Ion Exchange and Solvent Extraction

A Series of Advances

Volume 18

Ion Exchange and Solvent Extraction

A Series of Advances

Volume 18

Edited by
ARUP K. SENGUPTA

CRC Press is an imprint of the
Taylor & Francis Group, an **informa** business

CRC Press
Taylor & Francis Group
6000 Broken Sound Parkway NW, Suite 300
Boca Raton, FL 33487-2742

First issued in paperback 2019

ISBN-13: 978-0-8493-7397-8 (hbk)
ISBN-13: 978-0-367-40354-6 (pbk)

Visit the Taylor & Francis Web site at
http://www.taylorandfrancis.com

and the CRC Press Web site at
http://www.crcpress.com

Table of Contents

Preface vii
Editor xi
Contributors xiii

Chapter 1 Sorption of Solvent Mixtures in Ion Exchange Resins: Influence of Elastic Properties on Swelling Equilibrium and Kinetics 1

Tuomo Sainio, Markku Laatikainen, and Erkki Paatero

Chapter 2 Development of Simulated Moving Bed Reactor Using a Cation Exchange Resin as a Catalyst and Adsorbent for the Synthesis of Acetals 45

Viviana M.T.M. Silva, Ganesh K. Gandi, and Alírio E. Rodrigues

Chapter 3 Ion Exchange Resins in Drug Delivery 103

Sunil K. Bajpai, Manjula Bajpai, and Sutanjay Saxena

Chapter 4 Biopolymers as Supports for Heterogeneous Catalysis: Focus on Chitosan, a Promising Aminopolysaccharide 151

Eric Guibal, Thierry Vincent, and Francisco Peirano Blondet

Chapter 5 Ion Exchange Selectivity as a Surrogate Indicator of Relative Permeability of Homovalent Ions in Reverse Osmosis Processes 293

Parna Mukherjee and Arup K. SenGupta

Chapter 6 Chitosan: A Versatile Biopolymer for Separation, Purification, and Concentration of Metal Ions 339

Katsutoshi Inoue and Yoshinari Baba

Chapter 7 Short-Bed Ion Exchange 375

Craig J. Brown

Index 405

Preface

No specialty thrives in isolation. The field of ion exchange grew over decades by permeating into myriad areas from deionization to drug delivery. As a result, new knowledge was created in many seemingly disjointed scientific disciplines. Scientists and researchers — while using, applying, and moving the boundaries of ion exchange — are often separated by their professional fields and literally unknown to each other. In order to bridge the gap and break the barrier in the field of ion exchange, the first international workshop on Frontiers and Interfaces of Ion Exchange was recently held during June 11–15, 2006, in Antalya, Turkey. Nearly 120 attendees from 24 countries attended the meeting. Invited speakers were selected to cover a broad range of specialty research areas, namely, catalysis, synthesis, molecular imprinting, green processes, drug delivery, nanotechnology, and obviously, water treatment and environmental pollution control. Several invited speakers had never attended an "ion exchange" meeting previously but their ongoing research and interaction during the workshop inevitably enriched the field.

In accordance with the spirit of the first international workshop on ion exchange and Volumes 14 and 16 of the series earlier edited by myself, the current volume contains seven chapters encompassing a wide gamut of topics; they truly reflect the diversity in the field of ion exchange. As the editor, I must also mention to potential readers that the publisher of the series has changed. Marcel Dekker, the publisher of the Ion Exchange and Solvent Extraction Series for the 17 volumes spanning a period over 40 years, has recently been taken over by Taylor & Francis. I have personally been assured by the administration of Taylor & Francis that the present series will continue and they recognize the importance of this expanding field. One more change is also on the way. Professor Yizhak Marcus has decided to step down as the editor of the "Solvent Extraction" part of the series and is being replaced by Dr. Bruce Moyer of Oak Ridge National Laboratory. I salute Professor Marcus for his invaluable contribution to this series for more than four decades and welcome Dr. Moyer as a worthy successor to make this series move forward with renewed energy.

The breadth and synergy among many emerging areas in ion exchange serve as the primary theme of this current volume, which contains seven chapters written by professionals from academic institutions, research laboratories, and industries around the world. The volume indeed encompasses a wide range of topics.

It is recognized that the majority of the applications of ion exchange resins are geared toward the separation of ions from the aqueous phase. Polymer-based ion exchange resins can also serve as a medium (stationary phase) for carrying out chemical reactions and separating the reactants and products simultaneously. In such processes, sorption of solvent and mixture of solvents onto ion exchange resins influence the overall effectiveness of the process application. Chapter 1 discusses how solvent sorption equilibria and kinetics vary with the elastic properties of ion exchange resins, which, in turn, are dependent on type of functional groups, ionic forms, polymer matrix composition, and the degree of cross-linking. The thermodynamic modeling approach can be applied to the mixtures of polar and weakly polar solvents, and it explicitly takes into account the elasticity of the polymer network.

In synthesis reactions limited by chemical equilibria, the process conversion can be enhanced by selectively separating the products as they are formed. Ion exchange resins can serve as the heart of many such processes by simultaneously acting as a catalyst (or a reactant) and a separating agent. Development and modeling such a chromatographic reactor-separator and, more specifically, the simulated moving bed reactor (SMBR) for the synthesis of acetaldehyde dimethylacetal or 1,1-dimethoxyethane (DME) and acetaldehyde diethylacetal or 1,1-diethoxyethane (DEE) using a cation exchange resin (Amberlyst 15) is the primary objective of Chapter 2. In addition to the development and validation of models, the chapter provides experimental data from a pilot SMBR unit in Novasep, France, for the synthesis of acetals (DME and DEE) from acetaldehyde and alcohol as reactants and Amberlyst 15 resin as the catalyst and a selective adsorbent as well.

Ion exchange resin-based drug formulations have been the subject of intense research for nearly two decades and the resulting products are gradually moving into the marketplace. Besides oral drug delivery, ion exchange resins are being explored for transdermal, nasal, ophthalmic, and site-specific routes. Obviously, the pertinent question is: What is the advantage of getting a drug released from ion exchange resins? Controlling the rate of dissolution and improving the chemical stability and taste are some of the critical areas where drug delivery through ion exchange offers well-observed advantages. Chapter 3 discusses various scenarios of drug delivery for a combination of drugs and ion exchange resins. In addition, the chapter attempts to elucidate how the process variables, namely, temperature, ionic strength, pH, molecular weight of the drug, and cross-linking in the resin influence the overall process of drug delivery.

The importance of designing and preparing support materials for catalysts, especially for metal catalysts, is now well recognized. Metal binding capacity, chemical-thermal statibility, pore structure, physical morphology, and flexibility in imparting functional groups are some of the desirable attributes for the host materials. Also, there is now a new emphasis on biorenewable materials as hosts, wherever possible. Chitosan, a modified naturally available biopolymer with ion exchange properties for metals sorption, has of late come to the forefront as a support material for heterogeneous catalysis. Chapter 4 provides extensive coverage for preparation, usage, and performance evaluation of biopolymers as catalyst supports with particular emphasis on chitosan. The chapter includes many examples of reactions, namely, hydrogenation, oxidation, reduction, hydroxylation, and carbonylation catalyzed by biopolymer-supported catalysts.

Reverse osmosis (RO) and ion exchange are characteristically two different processes with no apparent similarity. RO is a nonselective pressure-driven membrane process applied primarily to separate dissolved solids from water. It is, however, well documented that RO membranes reject different electrolytes or ions to different extents, that is, permeation of different ions through RO membranes varies, all other conditions remaining identical. Also, the rejection of a specific ion is influenced by the accompanying electrolytes in feed water. With an increased application of RO and nanofiltration (NF) processes in the area of water treatment and wastewater reuse, there is now a greater need in predicting the relative degree of rejection or permeation of various ions, including environmentally regulated species. Chapter 5 provides convincing experimental evidence and elucidates underlying scientific reasons to confirm that ion exchange selectivity data for various ions can be used as surrogate parameters to predict the relative permeability of different ions in RO and NF processes. A simple ion chromatograph can provide the requisite information leading to the development of a quantitative model describing individual salt permeability. For trace ions of environmental significance, namely, perchlorate, nitrate, cesium, arsenate, chromate, and selenate, the ion exchange selectivity approach offers an insight with respect to their relative permeability in pressure-driven membrane processes.

Synthetic polymer-based materials constitute the majority of the ion exchange market and this trend is unlikely to change in the near future. Nevertheless, many naturally occurring biorenewable materials exhibit ion exchange properties resulting from the presence of a variety of chemical functional groups. In this regard, chitosan is probably a leading candidate due to the presence of both carboxylate and amino functional groups. Also, chitosan is amenable to chemical modification for improved chemical stability and mechanical strength. Chapter 6 provides a detailed account of how chitosan and its modified forms can find applications in separation and purification of metal ions.

The packed- or fixed-bed process where the mobile liquid passes through stationary ion exchange beads in a column is by far the most popular unit operation due to its simplicity of construction and operation. This method is routinely used for water softening, water demineralization, and removal of target-contaminating ions. Poor kinetics is one of the major limitations of the ion exchange process and intraparticle diffusion is often the rate-limiting step in the majority of the applications. During the exhaustion cycle, there are three specific zones for a solute in a fixed bed, viz., saturated, unused, and mass-transfer zone (MTZ). For intraparticle diffusion-controlled processes, the length of MTZ is proportional to the square of the diameter of the spherical ion exchange resin beads. Reducing the bead size reduces the length of the mass-transfer zone; however, the pressure drop across the bed increases with smaller particle sizes. "Short-Bed Ion Exchange" is a gainful compromise between the two: It offers faster kinetics with an acceptable pressure drop using skid-mounted ion exchange units. Chapter 7 provides specific advantages of short-bed units and presents many novel applications of ion exchange.

Arup K. SenGupta
Lehigh University, Pennsylvania

Editor

Arup K. SenGupta, Ph.D., is the P.C. Rossin Professor in the Department of Civil and Environmental Engineering, Lehigh University, Bethlehem, Pennsylvania. Dr. SenGupta is the author or coauthor of more than 100 scientific publications in peer-reviewed journals and conference proceedings, as well as the recipient of the 2004 International Ion Exchange Award from Cambridge University, Great Britain. He was the North American Editor of *Reactive and Functional Polymers* from 1996 to 2006. He is a member of the American Chemical Society (ACS), American Institute of Chemical Engineers (AIChE), Association of Environmental Engineering and Science Professors (AEESP), and American Water Works Association (AWWA). Dr. SenGupta received his B.S. degree (1972) in chemical engineering from Jadavpur University, Kolkata, India, and his Ph.D. degree (1984) in environmental engineering from the University of Houston, Texas.

Contributors

Yoshinari Baba
Department of Applied Chemistry
Miyazaki University Japan
Miyazaki, Japan

Manjula Bajpai
Polymer Research Laboratory
Government Model Science College (Autonomous)
Jabalpur, India

Sunil K. Bajpai
Polymer Research Laboratory
Government Model Science College (Autonomous)
Jabalpur, India

Hans-Jörg Bart
Lehrstuhl für Thermische Verfahrenstechnik
Technische Universität Kaiserslautern
Kaiserslautern, Germany

Francisco Peirano Blondet
Laboratoire Génie de l'Environnement Industriel
Ecole des Mines
Alès, France

Craig J. Brown
Chemionex, Inc
Pickering, Ontario, Canada

R. Chiarizia
Chemistry Division
Argonne National Laboratory
Argonne, Illinois

Ganesh K. Gandi
Laboratory of Separation and Reaction Engineering (LSRE)
University of Porto
Porto, Portugal

Bohumír Grüner
Institute of Inorganic Chemistry
Czech Academy of Sciences
Řež, Czech Republic

Eric Guibal
Laboratoire Génie de l'Environnement Industriel
Ecole des Mines
Alès, France

A.W. Herlinger
Department of Chemistry
Loyola University — Chicago
Chicago, Illinois

Katsutoshi Inoue
Department of Applied Chemistry
Saga University
Saga, Japan

Zdenek Kolarik
Karlsruhe, Germany

Markku Laatikainen
Laboratory of Industrial Chemistry
Lappeenranta University of Technology
Lappeenranta, Finland

Parna Mukherjee
California Polytechnic State University
San Luis Obispo, California

Erkki Paatero
Lappeenranta University of Technology
Laboratory of Industrial Chemistry
Lappeenranta, Finland

Michel Perrut
Separex
Champigneulles, France

Jiří Rais
Nuclear Research Institute Řež plc
Řež, Czech Republic

Alírio E. Rodrigues
Laboratory of Separation and Reaction Engineering (LSRE)
University of Porto
Porto, Portugal

Tuomo Sainio
Laboratory of Industrial Chemistry
Lappeenranta University of Technology
Lappeenranta, Finland

Sutanjay Saxena
Polymer Research Laboratory
Government Model Science College (Autonomous)
Jabalpur, India

Viviana M.T.M. Silva
Laboratory of Separation and Reaction Engineering (LSRE)
University of Porto
Porto, Portugal

Geoffrey W. Stevens
Department of Chemical and Biomolecular Engineering
The University of Melbourne
Victoria, Australia

Thierry Vincent
Laboratoire Génie de l'Environnement Industriel
Ecole des Mines
Alès, France

1 Sorption of Solvent Mixtures in Ion Exchange Resins: Influence of Elastic Properties on Swelling Equilibrium and Kinetics

Tuomo Sainio, Markku Laatikainen, and Erkki Paatero

CONTENTS

1.1 Introduction 2
1.2 Sorption and Swelling Equilibria 3
1.2.1 Model Development 3
1.2.1.1 General Condition for Phase Equilibrium 4
1.2.1.2 Gibbs Energy of Mixing: Liquid Lattice Model and Counterion Condensation Theory 6
1.2.1.3 Shear Modulus and Swelling Pressure 9
1.2.2 Experimental 11
1.2.2.1 Resin Properties 11
1.2.2.2 Mechanical Measurements 12
1.2.2.3 Measurement of the Swelling Ratio and Solvent Uptake 13
1.2.3 Results and Discussion 14
1.2.3.1 Elastic Properties 14
1.2.3.2 Equilibrium Swelling 22

1.3 Solvent Diffusion and Swelling Kinetics....29
1.3.1 Model Development....30
1.3.1.1 Maxwell–Stefan Approach....31
1.3.1.2 Fick's Law....32
1.3.2 Experimental....32
1.3.3 Results and Discussion....33
1.3.3.1 Fickian Diffusion Coefficients at Constant Swelling Ratio....34
1.3.3.2 Swelling and Shrinking Kinetics....35
1.4 Summary and Conclusions....39
Acknowledgments....41
References....41

1.1 INTRODUCTION

The behavior of strong acid cation exchangers in nonaqueous solvents and in aqueous-organic solvent mixtures has been extensively investigated during the past decades.[1–9] Modeling of multicomponent sorption equilibria and of diffusion of liquids in a swollen cross-linked polymer network plays a key role in understanding ion exchange resin catalysis and membrane separation processes. It is well known that the solvent sorption equilibria and resin swelling kinetics are influenced by the nature of the functional group, the counterion, and the polymer matrix, as well as the cross-link density of the resin. The purpose of the present work is to investigate the role of the elastic properties of the resin on these two phenomena in solvent mixtures. We also demonstrate the use of a thermodynamic modeling approach in describing sorption and swelling equilibria as well as swelling kinetics of ion exchange resins. The essence of this approach is that the elastic properties of the resin are explicitly taken into account.

The work is divided into two parts as follows. We begin by deriving a general condition for phase equilibrium in rubbery cross-linked polymers by using thermodynamics of polymer solutions and rubbery materials. An expression for the solvent activities in the liquid and resin phases is given. In this context we also present a number of models commonly used for describing the elasticity of gels and calculating the swelling pressure in the gel phase. To quantify the elastic properties of strong and weak acid cation exchange resins, we present shear modulus data obtained from direct mechanical measurement of single particles and swelling pressure data obtained from vapor sorption isotherms. The applicability of various elasticity models is discussed in the light of such data. Further, the thermodynamic phase equilibrium model is used as a semipredictive tool to describe sorption and swelling equilibria in some example systems.

For rubbery polymers, swelling and shrinking kinetics in solvent mixtures is controlled by diffusion of liquids inside the particle. This is the topic of the second part. We demonstrate how the elastic properties of the resin can be taken into account in a relatively simple manner by including them in the chemical potential driving force for diffusion. Literature data of solvent diffusion coefficients in strong cation exchange resins are scarce. To give an idea of the influence of the extent of resin swelling on the solvent diffusion coefficients of common solvents, we present some experimental data for strong cation exchange resins.

1.2 SORPTION AND SWELLING EQUILIBRIA

Conventional heterogeneous models based on analogies to surface adsorption are adequate in describing the distribution of compounds between a fluid and a rigid material, and are often used for ion exchange resins as well. However, the concept of fixed adsorption sites does not reflect the sorption phenomenon in swollen resins, which have a more or less flexible polymer structure. When the polymer network is expanded, only a fraction of the sorbed molecules are in the vicinity of the functional groups that are usually considered as the adsorption sites, and solvent–solvent interactions also have to be included. Therefore, statistical models derived from the thermodynamics of polymer solutions and gels[10,11] have recently been used to describe the observed phase equilibria in systems consisting of a solvent mixture and an ion exchange resin.[12–17] Similar models have also been tested in systems containing a binary solvent mixture, solutes, and a resin.[18,19] This approach has also been applied to the dynamic modeling of chemical reactors.[12,13,20,21]

Here we illustrate the use of a thermodynamic treatment in calculating the equilibrium states of systems containing solvent mixtures (or other neutral components) and moderately or densely cross-linked ion exchange resins. Only nonreactive systems are considered, but extension to reactive systems is straightforward.[22]

1.2.1 Model Development

The system under consideration consists of three parts: a homogeneous liquid phase, a homogenous polyelectrolyte solution, and an elastic cross-linked polymer network (Figure 1.1). A cross-linked polymer network may swell to a variable extent depending on the external liquid-phase composition, the affinity of the polymer for the solvents, and the number of cross-links. The configurational entropy of a cross-linked polymer network decreases as the extent of swelling increases. This brings about a tensile force that opposes the expansion of the network, which is observed as an increase in the pressure of the polymer phase. Consequently, there exists a pressure difference (swelling pressure) between the polymer phase and the external liquid phase.

In the present context, only the distribution of neutral liquid compounds between the liquid phase and ion exchange resin particles is relevant: the dissociated counterions cannot leave the resin phase due to the condition of electroneutrality. However, the counterions contribute to the free energy of the polymer phase and the effect is taken into account by the mixing entropy of the free counterions.

It should be noted that although electrolyte solutions are not considered here, the modeling approach can be extended to systems with ionic species by including

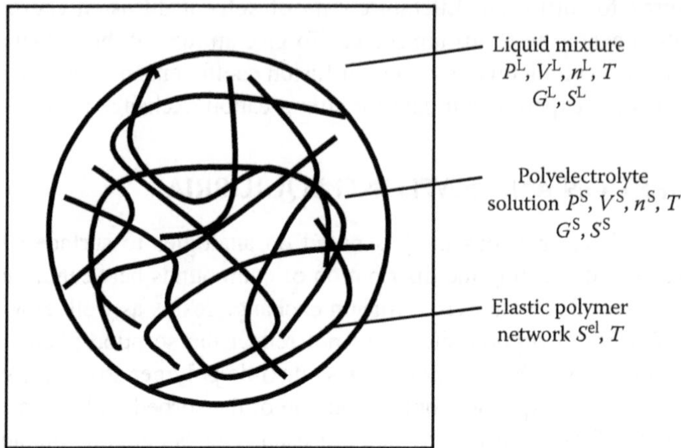

FIGURE 1.1 The system consists of three parts: multicomponent liquid mixture (L); a homogeneous polyelectrolyte solution (S); and an elastic polymer network (el). The thermodynamic quantities are explained in the text.

the condition of electroneutrality in each phase as an additional constraint.[11] This enables calculation of the equilibrium state of the system without employing the Donnan potential concept.

1.2.1.1 General Condition for Phase Equilibrium

The first and second laws of thermodynamics yield Equation (1.1) for an isothermal reversible transition under constant external pressure. In Equation (1.1), U^{sys} and S^{sys} are the total internal energy and the entropy of the system, and W is the work done by the system. Exact and inexact differentials are denoted with d and $đ$, respectively.

$$-\mathrm{d}U^{sys} - đW + T\mathrm{d}S^{sys} = 0 \tag{1.1}$$

According to the classical theory of rubber elasticity, deformation of the polymer network does not involve changes in the internal energy or volume of the polymer network, but only changes in its configurational entropy.[10] If only the expansion work is taken into account and if the liquid-phase pressure equals that of the surroundings, Equation (1.1) may be written as shown in Equation (1.2), where S^{el} denotes the configurational entropy of the polymer network, and T and p are temperature and pressure. The liquid and polymer phases are denoted by superscripts L and P.

$$-\mathrm{d}U^{L} - p^{L}\mathrm{d}V^{L} + T\mathrm{d}S^{L} - \mathrm{d}U^{P} - p^{L}\mathrm{d}V^{P} + T\mathrm{d}S^{P} + T\mathrm{d}S^{el} = 0 \tag{1.2}$$

Temperature and external pressure are assumed constant, and Equation (1.2) can be rearranged into the form shown in Equation (1.3), where G is the Gibbs energy, and n^{L} and n^{P} denote the number of moles in liquid and polymer phases, respectively. It should be noticed that the Gibbs energy of the polymer phase is evaluated at liquid-phase pressure because of cancellation of terms.

$$\mathrm{d}\left(G^{\mathrm{L}}\left(T,p^{\mathrm{L}},n^{\mathrm{L}}\right)+G^{\mathrm{P}}\left(T,p^{\mathrm{L}},n^{\mathrm{P}}\right)-TS^{\mathrm{el}}\right)=0 \tag{1.3}$$

Since reversible transitions pass through equilibrium stages, Equation (1.3) gives a condition for equilibrium in the system: *the expression in the parentheses has its minimum value at equilibrium.* However, it is worthwhile to elaborate the equilibrium condition somewhat further. As a first step, the effect of the elastic response of the polymer network can be expressed in terms of more easily measured quantities than $-T\mathrm{d}S^{\mathrm{el}}$. As a second step, the Gibbs energies in Equation (1.3) can be replaced with the Gibbs energy of mixing by using the material balance.

The stress in a cross-linked polymer network due to a deformation by an external force is obtained from the classical theory of rubber elasticity. The stress that opposes further swelling of a swollen spherical cross-linked polymer network (i.e., tensile force per unit area of swollen sample) is calculated as shown in Equation (1.4), where V^0 is the volume of the unswollen undeformed sample, and α is the deformation factor. The stress, denoted with π_{sw} in Equation (1.4), should be interpreted as an additional pressure exerted on the polymer phase, and is termed *swelling pressure.*[10] Consequently, the resin phase pressure is $p^{\mathrm{P}} = p^{\mathrm{L}} + \pi_{\mathrm{sw}}$. The factor $3\alpha^2$ in the denominator of Equation (1.4) originates from the spherical geometry and the assumption of isotropic swelling. Since $3\alpha^2\mathrm{d}\alpha V^0 = \mathrm{d}V^{\mathrm{P}}$, the change in the configurational entropy of the polymer network can be expressed in terms of the swelling pressure and the volume of the polymer phase, as shown in Equation (1.5).

$$\pi_{\mathrm{sw}} = -\frac{T}{3\alpha^2 V^0}\frac{\mathrm{d}S^{\mathrm{el}}}{\mathrm{d}\alpha} \tag{1.4}$$

$$-T\mathrm{d}S^{\mathrm{el}} = \pi_{\mathrm{sw}}\mathrm{d}V^{\mathrm{P}} \tag{1.5}$$

In order to introduce the reaction and mixing quantities, the Gibbs energies of the liquid and polymer phases at the liquid-phase pressure can be expressed as in Equation (1.6) and Equation (1.7), provided that system temperature and liquid-phase pressure are chosen as the reference state. The numbers of components in the liquid and solid phases are denoted by NL and NP, respectively. G_i^0 is the standard Gibbs energy of a pure component i, and $\Delta_{\mathrm{mix}}G$ is the Gibbs energy change due to mixing of the components.

$$G^{\mathrm{L}}\left(T,p^{\mathrm{L}},n^{\mathrm{L}}\right)=\sum_{NL}n_i^{\mathrm{L}}G_i^0\left(T,p^{\mathrm{L}}\right)+\Delta_{\mathrm{mix}}G^{\mathrm{L}}\left(T,p^{\mathrm{L}},n^{\mathrm{L}}\right) \tag{1.6}$$

$$G^{\mathrm{P}}\left(T,p^{\mathrm{L}},n^{\mathrm{P}}\right)=\sum_{NP}n_i^{\mathrm{P}}G_i^0\left(T,p^{\mathrm{L}}\right)+\Delta_{\mathrm{mix}}G^{\mathrm{P}}\left(T,p^{\mathrm{L}},n^{\mathrm{P}}\right) \tag{1.7}$$

By substituting Equation (1.5) through Equation (1.7) into Equation (1.3), and using the material balance $n_i^{\mathrm{P}}+n_i^{\mathrm{L}}=n_i^0$, the equilibrium condition can be expressed in terms of quantities that can be evaluated from experimentally measured data. The equilibrium compositions of both phases are then obtained by minimization of the objective function Y given in Equation (1.8).

$$Y=\Delta_{\mathrm{mix}}G^{\mathrm{L}}\left(p^{\mathrm{L}}\right)+\Delta_{\mathrm{mix}}G^{\mathrm{P}}\left(p^{\mathrm{L}}\right)+V^{\mathrm{P}}\pi_{\mathrm{sw}} \tag{1.8}$$

Although phase equilibrium calculation by numerical minimization of Equation (1.8) is straightforward, an alternative phase equilibrium condition can be derived from it. Firstly, an infinitesimally small amount $\mathrm{d}n_i^{\mathrm{L}}$ of component i is transferred across the phase boundary from the liquid phase to the resin phase. If the system is initially at equilibrium, the objective function changes, as shown in Equation (1.9). Because of the equilibrium assumptions, the derivative $(\partial Y/\partial n_i^{\mathrm{L}})_{TPnj}$ is equal to zero, and substituting Equation (1.6) and Equation (1.7) into Equation (1.9) results in a well-known relationship, shown in Equation (1.10), which states that the chemical potentials of each component are equal across the phase boundary. Here R_{g} is the gas constant, a is activity, and V_{m} is the partial molar volume.

$$\mathrm{d}Y=\frac{\partial G^{\mathrm{L}}}{\partial n_i^{\mathrm{L}}}\left(-\mathrm{d}n_i^{\mathrm{L}}\right)+\frac{\partial G^{\mathrm{P}}}{\partial n_i^{\mathrm{S}}}\mathrm{d}n_i^{\mathrm{L}}+\pi_{\mathrm{sw}}\frac{\partial V^{\mathrm{P}}}{\partial n_i^{\mathrm{S}}}\mathrm{d}n_i^{\mathrm{L}} \tag{1.9}$$

$$R_{\mathrm{g}}T\ln a_i^{\mathrm{L}}\left(T,p^{\mathrm{L}}\right)=R_{\mathrm{g}}T\ln a_i^{\mathrm{P}}\left(T,p^{\mathrm{L}}\right)+V_{\mathrm{m},i}\pi_{\mathrm{sw}} \tag{1.10}$$

1.2.1.2 Gibbs Energy of Mixing: Liquid Lattice Model and Counterion Condensation Theory

In order to calculate the equilibrium phase compositions by minimizing the objective function Y in Equation (1.8), an expression is required for the Gibbs energy change when mixing the solvents, the polymer, and the dissociated counterions. In the present work, the lattice theory by Flory and Huggins[10] is used (for other models, see, for example, Reference 14). The mixing Gibbs energy is given in Equation (1.11) for the polymer phase, but the same equation may also be used for the liquid phase when ϕ_p is set equal to zero. In contrast to the original model, the Flory–Huggins interaction parameter χ is allowed to depend on both composition and temperature. Only the entropic effect of the dissociated counterions is included explicitly; electrostatic effects were thus implicitly included in the solvent–polymer interaction parameters.

$$\frac{\Delta_{\text{mix}} G^{\text{P}}\left(T, n^{\text{P}}\right)}{R_g T} = \sum_{j=1}^{NP-1} n_j \ln \phi_j + n_c \ln \phi_c + n_p \ln\left(\phi_p - \phi_c\right) + \sum_{j=1}^{NP-1}\sum_{k=1}^{j-1} \chi_{kj}\left(u_j^{kj}, T\right) n_k \phi_j + \sum_{j=1}^{NP-1} \chi_{jp}\left(\phi_p, T\right) n_j \phi_p \tag{1.11}$$

In Equation (1.11), n denotes the number of moles, ϕ is the volume fraction, and χ is the Flory–Huggins interaction parameter. The subscripts j and k refer to the liquid components, subscript c to the dissociated counterion, and subscript p to the polymer. Simple linear relations, shown in Equation (1.12) and Equation (1.13), were chosen for the concentration and temperature dependencies of the interaction parameters.

$$\chi_{ij}\left(T, u\right) = \chi_{ij}^0 + b_{ij} u_j^{ij} + \left(a_{ij} + c_{ij} u_j^{ij}\right) / T \tag{1.12}$$

$$\chi_{ip}\left(T, \phi\right) = \chi_{ip}^0 + b_{ip} \phi_p + \left(a_{ip} + c_{ip} \phi_p\right) / T \tag{1.13}$$

By following Pouchly et al.,[23] a reduced volume fraction, defined as $u_i^{ij} = \phi_i/(\phi_i+\phi_j)$, was used as the concentration variable for the solvent–solvent interaction parameter, whereas ϕ_p was used for the solvent–polymer interactions. The following relationships hold for the parameters in Equation (1.12): $\chi^0_{ji} = m_{ji}(\chi^0_{ij}+b_{ij})$, $a_{ji} = m_{ji}(a_{ij}+c_{ij})$, $b_{ji} = -m_{ji}b_{ij}$, and $c_{ji} = -m_{ji}c_{ij}$. The molecular size ratio m_{ij} was calculated from the pure component molar volumes according to $m_{ij} = V_{m,i}/V_{m,j}$. The parameter m_{ip} may be taken as zero because the molecular size of the polymer is large compared to that of the solvents. It should also be noted that $\chi_{ii} \equiv 0$.

Phase equilibrium calculations are often carried out by equating the chemical potentials μ of the solvents across the phase boundary [cf. Equation (1.10)]. For this purpose, the expression for μ_i in the polymer phase [Equation (1.14)] is obtained by differentiating the Gibbs energy of mixing, Equation (1.11), with respect to n_i. Again, the same expression also applies for the liquid phase when ϕ_p is set to zero. The index NP+1 in the summations refers to the dissociated counterions. In the derivation of Equation (1.14), it has been assumed that the degree of dissociation is independent on solvent composition (see References 22 and 24 for an extended model).

$$\frac{\Delta\mu_i^{\mathrm{P}}\left(p^{\mathrm{L}}\right)}{R_{\mathrm{g}}T}=\ln a_i^{\mathrm{P}}\left(p^{\mathrm{L}}\right)=1+\ln\phi_i^{\mathrm{P}}-\sum_{j=1}^{NP+1}m_{ij}\phi_j^{\mathrm{P}}+\sum_{j=1}^{NP}\chi_{ij}\phi_j^{\mathrm{P}}-\sum_{j=1}^{NP}\sum_{k=1}^{j-1}m_{ik}\chi_{kj}\phi_k^{\mathrm{P}}\phi_j^{\mathrm{P}}$$
$$-\sum_{j=1}^{NP-1}u_i^{ij}u_j^{ij}\phi_j^{\mathrm{P}}\frac{\partial\chi_{ij}}{\partial u_j^{ij}}-\sum_{j=1}^{NP-1}m_{ij}\phi_j^{\mathrm{P}}\phi_p^2\frac{\partial\chi_{jp}}{\partial\phi_p} \quad (1.14)$$

The use of the Gibbs energy of mixing and activity models described above calls for a quantitative expression for the degree of dissociation of the charged functional groups. Here, this was approximated in the framework of Manning's *counterion condensation theory*.[25,26] According to the theory, the free energy of a polyelectrolyte solution attains its minimum value when a fraction of the counterions is dissociated and free, while the rest is "topologically condensed" close to the polymer matrix. Furthermore, there exists a well-defined minimum distance, equal to the Bjerrum length, between two adjacent dissociated groups in the polymer matrix. Although the condensed counterions are not covalently bound to the functional group, they renormalize the line charge density of the polyelectrolyte to a threshold value that depends on temperature and the dielectric properties of the solvent environment.[25] In the current context, the important property of the condensed counterions is that they do not contribute to the mixing entropy of the system but are an inseparable part of the polymer structure. According to the counterion condensation theory, the amount of dissociated (free) counterions is obtained by Equation (1.15), and their volume fraction was approximated by Equation (1.16).

$$n_c=n_m DF\left(z\psi\right)^{-1},\qquad \psi=e_0^2/4\pi\varepsilon_0\varepsilon k_{\mathrm{b}}T\delta \quad (1.15)$$

$$\phi_c\approx\phi_p m_{cm}\,n_c/n_m \quad (1.16)$$

In Equation (1.15) and Equation (1.16), *DF* denotes the degree of functionalization of the polyelectrolyte, *z* is the counterion valence, and ψ is the charge density parameter in Manning's theory. Furthermore, ε_0 is the contour length between two charged groups, e_0 is the elementary charge, $_0$ is the permittivity of the solvent mixture in the polymer-phase solution (excluding the polymer), and k_{b} is the Boltzmann constant. Subscript *m* refers here to the repeating unit of the polymer.

It should be recognized that the counterion condensation theory is a limiting law, and is used here only as an approximation for the amount of free counterions. The interactions between functional groups located in different branches of the polymer network are neglected for simplicity. Consequently, the degree of dissociation will probably be overestimated at least for highly cross-linked resins. In a homogeneous polystyrene-divinylbenzene (PS-DVB) ion exchange resin with cross-link density of 4 wt%, for example, the average number of monomers between the cross-links is approximately 30, and in the swollen state, the interactions between adjacent charged groups in a polymer branch are dominant. At cross-link density of 20 wt%, however, there are only approximately six monomers per branch, and interactions between different branches become much more probable.

1.2.1.3 Shear Modulus and Swelling Pressure

As already mentioned above, stresses in a swollen polymer network are of entropic origin and can be interpreted as an additional pressure, termed *swelling pressure*, exerted on the polymer phase. Models for π_{sw} are derived from the statistical analysis of cross-linked polymer networks (the theory of rubber elasticity). The entropy change in any process is linked to the probabilities of finding the system in the initial and final states according to the Boltzmann relation. Similarly, the entropy change in an elastic deformation of a polymer network is obtained by calculating the probabilities of finding the network in three-dimensional configurations corresponding to the undeformed and deformed states. Deformation under simple shear load without a volume change is usually employed for polymer gels and the material properties are thus characterized by the shear modulus G.

The amount and distribution of the cross-links in the network affect the number of ways the polymer network can be arranged into a given spatial configuration. The elastic models presented in the literature differ mainly in their mathematical description of the distribution of the cross-links in the polymer network (chain-length distribution). The shear modulus and swelling pressure of an ideal Gaussian network are given in Equation (1.17) and Equation (1.18),[27] where ϕ_p and $\phi_p{}^0$ are the polymer volume fractions at the experimental conditions and at the reference state, respectively.

$$G = K_{el}\phi_p^0(\phi_p / \phi_p^0)^{1/3} \tag{1.17}$$

$$\pi_{sw} = K_{el}\phi_p^0(\phi_p / \phi_p^0)^{1/3} \tag{1.18}$$

In the original theory, $K_{el} = R_gTn_c$ where n_c is the concentration of the chains in the dry polymer. "Chain" refers to the polymer segment between two consecutive cross-link points. Moreover, the dry polymer is taken here as the reference state, and therefore $\phi_p^0 = 1$ and it is omitted in the following equations. Normally ϕ_p^0 is defined at the unstrained conditions, where the cross-links are introduced in the network. No such well-defined reference state exists for commercial resins, and K_{el} is treated here as an adjustable parameter.

The Flory elasticity model,[10] shown in Equation (1.19), assumes a Gaussian chain length distribution, too. Gusler and Cohen[28] developed a modified expression for π_{sw} in which the Gaussian polymer chain length distribution was replaced with a near-Gaussian one. The Gusler–Cohen expression for π_{sw} is shown in Equation (1.20). In Equation (1.19) and Equation (1.20), M_c and M denote the molar masses of the polymer chains and the resin, respectively. In practice, the first term in parentheses is often neglected because $2M_c << M$.

$$\pi_{sw} = K_{el}\left(1 - \frac{2M_c}{M}\right)\left(\phi_p^{1/3} - \frac{1}{2}\phi_p\right) \tag{1.19}$$

$$\pi_{sw} = K_{el}\left(1 - \frac{2M_c}{M}\right)\left(\frac{5}{3}\phi_p^{1/3} - \frac{7}{6}\phi_p\right) \tag{1.20}$$

The Gaussian models assume that the average end-to-end distance of the polymer chains is much less than the length of a fully extended chain. Therefore, the models are valid only as long as the extension of the chains does not exceed about one third of the fully extended length.[29] For the densely cross-linked resins considered here, the assumption of the Gaussian distribution is probably not valid even in the unswollen state. Therefore, a model derived from the three-chain model of real elastic networks[16,30] was chosen, and the swelling pressure for the pressure correction term in Equation (1.10) can be calculated according to Equation (1.22). In a similar way, the shear modulus G is obtained from Equation (1.21). In this model, the chain-length distribution of the polymer is described using the inverse Langevin function $L^{-1}(\gamma)$, where γ is the fractional extension of the chains.

$$G = K_{el}(d_G\phi_p)^{1/3}\frac{1}{6}\left(\frac{L^{-1}(\gamma)}{\gamma} + \frac{dL^{-1}(\gamma)}{d\gamma}\right), \quad \gamma = \left(d_G\phi_p\right)^{-1/3} N_{stat}^{-1/2} \tag{1.21}$$

$$\pi_{sw} = \frac{1}{3}K_{el}(d_\pi\phi_p)^{2/3} N_{stat}^{1/2} L^{-1}(\gamma), \quad \gamma = (d_\pi\phi_p)^{-1/3} N_{stat}^{-1/2} \tag{1.22}$$

In Equation (1.21) and Equation (1.22), N_{stat} is the number of statistical segments and γ is defined as the ratio of the actual end-to-end distance of the network chain to its total length.[27] The correction factors d_G and d_π account for the imperfections of the network and they formally give the fraction of active chains. For perfect networks, both are equal to unity. The physical significance is less clear in the densely cross-linked resins considered here, and d_G and d_π are treated as empirical parameters. The form of Equation (1.21) is taken from Zaroslav et al.,[31] who have used d_G to improve the agreement between the experimental data of polyelectrolyte gels and the theoretical predictions.

TABLE 1.1
Properties of the Resins

Resin Name	Resin Type	Counterion	Cross-Linkage (wt% of DVB)	d_{av}[a] (mm)	Effective Capacity (mequiv/g)	Water Retention (wt%)	V_w/V^0
CS16G	SCE X8	H^+	8	0.85	5.1	55	2.2
	SCE X8	Na^+	8	0.35	4.7	47	2.1
	SCE X8	La^{3+}	8	0.35	4.2	40	1.7
CS11G	SCE X5.5	Na^+	5.5	0.35	4.6	56	2.7
	SCE X5.5	$(CH_3)_4N^+$	5.5	0.35	3.7	50	2.0
CS08G	SCE X4	Na^+	4	0.35	4.5	65	3.6
KEF76M[b]	SCE X7	H^+	7	0.30	5.1	57	
Amberlyst 15M[c]	SCE X20	H^+	20	0.4–0.7	5.1	32	
AS11G	SAE X5.5	Cl^-	5.5	0.35	3.6	49	1.7
CA06G	WCE X3	Na^+	3	0.30	9.9	81	8.0
CA08G	WCE X4	Na^+	4	0.30	9.6	76	6.2
CA12G	WCE X6	Na^+	6	0.30	9.3	70	4.8
CA16G	WCE X8	Na^+	8	0.30	9.1		
IRC86G	WCE NA	Na^+	NA	0.6–0.8	9.3	67	4.0

Note: SCE, strong cation exchanger; SAE, strong anion exchanger; WCE, weak cation exchanger; NA, cross-link density not available; G, gel type; M, mesoporous or macroporous.

[a] Average bead diameter in water.
[b] Mesoporous resin, porosity 7 vol%
[c] Macroporous resin, porosity 30 vol%

1.2.2 Experimental

1.2.2.1 Resin Properties

The strong cation exchangers (CS16G, CS11G, CS08G, and KEF76M), weak cation exchangers (CA06G, CA08G, CA12G, and CA16G), and the strong anion exchanger AS11G were delivered by Finex Oy (Finland). The macroreticular strong cation exchanger (SCE) Amberlyst 15 and the weak cation exchanger (WCE) Amberlyst IRC86 were obtained from Rohm & Haas. Some characteristics of the resins are shown in Table 1.1. The sulfonated PS-DVB resins are functionalized with sulfonic acid groups, the strong anion exchangers with quaternary ammonium groups, and the acrylic cation exchanger with carboxylic acid groups. The nominal cross-link density is given as a weight percentage of divinylbenzene (DVB) in the polymerization mixture. Before use in the experiments, the resins were pretreated and rendered to the desired ionic form by conventional methods.[14,24]

1.2.2.2 Mechanical Measurements

The shear modulus of the resin beads was measured using the static compression method originally applied to polymeric particles by Knaebel et al.[32] and Briscoe et al.[33] The method is based on the theoretical and experimental study of Hertz on glass spheres.[34] All measurements were carried out at room temperature (22±1°C). In a typical experiment, a single resin bead immersed in the equilibration solution was placed between two parallel plates that were attached to a micrometer screw and to a microload cell (Kyowa LTS-200GA or Kyowa LVS-20GA), respectively. Alternatively, the resin bead was first swollen with a given amount of water, and during the measurement it was immersed in *n*-heptane in order to maintain the same water content. The deformation was increased stepwise up to about 10% of the bead diameter. The volume changes of the bead were avoided by keeping the measurement cycle time as short as possible. One cycle took only a few minutes involving 30 to 150 points. The diameter of the undeformed resin bead was measured with an optical microscope.

The shear modulus, G, of the swollen resins was calculated from the Hertz equation, which under constant volume conditions can be written as Equation (1.23).[32,33]

$$\Delta d = \left[\frac{3F}{4G\sqrt{d_0}} \right]^{2/3} \tag{1.23}$$

Here Δd is the deformation of the bead under load F and d_0 is the original diameter of the swollen bead.

For all the resins, at least three beads were measured and at least two compression cycles were recorded for each bead. The moduli calculated from duplicate cycles differed less than 5%. The data corresponding to deformations of 1 to 3% were neglected because of uncertainties due to very small compressive force.[35] Typical data plotted in the form of Equation (1.23) are presented in Figure 1.2 and the shear modulus is obtained as the inverse of the slope. The linear dependence was excellent in water and dilute salt or alcohol solutions. For the resins equilibrated in concentrated solutions, on the other hand, the differences between duplicate cycles were larger because of slow restoration of the spherical shape after compression. This is due to the transition of the resin from rubbery to glassy state, and in such cases only the first cycle was used. Moreover, marked deviations from linearity were observed in concentrated alcohol solutions (Figure 1.2), and therefore the values measured in the transition region are only approximate. It may be noted that the linear dependence was again observed in dry or nearly dry states, where the modulus is much higher (i.e., 2 to 4 GPa). However, this region has no interest for the purposes of this study and it was not investigated systematically.

The variation between individual beads of a given resin was tested measuring 40 beads of NaCS16G, 17 beads of NaCS11G, and 14 beads of NaIRC86. At the confidence level of 99%, the moduli were 21.4±1.9 MPa, 9.6±0.6 MPa, and 8.9±0.7 MPa, respectively, and the deviations from the average value were in

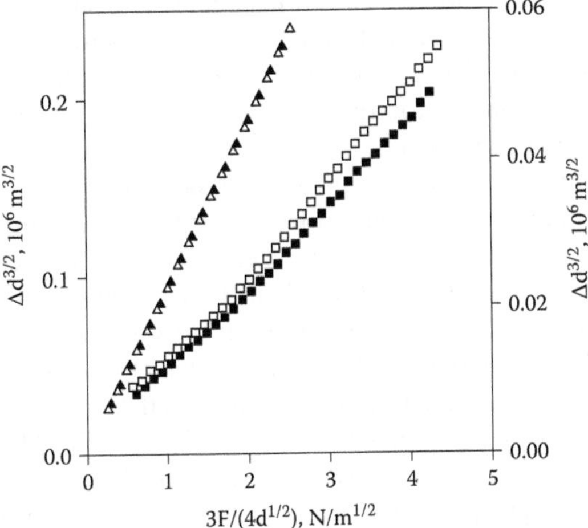

FIGURE 1.2 Typical compression cycles plotted in the form of Equation (1.23). Resin bead swollen in water (triangles, left *y* axis) and in a concentrated ethanol solution (squares, right *y* axis). Filled and open symbols represent the first and second cycles, respectively.

most cases less than 10%. The deviations were smaller for the less densely cross-linked resins, probably because of the more isotropic structure of the network.

1.2.2.3 Measurement of the Swelling Ratio and Solvent Uptake

The volume changes of the resin were measured for a resin bed placed in a glass column or for individual particles by means of an optical microscope. In the former method, the height of the resin bed was typically 0.15 m, and the resin was equilibrated with 15 to 25 bed volumes of the liquid mixture during 1 to 5 h.

The water-swollen state rather than the dry state was taken as the reference volume because of the discontinuity in resin properties at low degrees of swelling. As will be discussed later, the resins undergo a transition from the rubbery swollen state to a glassy state, and the properties change dramatically. If not indicated otherwise, the swelling ratio, q, is thus expressed in terms of the polymer volume fraction, ϕ_p, and it was calculated from Equation (1.24). Consequently, the non-swollen state ($q = \phi_p^{-1} = 1$) is a hypothetical dry rubbery polymer. If the actual volume of the dry resin, V^0, is used as the reference, somewhat smaller values are obtained for the extent of swelling.

$$q = \phi_p^{-1} = \frac{V}{V_w \phi_{p,w}} \tag{1.24}$$

The polymer volume fraction in the water-swollen resin, $\phi_{p,w}$, was calculated from the water retention and the density of the swollen resin. The density was measured by means of a pycnometer.

The liquid present in the pores of a macroporous resin is not regarded here as sorbed, and the water-retention data were corrected appropriately for the porous Amberlyst 15 and KEF 76M resins. The swelling-ratio data were interpolated with polynomials of third or fourth degree for calculating the extent of swelling in the batch sorption experiments.

Phase equilibria in ternary solvent–solvent–polymer systems were measured at 298 K using gravimetric[14,24] and dilatometric[36] methods. In the gravimetric method, the equilibrium distribution of water and ethanol between the resin and liquid phases was obtained as follows. Four to six grams of the water-swollen resin or three grams of the ethanol-swollen resin was weighed accurately in a test tube and a known amount of the desired water–ethanol solution was added. The test tube was tightly sealed and the sample was then agitated for 3 d at room temperature and finally equilibrated for 3 d at 298 ± 0.1 K. After equilibration, the solution was recovered and the resin was centrifuged and weighed. The solvent composition in the liquid phase was determined by precision density measurement (Anton Paar, DMA 55) and the absorbed amounts were obtained from mass balances. The deviation between the duplicate measurements at high water mole fractions was ± 0.1 mol/equiv and somewhat greater at low water contents.

In the dilatometric method, the amounts of the two solvents sorbed by the resin were calculated by combining two data sets.[22] Changes in the composition of the liquid phase due to selective solvent uptake of the resin were measured by means of gas chromatography (Agilent 6890N, HP INNOWax column). The swelling ratio of the resin corresponding to the equilibrium composition of the liquid phase was obtained from the changes in the bed volume.

1.2.3 Results and Discussion

In this section, the experimental data on solvent sorption and volume swelling of different commercial resins are discussed in view of the equilibrium model described in Section 1.2.1. The influence of the elastic properties is emphasized and, therefore, the experimental data and their correlation with the proposed models are considered first. Combination of the elastic and mixing models to explain the behavior of the cross-linked resins is discussed in Section 1.2.3.2. Details of the model calculations and parameter estimation are omitted here but can be found in References 15, 24, and 22.

1.2.3.1 Elastic Properties

As discussed in Section 1.2.1.3, the two elastic parameters, K_{el} and N_{stat}, describe the dependence of both the shear modulus and the swelling pressure on the swelling ratio. Experimental data of both quantities are available, and Equation (1.21) and Equation (1.22) can thus be used to estimate the values for the parameters. Determination of the shear modulus by mechanical measurements is straightforward and the data for the resins studied here have been reported previously.[15,16,24] At high cross-link densities, however, the rubbery region is relatively narrow and the interference due to the glass transition is difficult to eliminate. For such cases, the swelling pressures evaluated from water sorption isotherms may give more realistic estimates for the elastic parameters. The isotherm measurements are much more tedious, and high accuracy is required to obtain reliable values for the swelling pressures.

In practice, however, the parameter values obtained from the different data sets may vary appreciably and the swelling pressures tend to give higher values for K_{el} or lower values for N_{stat}. The consistency can be improved by adjusting the correction factors d_G and d_π appearing in Equation (1.21) and Equation (1.22). Because these quantities are purely empirical, it was assumed here that $d_G = 1$, and the value of d_π was then estimated from the experimental data.

Mechanical Measurements

Besides giving information on the behavior of the polymer network in the swollen state, mechanical measurements also detect the transitions taking place at low swelling ratios. As discussed earlier, the main factor determining the mechanical properties of a given polymer matrix in the swollen state is the cross-link density. The transition from rubbery to glassy state, on the other hand, is affected by the characteristics of the polymer network, the size of the ionic groups, and the counterion. Some examples are given below to illustrate both aspects.

The shear moduli of strong and weak cation exchangers having different cross-link densities are shown in Figure 1.3 and Figure 1.4. All these resins were in the sodium form. Only the data points corresponding to the rubbery and the transition region are depicted, while the much higher values of the glassy state are omitted. The data were correlated with Equation (1.21) and the calculated values are shown as solid lines. Because of the small number of data points and the interference of the glass transition, only manual adjustment was used. The estimated values of K_{el} and N_{stat} are given in Table 1.2. For comparison, the values obtained with the ideal Gaussian model [Equation (1.17)] and with the same K_{el} are shown for the lowest cross-link densities as dashed lines.

The experimental data indicate the presence of two phenomena, which tend to increase the modulus and thus a more or less distinct minimum appears at an intermediate swelling ratio. The minimum is most pronounced for the weak cation exchanger and the minimum is only about 40% of the value measured in water (Figure 1.4). The resins thus behave in a qualitatively similar way as weakly cross-linked hydrogels in brine solutions,[27,37–39] although the minimum occurs now at much lower swelling ratios. The hydrogels typically swell from 100 to over 1000 times from the dry volume, whereas the maximum swelling ratio of the resins is less than 10 (see Table 1.1). In analogy with the highly expanding gels, the increasing modulus at high swelling ratios can be explained by the finite expansibility of the polymer chains. In terms of the thermodynamic model presented in Section 1.2.1, the high modulus values are due to the entropy decrease caused by the large deviations from the most probable chain-length distribution.

This explanation finds support also from the data shown in Figure 1.5, where the shear moduli of a fully and partially neutralized weak cation

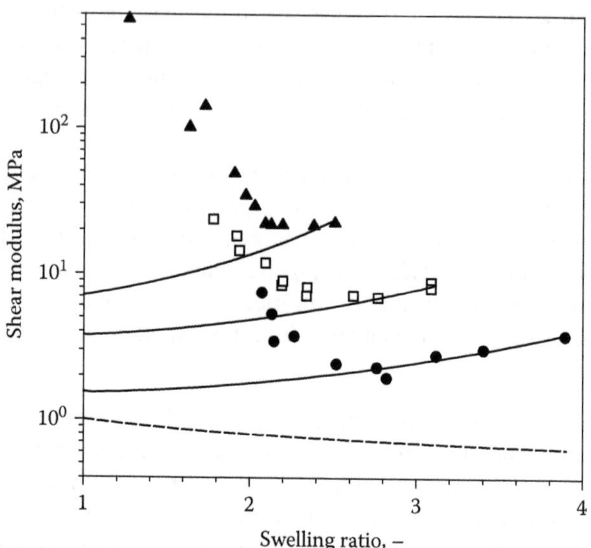

FIGURE 1.3 Effect of cross-linking on the shear modulus of Na-form strong cation exchangers (SCE). NaCS08G (filled circles); NaCS11G (open squares); NaCS16G (filled triangles). Solid and dashed lines represent the values calculated with the three-chain model [Equation (1.21)] and the ideal Gaussian model [Equation (1.17)], respectively.

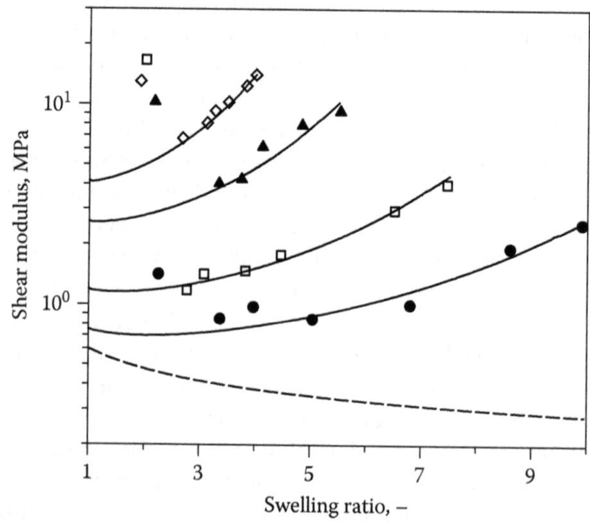

FIGURE 1.4 Shear moduli of Na-form weak cation exchangers (WCE). CA06G (filled circles); CA08G (open squares); CA12G (filled triangles); CA16G (open diamonds). Solid and dashed lines have the same meaning as in Figure 1.3.

TABLE 1.2
Elastic Parameters Estimated by Means of Equation (1.21) (Mechanical Measurements) and Equation (1.22) (Swelling Pressure); D_G = 1

Resin	K_{el} (MPa)	N_{stat},	d_{π},
Strong cation exchangers:			
CS08G, Na^+	1.0	3.7	—
CS11G, Na^+	2.3	3.3	—
CS16G, Na^+	3.3	2.5	—
CS16G, H^+	4.2	2.8	—
Dowex 50 X4, H^+	2.5	4.3	0.70
Dowex 50 X8, Na^+	3.3	2.5	0.68
Dowex 50 X8, H^+	4.2	2.8	0.70
Dowex 50 X16, H^+	8.5	2.3	0.70
KEF76M, H^+	3.0	3.0	—
Amberlyst 15M, H^+	5.7	2.1	—
Weak cation exchangers:			
CA06G, Na^+	0.6	6.3	—
CA08G, Na^+	0.9	5.2	—
CA12G, Na^+	1.8	4.2	—
CA16G, Na^+	2.6	3.5	—

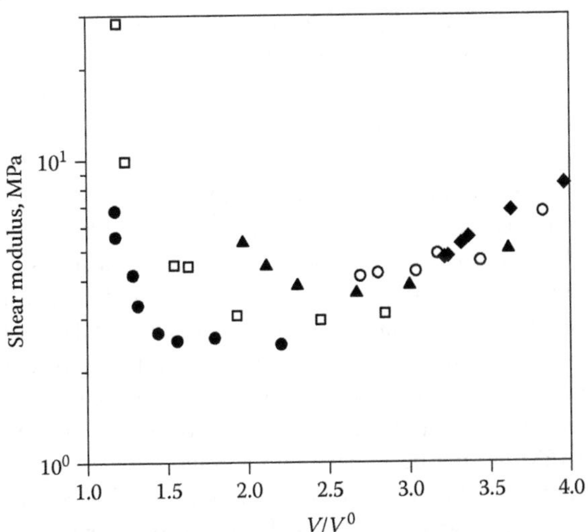

FIGURE 1.5 Shear moduli of fully and partially neutralized Amberlyst IRC86 (WCE). Nominal degree of neutralization α = 1 (filled diamonds); α = 0.75 (open circles); α = 0.5 (filled triangles); α = 0.25 (open squares); α = 0 (filled circles).

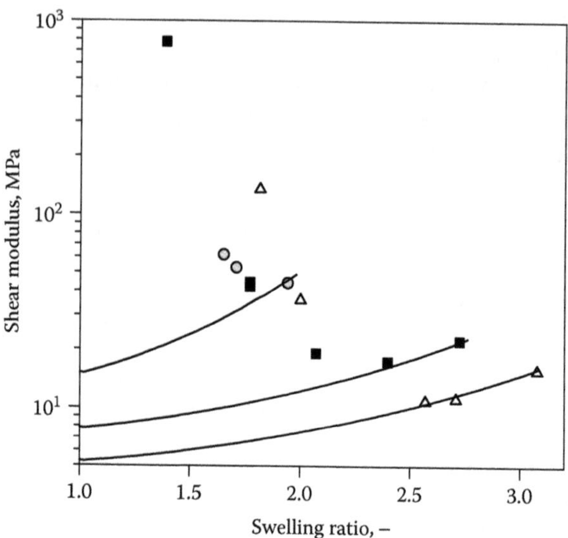

FIGURE 1.6 Shear moduli of KEF 76M (open triangles); CS16G (filled squares); and Amberlyst 15M (shaded circles) ion exchange resins in H^+ form at room temperature. Solid lines represent the values calculated with the three-chain model [Equation (1.21)] and the parameters given in Table 1.2.

exchanger (H/Na-IRC86) are depicted. Here α means the nominal degree of neutralization calculated from the added amount of NaOH. The swelling ratio is given as V/V^0 and the values are thus not directly comparable with Figure 1.4.

The data of the H-form resin ($\alpha = 0$) exhibit no minimum, because the driving force for swelling is small due to very limited dissociation of the carboxylic groups. Therefore, the fractional extension of the chains remains low and the system follows the predictions of the Gaussian model. As the number of ionic groups increases with increasing α, the resin swells more and, at the same time, the upturn of the modulus becomes more evident. The network expansion induced by neutralization thus seems to be the primary cause for the increasing modulus.

Qualitatively similar results of some strong cation exchangers in the acid form are shown in Figure 1.6. Because of the higher cross-link density, these data demonstrate the difficulty of obtaining reliable estimates for the elastic parameters. In case of Amberlyst 15M, no separate rubbery region is observed and application of the elastic model is thus questionable. Moreover, the absolute values of the resins should be compared with caution, because the heterogeneous structure of Amberlyst 15M and KEF 76M has not been taken into account in calculation of G.

The rapid upturn of the modulus at low swelling ratios is evident in all data shown in Figure 1.3 through Figure 1.6, and it indicates the onset of glass transition at the temperature of the measurement.[40] The modulus of the dry resins is in the range of 2 to 4 GPa, which is typical for glassy polymers. The glass transition temperature, T_g, of sulfonated polystyrenes and of polyacrylates in dry state is well

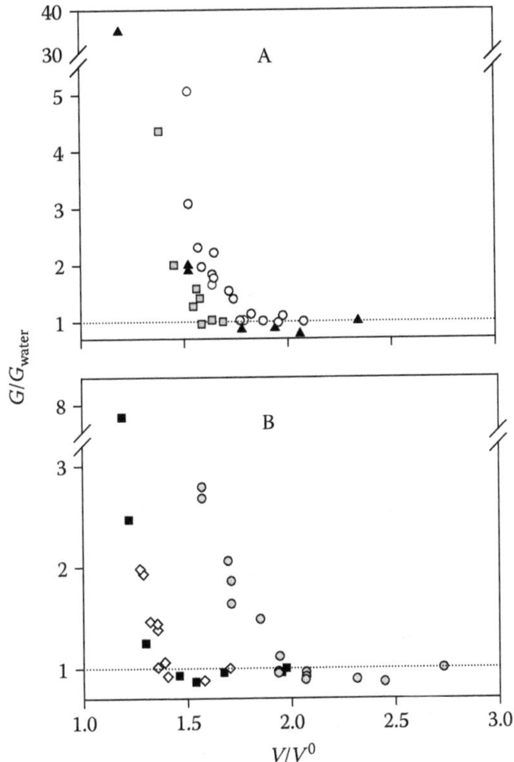

FIGURE 1.7 Effect of counterion on the transition from rubbery to glassy state. A: Strong cation exchanger (X8); H^+ (filled triangles); Na^+ (open circles); La^{3+} (shaded squares). B: Strong cation exchanger (X5.5); Na^+ (shaded circles); $(CH_3)_4N^+$ (filled squares); strong anion-exchanger (X5.5); Cl^- (open diamonds).

above room temperature[40] but decreases markedly due to the plasticizing effect of water or other absorbed solvents. The transition takes place at a characteristic swelling range depending on the resin backbone, ionic group, and counterion. The same factors also appear to influence the width of the transition region, and it is thus impossible to compare the systems in terms of exact numeric values.

The influence of polymer matrix and cross-linking on the transition was already shown in Figure 1.3 through Figure 1.6 and some additional examples are given in Figure 1.7. For convenience, the results are given here as relative values, G/G_{water}, where G_{water} is the modulus measured in pure water.

The weak cation exchanger has a transition located at slightly lower q values than the strong cation exchanger, as can be seen from Figure 1.3 and Figure 1.4. It is likely that the sulfonated PS-DVB resin with aromatic side chains is more rigid than the acrylic resin, having a much lower aromatic ring content (cross-links only).[41] The data of Figure 1.4 also indicate a shift to higher swelling ratios when the acrylic resin is rendered to sodium form. At the same time, the transition

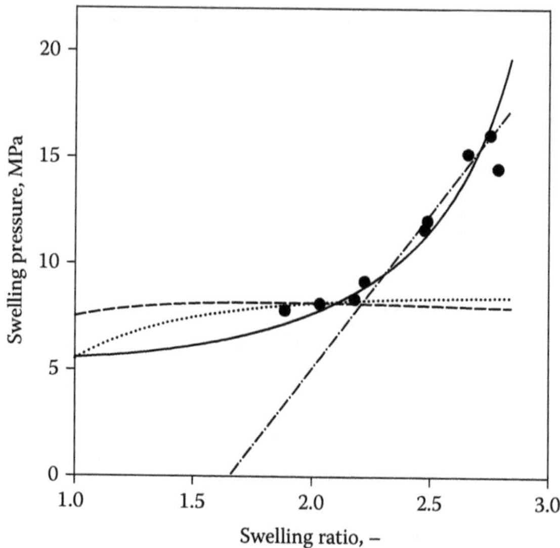

FIGURE 1.8 Correlation of the swelling pressures of a H-form sulfonated PS-DVB resins (X8)[42] with different elastic models. Flory model [Equation (1.19)] (dashed line); Gusler–Cohen model [Equation (1.20)] (dotted line); three-chain model [Equation (1.22)] (solid line); and linear model [Equation (1.25)] (dash-dotted line).

appears to become less definite. A similar trend from the acid form to the univalent salt form is observed also for the strong cation exchanger (Figure 1.7A).

According to Figure 1.3 and Figure 1.4, the transition region also moves toward lower swelling degrees when the cross-link density increases, although the effect is weak. The data in Figure 1.7B indicate that a more pronounced shift is produced by a large counterion [$(CH_3)_4N^+$ in the sulfonated resin] or by a bulky functional group [$–N^+(CH_3)_3$ in the strong anion exchanger]. All these cases can be considered equivalent to internal plasticization, where the glass transition temperature is lowered by introduction of bulky side chains in the polymer backbone. The analogous influence of the three-valent counterion (La^{3+} in Figure 1.7B) may be due either to the large size of the highly hydrated cation or to the electrostatic cross-linking effect.

Swelling Pressure

The predictions of different elastic models mentioned in Section 1.2.1.3 are compared in Figure 1.8. The experimental data for H-Dowex® 50 X8 are taken from Boyd and Soldano[42] and Nandan and Gupta.[43] The swelling ratios have been calculated from the reported water contents and resin equivalent volumes.

Figure 1.8 clearly demonstrates the inability of the Gaussian and near-Gaussian models [Equation (1.19) and Equation (1.20)] to correlate the swelling pressure data of a densely cross-linked resin. As expected, they cannot explain the upturn near the swelling maximum, where the fractional extension of the chains is large. The situation is thus analogous with the modulus data shown in Figure 1.3 and

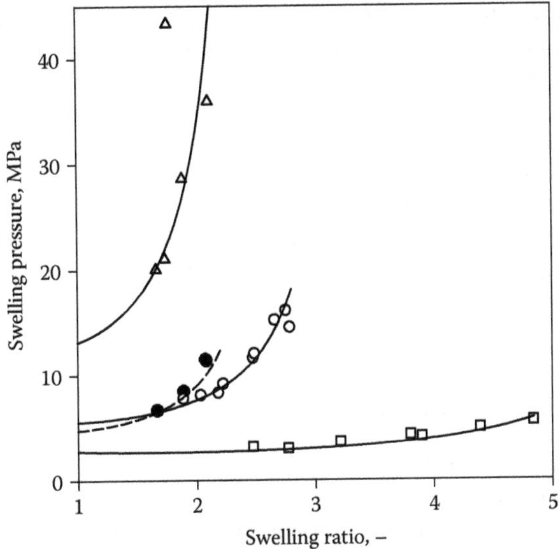

FIGURE 1.9 Swelling pressure in strong cation exchange resins in H^+ (open symbols); Na^+ (filled symbols) form. Dowex 50 X4 (squares); Dowex 50 X8 (circles); Dowex 50 X16 (triangles).[42,43] Solid lines have been calculated from Equation (1.22) with the parameters given in Table 1.2.

Figure 1.4. The inadequacy of the statistical models with a Gaussian chain distribution was pointed out already by Boyd and Soldano,[42] and therefore they introduced a linear model given in Equation (1.25). Here q^* is the swelling ratio extrapolated to zero swelling pressure and B is the average bulk modulus.

$$\pi_{sw} = \frac{B}{q^*}(q - q^*) \tag{1.25}$$

The linear model gives a good approximation for the steeply rising part of the curve but fails at the low and moderate swelling ratios. The straight line in Figure 1.8 was calculated with B = 24 MPa; by using a Poisson ratio of 0.35,[42] this value corresponds to a shear modulus of about 8 MPa. As shown in Figure 1.5, this is less than half of the value measured experimentally in this study.

However, Figure 1.8 also shows that the data near the swelling maximum can be correlated equally well with the non-Gaussian model [Equation (1.22)] and, at the same time, a more realistic behavior is predicted at lower swelling ratios. Correlations with Equation (1.22) for other strong cation exchangers in acid and sodium forms are shown in Figure 1.9. The experimental values were obtained from the water vapor sorption results of Boyd and Soldano[42] and Nandan and Gupta.[43] The calculated curves were obtained with the parameter values given in Table 1.2. As discussed earlier, the values for K_{el} and N_{stat} were taken from the mechanical measurements and only the value of d_π was adjusted.

Both the experimental and calculated curves indicate the enormous pressure increase as the cross-link density increases. It is important to remember, however, that the transition to the glassy state starts around $q = 2$, and below this value the swelling pressure becomes an ill-defined quantity. Figure 1.9 also demonstrates the suitability of Equation (1.22) in correlation of all the data sets. In particular, the swelling pressures of the X8 resins can be explained with the parameters derived from the mechanical measurements of the CS series resins, provided that a suitable correction factor is used. The actual numeric values of d_π are not important, because the data in Figure 1.9 have been measured with the Dowex 50 resins and the values probably include also a contribution from the different resin matrices.

1.2.3.2 Equilibrium Swelling

Swelling equilibrium can be considered as balance between the osmotic pressure trying to expand the resin and the elastic stress counteracting this tendency. As discussed in Section 1.2.1, these contributions can be formally expressed in terms of the mixing interactions and the swelling pressure. Quantitative estimation of the latter was considered in the previous section, and two different methods have been used to evaluate the mixing effects. Several authors[13,14,22] have used the experimental resin sorption data to estimate the parameters involved and then the volume changes are obtained additively from the amounts of individual components. Alternatively, the mixing parameters can be estimated from the data measured with an uncross-linked polymer having the same structure as in the resin.[15] This "predictive" approach is considered below in order to illustrate the role of mixing and elastic contributions in sorption and volume swelling of strong and weak cation exchangers.

Sorption of Solvents

In order to illustrate the application of the equilibrium model to ion exchange resins, sorption of individual solvent components is considered in two ternary systems: water–ethanol–SCE and water–ethanol–WCE. The experimental data consist of sorbed amounts of water and ethanol in the NaCS and NaCA resins of different cross-link densities (for the resin characteristics, see Table 1.1). Moreover, phase diagrams were determined for the noncross-linked model polymers Na-poly(styrenesulfonic acid) (NaPSS) and Na-poly(acrylic acid) (NaPA).[15] The experimental data were correlated with the proposed thermodynamic model using the following strategy. First, the interaction parameters for the water–ethanol and water–polyelectrolyte binaries were estimated from the literature data. As an example, correlation of the isopiestic and osmotic data of NaPSS[44–47] and NaPA[48] with Equation (1.14) is shown in Figure 1.10. Here, the water activities, a_w, of different data sets are expressed as reduced osmotic pressure $\pi_{osm}/RT = -\ln a_w/V_{m,w}$ and $V_{m,w}$ is the partial molar volume of water.

Next, the parameters for the ethanol–polyelectrolyte pair were obtained from the ternary phase equilibrium data of the linear polymers (see Figure 1.11 and

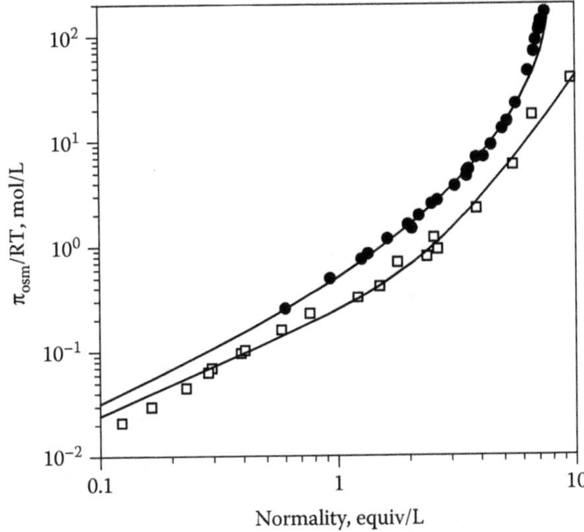

FIGURE 1.10 Correlation of the isopiestic and osmotic data of NaPSS (filled circles) and NaPA (open squares) with the mixing model given in Equation (1.14). The experimental data of NaPSS and NaPA are taken from References 44 through 47 and 48, respectively.

Figure 1.12). The interaction parameters were then used in simulation of the sorption data measured with the cross-linked resins (CS and CA series, see Table 1.1). The elastic contribution was calculated using the mechanical parameters of Table 1.2 as the starting point and adjusting the value of K_{el} if necessary.[15]

The phase diagrams of the linear and cross-linked systems are compared in Figure 1.11 and Figure 1.12. For clarity, only the experimental and calculated equilibrium curves are shown and the tie lines are omitted. The calculated values are depicted as solid lines and they were obtained with the parameters given in Reference 15. In the latter case, the equilibrium curves are located at high water contents and only the upper part of the triangle is shown.

As shown in Figure 1.11, the experimental data of the cross-linked NaCS resins indicate a tendency towards a common limiting curve, from which the equilibrium lines then deviate depending on the cross-link density. Moreover, the limiting curve appears to match with the binodal of the linear polymer. The close correlation is reasonable because of the chemical similarity of the linear and cross-linked materials; all monomers in the resins are structurally similar and they can be functionalized in the sulfonation step. Further evidence for the similarity between linear NaPSS and cross-linked NaCS resins has been found by replacing ethanol with 2-propanol.[15]

However, the phase equilibrium of the NaPSS/NaCS system is complicated by glass transition taking place at high polymer contents. In this region, comparison with the liquid–liquid equilibrium (LLE) data becomes meaningless, because the mechanisms controlling sorption in the glassy state are different from the processes in homogeneous solutions. According to the mechanical measurements,

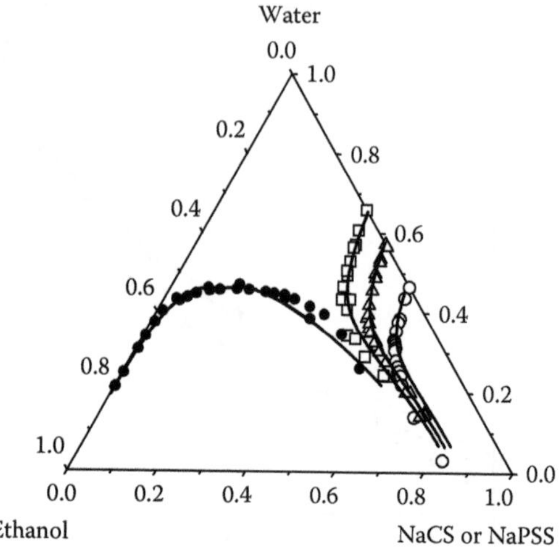

FIGURE 1.11 Phase diagrams for linear NaPSS (filled symbols) and for cross-linked NaCS resins (open symbols) in water–ethanol mixtures at 298 K. NaCS08G (squares); NaCS11G (triangles); and NaCS16G (circles). The solid lines are the equilibrium curves calculated with the parameter values of Reference 15. The compositions are in weight fractions.

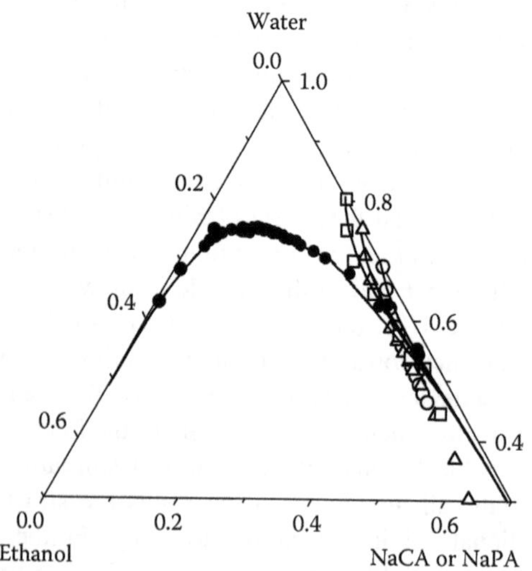

FIGURE 1.12 Phase diagrams for linear NaPA (filled symbols) and for cross-linked NaCA resins (open symbols) in water–ethanol mixtures at 298 K. NaCA06G (squares); NaCA08G (triangles); and NaCA12G (circles). The solid lines are the equilibrium curves calculated with the parameter values of Reference 15. The compositions are in weight fractions.

transition to the glassy state starts at room temperature at polymer weight fractions of about 0.65, which is not far from the point where the sorption data shown in Figure 1.11 reach the limiting curve. Above this limit, the system can be considered as homogeneous polymer solution and the basic assumptions of the thermodynamic model are valid. In terms of the model, the limiting curve is determined only by the activity terms of Equation (1.10), while the deviation from this curve is described by the pressure term. Exact separation of these two contributions is, of course, impossible, but in this case the simple additivity gives a reasonably good correlation. It is important to note, however, that some adjustment was needed for the elastic parameters obtained from the mechanical measurements.[15]

The phase diagram of the linear polyelectrolyte, NaPA, and the cross-linked NaCa resins (Figure 1.12) show a qualitatively similar behavior as in the previous case. However, in terms of ethanol compatibility the two systems are markedly different. It is evident from the binodal of NaPA that the ethanol content in the concentrated phase never exceeded 10 wt%, while the concentrated NaPSS phase contained more than 30 wt% of ethanol near the (estimated) critical point (Figure 1.11). The difference is even more obvious from the data of the cross-linked materials. The very narrow homogeneous region of the NaCA resins indicates that the water uptake selectivity is substantially higher than in the NaCS resins. Basically the ternary water–ethanol–resin system forms a salting-out system, where exclusion of ethanol from the resin phase depends on the concentration and type of the ionic groups. Because the cation is same in both cases and the ion concentrations also are similar, the selectivity difference should have its origin in the influence of the sulfonate and carboxylate anions on the absorbed species, mainly water. The salting-out effect is probably balanced to some extent by interactions between ethanol and the polymer matrix, especially when unfunctionalized segments are present.

Figure 1.12 shows that the interaction parameters of the linear NaPA and the elastic parameters given in Reference 15 explain satisfactorily the phase equilibria of the cross-linked resins at high water contents. As the water content decreases and the polymer content increases, deviations from the experimental values become more and more pronounced. The increasing compatibility of the NaCA resins with ethanol most probably stems from the hydrophobicity of the divinylbenzene cross-links.

The binodals of the linear polymers in Figure 1.11 and Figure 1.12 also imply the importance of the counterions in these systems. As discussed elsewhere,[15] the curves are located much more symmetrically than in case of neutral polymers, and the difference can be explained by the mixing entropy of the free counterions. The counterions are included explicitly in the swelling models of dilute polyelectrolyte gels (e.g., Reference 30); also, in the present model, they are treated as a separate component (see Section 1.2.1.2). The calculations indicate, however, that this effect becomes very small in the cross-linked resins of this study and can be safely neglected for any practical purposes.

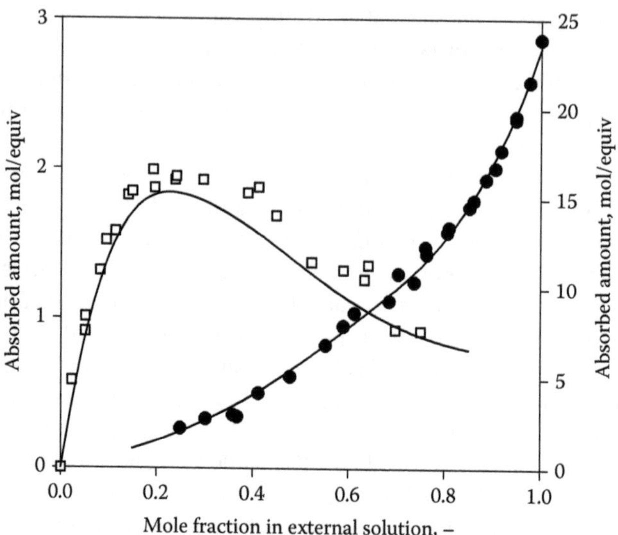

FIGURE 1.13 Sorption isotherms of water (filled circles, right axis) and ethanol (open squares, left axis) in the system water–ethanol–NaCS08G at 298 K. The solid lines represent the values calculated with the parameter values given in Reference 15.

Although the phase diagrams of Figure 1.11 and Figure 1.12 give a good overall picture of the ternary systems, the mutual interactions of water and ethanol are difficult to discern. Therefore, the experimental and calculated data of the lowest cross-link density in Figure 1.11 are replotted in the form of sorption isotherms (Figure 1.13). As shown in Figure 1.13, the mutual enhancement effect is clear, especially in sorption of the alcohol, and the maximum occurring in the isotherm is about three times the value attained in pure ethanol. The latter is estimated as 0.7 mmol/equiv, but because of the very slow equilibration and hysteresis effects[49] in the glassy state, the value is somewhat uncertain. The isotherm shape can be explained quite well with the thermodynamic model; in terms of the model parameters, the effect stems from the favorable interactions between ethanol and water in the resin phase.

The mutual enhancement or cosolvent effect is less pronounced in acid-form resins, because the difference in sorption of water and organic solvents is smaller. For example, the acid-form strong cation exchanger swells nearly equally well in ethanol and water. If, however, the polarity of the absorbing components differs markedly, a clear effect is observed as shown for acetic acid and ethyl acetate in Figure 1.14. Activities of the absorbing components rather than the external solution compositions are used here to facilitate comparison between the widely different systems.

As seen in Figure 1.14, the amount of acetic acid sorbed from an aqueous mixture may be up to twice as large as the amount sorbed from a mixture with ethyl acetate at the same liquid phase activity. The increased sorption stems partly

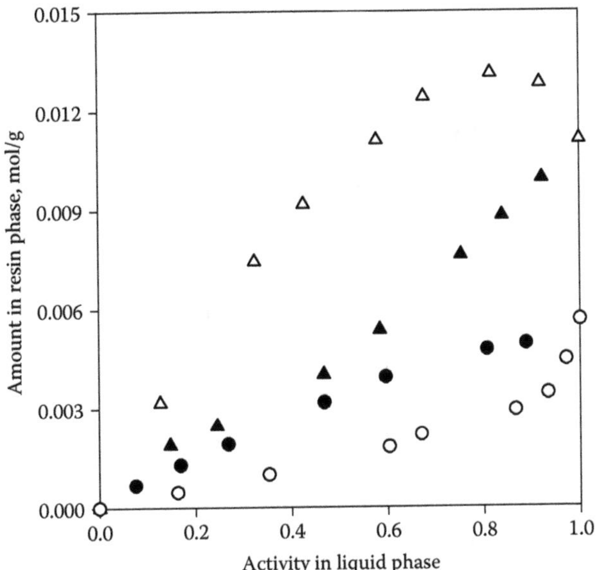

FIGURE 1.14 The cosolvent effect in the sorption of acetic acid and ethyl acetate in H-KEF 76M ion exchange resin at 298 K.[22] Symbols: (△) HOAc from HOAc–water; (▲) HOAc from HOAc–EtOAc; (○) EtOAc from EtOH–EtOAc; (●) EtOAc from HOAc–EtOAc. Liquid phase activities were calculated according to Reference 22.

from the larger extent of swelling in presence of water and partly from the favorable interactions between water and acetic acid in the resin phase. In a similar way, ethyl acetate is more effectively sorbed with acetic acid because, according to vapor–liquid equilibrium (VLE) data,[22] the mutual compatibility is far better for this pair than for the ethyl acetate–ethanol mixture.

Volume Changes in Solvent Mixtures

It was shown in Section 1.2.3.1 that the common ion exchange resins with a PS-DVB or acrylic backbone undergo a glass transition at a specific swelling range. Because of the "frozen" structure in the glassy state, the volume additivity observed in the rubbery region is expected not to hold. The nonlinear behavior at low water contents was first reported by Boyd and Soldano[42] and it has been ascribed to variation of the molar volumes due to electrostriction.[50] The volumetric behavior should also indicate the presence of a gel transition, which in contrast to the glass transition can be considered as a first-order phase transition.[51]

The swelling ratios of the strong and weak Na-form cation exchangers, which were equilibrated in aqueous ethanol solutions, are shown in Figure 1.15. The calculated values obtained additively from the predicted absorbed amounts are given as solid curves. Constant molar volumes of 18.0 and 58.5 mL/mol were used for water and ethanol, and the equivalent volumes of the strong and weak cation exchangers were 130 to 140 mL/equiv and 45 to 50 mL/equiv, respectively.

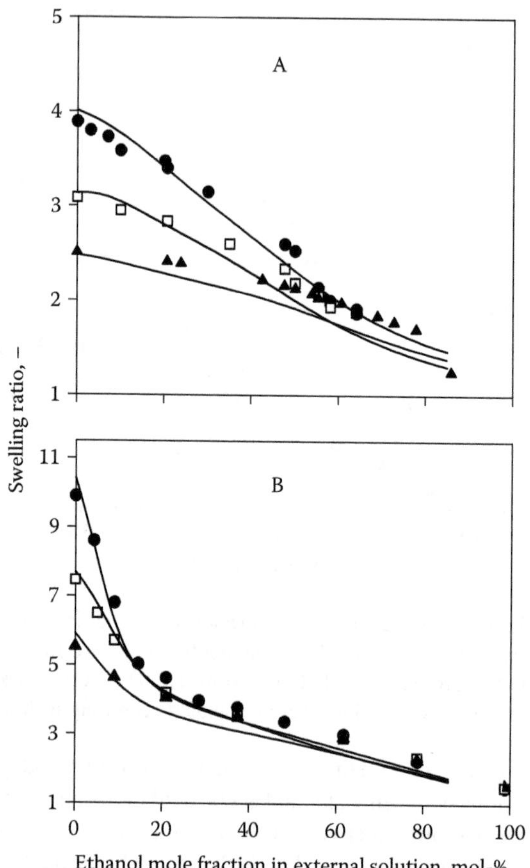

FIGURE 1.15 Volume swelling of strong (A) and weak (B) Na-form cation exchangers in water–ethanol mixtures. A: CS08G (filled circles); CS11G (open squares); CS16G (filled triangles). B: CA06G (filled circles); CA08G (open squares); CA12G (filled triangles). The solid lines represent the calculated values.

The data of the NaCS resins (Figure 1.15A) suggest a close correlation between the glass transition and the swelling behavior. The glass transition occurs at the same external liquid composition range (around 60 mol% of ethanol) where the influence of cross-linking on the volume swelling disappears. The latter phenomenon is indicative of a transition to a gel, and the PS-DVB matrix seems to undergo a direct change from the glassy state to a swollen gel within a relatively narrow composition range. The calculated and experimental curves agree qualitatively, but in the transition region the measured swelling ratios tend to be higher. This is mainly due to the inability of the model to take into account the hydrophobic interactions of ethanol and the PS-DVB matrix.

However, the glass and gel transitions need not coincide, as is shown in Figure 1.15B for the NaCA resins. The glass transition of this polymer matrix starts at an ethanol concentration of about 75 mol% (Reference 40), and this transition seems to be reflected in the swelling curves as a bend. A much stronger volume increase occurs, however, at much more dilute ethanol solutions and the transition to a true gel starts at 20 mol% of ethanol. This behavior stems from the sorption properties of the weak acid cation exchange (WCE) resin: the influence of the cross-link density on the water and ethanol isotherms vanishes around 20 mol% of ethanol; at higher ethanol concentrations, all values lie on a single curve. The fact that the glass transition and the gel transition are widely separated in the WCE resin, while they practically coincide in the strong acid cation exchange (SCE) resin, suggests stronger chain–chain interactions in the former material. The exact interaction mechanisms, however, are unknown. The model calculations give again a rather good correlation with the experimental data in the gel region, but otherwise the calculated values are too low because of underestimation of the ethanol sorption.

1.3 SOLVENT DIFFUSION AND SWELLING KINETICS

We now turn our attention to dynamics of the sorption phenomena in gel-type ion exchange resin beads. Diffusion of liquids in weakly or moderately cross-linked resins is usually coupled with changes in the total volume and density of the particle, which makes diffusion in resins an intriguing problem. Dependency of the solvent diffusion coefficients on the density of the polymer network must be adequately described, and the mathematical modeling approach (including the numerical solution scheme) must be capable of handling dimensional changes of the particle.

On the molecular level, diffusion in liquids is usually described as a series of thermally activated jumps into cavities formed as a result of molecular vibrations and motion. This framework lends itself readily at describing the diffusion of solvents in a cross-linked polymer network as well. The diffusion rate in a swollen cross-linked polymer network is known to depend, on one hand, on the size of the diffusing species relative to the mesh width of the network and, on the other hand, on the interactions between the polymer chains and the diffusing species. The situation becomes more complex when the transport of solvents requires rearrangement of the polymer network. The rate at which polymer segments relax and react to changes in the solvent composition, temperature, or other properties of the system can be low enough to affect the mass transport during swelling and shrinking of a cross-linked polymer.

Alfrey et al.[52] classify diffusion of solvents in polymers into three types: Type I (Fickian), Type II (relaxation-controlled), and Type III (anomalous or non-Fickian). In Type I behavior, the rate of molecular diffusion is much less than that of the relaxation processes of the polymer segments, and the polymer network reacts rapidly to changes in the solvent composition or temperature. This is also the case when the change in the solvent composition is small enough not to affect the structure of the polymer network. In Type II diffusion, the relaxation processes are very slow compared to the rate of diffusion. This is the case when a glassy cross-linked polymer is placed in contact with a good solvent; a sharp traveling front between the swollen rubbery and the nonswollen glassy regions is developed. The velocity of this front controls the sorption rate. Type III diffusion is observed when the characteristic times of diffusion and relaxation are of the same order of magnitude. Since gel-type ion exchange resins are in the rubbery state in separation processes, the mathematical models in the following do not include time-dependent relaxation processes, and are intended for describing Type I diffusion phenomena only.

1.3.1 Model Development

Coupled diffusion of solvents and volume changes of particles are conveniently modeled by using a Lagrangian coordinate system where the mass of the polymer is used as the spatial variable.[53] Such a frame of reference will be here referred to as *mass coordinate system*. The coordinate transformation between radial coordinates and mass coordinates is defined by Equation (1.26) and an arbitrary reference point, for example, $W(R{=}0){=}0$. In Equation (1.26), W and R are the spatial variables in the mass and radial coordinate systems, respectively, A is the cross-sectional area of the particle, and ρ_m is the density of the polymer.

$$\frac{\partial W}{\partial R} = \frac{A(W)\rho_m}{q(W)} \tag{1.26}$$

Several approaches to modeling diffusion of solvents in cross-linked polymers have been presented in the literature. The essence of the Tanaka et al.[54] and Tanaka and Fillmore[55] approach is that liquids do not diffuse into the polymer network but the polymer network diffuses into a stagnant fluid. The diffusion coefficient of the polymer network is defined using the bulk modulus of the network and a solvent–polymer friction coefficient. Modifications of the approach have also been presented[56] to include the shear modulus of the polymer in the definition of the network diffusion coefficient. Also, the swelling kinetics of strong acid PS-DVB ion exchange resins in water have been interpreted in terms of the Tanaka–Fillmore approach.[57] Here we are interested in diffusion-induced swelling of gels in liquid mixtures with two or more components. Unfortunately, no extension of the Tanaka–Fillmore approach to such systems is available. In rigorous treatments of the coupled diffusion and swelling problem, the presence of the cross-links and the resulting limited expansibility of the polymer network are taken into account by introducing stress tensors in equations of motion.[53,58,59] The volumetric flux into a volume element of the particle is reduced if it results in an increase in the stress of the polymer network.

Here we adopt a simpler approach that does not involve stress tensors: the effect of the cross-links is embedded into the diffusion driving force through the solvent activity, which depends on the extent of swelling, as already discussed in connection with the phase equilibrium model. The resins will be again treated as homogeneous, cross-linked polymer solutions. The solvent concentrations are thus based on the volume of the swollen network, including the polymer chains (nonstructured model).

1.3.1.1 Maxwell–Stefan Approach

An extended Flory–Huggins model for the chemical potential (activity) of the solvents in a cross-linked polymer solution was presented in Section 1.2.1.2 [Equations (1.14) and (1.22)]. The same model is now applied in calculating the diffusive fluxes using the generalized Maxwell–Stefan approach. Neglecting viscous flow within the polymer network, which can be expected to be very small in moderately cross-linked resins, the force balance in mass coordinates reads

$$-\frac{\partial \mu_i}{\partial W} = \frac{q^2}{A^2 \rho_{\mathrm{m}}^2} \sum_{\substack{j=1 \\ j \neq i}}^{NP} \frac{R_{\mathrm{g}} T}{Ð_{ij}} \phi_j \left(u_i^{\mathrm{W}} - u_j^{\mathrm{W}} \right) \tag{1.27}$$

where μ is the chemical potential, $Ð$ is the Maxwell–Stefan diffusion coefficient, and u^{W} denotes species velocity in the frame of reference. As before, the index NP refers to the polymer. Since we focus on nonelectrolyte solutions, the counterions (index NP+1 in Section 1.2.1) are not considered mobile.

Velocity differences in the force balances are somewhat impractical, and they can be replaced with molar fluxes N_i, as shown in Equation (1.28).

$$-\frac{\partial \mu_i}{\partial W} = \frac{q}{A\rho_{\mathrm{m}}} \sum_{\substack{j=1 \\ j \neq i}}^{NP} \frac{R_{\mathrm{g}} T}{Ð_{ij}} \frac{V_{\mathrm{m},i} \phi_j}{\phi_i} N_i - \frac{q}{A\rho_{\mathrm{m}}} \sum_{\substack{j=1 \\ j \neq i}}^{NP} \frac{R_{\mathrm{g}} T}{Ð_{ij}} V_{\mathrm{m},j} N_j \tag{1.28}$$

Unique values for the molar fluxes cannot be determined using a set of force balance equations alone because there is at least one unknown variable more than there are linearly independent equations. A particularly simple bootstrap relation, namely $N_i = 0$, can be applied in the present case because the polymer does not move with respect to itself, and the molar fluxes are readily obtained by employing the usual methods of linear algebra.

Although the concentration dependency of the Maxwell–Stefan diffusion coefficients in nonideal solvent mixtures is remarkably weaker than that of those defined using a concentration gradient driving force, this effect should be taken into account. The interpolation formula shown in Equation (1.29), which employs only diffusion coefficients at infinite dilution D_i^0,[60] was chosen to approximate the $Ð_{ij}$ in the multicomponent mixture.

$$Ð_{ij} = \left(D_i^0 \right)^{\frac{1-\phi_i+\phi_j}{2}} \left(D_j^0 \right)^{\frac{1+\phi_i-\phi_j}{2}} \tag{1.29}$$

Solvent–polymer diffusion coefficients were calculated with the simple obstruction model shown in Equation (1.30), which is based on that proposed by Mackie and Meares.[61] The reference state in the original model is free liquid. In this work, however, the reference state was chosen as a hypothetical, infinitely dilute solution with respect to the polymer, and the diffusion coefficients in the reference state, $Đ_i^{S,ref}$, are regarded as adjustable parameters.

$$Đ_i^{S} = Đ_i^{S,ref}\left[\left(1-\phi_p\right)/\left(1+\phi_p\right)\right]^2 \tag{1.30}$$

In the case of molecules that are large relative to the free interchain distance in the swollen resin, such a geometric obstruction model is clearly an oversimplification. However, nearly all diffusion coefficient correlations proposed in the literature share a common drawback: they contain one or more undefined structural parameters that have to be estimated from experimental diffusion coefficient data for each system. Comparison of various correlations is outside the scope of the present work, and can be found elsewhere.[62]

1.3.1.2 Fick's Law

In order to compare predictions of the Maxwell–Stefan approach with a model that does include the effect of cross-links, we introduce Fick's first law in the mass coordinate system

$$N_i = -D_i^{S}\frac{A\rho_m}{q}\frac{\partial C_i}{\partial W} \tag{1.31}$$

where D_i is a Fickian diffusion coefficient and C is concentration. It should be noted that, owing to our choice of the frame of reference, there is no convection during swelling or shrinking with respect to the reference plane across which the diffusive fluxes are evaluated. In addition, the numerical value of D_i^S is not affected by the coordinate transformation, and values defined in the conventional manner using laboratory-fixed coordinates can be used. However, it should also be noted that the usual assumption of negligible molar average velocity of the solvent mixture with respect to the coordinate system has not been removed, and D_i^S should be regarded as an apparent quantity.

The geometric obstruction model in Equation (1.30) is used in the case of Fick's law as well by replacing D_i^S and $Đ_i^{S,ref}$ with D_i^S and $D_i^{S,ref}$. The interaction between the solvent molecules was neglected.

1.3.2 Experimental

Solvent diffusion in gel-type ion exchange resins was studied both at constant swelling ratio and under conditions where diffusion is coupled with volume changes of the particle. Apparent Fickian diffusion coefficients of water, ethanol, acetic acid, and ethyl acetate in SCE X8 resin (CS16G, see Table 1.1) at different swelling ratios were estimated from batch sorption kinetics data. The extent of swelling of the resin was varied by changing the ionic form of the resin (H^+, Na^+) and the solvent environment (water, ethanol). The average diameter (averaged by using the surface area to volume ratio) of the resin particles was determined for each solvent environment and ionic form using an optical microscope and image analysis. In the H^+ form, the average diameter of water-swollen particles was 845 μm.

The amount of dry resin was 6 to 17 wt% of the liquid phase mass, the volume of which was typically 300 cm^3 (100 cm^3 in the case of sorption of water from ethanol and D_2O (99.8 wt%) from water). The mole fraction of the solute (water, ethanol, acetic acid, or ethyl acetate) after injection into the main solvent was typically 0.03 to 0.10. Such large amounts were required because the sorption of liquids into the resin is very low at low swelling ratios. In order to ensure efficient mixing of the solute and the liquid phase, the solute was not injected directly into the adsorber but mixed with an approximately 100 cm^3 sample of the main solvent withdrawn from the adsorber. During the sorption kinetics experiments, samples of approximately 0.5 cm^3 were withdrawn from the vessel, and analyzed offline with a gas chromatograph (Agilent 6890N) fitted with a thermoconductive detector. In the case of water and D_2O sorption, the liquid phase composition was monitored continuously by using a differential refractive index detector (Schambeck RI2000-F) in a side loop.

Coupled solvent diffusion and volume change of the particle was investigated in a specially constructed flow-through cell. One or two resin beads were placed in the cell and equilibrated in a given solvent mixture. A step change was made in the feed composition, and the resulting dimensional changes of the particles were recorded with a digital camera attached to an optical microscope. The sampling interval was varied between 1 and 30 s, depending on the volume change rate of the particles. The volume of the cell was approximately 0.1 cm^3 and the feed volumetric flow rate was typically 6.0 cm^3 min^{-1}. The particle size was determined offline by using image analysis. The swelling ratio at the initial and final states of the swelling and shrinking experiments was calculated from experimental sorption isotherms.[22] These values were then used to convert the particle size data to swelling ratios.

Shrinking and swelling kinetics of SCE X4 (Na^+) and SCE X8 (H^+) were studied using water–ethanol and water–acetic acid mixtures. In the former case, the particles were initially equilibrated in water, which was then displaced by aqueous ethanol solutions (5 to 90 mol% EtOH) resulting in shrinking of the particles. In swelling kinetics experiments with SCE X8 resin, the initial state was a 50 mol% aqueous acetic acid solution that was displaced with water. Shrinking kinetics were studied by restoring the original feed concentration.

1.3.3 Results and Discussion

In this section, experimental data on solvent diffusion in strong acidic cation exchange resins as well as kinetics of the accompanying particle swelling and

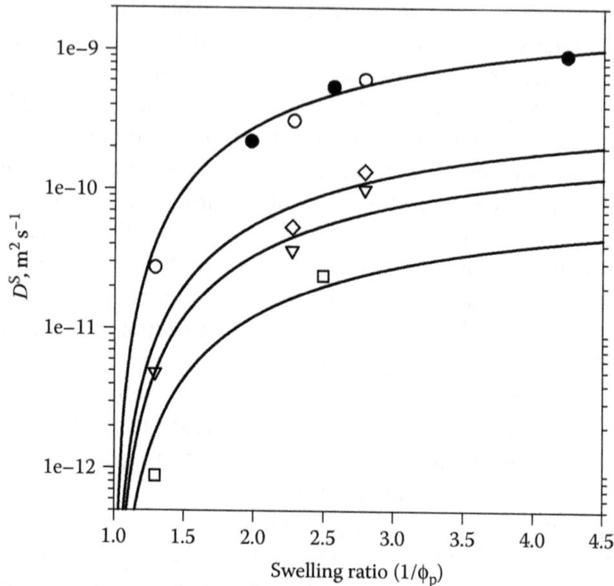

FIGURE 1.16 Apparent Fickian diffusion coefficients of water (circles); ethanol (diamonds); acetic acid (triangles); and ethyl acetate (squares) in H^+ form SCE X8 resin. Open symbols: data from Reference 22; filled symbols: data from Reference 64. Solid lines: calculated using Equation (1.30) with $D_i^{S,ref}$ values given in the text.

shrinking kinetics are discussed. Predictions of two modeling approaches, generalized Maxwell–Stefan model and Fick's law described in Section 1.3.1, are compared with the experimental results. Details of the calculation methods can be found in References 22 and 63.

1.3.3.1 Fickian Diffusion Coefficients at Constant Swelling Ratio

The relationship between the diffusion rate and the extent of swelling is illustrated in Figure 1.16. It should be noted that the use of the mass coordinate system instead of the conventional radial coordinates in the parameter estimation does not influence the numerical value of the apparent diffusion coefficient, D^S. As seen in Figure 1.16, the diffusion coefficients of all solvents decrease rapidly as the resin shrinks. This is due to increasing tortuosity of the medium and decreasing cross-sectional area available for diffusion between the polymer chains. Moreover, as the volume fraction of solvents in the resin decreases, the interactions between the solvents and the functional groups of the resin become more significant. In the present case, however, these interactions are not very strong, except perhaps for sorption of water from ethanol, and the last effect is considered to be of minor importance.

It is also seen in Figure 1.16 that the diffusion coefficients of D_2O and H_2O in water-swollen and ethanol-swollen GX8 resins are in accordance with those given by Meares[64] for resins of the same type but with different cross-link densities.

As discussed in Section 1.2.3.1 (see Figure 1.6), the increase in the shear modulus at swelling ratios below 2.2 indicates the transition from rubbery to glassy state, and below 1.5 the shear modulus approaches values that are typical for glassy polymers (see Figure 1.6). The experimental data in Figure 1.16 imply, however, that the glass transition has no drastic effect on diffusion coefficients. This is in contrast with results obtained by Wang et al.[65] with nonionic poly(methyl methacrylate) (PMMA) membranes. Also the diffusion coefficients of small molecules, such as diacetyl, in nonswollen PMMA collapse due to glass transition.[66] The main difference in the chemical structure of PS-DVB resins and PMMA membranes is that the former contains bulky aromatic rings and large sulfonic acid moieties.

The observation that diffusion coefficients in PS-DVB ion exchange resins are rather insensitive to glass transition could mean that thermal fluctuations of the cross-linked polymer network do not control the diffusion rates even in the rubber-like state. In addition, it is possible that the polymer network in the glassy state remains somewhat open even when the amount of absorbed solvents is very low. This may be related to the fact that the equivalent volume of linear polystyrene sulfonate has been found to decrease linearly with decreasing amount of sorbed water except at low water contents where it decreases at a lower rate.[46] The equivalent volume of dry NaPSS is approximately 8% larger than the equivalent volume of the polymer matrix obtained by extrapolating from data at large swelling ratios. As mentioned earlier, the classical explanation for this phenomenon is that the molar volume of water changes due to strong electric field inside the resin (electrostriction).[46] Despite these speculations, it should be recalled that the glassy state of the polymer is known to influence the sorption of liquids when diffusion assumes reorientation of the polymer segments.[67]

The solid lines in Figure 1.16 are calculated by using the Mackie and Meares correlation, Equation (1.30). The reference values $D_i^{S,ref}$ for water, ethanol, acetic acid, and ethyl acetate were 2.5×10^{-9}, 0.4×10^{-9}, 0.3×10^{-9}, and 0.11×10^{-9} $m^2\ s^{-1}$, respectively. The reference value used for water is close to the diffusion coefficient in free liquid, but those of the other components are significantly lower. This indicates that the mesh width of the polymer network is too small for free Brownian motion of molecules larger than water.[68] In the present work it is assumed that the accessible volume for diffusion depends on the extent of swelling only and not on the solvent environment. As pointed out by Freger et al.,[69] solvation of the counterions by a polar major solvent may decrease the effective cross-sectional area of the apertures in the polymer network. More data at low swelling ratios are required before quantitative theoretical correlations between the solvent molecule size and the swelling ratio can be applied.

1.3.3.2 Swelling and Shrinking Kinetics

The shrinking kinetics of SCE X4(Na^+) resin when water is displaced with aqueous ethanol solutions of increasing alcohol concentration are shown in Figure 1.17. The swelling ratio is plotted against a modified time variable in order

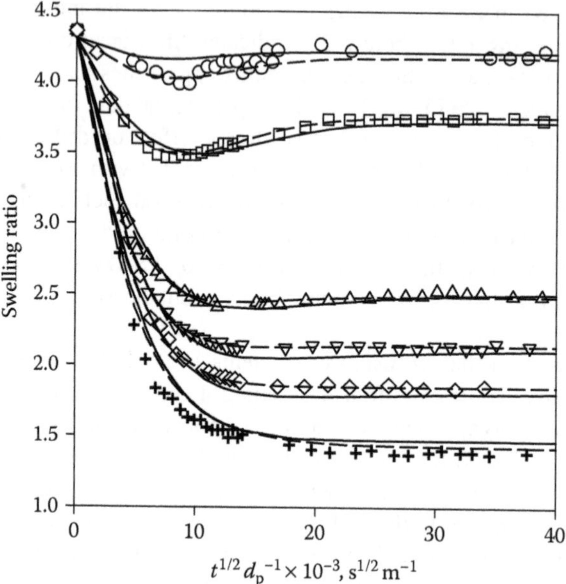

FIGURE 1.17 Shrinking kinetics of SCE X4(Na^+) in ethanol–water mixtures at 298 K. Initial state: resin swollen by water. Ethanol mole fraction in liquid phase and initial particle size: (○) 0.05 mol%, 521 × 10^{-6} m; (□) 0.20 mol%, 622 × 10^{-6} m; (△) 0.50 mol%, 622 × 10^{-6} m; (▽) 0.60 mol%, 597 × 10^{-6} m; (◇) 0.70 mol%, 613 × 10^{-6} m; (+) 0.90 mol%, 612 × 10^{-6} m. Calculated results: Maxwell–Stefan approach (solid line); Fick's law (dashed line).

to enable comparison of the results for particles of different size. A distinct minimum in the swelling ratio is observed at low concentrations of the organic solvent, but it diminishes as the ethanol concentration increases and is not observed above 60 mol% of ethanol. Such a minimum is due to differences in the diffusion coefficients, sorption isotherms, and molar volumes of the diffusants. The resin shrinks rapidly at the early stages because the volumetric flux of water out of the resin is larger than that of ethanol into the resin. At the minimum, the volumetric fluxes have become equal. When no minimum is observed, the volumetric fluxes are not equal until they are very small.

According to mechanical measurements, the SCE X4(Na^+) resin enters the glass transition region as the swelling ratio decreases below 2.5.[16] Thus the polymer is not completely rubbery at the final state in high ethanol concentrations. Nevertheless, the shape of the shrinking curves indicates that the diffusion coefficients of the solvents are not greatly influenced by the glass transition. The behavior is similar to that of the solvent diffusion coefficients in the SCE X8(H^+) resin (Figure 1.16), and thus supports the conclusion that the free space available for diffusion in the glassy resin is large enough to abate the effect of reduced mobility of the polymer segments.

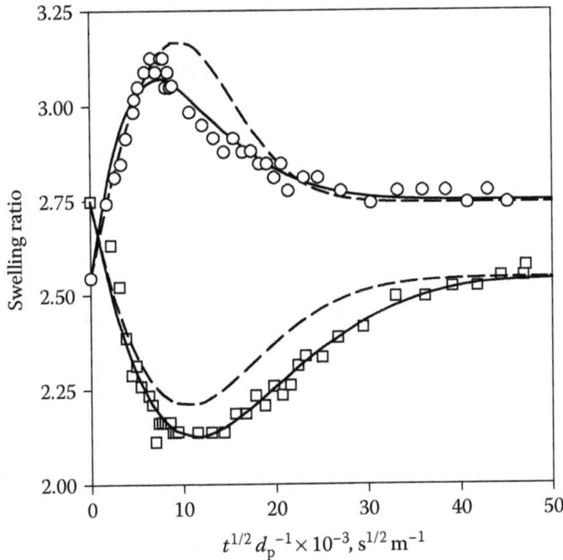

FIGURE 1.18 Swelling and shrinking kinetics of SCE X8(H^+) at room temperature. Initial and final states: water and 50 mol% HOAc–water mixture. Initial particle sizes: 984 μm (○); 903 μm (□). Lines were calculated with Maxwell–Stefan approach (solid line) and Fick's law (dashed line) by using diffusion coefficients given in the text and activity model parameters given in Reference 22.

The shrinking and swelling of SCE X8(H^+) resin as the external phase is changed from water to a 50 mol% aqueous acetic acid mixture and vice versa are displayed in Figure 1.18. In addition to a minimum in the shrinking curve, a distinct maximum in the swelling curve is observed. In this case, the volumetric flux of water into the resin exceeds that of acetic acid out of the resin. The data in Figure 1.18 also show that the final state is attained more slowly during shrinking than swelling. This phenomenon was already observed in early works with ion exchange resins and other cross-linked polymers.[1] Since shrinking starts from the surface of the particle, the solvents have to diffuse through a dense region, the thickness of which only increases with time.

Calculated shrinking and swelling kinetics curves obtained by using the Maxwell–Stefan approach and the Fick's law are plotted in Figure 1.17 and Figure 1.18. The diffusion coefficients at the reference state used in the calculations are given in Table 1.3. It is worthwhile noting that the $D_i^{S,ref}$ value obtained for water is close to that estimated from the shrinking and swelling data of the SCE X8(H^+) resin. The Flory–Huggins interaction parameters and solvent–solvent Maxwell–Stefan diffusion coefficients used are given in Reference 22.

If only shrinking kinetics data are used (Figure 1.17), both diffusion models can explain the coupled diffusion and volume changes of the resin when combined with the diffusion coefficient correlation given in Equation (1.30). However, if

TABLE 1.3
Diffusion Coefficients at the Reference State in SCE X4(Na^+) and SCE X8(H^+)

Resin	Component	Fick's Law $D_i^{S,ref}$, m² s⁻¹	Maxwell–Stefan $Đ_{i,p}^{S,ref}$, m² s⁻¹
SCE X4(Na^+)	Water	3.2×10^{-9}	3.6×10^{-9}
	Ethanol	1.0×10^{-9}	2.2×10^{-10}
SCE X8(H^+)	Water	3.3×10^{-9}	2.6×10^{-9}
	Acetic acid	8.5×10^{-10}	1.9×10^{-10}

Note: The values are estimated from swelling and shrinking data in Figure 1.17 and Figure 1.18.

both shrinking and swelling kinetics are to be explained simultaneously, the Maxwell–Stefan model outperforms the Fickian model (Figure 1.18). To improve the fit to the shrinking data in Figure 1.18, the water to acetic acid diffusion coefficient ratio should be increased, which would lead into an even higher overshooting in the swelling kinetics curve. The Maxwell–Stefan model, on the other hand, accounts for the finite expansibility of the polymer network through the chemical potential of the diffusing species, and the swelling behavior of the resin is predicted more accurately.

This point can be further illustrated by comparing the intraparticle swelling ratio profiles calculated with the two diffusion models shown in Figure 1.19. The maximum in the local swelling ratio inside the particle grows higher as it advances towards the center of the particle. To the left of the maximum, the extent of swelling increases due to diffusion of water towards the particle center. To the right of the maximum, on the other hand, the swelling ratio decreases due to diffusion of acetic acid out of the resin. Both the Maxwell–Stefan model and the Fickian model predict that the center of the particle will undergo the largest changes in the extent of swelling, but only Fick's law predicts unrealistically large swelling ratios inside the particle. This is attributed to the fundamental difference in these models: incorporation of the swelling pressure in the chemical potential term in the Maxwell–Stefan model properly accounts for the finite extensibility of the polymer chains. An increase in the extent of swelling increases the chemical potential as discussed in Section 1.2.1, and the diffusive flux to a highly swollen region in the particle is reduced. Even if the difference in the diffusion coefficients of the two liquids is very large — three orders of magnitude, for example — the maximum swelling ratio inside the particle as predicted by the Maxwell–Stefan model does not exceed a value of 3.35.

Finally, it should be noted that a central assumption of these diffusion models is that the polymer network is isotropic with respect to the cross-link density. This is not true in reality, and swelling is long known to cause internal mechanical

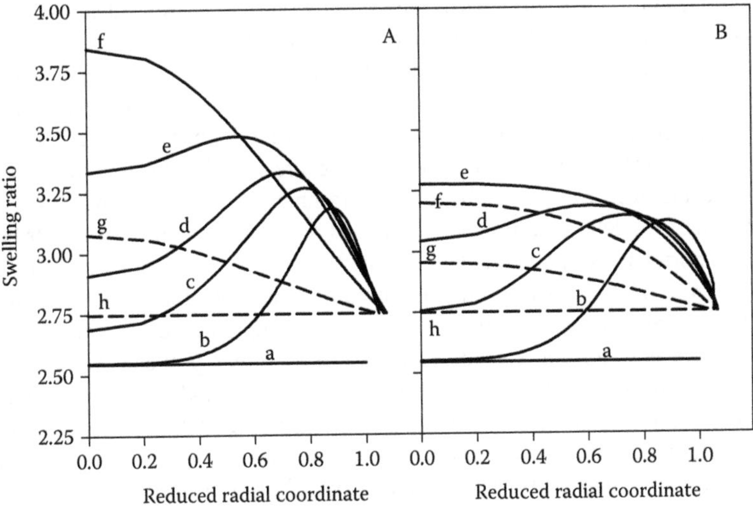

FIGURE 1.19 Local swelling ratios in a SCE X8(H^+) particle during swelling. Calculated with Fick's law (A) and the Maxwell–Stefan approach (B). Initial state: 50 mol% aqueous acetic acid mixture; final state: water. Solid lines represent the situation before and dashed lines after reaching the maximum swelling ratio. Time values (in minutes): a = 0; b = 0.25; c = 0.55; d = 0.85; e = 1.30; f = 2.90; g = 6.90; and h = ∞. Reduced radial coordinate = $R/R(t{=}0\ \text{min})$.

stresses in ion exchange resin particles.[1] These stresses often lead to cracking of the particles, which was also encountered in the experiments carried out for this work.

1.4 SUMMARY AND CONCLUSIONS

In the present work, the role of the elastic properties of ion exchange resins on solvent sorption equilibria and kinetics was investigated experimentally. Several strong and weak acid cation exchange resins with different ionic forms, polymer matrices, and a wide range of cross-link densities were studied. Mixtures of polar and weakly polar solvents were used as the swelling media. Moreover, the use of a thermodynamic modeling approach that explicitly takes into account the elasticity of the polymer network was demonstrated. A general equilibrium condition for a swollen resin was derived, and explicit expressions were given for the entropy of mixing, the enthalpic interactions, and the elastic reaction of the cross-linked network.

The elastic properties of the resins were quantified by direct mechanical measurements of shear modulus of swollen resin particles. Also, swelling pressure data extracted from vapor sorption isotherms were used. The shear modulus measurements give valuable information about the state of the polymer network;

a steep rise in the shear modulus as the swelling ratio decreases indicates the onset of isothermal glass transition. Moreover, such data are better suited for evaluating network elasticity correlations (swelling pressure models) than phase equilibrium data because the effects of solvent–polymer interactions and the resin elasticity are decoupled. It was shown that a non-Gaussian model based on the three-chain concept can accurately describe the swelling pressure inside moderately cross-linked resins over a relatively large swelling range. On the other hand, the ideal or modified models assuming an essentially Gaussian chain-length distribution failed to correlate even qualitatively the shear modulus and swelling pressure data.

The role of the elastic properties in equilibrium swelling was illustrated by two ternary model systems, and the required mixing parameters were estimated from phase equilibrium data of uncross-linked linear polyelectrolytes with a similar chemical structure. The additivity of the elastic and mixing contributions implied in the model was shown to give a reasonable good agreement with the experimental data at large swelling ratios. The failure of the model at low swelling ratios was attributed partly to the isothermal glass transition and partly to the heterogeneous structure of the commercial resins. The latter factor was important in acrylic WCE resins cross-linked with divinylbenzene, because the nonfunctionalized cross-links render the material more hydrophobic.

Swelling kinetics was studied by two methods. The pronounced effect of swelling ratio on the diffusion coefficient observed in batch sorption experiments at constant swelling ratio was described with a simple geometric obstruction model. Experiments of coupled diffusion and volume changes showed that local minima and maxima in the swelling ratio occur during shrinking and swelling of the resins, respectively. Correlation of these data with the proposed diffusion models indicated that the elastic properties of the resin must be taken into account when describing the dynamics of resin swelling. The elastic influence was incorporated in the Maxwell–Stefan diffusion model by evaluating the thermodynamic driving force with the same expression as in the phase equilibrium calculations. The Fickian diffusion model that does not contain the elastic reaction of the polymer network was found to result in unrealistically large intraparticle swelling ratios, especially close to the particle center. Resin shrinking is less influenced by the elasticity and both diffusion models describe the dynamics with practically the same accuracy, when the diffusion coefficients are used as adjustable parameters.

The elastic stress of the swollen network thus poses an additional constraint for the diffusion in the resin particle. In the present model, this effect is described by the swelling pressure, and diffusing species of different size are thus affected to different extent. The situation may be compared with diffusion of ions in an ion exchange resin, where the condition of local electroneutrality adjusts the diffusion rates of individual ions. Here the limiting factor is the local electrical potential and the effect depends on the charge of the diffusing ion. In fact, the present model can be easily generalized for charged species as well by including the electrochemical potential as an additional driving force.

ACKNOWLEDGMENTS

The authors are indebted to all of our coworkers at the LUT laboratory of Industrial Chemistry who have participated in the studies considered in this paper. In particular, the contribution of Dr. Jari Tiihonen is gratefully acknowledged. Finex Oy (Finland) is thanked for providing the ion exchange resins.

REFERENCES

1. Helfferich, F., *Ion-Exchange*, Dover Publications, New York, 1995.
2. Dorfner, K., *Ion Exchangers*, Walter de Gruyter & Co., Berlin, 1991.
3. Samuelson, O., *Ion Exchange Separation in Analytical Chemistry*, John Wiley & Sons, New York, 1963.
4. Moody, G.J. and Thomas, J.D.R., Inorganic ion exchange in organic and aqueous-organic solvents, a review, *Analyst*, 93, 557, 1968.
5. Heumann, W.R., Ion exchange in nonaqueous and mixed media, *CRC Crit. Rev. Anal. Chem.*, 2, 425, 1971.
6. Gregor, H.P., Nobel, D., and Gottlieb, M.H., Studies on ion-exchange resins. XII. Swelling in mixed solvents, *J. Phys. Chem.*, 59, 10, 1955.
7. Rückert, H. and Samuelson, O., Die Verteilung von Äthylalkohol und Wasser bei Ionenaustauschern auf Harzbasis, *Acta Chem. Scand.*, 11, 303, 1957.
8. Pietrzyk, D.J., Ion-exchange resins in non-aqueous solvents. III. Solvent-uptake properties of ion-exchange resins and related adsorbents. *Talanta*, 16, 169, 1969.
9. Sinegra, J.A. and Carta, G., Sorption of water from alcohol-water mixtures by cation-exchange resins, *Ind. Eng. Chem. Res.*, 26, 2437, 1987.
10. Flory, P.J., *Principles of Polymer Chemistry*, Cornell University Press, Ithaca, New York, 1953.
11. Maurer, G. and Prausnitz, J.M., Thermodynamics of phase equilibrium for systems containing gels, *Fluid Phase Equilib.*, 115, 113, 1996.
12. Lode, F., A Simulated Moving Bed Reactor (SMBR) for Esterifications, Ph.D. dissertation, Shaker Verlag, Aachen, 2002.
13. Mazzotti, M., Neri, B., Gelosa, D., Kruglov, A., and Morbidelli, M., Kinetics of liquid-phase esterification catalyzed by acidic resins, *Ind. Eng. Chem. Res.*, 36, 3, 1997.
14. Tiihonen, J., Laatikainen, M., Markkanen, I., and Paatero, E., Sorption of neutral components in ion-exchange resins. 1. Effect of cross-link density and counterion on selective sorption of water–ethanol mixtures in sulfonated PS–DVB resins, *Ind. Eng. Chem. Res.*, 38, 4832, 1999.
15. Laatikainen, M., Markkanen, I., Tiihonen, J., and Paatero, E., Liquid-liquid equilibria in ternary systems of linear and cross-linked water-soluble polymers, *Fluid Phase Equilib.*, 201, 381, 2002.
16. Tiihonen, J., Markkanen, I., Kärki, A., Äänismaa, P., Laatikainen, M., and Paatero, E., Modelling the sorption of water–ethanol mixtures in cross-linked ionic and neutral polymers, *Chem. Eng. Sci.*, 57, 1885, 2002.
17. Anasthas, H.M. and Gaikar, V.M., Adsorption of acetic acid on ion-exchange resins in non-aqueous conditions, *React. Funct. Polym.*, 47, 23, 2001.

18. Tiihonen, J., Laatikainen, M., Markkanen, I., and Paatero, E., Sorption of neutral components in ion-exchange resins. 2. Sorption of d-xylose in sulfonated PS–DVB resins from water–ethanol mixtures, *Ind. Eng. Chem. Res.*, 38, 4843, 1999.
19. Tiihonen, J., Sainio, T., Kärki, A., and Paatero, E., Co-Eluent effect in partition chromatography: rhamnose–xylose separation with strong and weak cation exchangers in aqueous ethanol, *J. Chromatogr. A*, 982, 69, 2002.
20. Mazzotti, M., Neri, B., Gelosa, D., and Morbidelli, M., Dynamics of a chromatographic reactor: esterification catalyzed by acidic resins, *Ind. Eng. Chem. Res.*, 36, 3163, 1997.
21. Musante, R.L., Grau, R.J., and Baltanás, M.A., Kinetic of liquid-phase reactions catalyzed by acidic resins: the formation of peracetic acid for vegetable oil epoxidation, *Appl. Catal. A: Gen.*, 197, 1653, 2000.
22. Sainio, T., Ion-exchange Resins as Stationary Phase in Reactive Chromatography, Acta Universitatis Lappeenrantaensis 218, Dissertation, Lappeenranta University of Technology, Lappeenranta, 2005.
23. Pouchly, J., Zivny, A., and Solc, K., Thermodynamic equilibrium in the system macromolecular coil–binary solvent, *J. Polym. Sci. Part C*, 23, 245, 1968.
24. Sainio, T., Laatikainen, M., and Paatero, E., Phase equilibria in solvent mixture–ion-exchange resin catalyst systems, *Fluid Phase Equilib.*, 218, 269, 2004.
25. Manning, G.S., Limiting laws and counterion condensation in polyelectrolyte solutions. I. Colligative properties, *J. Chem. Phys.*, 51, 924, 1969.
26. Manning, G.S. and Ray, J., Counterion condensation revisited, *J. Biomol. Struct. Dyn.*, 16, 461, 1998.
27. Dubrovskii, S.A., Compressional and shear behavior of weakly ionic polyacrylamide gels, *Polym. Gels Networks*, 4, 467, 1996.
28. Gusler, G.M. and Cohen, Y., Equilibrium swelling of highly cross-linked polymeric resins, *Ind. Eng. Chem. Res.*, 33, 2345, 1994.
29. Treloar, L.R.G., *The Physics of Rubber Elasticity*, 3rd ed., Clarendon Press, Oxford, 1975.
30. Hasa, J., Ilavsky, M., and Dusek, K., Deformational, swelling, and potentiometric behavior of ionized poly(methacrylic acid) gels. I. Theory, *J. Polym. Sci., Polym. Phys. Ed.*, 13, 253, 1975.
31. Zaroslov, Y.D., Philippova, O.E., and Khokhlov, A.R., Change of elastic modulus of strongly charged hydrogels at the collapse transition, *Macromolecules*, 32, 1508, 1999.
32. Knaebel, A., Rebre, S.R., and Lequeux, F., Determination of the elastic modulus of superabsorbent gel beads, *Polym. Gels Networks*, 5, 107, 1997.
33. Briscoe, B.J., Liu, K.K., and Williams, D.R.J, Adhesive contact deformation of a single microelastomeric sphere, *Colloid Interface Sci.*, 200, 256, 1998.
34. Hertz, H., Ueber die Beruhrung Fester Elastischer Körper, *Z. Reine Angew. Math.*, 92, 156, 1882.
35. Liu, K.K., Williams, D.R., and Briscoe, B.J., The large deformation of a single micro-elastomeric sphere, *J. Phys. D: Appl. Phys.*, 31, 294, 1998.
36. Christensen, S.G. and Thomsen, K., Experimental measurement and modeling of the distribution of solvent and ions between an aqueous phase and an ion-exchange resin, *Fluid Phase Equilib.*, 228–229, 247–260, 2005.
37. Oppermann, W., in *Swelling Behavior and Elastic Properties of Ionic Hydrogels, in Polyelectrolyte Gels*, ACS Symp Ser 480, Harland, R.S. and Prud'homme, R.K., Eds, American Chemical Society, Washington, DC, 1992, pp. 159–170.

38. Schröder, U.P. and Oppermann, W., Mechanical and stress-optical properties of strongly swollen hydrogels, *Makromol. Chem., Macromol. Symp.*, 76, 63, 1993.
39. Nisato, G., Skouri, R., Schosseler, F., Munch, J.P., and Candau, S.J., Elastic behavior of salt-free polyelectrolyte gels, *J. Faraday Discuss.*, 101, 133, 1995.
40. Tiihonen, J., Markkanen, I., Laatikainen, M., and Paatero, E., Elasticity of ion-exchange resin beads in solvent mixtures, *J. Appl. Polym. Sci.*, 82, 1256, 2001.
41. Mark, J.E., Eisenberg, A., Graessley, W.W., Mandelkern, L., and Koenig, J.L., *Physical Properties of Polymers*, American Chemical Society, Washington, DC, 1984, pp. 55–94.
42. Boyd, G.E. and Soldano, B.A., Osmotic free energies of ion exchangers, *Z. Electrochem.*, 57, 162, 1953.
43. Nandan, D. and Gupta, A.R., Solvent sorption isotherms, swelling pressures, and free energies of swelling of polystyrenesulfonic acid type cation exchangers in water and methanol, *J. Phys. Chem.*, 81, 1174, 1977.
44. Waxman, M.H., Sundheim, B.R., and Gregor, H.P., Studies in ion exchange resins. VI. Water vapor sorption by polystyrenesulfonic acid, *J. Phys. Chem.*, 57, 969, 1953.
45. Reddy, M. and Marinsky, J.A., Further investigation of the osmotic properties of hydrogen and sodium polystyrenesulfonates, *J. Phys. Chem.*, 74, 3884, 1970.
46. Timmermann, E.O., Untersuchungen über konzentrierte Polyelektrolytlösungen und Ionenaustauscher. II. Quellung und partielle molare volumina, *Z. Phys. Chem. Neue Folge*, 72, 140, 1970.
47. Gluekauf, E.A., Theoretical treatment of cation exchangers I. The prediction of equilibrium constants from osmotic data, *Proc. Royal Soc.*, 214A, 207, 1952.
48. Hiraoka, K. and Yokoyama, T., Analyses of sorption isotherms of poly(acrylic acid) salts according to sorption theories, *Kenkyu Hokoku-Nagasaki Daigaku Kogakubu*, 20, 51, 1990 (in Japanese).
49. Mattisson, M. and Samuelson, O., The sorption of polyalcohols from aqueous alcohol by cation exchange resins, *Acta Chem. Scand.*, 1395, 1958.
50. Boyd, G.E., The nature of the selective binding of ions by polyelectrolyte gels: volume and entropy change criteria, in *Charged Gels and Membranes*, Vol. I, Selegny, E., Ed., Reidel Publishing Co., Dordrecht, 1976, pp. 73–89.
51. Tanaka, T., Phase transitions of gels, in *Polyelectrolyte Gels*, ACS Symp. Ser. 480, Harland, R.S. and Prud'homme, R.K., Eds., American Chemical Society, Washington, DC, 1992.
52. Alfrey, T., Gurnee, E.F., and Lloyd, W.G., Diffusion in glassy polymers, *J. Polym. Sci.*, C12, 249, 1966.
53. Billovits, G.F. and Durning, C.J., Polymer material coordinates for mutual diffusion in polymer–penetrant systems, *Chem. Eng. Commun.*, 82, 21, 1989.
54. Tanaka, T., Hocker, L., and Benedek, G.B., Spectrum of light scattered from a viscoelastic gel, *J. Chem. Phys.*, 59, 5151, 1973.
55. Tanaka, T. and Fillmore, D.J., Kinetics of swelling of gels, *J. Chem. Phys.*, 70, 1214, 1979.
56. Peters, A. and Candau, S.J., Kinetics of swelling of polyacrylamide gels, *Macromolecules*, 19, 1952, 1986.
57. Ikkai, F. and Shibyama, M., Microstructure and swelling behavior of ion-exchange resin, *J. Polym. Sci. B, Polym. Phys.*, 34, 1637, 1996.
58. Carbonell, R.G. and Sarti, G.C., Coupled deformation and mass-transport processes in solid polymers, *Ind. Eng. Chem. Res.*, 29, 1194, 1990.

59. El Afif, A. and Grmela, M., Non-Fickian mass transport in polymers, *J. Rheol.*, 46, 591, 2002.
60. Xu, X. and Cui, Z.F., Modeling of the co-transport of cryoprotective agents in porous medium as a model tissue, *Biotechnol. Progr.*, 19, 972, 2003.
61. Mackie, J.S. and Meares, P., The diffusion of electrolytes in a cation-exchange resin membrane. I. Theoretical., *Proc. Roy. Soc. (London)*, A232, 498, 1955.
62. Amsden, B., Solute diffusion within hydrogels, mechanisms and models, *Macromolecules*, 31, 8382, 1998.
63. Sainio, T., Tiihonen, J., and Paatero, E., Mass Transfer Coupled with Volume Changes in Ion-Exchange Resin Beads, IEX 2004 conference, July 4–7, 2004, Cambridge, England.
64. Meares, P., Transport in ion-exchange polymers, in *Diffusion in Polymers*, Crank, J. and Park, G.S., Eds., Academic Press, London, 1968.
65. Wang, B-G., Yamaguchi, T., and Nakao, S., Prediction of solvent solubility, diffusivity, and permeability in glassy polymeric membranes, *Polymer*, 42, 5225, 2001.
66. Karlsson, O.J., Stubbs, J.M., Karlsson, L.E., and Sundberg, D.C., Estimating diffusion coefficients for small molecules in polymers and polymer solutions, *Polymer*, 42, 4915, 2001.
67. Kuipers, N.J.M. and Beenackers, A.A.C.M., Non-Fickian diffusion with chemical reaction in glassy polymers with swelling induced by the penetrant: a mathematical model, *Chem. Eng. Sci.*, 48, 2957, 1993.
68. Crank, J., *The Mathematics of Diffusion*, Oxford University Press, Oxford, 1975.
69. Freger, V., Korin, E., Wisniak, J., Korngold, E., Ise, M., and Kreuer, K.D., Diffusion of water and ethanol in ion-exchange membranes: limits of the geometric approach, *J. Membr. Sci.*, 160, 213, 1999.

2 Development of Simulated Moving Bed Reactor Using a Cation Exchange Resin as a Catalyst and Adsorbent for the Synthesis of Acetals

Viviana M.T.M. Silva, Ganesh K. Gandi, and Alírio E. Rodrigues

CONTENTS

2.1 Introduction 46
2.2 Applications 47
  2.2.1 Perfumes and Flavors 47
  2.2.2 Additives to Fuel 49
    2.2.2.1 Oxygenated Additives to Gasoline 49
    2.2.2.2 Oxygenated Additives to Diesel 52
2.3 Synthesis of Acetals 55
  2.3.1 Acid Catalysis 57
  2.3.2 Equilibrium Displacement 57
  2.3.3 Reactive Separations 60
2.4 Chemicals, the Catalyst, and Methods of Analysis 60
2.5 Acetalization Reaction in Batch Reactor 61
  2.5.1 Experimental Setup 61
  2.5.2 Thermodynamic Results 61
  2.5.3 Kinetic Model 65
    2.5.3.1 Batch Reactor Model 68
    2.5.3.2 Optimization of the Kinetic Parameters 70

2.6 Acetalization Reaction in Fixed Bed Reactor 71
2.6.1 Experimental Setup 75
2.6.2 Modeling of Fixed Bed 76
2.6.3 Binary Adsorption Experiments 77
2.6.4 Reaction Experimental Procedure 79
2.7 Acetalization Reaction in Simulated Moving Bed Reactor 81
2.7.1 Simulated Moving Bed Concept 82
2.7.2 Simulated Moving Bed Reactor (SMBR) 84
2.7.3 SMBR Experimental Unit (LICOSEP 12-26) 85
2.7.4 Mathematical Model 87
2.7.4.1 Bulk Fluid Mass Balance to Component i and in Section j 87
2.7.4.2 Pellet Mass Balance to Component i and in Section j 88
2.7.4.3 SMBR Performance Criteria 90
2.7.5 Experimental Results and Discussion 91
2.8 Conclusions 92
2.9 Nomenclature 93
Greek Letters 95
Subscripts 95
References 96

2.1 INTRODUCTION

The main purpose of this work is to study the synthesis of acetaldehyde dimethylacetal and acetaldehyde diethylacetal [considered here as 1,1-dimethoxyethane (DME) and 1,1-diethoxyethane (DEE), respectively], focusing on the kinetic and thermodynamic aspects, along with the adsorption on acid ion exchange resins to support the development of a new technology (simulated moving bed reactor). DME and DEE can be produced by a reversible liquid phase reaction between acetaldehyde and methanol or ethanol, respectively, in acid medium. In reactions limited by chemical equilibrium where more than one product is formed, conversion can be enhanced in a hybrid apparatus where the products are separated as they are formed. Since the resin used in acetal synthesis can act both as a catalyst and as well as a selective adsorbent, the use of a chromatographic reactor-separator — in particular, the simulated moving bed reactor (SMBR) — is a feasible process. Hence, here the process development for the synthesis of DME and DEE synthesis in a SMBR using ion exchange Amberlyst 15 resin as catalyst and as well as selective adsorbent was considered.

The chapter first reviews applications for acetals in various areas with emphasis in perfumes and flavors and as additives to fuel diesel. Then, the methodology for process development including kinetics in a batch reactor, fixed bed adsorptive, and SMBR will be addressed. All these steps are supported by both modeling/simulation and experimental studies.

2.2 APPLICATIONS

Acetals are oxygenated organic compounds represented by the following structure R_2–CH–(O–R_1)$_2$, wherein R_1 and R_2 each independently represents alkyl groups derived from an alcohol and an aldehyde, respectively. Acetals are used in perfumes to impart a fruity odor, in the flavoring of food,[1] and in the aroma of alcoholic beverage, such as spirit drinks.[2] Acetals are also used as starting materials for agricultural chemicals[3] and intermediates for pharmaceuticals products, such as vitamins[4] and analgesics and antifungal.[5] DME and DEE are used as solvent, especially for cellulose and its derivatives,[6] plasticizers, and fuel for internal combustion engines and blenders.[7] Acetals also find widespread use as an intermediate for the synthesis of various industrial chemicals where it is necessary to protect the carboxylic group of aldehydes and ketones, especially alkyl vinyl ethers[8] and synthetic perfumes. It is a minor component of acetal resin,[9] used as adhesives.[10] The main area of polyacetal resins consumption in Western Europe is the automotive sector (39%); domestic, household, and office equipment (14.7%); and the electrical engineering industry (10.8%).[11] The annual consumption of polyacetals in the United States, Japan, and Europe was estimated to be 388,000 tons, but the polyacetal resins are expensive and are being replaced by other cheaper polymers (polypropylene, polystyrene).

In past years, the development of more environmentally acceptable gasoline and fuel has been studied intensively. Reduction of pollutants in exhaust emissions of diesel engines represents a challenge to modern society. Acetals have been under consideration as oxygenated additives to diesel fuel because they reduce drastically the emission of particles and NO_x.[12] The NO_x emissions can be reduced by decreasing the combustion temperature. One way to achieve this lower combustion temperature is to add water to the fuel. However, by adding water, phase separation will occur in most fuel systems, especially at lower temperatures, for example, below 0°C. Some patents address the problem of water in the fuel composition by including in the fuel polyethers or acetals with or without methanol or ethanol. It is known that alcohol-containing fuels provide relatively low emissions of carbon, carbon oxide, and nitrogen oxide.[13,14]

2.2.1 Perfumes and Flavors

Since early antiquity, resins from animal and plant sources have been used extensively for perfumery and flavor purposes. One of the most important steps in the history of natural fragrance materials was the industrialization of essential oils production. With the industrial synthesis of vanillin (1874) and coumarin (1878), a new branch of the chemical industry was founded.[15] The number of synthetically produced fragrance and flavor chemicals has since expanded continuously. Modern perfumery began at the end of the 19th century when the first synthetic fragrance substances became available. The importance of perfumery has increased greatly since then. It includes not only the manufacture of fine products (perfume and eau de cologne) but also the use of perfumes in cosmetic,

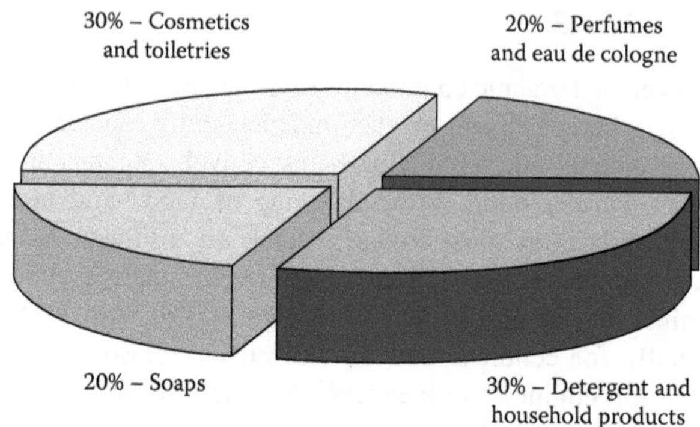

FIGURE 2.1 Applications distribution of perfume oils. (From Sturm, W. and Peters, K., Perfumes, in *Ullmann's Encyclopedia of Industrial Chemistry*, Wiley-VCH, Weinheim, 1999. With permission.)

toiletries, soaps, and household or industrial products. Worldwide consumption of perfume oils for all applications in 1984 was estimated at ca. 1.8×10^9 Euro, as shown in Figure 2.1.[16]

Aliphatic aldehydes are among the most important components used in perfumery. Acetals derived from aliphatic aldehydes have odor characteristics that resemble those of the aldehydes but are less pronounced. In spite of the aroma of acetals being weaker and slightly different from the correspondent aldehyde, they are used as stabilizers in the industry of perfumes. These acetals are used for the design of synthetic perfumes to increase the resistance to oxidation and therefore the lifetime of perfumes. Because of their stability toward alkali, acetals are used as soap perfume oils.[1]

DEE and DME are important fragrances in their own right, as well as raw materials for fragrances and pharmaceutical products. Elimination of alcohol from acetal easily gives, for example, vinyl ether, which is a reactive building block for pharmaceutical and fragrance synthesis.[17] These are higher impact chemicals used to top note fruit flavors (apple, peach) and spirits (whiskey) and contribute a significant strong, sweet, fruity odor to the typical aroma of sherry wines.[18] DEE is referred to as a basic fragrance in perfume and aromatic compositions, by the Scientific Committee on Cosmetic Products and Non-Food Products (SCCNFP) and by the European Flavour and Fragrance Association (EFFA).[19]

Acetals are also used to prepare aqueous liquids since they are extraordinarily suitable as solubility promoters in microemulsions and liquid–crystalline phases. Applications of microemulsions and liquid–crystalline phases are also known as formulations for crop protection, cosmetics, and pharmaceuticals as a new type of solvents and in emulsion polymerization. Particularly suitable are acetals having a total of 5 to 15 carbon atoms.[20] DEE appears in the composition of many cosmetically acceptable aqueous mediums (compatible with keratinous materials such as the skin or hair), such as shampoos[21] and conditioners, styling gels, and mousses,[22] for example.

TABLE 2.1
Environmental Compatibility and Availability of Various Fuels

Fuel	Environmental Problems	Availability
Gasoline (with catalyst)	CO_2, ozone among other	Worldwide
Diesel fuel	CO_2, particle NO_x ozone	Worldwide
Liquefied petroleum gas (LPG)	CO_2, among others	Some countries
Methanol		
Fossil	CO_2, formaldehyde	Locally possible in short time
Biomass	Formaldehyde	—
Ethanol (biomass)	Acetaldehyde	Brazil
Vegetable oils	Particle, among others	Possible locally to some extent

Source: From Fabri, J., et al., Motor fuels, in *Ullmann's Encyclopedia of Industrial Chemistry*, Wiley-VCH, Weinheim, 1999. (With permission.)

2.2.2 Additives to Fuel

In recent years there has been a considerable increase in research and development in the area of environmentally acceptable alternative fuels as well as the use of alternative feedstock as building blocks for applied chemicals and petrochemicals (Table 2.1). Conventional process industries are also faced with restricted emission control requirements. This trend is going to intensify for the next several decades because of the dwindling supply of petroleum reserves, emphasis on environmentally safe chemical technologies, and fierce competitiveness in the global petrochemical market. However, the utilization of alternative fuels requires a separate fuel distribution infrastructure and vehicle fuel systems need to be modified for these fuel properties. In contrast, blending oxygenates in existing gasoline fuel provides an energy equivalence higher than conventional gasoline. Its distribution cost is minor compared with alternative fuels, and the automotive fuel system modification is not needed. In essence, the high octane and environmental benefits make oxygenates more attractive.[23]

2.2.2.1 Oxygenated Additives to Gasoline

At the end of the 1950s, petroleum refinery seemed to have reached a satisfactory technological level representing a good energy-economy balance. However, a great increase in energy consumption and deterioration of the environment led to the recognition during the 1960s that energy production and automotive transportation were major causes of air-quality deterioration. By the 1970s, it became clear that petroleum refining had to meet a new goal: energy, economy, and environment together. Whereas this objective was recognized broadly, in the United States there was a succession of legislative acts and technical interventions (Table 2.2). Since 1970, the U.S. Environmental Protection Agency (EPA) has had the responsibility for setting and ensuring National Ambient Quality Standards (NAAQS).[24]

TABLE 2.2
Legislative and Technical Interventions on U.S. Fuel Market

1955	First congressional law on clean air (Public Law 159)
1970	The Clean Air Act (CAA)
1975	Catalytic converters installed on new cars, lead phase-down
1978	EPA grants waiver for ethanol blending (up to 10 vol%) "gasohol"; ethanol blended into gasoline for the first time
1979	EPA grants waiver for MTBE blending up to 7 vol%
1980	Federal income tax credit established for ethanol blending
1981	EPA defines "sub-sym." to gasoline, fuel containing 2 wt% oxygen (11 vol% MTBE)
1985	EPA sets the allowed lead to 0.5 gplg. Lead banking and trading begin (first environmental credit trading programs)
1987	Colorado oxygenated fuels program for carbon monoxide control begins (first in the nation)
1988	EPA grants "sub-sym." waiver for MTBE blending (up to 15 vol%)
1989	EPA grants "sub-sym." waiver for oxygenate blending (up to 2.7 wt% oxygen); federal Phase 1 Reid Vapor Pressure (RVP) takes effect. ARCO Products introduces EC-1, the first gasoline designed for emission reduction
1990	Clean Air Act Amendments of 1990 introduce the federal oxy-fuel and reformulated gasolines (RFG) program (Tier 1 standards); Internal Revenue Service (IRS) extends to ETBE the federal income tax credit for ethanol blending
1991	ARCO introduces EC-X reformulated gasoline
1992	Federal oxy-fuel program begins federal Phase 2; RVP standards take effect
1993	EPA regulations controlling the sulfur content of diesel fuel extended to 0.05 wt% take effect
1995	Federal RFG program begins
1999	National LEV program begins
2004	Federal Tier 2 standards begins

Source: From Ancillotti, F. and Fattore, V., Oxygenate fuels: market expansion and catalytic aspect of synthesis, *Fuel Process. Technol.*, 57, 163, 1998. (With permission.)

The most publicized environmental measure adopted in the 1970s was the progressive phasing out of lead additives in motor fuels and the introduction of severe emission limits for 1975 that could be met only by the application of catalytic converters for the oxidation of CO and HC in automobile exhaust.

By 1978 even NO_x emissions had to be reduced drastically. This requirement brought about the development of the three-way catalytic converter, allowing a simultaneous reduction of CO, HC, and NO_x. The complete removal of lead additives from gasoline was necessary for the technical operability of catalytic converters because the catalysts are intolerant of lead, which causes rapid deactivation. An important technical consequence of the reduction of lead concentration in gasoline was the notable drop in gasoline octane number that could not be tolerated by millions of automobiles designed for high-octane fuel. The immediate solution for the problem was to raise the octane number with increasing concentrations of butanes and aromatics in gasoline.[24] A second and more innovative solution was the use of a new class of high-octane oxygenated components, called oxygenates.

Oxygenates are liquid organic compounds that are blended into gasoline to increase its oxygen content. During combustion, the additional oxygen in the gasoline reduces the output of carbon monoxide and limits emissions of ozone-forming materials. Oxygenates began to be added to gasoline and to play an increasingly important role in gasoline formulation as octane boosters replaced alkyl lead after 1973. During the oil crises of the early 1970s, oxygenates originating from sources partially alternative to crude oil also played a role as volume extenders.

The oxygenated market was dominated by methyl *tert*-butyl ether (MTBE) due to its octane blending value, relatively low volatility, and complete miscibility with gasoline, low susceptibility to phase separation in the storage and distribution system, and low tendency to undergo peroxidation. MTBE manufacture had a high commercial growth.[23] However, the unleaded gasoline–catalytic converter solution did not completely solve the car emission abatement. At the end of the 1980s, public concern in the United States about the winter's high-CO concentration and summer's ozone levels brought into consideration a deep change in gasoline and diesel fuel composition (reformulation) as an essential means to less road pollution.

In 1990, the Clean Air Act Amendments (CAAA) approved two new programs called the oxygenated fuel program and the reformulated gasoline program (RFG). For the first time, the U.S. government regulated the composition and quality of motor fuels as a matter of environmental and health policy. Specifications for gasoline and diesel fuel were enforced by CAAA because the two new fuels, oxygenated gasoline and RFG, required certain oxygen content. The CAAA gave the oxygenation option a new role with clean air additives, advancing that of the octane supplier. Oxygenated additives as aliphatic alcohols and ethers, in specific volumes and combinations, have similar octane values as conventional gasoline, and therefore they were able to provide the oxygen that gasoline needed under the oxygenated fuels program. It was in this context that MTBE and other ethers were considered oxygenates.

The use of MTBE in reformulated gasoline has resulted in contamination of water resources across the United States, and the introduction of others ethers [ethyl tertiary butyl ether (ETBE), tertiary amyl ether (TAME), di-isopropyl ether (DIPE), and *tert*-butyl isopropyl ether (IPTBE)] as oxygenated fuels are being considered. Ethers like MTBE and ETBE have advantages over lighter alcohols (for example, methanol, ethanol, etc.), especially with lower Reid vapor pressures and lower vaporization latent heats, and so forth (see Table 2.3).

In Europe, unleaded gasoline had been introduced in the mid-1980s, but at different paces in different countries. It anticipated the sales of catalyst-equipped cars, but legislated that all new cars must be capable of operating on the Euro-premium unleaded petrol even if they were actually run on leaded gasoline. Some countries achieved an early introduction by encouraging sales of unleaded petrol and low-emission catalyst-equipped vehicles by using tax concessions; by 1995, Austria, Sweden, and Finland were totally unleaded whereas the share of unleaded petrol was only 22% in Spain. Another important market difference was that,

TABLE 2.3
Oxygenates for Reformulated Gasoline

	Blending Octane (*R*+*M*)/2	Reid Vapor Pressure (psi)	Boiling Point (°C)	Oxygen (wt%)
Ethers				
MTBE	109	8–10	55	18.2
TAME	104	3–5	86	15.7
ETBE	110	3–5	72	15.7
TAEE	105	0–2	101	13.8
DIPE	105	4–5	68	15.7
Alcohols				
Methanol	116	50–60	65	49.9
Ethanol	113	17–22	78	34.7
TBA	101	10–15	83	21.6

Source: From Ancillotti, F. and Fattore, V., Oxygenate fuels: market expansion and catalytic aspect of synthesis, *Fuel Process. Technol.*, 57, 163, 1998. (With permission.)

TABLE 2.4
Legislative and Technical Interventions on European Fuel Market

1983	European Directive 70/220/EEC: First vehicle emissions requirements
1988	European Directive 88/77/EEC: Catalyst first required
1991	European Directive 93/59/EC: Euro 1 limits (3-way catalysts)
1994	European Directive 93/59/EC: Euro 2 limits
2000–2005	European Directive 98/69/EC: Euro 3/4 limits

compared to the United States, diesel fuel played a more important role in the market. Because the main concerns in Europe regarding air quality are ozone, NO_x, and particulate emissions, diesel vehicles (passenger cars as well as heavy-duty diesel engines) were included in the study. European Union legislative acts and technical interventions are presented in Table 2.4.

The auto and oils programs conducted in the United States, Europe, and Japan are the scientific basis for exhaust emission legislation and fuel quality regulations in these countries. Figure 2.2 shows the evolution of emissions legislation.

2.2.2.2 Oxygenated Additives to Diesel

Diesel engines are more efficient than gasoline engines, but they emit 10 to 20 times more particulate matter (PM) than gasoline engines, so the environmental and health effects of diesel engine emissions continue to be a significant concern (Table 2.5). It is known that addition of oxygenates to diesel from petroleum lowers particulate matter emissions. Just adding oxygenated compounds to

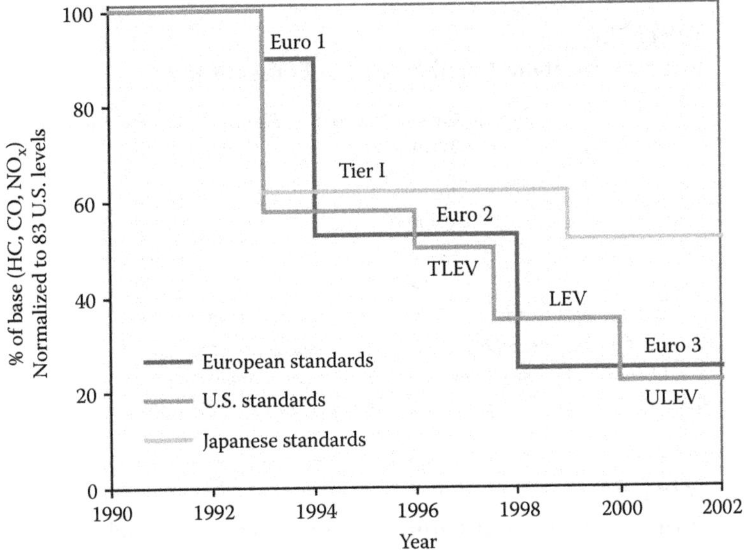

FIGURE 2.2 Evolution of emissions legislation. (From Homeister, N.L., Vehicle emissions standards around the globe in *The New Zealand Road to Cleaner Air*, Ed., World Energy Council, New Zealand, 2001. With permission.)

TABLE 2.5
Emissions from Conventional Light Duty Vehicles Made between 1981 and 1984 under Typical Urban Driving Conditions

	PM gr/km	SO_2 gr/km	NO_x gr/km	CO gr/km	VOC gr/km
Diesel oil	0.3	0.35	0.63	1.1	1.5
Gasoline		0.04	1.6	23.4	2.8
LPG		0.00	1.2	3.3	1.4

Source: From Economopoulos, A.P., *Assessment of Sources of Air, Water and Land Pollution. 2. Approaches for Consideration in Formulating Environmental Control Strategies,* World Health Organization, Geneva, 1993. (With permission.)

diesel fuel involves problems with respect to availability, toxicity, cost, solubility, and biodegradability.

The SO_2 emissions from diesel-powered cars are directly related to the sulfur content of the fuel. The quality of the diesel fuel — as expressed through the cetane number (related to the ignition quality), the distillation range, and the aromatic content — considerably affects the emissions of all major pollutants. It is believed that cetane numbers below 50 tend to increase all pollutants. A rough quantitative idea about the impact of the cetane number on the emission rates is

TABLE 2.6
Impact of Cetane Number on the Emission Rates

	Change in Cetane Number from 52 to 47.5	Change in Cetane Number from 52 to 44.5
CO	+19.0%	+31.5%
VOC	+38.0%	+63.0%
NO_x	+5.5%	+9.0%
Smoke	+23.0%	+38.0%

Source: From Economopoulos, A.P., *Assessment of Sources of Air, Water and Land Pollution. 2. Approaches for Consideration in Formulating Environmental Control Strategies*, World Health Organization, Geneva, 1993. (With permission.)

TABLE 2.7
Europe: Current and Future Specifications for Diesel

	Euro 2 Pre-2000	Euro 3 2000	Euro 4 2005
Cetane number (min)	49	51	
Sulfur, ppm (max)	500	350	50
Density (kg/m^3)	820–860	820–845	
Polyaromatics, % (max)	—	11	
*T*95%, °C (max)	370	360	

Source: From Becker, R.F., The auto/oils programmes: petrol and diesel, in *The New Zealand Road to Cleaner Air*, Ed., World Energy Council, New Zealand, 2001. (With permission.)

given in Table 2.6, but it should be stressed that the data from the different sources vary widely.

For diesel, European Community (EC) specification in 2000 was constrained by increasing the minimum cetane number to 51 (from 49), reducing the maximum density to 845 (from 860), reducing maximum 95% point of the distillation (*T*95%) to 360°C (from 370), reducing the maximum sulfur content to 350 ppm (from 500), and introducing a polycyclic aromatics limit of 11%. For the year 2005, a further reduction to 50 ppm sulfur was imposed (Table 2.7).

Reduction of diesel engine emissions has traditionally been achieved through a combination of fuel system, combustion chamber, and engine control modifications. Another method of reformulating diesel fuel to reduce emissions is to incorporate oxygen in the fuel. Oxygenation of diesel fuel offers the possibility of reducing particulate matter emissions significantly, even for the current fleet of diesel vehicles. For instance, the replacement in motor fuel of 15% of the diesel by the alcohol provides cleaner exhaust and provides acceptable power without modification of the existing diesel engines.[25] However, the most widely available and inexpensive alcohols (methanol and ethanol) used as a part of motor fuel have the problem of immiscibility with diesel.

In spite of the restrictive emissions legislation in the European countries and the United States, modern society is concerned with the damage to the global balance of carbon dioxide in the atmosphere, which is linked to the intensive burning of petroleum, coal, and fossil gas. The damage to the carbon dioxide balance in the atmosphere causes global climate warming and has a negative effect on the nature of Earth. Because of this, the development of alternative motor fuel obtained from renewable plant resources is of real significance. However, the existing global inventory of vehicles and machinery with diesel engines does not allow the complete elimination of diesel oil as a motor fuel. Therefore, the research and development of additives to diesel in order to decrease the emissions is still necessary, but should obey the following physical and chemical properties[13]: (i) density at 20°C not less than 0.775 g/cm^3; (ii) the cloud temperature is not higher than 0°C at atmospheric pressure; (iii) stable at atmospheric pressure from cloud temperature of 0°C to initial boiling point of 50°C; (iv) amounts of liquid evaporated by boiling at atmospheric pressure not less than 98% of the total volume of the motor fuel composition distills at temperature no higher than 400°C, suitable no higher than 370°C, and preferably no higher than 280°C; (v) heat of combustion on oxidation by oxygen of not less than 39 MJ/kg; (vi) self-ignition temperature from 150°C to 300°C; and (vii) ability to accommodate at least 1% of water by volume.

DME and DEE can be proposed as one of the oxygenated components to be added to diesel fuel, resulting in cleaner exhaust emissions that do not adversely affect engine performance.[13,14,25]

2.3 SYNTHESIS OF ACETALS

The primary process for the preparation of acetals is the reaction of an aldehyde with an alcohol, accordingly

$$2\text{ Alcohol} + \text{Aldehyde} \leftrightarrow \text{Acetal} + \text{Water}$$

The lower acetals have boiling points distinctly higher than those of the corresponding aldehydes. Some physical data of lower acetals and respective reactants are presented in Table 2.8. The most well-known methods for the production of DEE or DME involve the reaction of acetaldehyde and ethanol or acetaldehyde and methanol, respectively, in acid medium[26,27]; however, acetaldehyde could be replaced by vinyl ether, acetylene, or ethylene.[1]

Traditionally, the acetalization reaction is catalyzed by mineral or carboxylic acids.[28] The disadvantage of using soluble catalysts is that they must be neutralized after reaction. Therefore, heterogeneous catalysts such as ion exchange resins (acid type) or zeolites are used, which have the advantage of being easily separated from the reaction medium and of having a long lifetime. Resins are of particular

TABLE 2.8
Different Acetals and Their Physical Properties

Acetal Name	Molecular Formula	MW	Density	BP (°C)	RI
Formaldehyde dimethylacetal	$CH_2(OCH_3)_2$	76	0.8593	41.6	1.3589
Formaldehyde diethylacetal	$CH_2(OCH_2CH_3)_2$	104	0.832	88.0	—
Acetaldehyde dimethylacetal	$CH_3CH(OCH_3)_2$	90	0.8516	64.5	1.3665
Acetaldehyde diethylacetal	$CH_3CH(OCH_2CH_3)_2$	118.18	0.8254	102.7	1.3819
Acetaldehyde dipropylacetal	$CH_3CH(OCH_2CH_2CH_3)_2$	146	—	—	—
Acetaldehyde dibutylacetal	$CH_3CH(OCH_2CH_2CH_2CH_3)_2$	174	—	—	—
Propionaldehyde dimethylacetal	$CH_3CH_2CH(OCH_3)_2$	101.15	0.8649	84–86	—
Propionaldehyde diethylacetal	$CH_3CH_2CH(OCH_2CH_3)_2$	132.21	0.8232	124	1.3924
Butyraldehyde dimethylacetal	$CH_3CH_2CH_2CH(OCH_3)_2$	118.18	—	112–113	1.3888
Butyraldehyde diethylacetal	$CH_3CH_2CH_2CH(OC_2H_5)_2$	146.22	0.8320	145–145.5	1.3970
Pentanal diethylacetal	$CH_3(CH_2)_3CH(OCH_2CH_3)_2$	160.25	0.8304	162.5–163.5	1.4021
Hexanal diethylacetal	$CH_3(CH_2)_4CH(OCH_2CH_3)_2$	174.28	—	155–158	—

Source: Kohlpaintner, C. et al., Aldehydes, aliphatic and araliphatic, in *Ullmann's Encyclopedia of Industrial Chemistry*, Wiley-VCH, Weinheim, 1999. (With permission.)

interest since they act like a catalyst for the acetalization reaction and as well as a selective adsorbent to the species involved in the process.[29,30]

The synthesis of acetals is a reversible reaction. In order to obtain acceptable acetal yields, the equilibrium must be displaced in the direction of acetal synthesis. Several methods are used to displace equilibrium towards acetal formation, such as:

1. To use a large excess of one of the reactants — in general, the alcohol — which then requires elimination of that excess in a step of purification of the desired product
2. To eliminate water by azeotropic distillation between a solvent and water; the solvent and water must be partially miscible and the boiling points of the different constituents in the reaction medium must be compatible with that azeotrope
3. To use reactive separations (such as reactive distillation, simulated moving bed reactor, pervaporation reactor, etc.) in order to remove the products from the reaction medium

At present, the process of acetals production (DEE, DME, or others) consists of a reaction followed by separation, usually consecutive distillations to separate the mixture.[17,31]

In reactions limited by chemical equilibrium where more than one product is formed, conversion can be enhanced in a multifunctional reactor where the products are separated as they are formed. Novel reactor configurations and the choice of operating conditions can be used to maximize the conversion of reactants and improve selectivity of the desired product, thereby reducing the costs associated with the separation step. Recently, reactive separations such as reactive distillation, chromatographic reactors, and membrane reactors have been intensively applied to esterifications processes, mainly to the production of oxygenates to fuel. There are a great number of patents related to acetal production, summarized in Table 2.9. The reaction is usually catalyzed by a strong acid, such as sulfuric acid. In order to overcome equilibrium limitations, an excess of one reactant is commonly applied. Another technique is the use of a solvent (*n*-hexane, kerosene, benzene), substantially immiscible with water in order to extract the acetal. Distillation is the most used separation technology; few patents mention liquid–liquid extraction and pervaporation as an alternative.

2.3.1 Acid Catalysis

The selection of catalysts for the synthesis of DEE and DME was, considering the nature of the reaction, oriented to acid substrates. Almost all industrial applications to oxygenates synthesis have adopted acidic ion exchange resins as catalysts. The first synthesis of ethers (MTBE, ETBE), esters, acetals, and ketals were done on inorganic catalysts such as mineral acids, with sulfuric acid, for example. But these catalysts presented low selectivity, favoring the formation of dehydrated by-products. Moreover, the use of soluble inorganic acids makes the separation and purification of the desired product a rather complex operation. Solid catalysts such as alumino silicates (mortmorillonites and bentonites, for example), and zeolites (Y, mordenite, ZSM-5, ferrierites) were demonstrated to be good catalysts, but catalytic activity or selectivity of these catalysts always had lower results when compared to macroreticular acid resins.

Ion exchange resins have functional groups tied to insoluble molecules; those groups contain hydrogen or hydroxyl ions to act as acid or alkaline catalysts, respectively, and are able to catalyze many reactions normally carried out in homogeneous systems. The structure of these heterogeneous catalysts obtained by polymerization can be of gel-type resins or macroreticular-type resins. Macroreticular resins have macropores that allow a significant increase in the reaction rate between nonpolar molecules or compounds with a high molecular weight. Ion exchange resins conventionally used to catalyze etherification and esterification reactions are Dowex 50 (Dow Chemical); Amberlite IR120, Amberlyst 15 (A15), and Amberlyst A36 (Rohm & Haas); and Lewatit (Bayer), as shown in Table 2.10.

2.3.2 Equilibrium Displacement

Similar to etherification and esterification reactions, acetalization has a limited conversion by nature of the equilibrium reaction. In order to increase the acetal yield, therefore, it would be necessary, for example, to use one of the reactants in large excess or to remove one product from the reaction system. The use of a large excess of one of the reactants implies another step of separation in order to recover and recycle the excessive reactant.

TABLE 2.9
Patented Processes for Acetal Production

Distilleries Des Deux-Sevres Henri Martin Guiton, 1932	Alcohol and aldehydes are placed together, the reaction being produced by means of a certain quantity of a catalyst, such as hydrochloric acid, sulfuric acid, *p*-toluene sulfonic acid, or acetic acid, hydrogen phosphide, calcium chloride, various other metallic salts, or the like. Acetal is formed, and the production stops at the state of equilibrium, where the reaction medium contains the four components. Due to the difficulty of separating acetal from the reaction mixture, two principal methods have been proposed. One of these consists in the use of certain salts, such as calcium chloride, to adsorb the water which is formed. And the other proposed method consists in adding liquid that is insoluble in water (such as gasoline) to the mixture in the reaction.
British Celanese 1952	The production of formals, acetals, and ketals were studied in order to improve the process for the production of the same in increase yields. For removing water from the reaction mixture, an entraining agent was adapted in a continuously operated azeotropic distillation.
Union Rheinische Braunkohlen Kraftstoff Aktiengesellschaft Korff et al., 1981	The improved process for the production of acetaldehyde dimethylacetal by reacting methanol with carbon monoxide and hydrogen in the presence of a cobalt-containing catalyst, halogen or halide as promoter, and a 3-valent phosphorus compound as ligand using a nickel compound as a cocatalyst was presented.
Union Carbide Corporation Wegman et al., 1984	Process for selective production of acetaldehyde and dimethylacetal by the reaction of methanol and carbon monoxide in contact with a catalyst containing cobalt atom, halide atom, and a mixture of a trivalent phosphorus compound and a trivalent nitrogen compound. The amount of trivalent nitrogen compound in its mixture varies from 5 mol% to 50 mol% of such mixture.
Kuraray Co. Ltd. Iwasaki et al., 1998	Description of producing acetals comprising reacting an aldehyde or a ketone with an alcohol in the presence of a titanium compound having an acetylacetone as a ligand, or in the presence of a compound selected from the group consisting of stannous chloride dihydrate, cerium chloride hexahydrate, and bismuth chloride. The process can be used in the synthesis of unstable acetals or when water exists in the reaction mixture.
Catalytic Distillation Technologies Smith et al., 2000	Acetals are produced from the reaction of aldehydes and alcohols, e.g., methylal is produced by the reaction of formaldehyde and excess methanol. By the reaction in a reactive distillation column, the products acetal and water formed through the reaction in presence of catalyst were separated through the fractional distillation of the reaction mixture.
BASF Aktiengesellschaft Therre et al., 2003	The preparation of acetals relates to a continuous process for the preparation of unsaturated acetals by reacting olefinically unsaturated aliphatic compounds with allyl alcohol in a reaction column, the resulting acetal is concentrated in at least two successive evaporation stages, and the recovered reactants are returned to the reaction column.
Roche Vitamins, Inc. Boesch et al., 2004	This process for the preparation of acetals and ketals by reacting an aldehyde or ketone with excess alcohol in the presence of an acidic catalyst and removing water formed by pervaporation. Manufacturing equipment for the preparation of acetals and ketals is also provided.

TABLE 2.10
Applications of Solid Catalysts to Oxygenates Production

Resin	Product	Reference
Amberlyst, A15	MTBE	52, 92, 93, 94
	ETBE	95
	TAME	96
	Acetic acid	97
	Methyl acetate	98
	Ethyl acetate	29
	-phenyl acetate	43
Amberlyst, A16	TAME	99
	TAEE	100
Amberlyst, A18	MTBE	48, 50
	ETBE	101
Amberlyst, A31	Bisphenol a	46
Lewatit SPC 118	TAME	102
Bayer K2631	MTBE	103
	ETBE	104
	IPTBE	105
ZSM-5	ETBE	106

When synthesizing acetals from aldehyde and alcohols, the best acetal yield is obtained when the water formed is eliminated as the reaction advances. A number of methods can be used to eliminate the water: distillation of an azeotrope between a solvent and water (the solvent and the water must be at least partially miscible and the boiling points of the different constituents in the reaction medium must be compatible with that azeotrope) or the water can be captured by adsorption on a dehydrating solid, which may be a molecular sieve or any other compound that can adsorb water without interfering in the reaction of acetal formation. Again, the use of a solvent to form an azeotrope with water leads to an additional separation step. The complexity of the vapor–liquid and liquid–liquid equilibrium makes the design and the synthesis of separation sequences a difficult task.

Innovative reactor configurations and choice of operating conditions of reactants improve the selectivity of desired product and thereby reduce the costs associated with the separation step. In recent years, there has been considerable academic and industrial interest in the area of reactive separation units that combine reaction and separation of species as a single step in a process. Reactive separations offer the following advantages:

- Increase conversion of reactants beyond thermodynamic limit
- Improved selectivity and yields of desired product
- Reduced requirements for external energy supply or cooling capacity
- Reduced capital costs by process intensification

2.3.3 Reactive Separations

Reaction and separation are traditionally viewed as separate steps in a chemical process. This paradigm is still part of mainstream practice in the chemical industry. However, the combination of reaction and separation in a single unit has aroused considerable interest in chemical engineering research and in industry. The term *multifunctional reactor* is often used to embrace reactive separations technology, which promises a reduction in capital costs, increased conversion, and reduced by-product formation; a variety of separation principles and concepts can be incorporated into the reactor.

The advantages of coupling chemical reaction and separation have been exploited since 1921 for methyl acetate production[32] with reactive distillation processes, and have been applied in the petrochemical industry. In this case, the multifunctional reactor is a distillation column filled with catalytically active packing. In the column, chemicals are converted on the catalyst while reaction products are continuously separated by fractioning (thus overcoming equilibrium limitations). The advantages of catalytic distillation units, in addition to the continuous removal of reaction products and higher yields due to the equilibrium shift, consist mainly of reduced energy requirements and lower capital investments.[33]

2.4 CHEMICALS, THE CATALYST, AND METHODS OF ANALYSIS

The chemicals used were ethanol (> 99.9% pure), methanol (> 99.9% pure), acetaldehyde (> 99.5% pure), DEE (> 99.0%), and DME (> 97.0%) from Sigma-Aldrich, U.K. The water used for calibration of gas chromatography was deionized. The use of these chemicals, mainly acetaldehyde and acetals, requires extra precautions.

A commercial sulfonic acidic ion exchange resin (Amberlyst 15, Rohm & Haas, France) was used as a catalyst as well as a selective adsorbent. The resin is in bead-form macroreticular polymer of styrene and divinylbenzene, with particle diameters varying from 300 to 1200 μm. The ion exchange capacity and other properties of the resin used in this work are presented in Table 2.11.

TABLE 2.11
Physical and Chemical Properties of Amberlyst 15 Resin

Properties	Amberlyst 15
Moisture content	52–57%
Shipping weight	770 g/L
Particle size	300–1200 μm
Concentration of acid sites	1.7 meq/mL
Surface area	45 m^2/g
Porosity	0.36
Average pore diameter	24 nm

The analysis of reactants and products was carried out using gas chromatography. The gas chromatograph used for analysis was from Chrompack, model Chrompack 9100 (Netherlands). The compounds were separated in a fused silica capillary column (Chrompack CP-wax 57 CB), 25 m ∞ 0.53 mm ID, using a thermal conductivity detector (TCD 903 A) for detection of peak separation of reactants and products. The column temperature was programmed with a 2-min initial holdup at 50°C, followed by 40°C/min increase in temperature up to 100°C and then held again for 1 min. The carrier gas used was helium.

2.5 ACETALIZATION REACTION IN BATCH REACTOR

2.5.1 Experimental Setup

Experiments were carried out in a glass-jacketed 1 dm^3 capacity autoclave (Buchi, Switzerland) with batch mode operations. The reactor was provided with a mechanically agitated stirrer in the range of 0 to 2900 rpm. The reactor was equipped with pressure sensor (pressure transducer Lucas P1231-0005-15 BAR A and manometer), and a temperature sensor [thermocouple K type, Omega, TJ36-CAIN-116(G)-12], along with safety blow-off valve as shown in Figure 2.3. The temperature in the reactor was controlled by a thermostatic water pump, model RE104 (Lauda, Germany) with a 4-liter capacity and temperature range from –10°C to 120°C; water flows through the glass-jacketed autoclave in order to maintain the desired reaction temperature.

In order to maintain the reaction mixture in the liquid phase over the whole temperature range, the pressure was set at 6 atm to 10 atm with the helium gas. A stainless steel meshed (10-μm mesh) basket was arranged to place the catalyst; the basket was initially placed at the top of the agitator with the help of the agitator's shaft support. The catalyst basket was introduced to the reactant's mixture at the beginning of the agitation (time zero) and after attaining the desired reaction conditions. One of the outlets of the reactor was connected directly to the liquid sampling valve (Valco, United States), which injected 0.1 μL of pressurized liquid in the reactor to the gas chromatograph (GC).

2.5.2 Thermodynamic Results

Some physical and thermodynamic properties of acetaldehyde, DME, DEE, ethanol, methanol, and water, available from literature, are presented in the Table 2.12. Thermodynamic properties, such as the enthalpy change of the reaction and equilibrium constant, are important for process design. Hence, to determine the above essential thermodynamic quantities, for both acetalization reactions, acetaldehyde–ethanol and acetaldehyde–methanol kinetic experiments were performed. Here, the experiments to measure the equilibrium constant for the acetalization reversible reaction were carried out in the batch reactor shown in Figure 2.3. The temperature dependency of equilibrium constant was determined from the Van't Hoff equation.

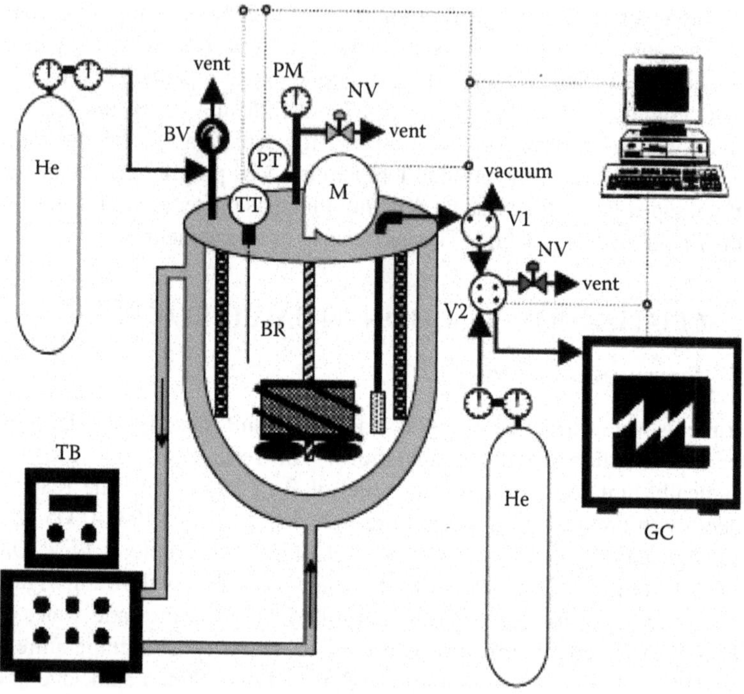

FIGURE 2.3 Laboratory-scale experimental setup. BR: batch reactor; M: motor; TT: temperature sensor; PT: pressure sensor; PM: manometer; BV: blow-off valve; V1: sampling valve; V2: injection valve; NV: needle valve; GC: gas chromatograph; TB: thermostatic bath.

TABLE 2.12
Basic Properties of Chemicals Used

Properties	Acetal-dehyde	DEE	DME	Ethanol	Methanol	Water
Molecular weight — M (g/mol)	44.054	118.17	90.12	46.069	32.04	18.015
Melting temperature — T_f (K)	150.2	173.0	163.8	159.1	175.5	273.2
Normal boiling temperature — T_b (K)	294.0	375.4	337.5	351.4	337.7	373.2
Critical temperature — T_c (K)	461.0	543.0[a]	507.8[a]	513.9	512.6	647.3
Critical pressure — P_c (bar)	55.7	30.2[a]	37.7[a]	61.4	80.9	221.2
Critical volume — V_c (cm^3/mol)	154.0	401.5[a]	289.5[a]	167.1	118.0	57.1
Acentric factor — ω	0.303	0.178[a]	0.403[a]	0.644	0.556	0.344

[a] Calculated for this work

Sources: Reid, R.C. et al., *The Properties of Gases and Liquids,* McGraw-Hill, New York, 1987; Perry, R.H., *Perry's Chemical Engineers Handbook*, McGraw-Hill, New York, 1984. (With permission.)

$$\ln K_{eq} = \frac{\Delta S^o}{R} - \frac{\Delta H^o}{R}\frac{1}{T} \tag{2.1}$$

All the experiments to measure the equilibrium constant for this reversible reaction were performed in a temperature range of 293 to 333 K at 0.6 MPa with Amberlyst 15 resin as the catalyst in a batch reactor. The reactant mixture volume is about 600 mL, and the reactant initial molar ratio is the stoichiometric $r_{A/B} = 2.0$. All experiments lasted long enough for the reaction to reach equilibrium.

The thermodynamic equilibrium constant for the liquid-phase reaction, considered here as a nonideal system, is given by

$$K_{eq} = \frac{a_{C_e} \times a_{D_e}}{a_{A_e}^2 \times a_{B_e}} = \frac{x_{C_e} \times x_{D_e}}{x_{A_e}^2 \times x_{B_e}} \times \frac{\gamma_C \times \gamma_D}{\gamma_A^2 \times \gamma_B} = K_x \times K_\gamma \tag{2.2}$$

where A is the alcohol (methanol or ethanol), B is the acetaldehyde, C is the acetal (DME or DEE), and D is the water.

The equilibrium constant can also be evaluated in terms of molar fractions, if the mixture has an ideal behavior and is given in the following Equation (2.3):

$$K_x = \frac{x_{C_e} \times x_{D_e}}{x_{A_e}^2 \times x_{B_e}} \tag{2.3}$$

The equilibrium constant expressed in terms of the molar concentration was also reported and it is given as shown in Equation (2.4):

$$K_C = \frac{C_{C_e} \times C_{D_e}}{C_{A_e}^2 \times C_{B_e}} \tag{2.4}$$

The concentrations were determined from the liquid molar volumes, estimated with the Gunn-Yamada method[34] and are given according to

$$C_i = \frac{x_i}{\sum_j x_j V_{ml,j}} \tag{2.5}$$

The activity coefficients of compounds, γ, were computed by the universal quasichemical functional group activity coefficients (UNIFAC) method.[35] The UNIFAC parameters of pure species, shown in Table 2.13(A) and Table 2.13(B), are the relative group volume (R_k) and surface area (Q_k) parameters and the group-group interaction parameters ($a_{m,n}$), respectively. The activity coefficients of all species in the reaction mixture at equilibrium were determined.

TABLE 2.13A
Relative Molecular Volume and Surface Area of Pure Species Parameters

(*i*)	Molecule	Group Identification					
		Name	No. Main	No. Sec	$v_k^{(i)}$	R_k	Q_k
1	Ethanol	CH_3	1	1	1	0.9011	0.848
		CH_2	1	2	1	0.6744	0.540
		OH	5	15	1	1.0000	1.200
2	Methanol	CH_3OH	6	16	1	1.4311	1.432
3	Acetaldehyde	CH_3	1	1	1	0.9011	0.848
		CHO	10	21	1	0.9980	0.948
4	DEE	CH_3	1	1	3	0.9011	0.848
		CH	1	3	1	0.4469	0.228
		CH_2O	13	26	2	0.9183	0.780
5	DME	CH_3	1	1	1	0.9011	0.848
		CH	1	3	1	0.4469	0.228
		CH_3O	13	25	2	1.1450	1.088
6	Water	H_2O	7	17	1	0.9200	1.400

Sources: Reid, R.C. et al., *The Properties of Gases and Liquids,* McGraw-Hill, New York, 1987; Perry, R.H., *Perry's Chemical Engineers Handbook*, McGraw-Hill, New York, 1984. (With permission.)

TABLE 2.13B
Interaction Parameters

$a_{m,n}$ (K)	1	5	6	7	10	13
1	0.0	986.5	697.2	1318.0	677.0	251.5
5	156.4	0.0	—	353.5	–203.6	28.06
6	16.510	—	0.0	–181.0	306.4	–128.6
7	300.0	–229.1	289.6	0.0	–257.3	540.5
10	505.7	529.0	–340.2	232.7	0.0	304.1
13	83.36	237.7	238.4	–314.7	–7.838	0.0

Source: Reid, R.C. et al., *The Properties of Gases and Liquids,* McGraw-Hill, New York, 1987. (With permission.)

The equilibrium constants in terms of molar fraction, molar concentrations, and activity coefficients are given in Table 2.14 for both systems.

The dependence of equilibrium constant on temperature can be estimated by fitting experimental values of $\ln K_{eq}$ vs. $1/T$ to Equation (2.1).

The fitting of experimental data by Equation (2.1), where the equilibrium constant was calculated with respect to activity coefficients, is given as follows in Equation (2.6) and Equation (2.7) for acetalization reactions of the acetaldehyde-ethanol system and acetaldehyde-methanol system, respectively:

TABLE 2.14A
Equilibrium Constants at Different Temperatures for Acetaldehyde–Ethanol System

T(K)	293	303	313	333
K_x	2.198	1.921	1.655	1.211
K_γ	1.580	1.606	1.636	1.711
K_C (L/mol)	0.150	0.132	0.115	0.086
K_{eq}	3.472	3.086	2.708	2.072

TABLE 2.14B
Equilibrium Constants at Different Temperatures for Acetaldehyde–Methanol System

T(K)	293	303	313	323	333
K_x	5.3532	4.0305	4.0062	2.9191	2.5547
K_γ	4.0974	3.9689	3.8052	3.7225	3.6177
K_C (L/mol)	0.5178	0.3856	0.3836	0.2756	0.2402
K_{eq}	21.9344	15.9965	15.2441	10.8663	8.9092

$$\ln K_{eq} = \frac{1270.1}{T(K)} - 3.0748 \quad (2.6)$$

$$\ln K_{eq} = \frac{2142.5}{T(K)} - 4.2475 \quad (2.7)$$

Therefore, ΔH° and ΔS° are obtained from the slope and the intercept, respectively, for both acetalization reactions, as shown in the Table 2.15. The standard free-energy change for the liquid-phase reaction can be related to the standard enthalpy and entropy changes by

$$\Delta G^o = \Delta H^o - T \times \Delta S^o \quad (2.8)$$

The standard molar properties for formation of acetals in the liquid phase at 298.15 K were estimated for both systems and are given in the Table 2.16.

2.5.3 Kinetic Model

The kinetic model for the production of acetals from alcohols and aldehydes is given on the basis of the below-mentioned steps. Once the species alcohol and

TABLE 2.15
Standard Reaction Properties Determined from Experimental Data at 298 K for Acetaldehyde–Ethanol System and Acetaldehyde–Methanol System

Properties	ΔH^o (kJ mol^{-1})	ΔS^o (J mol^{-1} K^{-1})	ΔG^o (kJ mol^{-1})
Acetaldehyde–ethanol system	−10.56	−25.56	−2.94
Acetaldehyde–methanol system	−17.81	−35.314	−7.28

TABLE 2.16
Standard Formation Properties at 298.15 K for Both Acetals

Properties	ΔH_f^o (kJ mol^{-1})	S_f^o (J mol^{-1} K^{-1})	ΔG_f^o (kJ mol^{-1})
Diethylacetal	−476.5	343.23	−243.5
Dimethylacetal	−402.29	308.71	−230.95[1]

aldehyde diffuse into the catalyst pore space, various different mechanisms can be considered, such as adsorption, surface reaction, and desorption. Typically, the Langmuir-Hinshelwood-Hougen-Watson (LHWW), Eley-Rideal, or Power Model is used. In this work, the Langmuir model was tested with experimental data, considering the following steps in the model:

- Adsorption of alcohol (a): (A – ethanol or methanol):

$$A + S \xleftrightarrow{K_{s,A}} AS \tag{2.9}$$

- Adsorption of acetaldehyde:

$$B + S \xleftrightarrow{K_{s,B}} BS \tag{2.10}$$

- Surface reaction between the adsorbed species of alcohol and acetaldehyde to give adsorbed hemi-acetal (I_1S):

$$AS + BS \xleftrightarrow{K_1} I_1S + S \qquad K_1 = \frac{\theta_{I_1}\theta_0}{\theta_A\theta_B} \tag{2.11}$$

- Surface reaction to obtain adsorbed water (DS):

$$I_1S + S \underset{\overleftarrow{k}_2}{\overset{\vec{k}_2}{\longleftrightarrow}} I_2S + DS \qquad K_2 = \frac{\vec{k}_2}{\overleftarrow{k}_2} = \frac{\theta_{I_2}\theta_D}{\theta_{I_1}\theta_0} \tag{2.12}$$

- Surface reaction of formation of adsorbed acetal (CS):

$$I_2S + AS \overset{K_3}{\longleftrightarrow} CS + S \qquad K_3 = \frac{\theta_C\theta_0}{\theta_{I_2}\theta_A} \tag{2.13}$$

- Desorption of acetal:

$$CS \overset{K^*_{s,C}}{\longleftrightarrow} C + S \tag{2.14}$$

- Desorption of water:

$$DS \overset{K^*_{s,D}}{\longleftrightarrow} D + S \tag{2.15}$$

$$\Re = \vec{k}_2\theta_{I_1}\theta_0 \quad - \quad \overleftarrow{k}_2\theta_{I_2}\theta_D \tag{2.16}$$

where $\Re$ is the rate of reaction and θ_0 is the fraction of vacant sites.

Langmuir adsorption isotherms are assumed to describe the adsorption behavior of the compounds of the reaction mixture in the surface of the resin. The model is based on activities (a_i), and therefore the adsorption equilibria of the species A_i is given by

$$\theta_i = \frac{K_{s,i}\, a_i}{1 + \sum_{j=1}^{N} K_{s,j}\, a_j} \tag{2.17}$$

where θ_i is the fraction of sites occupied by A_i, and $K_{s,i}$ is the equilibrium adsorption constant.

Combining the above equations, the reaction rate becomes

$$\Re = k_c \frac{a_A a_B - \dfrac{a_C a_D}{K_{eq}\, a_A}}{\left(1 + \sum_{j=A}^{D} K_{s,i}\, a_i + K_{s,I_1}\, a_A a_B + K_{s,I_2}\dfrac{a_C}{a_A}\right)^2} \tag{2.18}$$

In order to reduce the number of optimization parameters, several rate equations derived from Equation (2.17) were tested. The best fittings were obtained by considering that the products (acetal and water) and the intermediate I_2 are more adsorbed than the other species. The simplified rate equations are then:

- Model 1: Acetal is more adsorbed than other species:

$$\Re = k_c \frac{a_A a_B - \dfrac{a_C a_D}{K_{eq}\ a_A}}{\left(1 + K_{s,C}\ a_C\right)^2} \tag{2.19}$$

- Model 2: Water is more adsorbed than other species:

$$\Re = k_c \frac{a_A a_B - \dfrac{a_C a_D}{K_{eq}\ a_A}}{\left(1 + K_{s,D}\ a_D\right)^2} \tag{2.20}$$

- Model 3: The intermediate I_2 is more adsorbed than other species:

$$\Re = k_c \frac{a_A a_B - \dfrac{a_C a_D}{K_{eq}\ a_A}}{\left(1 + K_{s,I_2} \dfrac{a_C}{a_A}\right)^2} \tag{2.21}$$

Each kinetic model contains two parameters, the kinetic constant (k_c) and the adsorption parameters ($K_{s,C}$, $K_{s,D}$, or $K_{s,I2}$).

2.5.3.1 Batch Reactor Model

It is not possible to conclude that the experiments performed with the catalyst particles of lower fraction were conducted in absence of internal mass transfer reactions. Therefore, the model of the batch reactor considers internal pore diffusion of the species inside the resin. The fitting between the experimental kinetic data and the model of the batch reactor leads to the true kinetics parameters.

The macroreticular ion exchange resins show bidisperse pore distribution. The heterogeneous catalysis processes are regulated by transport phenomena, the adsorption, and the reaction at the solid surface. The mass transfer effects are due to three main mechanisms: the mass transfer of species between the bulk fluid phase and the external surface of the stationary phase particles (external mass transfer); the diffusive migration through the pores inside the particles (internal pore diffusion); and the surface diffusion. The internal pore diffusion may occur by molecular diffusion and Knudsen diffusion; for liquid systems, the Knudsen diffusion is negligible and, therefore, the proposed model just considers molecular diffusion as the only mechanism of internal pore diffusion. The macropores and micropore diffusion resistances should be taken into consideration; however, a monodisperse pore structure approximation can be made for Amberlyst 15 since the micropore diffusion had an insignificant contribution in the total diffusion flux into a macroreticular resin catalyst.[36] The transport phenomena through the micropores and the surface diffusion on the gel microspheres were not taken into account, but they could be lumped in the tortuosity factor.[37]

Therefore, the mathematical model for the isothermally operated batch reactor considers the diffusion of species, initially at the bulk external film, and secondly molecular diffusion through the macropores of the particle until the species reach the surface of the solid, where the reaction occurs[38]:

(a) Mass balance in the bulk fluid,

$$\frac{dC_{b,j}}{dt} = -\frac{A_p}{V_{liq}} D_j \left.\frac{\partial C_{p,j}}{\partial r}\right|_{r=r_p} \qquad (j = 1,2,3 \;\&\; 4) \tag{2.22}$$

$$A_p = \frac{3}{r_p} V_p \tag{2.23}$$

where $C_{b,j}$ is the bulk concentration for species j, $C_{p,j}$ is the concentration of species j inside the particle pores, A_p is the external area between the bulk fluid and the particle, V_{liq} is the total volume of reactant mixture, D_j is the effective diffusivity of species j inside the catalyst pores, r_p is particle radius, V_p is the total volume of the particles, r is the particle radial position, and t is the time coordinate.

(b) Mass balance in the particle,

$$\varepsilon_p \frac{\partial C_{p,j}}{\partial t} = \frac{1}{r^2}\frac{\partial}{\partial r}\left[D_j r^2 \frac{\partial C_{p,j}}{\partial r}\right] + (1-\varepsilon_p)\nu_j \rho_{solid} \Re \qquad (j = 1,2,3 \;\&\; 4) \tag{2.24}$$

Initial conditions:

$$t = 0 \qquad C_{b,j} = C_{b0,j}\,; \quad C_{p,j} = C_{p0,j} \tag{2.25}$$

Boundary conditions:

$$r = 0 \qquad \frac{\partial C_{p,j}}{\partial r} = 0 \tag{2.26}$$

$$r = r_p \qquad C_{b,j} = C_{p,j}\Big|_{r=r_p} \tag{2.27}$$

It is important to observe that for $r = 0$, Equation (2.24) is not defined and should be substituted by a limiting expression that using the L'Hopital's rule is transformed in[38]

$$\varepsilon_p \frac{\partial C_{p,j}}{\partial t} = 3\frac{D_j}{r_p^2}\frac{\partial^2 C_{p,j}}{\partial \rho^2} + (1-\varepsilon_p)\nu_j \rho_{solid} \Re \tag{2.28}$$

where $\rho = r/r_p$ is the normalized radial variable.

The effective diffusivity of the compound j $(D_j = \varepsilon_p D^o_{j,m} / \tau)$ was evaluated for the multicomponent liquid mixture; the diffusion coefficient $(D^o_{j,m})$ for a dilute solute j into a mixture of n components was determined by using the Wilke-Chang equation[34]

$$D^o_{j,m} = 7.4\times10^{-8}\frac{(\phi M)^{1/2} T}{\eta_m V^{0.6}_{ml,j}} \qquad \phi M = \sum_{\substack{k=1\\k\neq j}}^{n} x_k \phi_k M_k \tag{2.29}$$

where ϕ_k is the association factor of component k (ϕ_k is chosen as 2.6 for water, 1.9 for methanol, and 1.0 for unassociated compounds), M_k is molecular weight of component k, T is the temperature, η_m is the mixture viscosity and was predicted by the generalized corresponding states method, $V_{ml,j}$ is the molar volume of solute j, and x_k is the molar fraction for component k.

2.5.3.2 Optimization of the Kinetic Parameters

In order to determine the true kinetic parameters, the optimization of the kinetic data was performed. The kinetic parameters were estimated with a nonlinear regression subroutine, which uses the Levenberg–Marquardt method to minimize the sum of residual squares (SRS) between experimental and calculated molar fraction of all components.

$$SRS = \sum\sum_{i=1}^{4}(x_{i,\exp} - x_{i,theo})^2 \tag{2.30}$$

The theoretical molar fraction values ($x_{i,theo}$) were calculated by the proposed model. The values of true kinetic parameters in Equation (2.20) are $\ln k_c^{DME} = 58.6 - 16283/T(K)$; $\ln K_{s,D}^{DME} = 143.4 - 42215/T(K)$; $\ln k_c^{DEE} = 26.5 - 7817/T(K)$; and $\ln K_{s,D}^{DEE} = 13.4 - 3999/T(K)$. The average standard error between the experimental and simulated molar fractions of all species is 1.5%. The temperature dependence of the true kinetic constants was fitted to the Arrhenius equation, and the corresponding activation energy of both acetalization reactions are determined as 65.1 kJ mol^{-1} for the acetaldehyde–ethanol system, whereas we find activation energy of 72.4 kJ mol^{-1} for the acetaldehyde–methanol system.

The kinetic model is accurate in the range of initial molar ratio of reactants (ethanol or methanol to acetaldehyde) from 1.5 to 3.5 and at the temperature range 293 to 333 K. The comparison of experimental and theoretical data plots for different initial molar ratio of reactants for the acetaldehyde–ethanol system and the acetaldehyde–methanol system are shown in Figure 2.4 and Figure 2.5, respectively.

2.6 ACETALIZATION REACTION IN FIXED BED REACTOR

Strong acid-type resins have the advantage of being catalysts for the acetalization reaction and, simultaneously, a selective adsorbent for water.[29,30] Lately, the development of multifunctional reactors using acid resins as catalysts is being implemented in the production of oxygenated fuel additives (MTBE, ETBE, TAME, etc.), where reaction is of the type $A + B \rightarrow C$. Among all available technologies, the reactive distillation is the most used in that field when the boiling point of the desired product, C, is different from the reactants' boiling points.[39,40] In the case of acetalization reaction of acetaldehyde–ethanol, the two products, DEE and water, have boiling temperatures of 102°C and 100°C, respectively, where in the case of acetalization reaction of acetaldehyde–methanol, the product acetal (DME) and reactant methanol have boiling temperatures of 64.5°C and 64.0°C, respectively, and hence the use of an adsorptive chromatographic reactor seems to be a feasible and economical solution. The application of chromatographic reactors to equilibrium controlled reactions[41,42] leads to conversions higher than the equilibrium, since one of the products is being removed from the reaction zone. Perhaps one of the most interesting chromatographic reactors is the SMBR. This technology has been applied to reversible reactions catalyzed by acid resins[30,43] and also to biochemical reactions.[44,45] However, some authors first studied the dynamic behavior of the fixed bed adsorptive reactor to validate kinetic and adsorption data and to provide a better understanding of the performance of chromatographic reactors.[29,46,47] A number of mathematical models have been developed to explain the kinetic behavior of the fixed bed adsorptive reactor and to estimate the breakthrough curves. The mechanism of adsorption and reaction on a catalyst includes external diffusion, internal diffusion, the adsorption, and the reaction at the solid surface. The intraparticle diffusion may occur by molecular diffusion, Knudsen diffusion, and the surface diffusion, depending on the pore sizes, the adsorbate concentrations, and other conditions. The mass-transfer effects are due to four main mechanisms: axial mixing in the bulk mobile phase percolating between the stationary phase particles in the column (axial dispersion); the mass transfer of molecules between the bulk mobile phase and the external surface of the stationary phase particles (external mass transfer); the diffusive migration through the pores inside the particles (internal diffusion); and the surface diffusion.

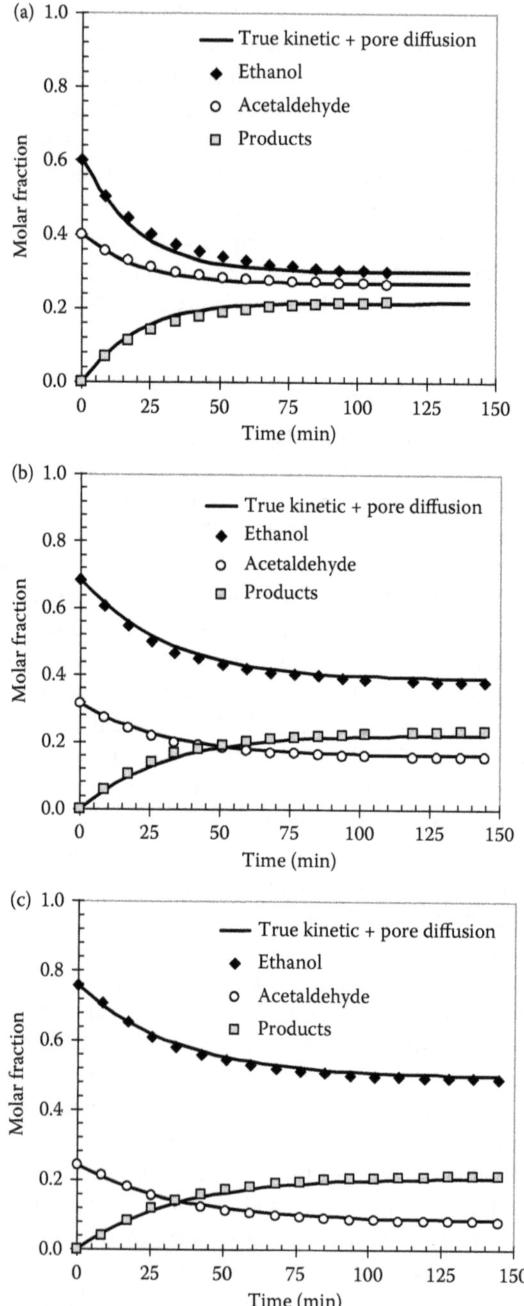

FIGURE 2.4 Comparison between experimental and simulated curves at T = 299 K; P = 6 atm; V = 600 mL; W_{cat} = 0.5 g; speed of agitation (*soa*) = 800 *rpm*; (a) $r_{A/B}$ = 1.5; (b) $r_{A/B}$ = 2.0; (c) $r_{A/B}$ = 3.5.

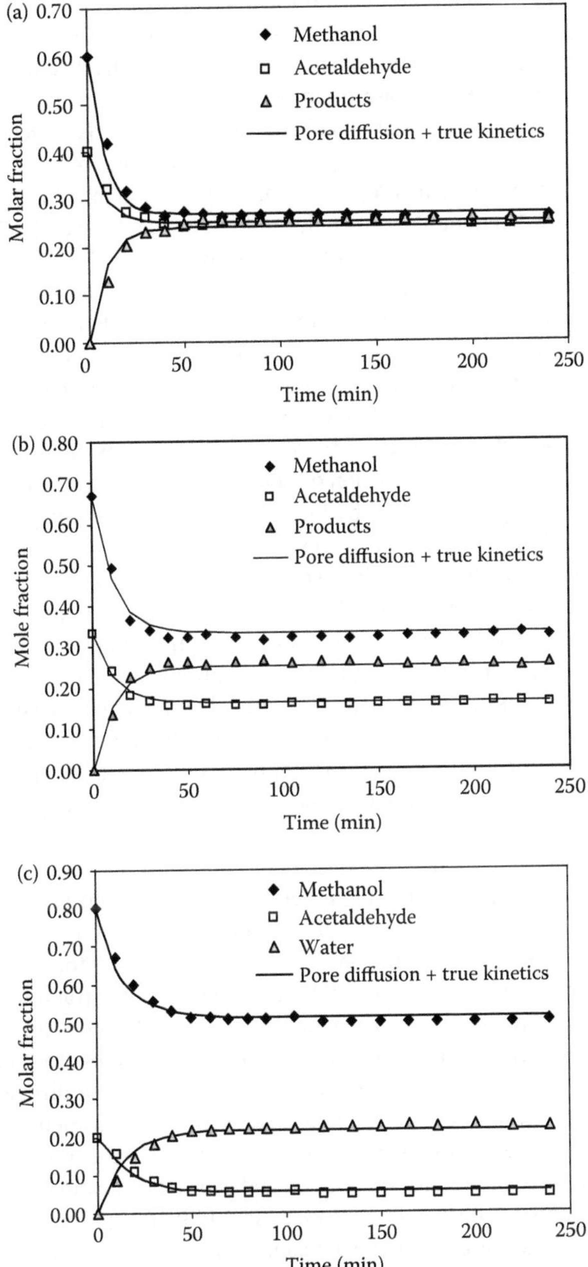

FIGURE 2.5 Comparison between experimental and simulated curves at T = 313 K; P = 6 atm; V = 600 mL; W_{cat} = 0.5 g; *soa* = 600 *rpm*; (a) $r_{A/B}$ = 1.5; (b) $r_{A/B}$ = 2.0; (c) $r_{A/B}$ = 3.5.

Ion exchange resins are produced through a copolymerization procedure with styrene and divinylbenzene used as a cross-linking agent. The acid groups, containing hydrogen ions in the form of sulfonic groups, are attached to the polymeric matrix developed in the gel phase by long polystyrene chains fixed by bridges of divinylbenzene, leading to a stable and rigid structure.[48] This structure can be of gel-type resins or macroreticular-type resins. Macroreticular resins, such as Amberlyst 15 (A15), have macropores, which allow a significant increase in the reaction rate between nonpolar molecules or high molecular weight.[49] The macroreticular ion exchange resins show bidisperse pore distribution.[50,51] The resin particles are considered to be an ensemble of gellular microspheres and most of the active sites are reported to lie within these microspheres.[51] The reactant species should first diffuse through the macropores to the external surface of microspheres and then penetrate into the gel phase. The pore-size distribution of the ion exchange resin depends upon the degree of cross-linking of the polymer matrix. The sulfonic groups constitute a dense network structure by forming hydrogen bonds between the sulphonic groups; therefore, the diffusion of polar compounds into the microspheres is expected to be strongly influenced by the interaction with the catalyst structure.[52]

The adsorption and reaction processes over catalysts with bidisperse pore-size distribution have been widely studied in the literature. The first and most used model was developed by Ruckenstein, who considers a spherical macroporous pellet to be an assembly of small microspheres.[53] The adsorbate diffuses into macropores, adsorbs on the macropore walls, and also diffuses into the micropores and is adsorbed there. Recently, this model was used to determine the effectiveness of bidisperse catalysts,[54] to study adsorption induced convection in the macropores of a bidisperse adsorbent particle.[55] Latter, an equivalent model was adopted that considered microspheres as homogeneous gel particles, where the adsorbed phase diffuses.[56]

Turner proposed a model structure where the solid network is described by the branched micropore–macropore model, including macropores for the transport and micropores to provide capacity of adsorption or reaction,[57] which was also used later by Villermaux.[58] This model was also applied for the analysis of diffusion and reaction in a catalyst with a bidisperse pore structure.[59] Recently, the Turner structure of a bidisperse model was adopted to study adsorption; because of its simplicity, the geometry of macro- and micropores and the diffusivities within pores is then well defined.[27,60] The ratio of the diffusion times in the macro- and micropores obtained from adsorption experiments of methanol and isobutylene in gas phase over A15 was found to be of the order of magnitude of unity. Therefore, in the analysis of rate data for etherification reactions, both macro- and micropore diffusion resistances should be taken into consideration.[52] The analysis of the pure component batch adsorption experiments of ethanol, methanol, and 2-methyl-2-butene on Amberlyst 15, using *n*-heptane as an inert solvent, showed that the surface diffusion had a significant contribution in the total diffusion flux into a macroreticular resin catalyst.[36]

The aim of this present section is to study the synthesis of acetals in a fixed bed adsorptive reactor, in view of future implementation of the process in a

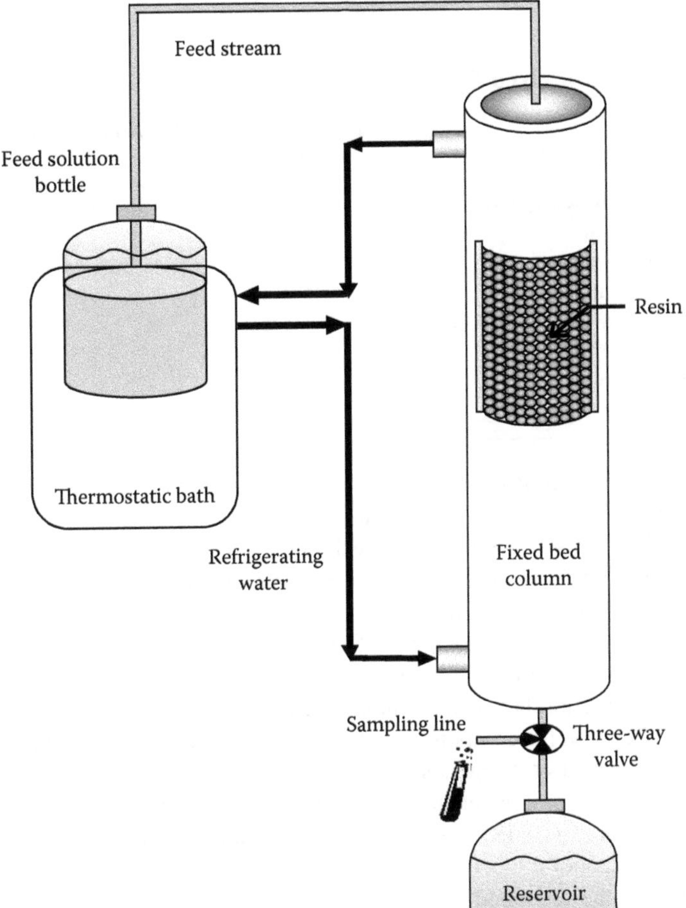

FIGURE 2.6 Fixed-bed reactor experimental setup.

simulated moving bed adsorptive reactor. Adsorption and reaction experiments were carried out at constant temperature and atmospheric pressure. The multicomponent adsorption parameters were obtained by performing binary adsorptive experiments in the absence of reaction. The model for the adsorptive reactor includes axial dispersion, external and internal mass-transfer resistances, constant temperature, and multicomponent Langmuir isotherms, and it was solved by orthogonal collocation by finite elements using the measured model parameters.

2.6.1 Experimental Setup

The experiments have been performed in a laboratory-scale jacketed glass column that was maintained at constant temperature, through a thermostatic bath, at atmospheric pressure, as shown in the Figure 2.6. The experimental breakthrough curves were obtained by analyzing with a gas chromatograph the small samples

TABLE 2.17
Characteristics of Fixed Bed

Properties	Acetaldehyde–Ethanol System	Acetaldehyde–Methanol System
Solid Weight (Amberlyst 15 Resin), Kg	40.0×10^{-3}	47.0×10^{-3}
Length of Bed (L), m	23.0×10^{-2}	11.0×10^{-2}
Internal diameter (D_i), m	2.6×10^{-2}	2.6×10^{-2}
Average radius of Resin Beads (r_p), μm	[300–400]	[300–400]
Bulk Density of Resin Beads (ρ_b), Kg/m^3	792	792
Bed Porosity (ε)	0.40	0.40
Resin Particle Porosity (ε_p)	0.36	0.36
Surface Area of Resin (S_A), m^2/g	45	45

that were withdrawn at different times from the column exit. The column was packed with a sulfonic acid ion exchange resin.

Amberlyst 15 is an ion exchange resin with a copolymer structure of styrene divinylbenzene that swells selectively when it is brought in contact with the liquid phase. The swelling ratio (ratio between the swollen and dry volume of the resin) depends upon the fluid composition. The swelling is due to the sorption of the different components of the mixture, depending on their relative affinities to the resin. The interaction between component and polymer matrix is caused by the interaction between the component and the acid site, the sulfonic acid group. The penetrability to acid sites depends on interactions between the pairs of components and the pairs of component and polymer matrix. Diffuse reflectance Fourier transform infrared (FT-IR) spectra of methanol and ethanol on Amberlyst 15 indicate that alcohol molecules are adsorbed by forming hydrogen bridges with SO_3H sites of the resin and among themselves. Some of the alcohol molecules were found to be strongly chemisorbed by dissociation of one or two hydrogen atoms.[36] The character of the resin is mainly determined by its structure (e.g., cross-linking degree and functional groups). The polymeric resins that contain sulfonic acid functional groups exhibit a strong selectivity for polar species. The effect of swelling yields changes in the length and in the bulk porosity of the fixed bed reactor, but during its operation, no variation in the bed length was noticed. Table 2.17 shows the characteristics of the fixed bed column.

2.6.2 Modeling of Fixed Bed

Here, it is considered an isothermal fixed bed packed with Amberlyst 15 resin. Axial dispersion as well as external and internal mass-transfer resistances in adsorbing species were considered in order to determine the model parameters using experimental data collected.

The adsorption isotherm is the equilibrium relationship between the concentration in the fluid and the concentration in the adsorbent particles at a given temperature and pressure. Particularly for liquid adsorption, the role of pressure is not that effective, as in the case of gas adsorption. The concentration of the adsorbate on the solid is given as number of moles or mass adsorbed per unit mass or unit volume of adsorbent (solid phase). The Langmuir isotherm model for multicomponent adsorption was considered, which can be expressed as follows:

$$q_i = \frac{Q_i K_i C_i}{1 + \sum_{j=1}^{n} K_j C_j} \tag{2.31}$$

where q_i is the solid concentration for species i in equilibrium with the fluid concentration inside the pores of the resin (C_i), Q_i is the adsorption capacity of species i, and K_i is the adsorption constant of species i.

The mathematical model used to describe the dynamic behavior of the fixed bed reactor considers the following assumptions[61,62]

1. The flow pattern is described by the axial dispersion plug flow model.
2. External and internal mass-transfer for adsorbable species is combined in a global resistance.
3. Isothermal process.
4. Constant column length and packing porosity of bed were assumed.

The model equations consists in the following system of four second-order partial differential equations in the bulk concentration of the *i*th component, C_i, and four ordinary differential equations in the average concentration of *i*th component into the particle pores, $\bar{C}_{P,i}$. The system of partial differential equations were solved by the method of lines (MOL) using orthogonal collocation in finite elements, with *B* splines as base functions through the numerical package PDECOL. Fifty subintervals for spatial discretization along the *z* axis were used, with two internal collocation points in each subinterval, resulting in 200 to 400 time-dependent ordinary differential equations for adsorption–reaction simulations. A tolerance equal to 10^{-7} was fixed for all simulations.

2.6.3 Binary Adsorption Experiments

The breakthrough curves of ethanol, methanol, acetals (DEE, DME), and water were measured in the absence of reaction. The resin was saturated with a certain component *A* and then the feed concentration of component *B* was changed stepwise. The adsorption parameters were optimized by minimizing the difference between the experimental and theoretical number of moles adsorbed or desorbed for all adsorption experiments. The acetaldehyde binary adsorption experiments cannot be performed in the presence of acidic catalysts such as resins or zeolites, due to their polymerization. Minimizing the difference between the experimental and simulated stoichiometric times of breakthrough curves, the adsorption parameters were optimized. The

TABLE 2.18A
Adsorption Equilibrium Isotherms for Acetaldehyde–Ethanol System

Component	Q ($\times 10^{-3}$mol/m$^3_{\text{real solid}}$)	K ($\times 10^3$ m^3/mol)	Density (kg/m^3)
Ethanol (A)	14.30	1.43	795
Acetaldehyde (B)	20.10	1.58	785
Acetal, DEE (C)	11.45	0.09	836
Water (D)	44.70	2.01	1003

TABLE 2.18B
Adsorption Equilibrium Isotherms for Acetaldehyde–Methanol System

Component	Q ($\times 10^{-3}$ mol/m$^3_{\text{real solid}}$)	K ($\times 10^3$ m^3/mol)	Density (kg/m^3)
Methanol (A)	14.52	0.60	791
Acetaldehyde (B)	18.65	0.38	785
Acetal (C)	7.67	0.32	852
Water (D)	28.49	1.15	1003

stoichiometric time can be determined from the mass balance over the adsorbent bed in the column, as shown below:

$$t_{st} = \frac{L}{u}\left[\varepsilon + (1-\varepsilon)\varepsilon_p + (1-\varepsilon)(1-\varepsilon_p)\frac{\Delta q}{\Delta C}\right] \tag{2.32}$$

where

$$\frac{\Delta q}{\Delta C} = \frac{q(C_F) - q(C_0)}{C_F - C_0} \tag{2.33}$$

Experimental stoichiometric times were calculated from the experimental outlet concentration as a function of time. In the breakthrough experiments, where component A displaces component B, the total amount necessary to saturate the column is given by the product between the volumetric flow rate and the area over the breakthrough curve limited by the feed concentration. The product of the flow rate and the area under the elution curve gives the total amount of component B that was initially saturating the column. The adsorption parameters were determined by optimizing the reaction experimental data. The determined adsorption parameters for all components are shown in Table 2.18 for both the ethanol–acetaldehyde system and the methanol–acetaldehyde system.

2.6.4 Reaction Experimental Procedure

Reaction experiments were performed by feeding continuously the mixture of alcohol and acetaldehyde to the fixed bed chromatographic reactor that was initially saturated with alcohol. Because the feed mixture (ethanol or methanol with acetaldehyde) is less dense than the pure corresponding alcohol, the flow direction was operated in a top-down fashion. However, we noted hydrodynamic effects due to axial back-mixing driven by natural convection of water. The concentration fronts moving within the column are hydrodynamically stable if the component in the liquid above the front is less dense than the component below the front. Since the products are denser than the reactants, the reaction mixture is denser than alcohol (initially saturating the column); therefore, the reactants mixture should be fed from the bottom. The reactants mixture, alcohol (ethanol or methanol), and acetaldehyde at the stoichiometric ratio was fed to the column at a certain known flow rate wherein the outlet concentration of the reaction mixture was indicated as a function of time. As the reaction mixture enters the column, acetaldehyde in the reactants mixture is adsorbed and reacts with adsorbed alcohol (ethanol or methanol) in the adsorbent phase (resin). Acetal (DEE or DME) and water are formed in the stoichiometric amounts, but the resin adsorbs preferentially water, whereas acetal is soon desorbed from the resin phase and carried out by the fluid stream along the column. As water is removed from the reaction medium, the acetalization reaction lasts until the acetaldehyde is consumed. This process of reaction in the fixed bed will continue until the resin in the column is completely saturated with acetaldehyde and water. Once the equilibrium is reached in the resin (solid phase), the selective separation of acetal and water is no longer possible. The concentration of the reaction mixture in the column remains constant as the steady state is achieved, and the reaction mixture at the equilibrium position consists of the concentrations in the outlet stream. The difference under the outlet concentration curves of the two products is relative to the difference in the adsorbed amount of acetal and water in the resin, once both products are formed at the same stoichiometric amounts. In this case, the model predictions are in agreement with the experimental data for the ethanol–acetaldehyde system and the methanol–acetaldehyde system.

When the steady state is achieved, the reaction experiment ends and it is necessary to regenerate the resin with a desorbent before starting the new experiment. During the regeneration of the fixed bed reactor, the adsorbate is removed from the adsorbent by displacement from the active sites, which is due to adsorption of a third material (the displacing agent). The choice of a suitable displacement agent depends both on the equilibrium of the system and on the kinetics of the adsorption and desorption. Here, methanol was used as a desorbent for regeneration of the fixed bed, as acetal (DEE or DME) and acetaldehyde are weakly adsorbed species and are the first to be desorbed. Alcohol (ethanol or methanol) was fed to the column in the top-down direction, as the reaction mixture present within the column is heavier than pure alcohol (ethanol or methanol).

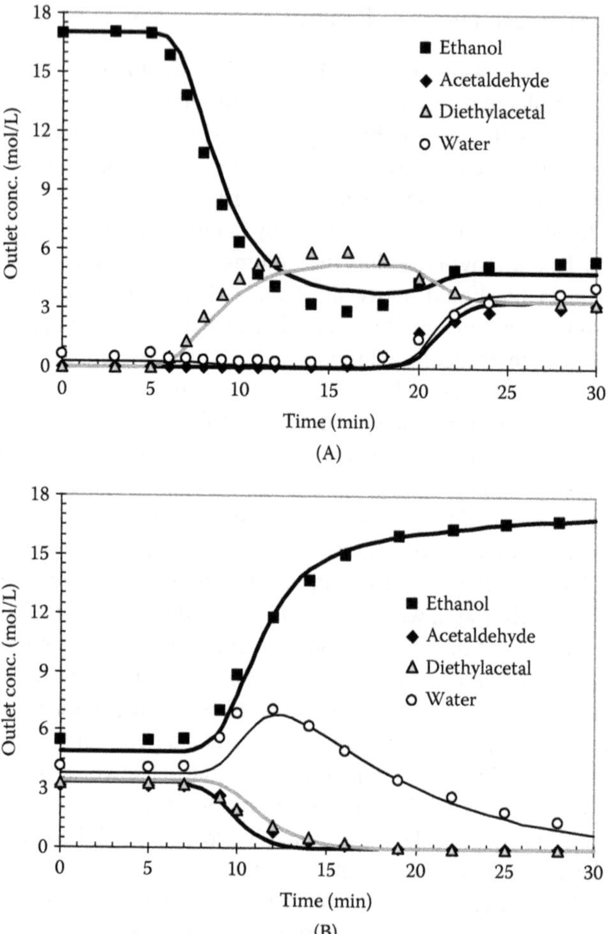

FIGURE 2.7A Concentration profiles in a fixed-bed adsorptive reactor initially saturated with ethanol and then fed with a mixture of ethanol (63%) and acetaldehyde (37%). Experimental conditions: Q = 9.5 mL/min (16 × 10^{-8} m^3/s), $C_{A,F}$ = 11.7 mol/L and $C_{B,F}$ = 6.8 mol/L; bed length: 23.0 cm; bed diameter: 2.6 cm; bed temperature: 10°C.

FIGURE 2.7B Concentration profiles in the regeneration step of a fixed-bed adsorptive reactor. The initial profiles in the bed are those at the final steady state of the run in Figure 2.7A. Experimental conditions: Q = 10.2 mL/min (17 × 10^{-8} m^3/s), $C_{A,F}$ = 17.0 mol/L, and $C_{B,F}$ = 0.0 mol/L.

In order to evaluate the effect of length of the reactor bed on the behavior of the fixed bed reactor or adsorptive reactor, other reaction and regeneration experiments were also performed with the longer bed in the column. It is expected that for a longer bed reactor that the same qualitative behavior will be found. The behavior of the longer fixed bed reactor (both reaction and regeneration) is shown in the Figure 2.7 and Figure 2.8 for the ethanol–acetaldehyde and methanol–acetaldehyde

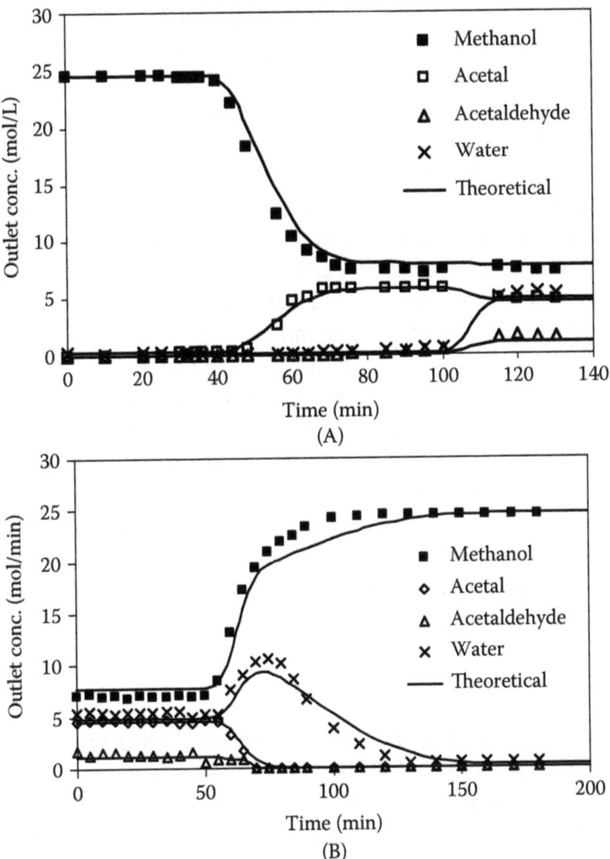

FIGURE 2.8A Concentration profiles in a fixed-bed adsorptive reactor initially saturated with methanol and then fed with methanol and acetaldehyde mixture. Experimental conditions: Q = 6.0 mL/min (10×10^{-8} m^3/s), $C_{A,F}$ = 16.88 mol/L, and $C_{B,F}$ = 5.63 mol/L; bed length: 85.0 cm; bed diameter: 2.6 cm; bed temperature: 20°C.

FIGURE 2.8B Concentration profiles in the regeneration step of a fixed-bed adsorptive reactor. The initial profiles in the bed are those at the final steady state of the run in Figure 2.8A. Experimental conditions: Q = 6.0 mL/min (17×10^{-8} m^3/s), $C_{A,F}$ = 24.7 mol/L, and $C_{B,F}$ = 0.0 mol/L.

systems, respectively. The experimental outlet concentrations of the reactor along the time are in agreement with the simulated results.

2.7 ACETALIZATION REACTION IN SIMULATED MOVING BED REACTOR

The combination of a chemical or biochemical reaction and a chromatographic separation process in a single-unit operation may improve reaction conversion as well as the separation efficiency. In addition, performing two unit operations in a single piece of equipment may reduce capital costs. For the past 30 years, the concept of chromatographic reactors has been extended to continuous processes in order to benefit from those features as enhanced productivity and decreased solvent consumption. Continuous mode of operation has been achieved by means of moving the adsorbent or simulating its motion. This is possible by such equipment configurations as a simulated moving bed,[63] a countercurrent moving bed,[64,65] a rotating annular chromatograph,[66] and a semicontinuous countercurrent chromatographic reactor-separator.[67] Perhaps one of the most interesting physical implementations of a chromatographic reactor-separator is the simulated moving bed reactor (SMBR). The countercurrent motion of the stationary phase is simulated by applying an intelligent scheme of valve switchings on a set of fixed beds. In such a reactor, reaction occurs either in the mobile or stationary phase. In the latter, the catalyst is supported or immobilized in the solid adsorbent, which promotes the separation of the reaction products. In the case of reactions catalyzed by acid sites, the use of acid ion exchange resins, such as Amberlyst 15, is very interesting since the resin acts both as catalyst as well a selective adsorbent. Moreover, since the acetalization is chemical-equilibrium controlled, the removal of the reaction products leads to reactants' conversion above the equilibrium limit. Because of this, the DEE production and separation in a SMBR seems to be a feasible process.

2.7.1 Simulated Moving Bed Concept

The simulated moving bed (SMB) is a continuous chromatographic countercurrent process developed in the 1960s by Universal Oil Products.[107] The first patent was licensed as the SORBEX process and issued for a number of large-scale separations in the petrochemical and sugar industries. Nowadays, the SMB technology has a wide range of applications, including the recently developed laboratory-scale pharmaceutical, fine chemistry, and bioseparation applications. The SMB is based on the principles of the true moving bed (TMB) concept. In the TMB (Figure 2.9), the liquid and the solid phases flow in opposite directions. The inlet (feed and eluent) and outlet (extract and raffinate) ports are fixed along the unit. According to the position of the inlet and outlet streams, four different sections could be distinguished:

Section I: located between the eluent and extract streams
Section II: between the extract and feed streams
Section III: placed between feed and raffinate streams
Section IV: located between raffinate and eluent streams

The major problem of TMB operation connected with the solid movement was overcome by the introduction of the SMB technology. The SMB unit consists of a set of interconnected columns in series; the countercurrent flow of solid and liquid phases is simulated by the periodic shifting of inlets and outlets in the direction of the fluid flow (Figure 2.10).

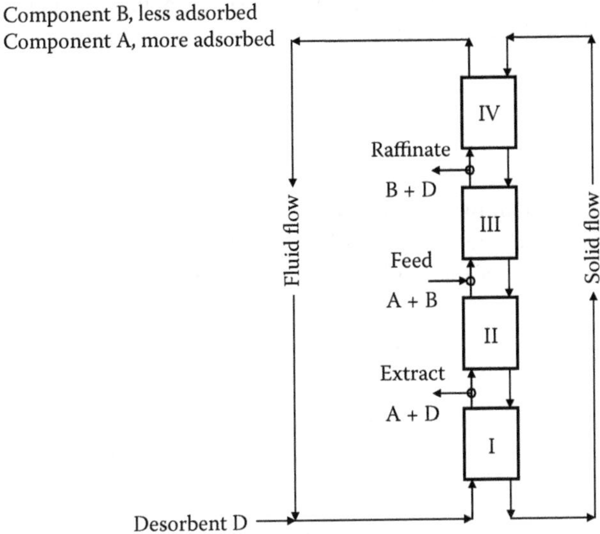

FIGURE 2.9 Schematic diagram of four-section true moving bed (TMB).

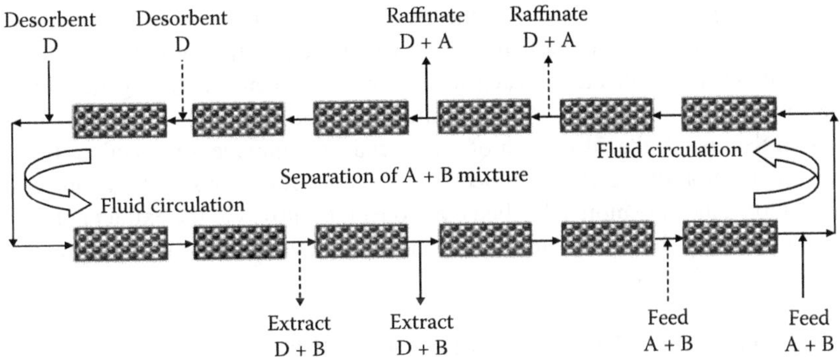

FIGURE 2.8 Schematic diagram of simulated moving bed (SMB).

Due to the switch of the inlet and outlet lines, each column changes the boundary conditions after the end of each switching time interval, depending on its location (section). This time dependence of the boundary conditions leads to a cyclic steady state for the SMB system, instead of a real steady state present in the TMB approach; it means that, after the cyclic steady state is achieved, the internal concentration profiles vary along the switch time interval, but they are identical at the same time for two consecutive cycles. The cyclic steady state of the SMB is equivalent to the real steady state of the TMB for a high degree of subdivisions of the SMB adsorbent bed[68,69]; hence, it is better to simulate and obtain the optimum operating conditions using the TMB model since it requires lower computing time. The cyclic behavior of the SMB can be predicted from the steady-state model of the TMB by considering the relation between the interstitial solid velocity U_s and the switching time t^* in SMB operation, that is,

$$U_s = \frac{L_C}{t^*} \tag{2.34}$$

where L_C is the length of the column.

The equivalence between the SMB and TMB systems is made by keeping constant the net flow of the liquid relative to the solid,

$$U_j^{SMB} = U_j^{TMB} + U_s \tag{2.35}$$

where U_j is the interstitial fluid velocity in the j section of the moving bed. The ratio between the fluid and solid interstitial velocities in each section is given by

$$\gamma_j = \frac{U_j^{TMB}}{U_s} \tag{2.36}$$

The design of SMB involves the right choice of the operating conditions, such as switching time period and flow rates in each section of the unit. The net flow rate has to be selected in each section in order to ensure the regeneration of the adsorbent in the Section I (Zone I), the desorption of the less adsorbed component in Section II, the adsorption of more retained component in Section III, and the regeneration of the eluent in Section IV. These conditions will guarantee the separation, since the more adsorbed component, B, moves to the extract port with solid phase, and the less retained component, A, moves to the raffinate port along with the liquid phase.

2.7.2 Simulated Moving Bed Reactor (SMBR)

Commonly, chemical processes were designed with reaction and separation units in series. Process integration by combining both steps of reaction and separation in one single unit may significantly reduce the costs for the whole process. Besides reactive distillation, reactive extraction, or membrane reactors, the interconnection of chemical or biochemical reaction with chromatographic separation forms an attractive integrated process for producing high-purity products. For reactions in series or in parallel, it may be possible the selective separation of desired intermediate species. In addition, the SMBR also enhances the conversion, yield, selectivity, and purity of the desired product beyond the levels predicted by thermodynamics for an equilibrium-limited reversible reaction, since the products are continuously removed from the reaction zone. The first applications of the SMBR were for gas phase as the hydrogenation of 1,3,5-trimethylbenzene[70] and the oxidative coupling of methane.[71] Several applications for the liquid phase are, for biochemical reactions, the isomerization of glucose by action of glucose isomerase,[63] the inversion of sucrose by action of invertase,[44,72] and the production of lactosucrose by action of -fructofuranosidase.[73] For chemical reactions, major examples are reversible reactions catalyzed by ion exchange resins. Examples are the esterification of acetic acid with -phenethyl alcohol,[43] the ethyl acetate synthesis on Amberlyst 15,[30,74] the synthesis of bisphenol *A* from acetone and phenol,[46] esterification of acetic acid with methanol,[47] and the synthesis of MTBE.[75] Another example is the dismutation of toluene into benzene and xylenes.[76] The design and optimization of SMBR to carry out simultaneous and continuous reaction and separation are essential to define the feasibility of the process at an industrial scale. The number of publications focusing on the design of nonreactive SMBs is quite large. Nevertheless, the same may not be said of the SMBR design,[74,77] and about the optimization of the SMBR process.[78,79] The design will define geometric and operating parameters that should lead not only to product separation but also to high reactant conversion.

2.7.3 SMBR Experimental Unit (licosep 12-26)

The SMBR unit is a pilot unit licosep 12-26 (Novasep, France) in which all SMBR experiments were performed in this present work. The SMBR unit is constructed with 12 columns of Superformance SP 300-26 (300 mm length of the column and internal diameter of 26 mm), by Göteo Labartechnik (Mühltal, Germany). These columns can withstand up to 60°C and 60 bar pressure. All 12 columns were packed with the acid resin Amberlyst 15 (Rohm & Haas, France). Each column is jacketed to ensure the desired temperature in the column; the jackets of each column are connected to one another with the help of silicon houses through a thermostat bath (Lauda, Germany). The thermostat bath can be operated for the circulation of water through the column's jacket in order to maintain the desired temperature of the packed bed in the column. Between every two columns, there exists a four-port valve (Top-Industrie, France) actuated by the control system. According to the operating conditions of the SMBR, when required, the valves allow either pumping of the feed or eluent into the columns or withdrawal of extract or raffinate from the system. Each of the inlet streams (feed or eluent) and outlet streams (extract or raffinate) is pumped by means of high performance liquid chromatography (HPLC) pumps. The recycling pump is arranged for the SMBR unit, and this is a positive displacement three-headed membrane pump (Dosapro Milton Roy, France) that can deliver flow rates as low as 20 mL/min up to 120 mL/min, and it can withstand up to 100 bar pressure. The calibration of the recycling pump should be performed every time whenever eluent or desorbent is changed. The other flow rates (eluent, extract, feed, and raffinate) are controlled by four Merck-Hitachi pumps (Merck-Hitachi L-6000 model for extract and raffinate, whereas for eluent and feed they are a Merck-Hitachi L-6200 model, Japan). All the pumps (eluent, extract, feed, raffinate, and recycling) are controlled by a microcomputer. The flow rates in the eluent and extract pumps can be operated up to 30 mL/min, whereas the feed and raffinate

FIGURE 2.9 SMB pilot unit at LSRE (Laboratory of Separation and Reaction Engineering).

pumps can be operated only up to 10 mL/min. In all the above pumps, the maximum allowable pressure is 60 bar. Between the 12th column and the 1st column of the system, there is a six-port valve used to perform internal tests. Figure 2.11 shows a front view of the SMB pilot unit, where the other six columns are fixed on the other side of the system. The SMBR apparatus is equipped with its own process control software (LICOSEP).

All tubing between the columns consists of 1/16-inch external diameter and 1 mm internal diameter. The recycling pump introduces a dead volume and its adverse effects in delaying the concentrations, leaving the last column (12th column) and entering to the first column, and this has been overcome by desynchronizing the switches of the ports that are about to or have been shifted across the pump at each cycle. In such cases, the switching time is delayed by t_d minutes, which is calculated from the following equation:

$$t_d \text{ (min)} = \frac{dead\ volume}{Q\ (mL/\text{min})} \tag{2.37}$$

TABLE 2.19
Characteristics of the SMB Column Packed with Amberlyst 15

Length of the packed bed (L)	0.23 m
Internal diameter of the column	0.026 m
Radius of the particle (r_p)	400 μm
External void fraction (ε)	0.4
Internal void fraction (ε_p)	0.4
Peclet number	300
Bulk density (ρ_b)	390 kg/m^3

where

$$\bar{Q} = \frac{\sum_{1}^{4} Q_j}{4} \quad (2.38)$$

and Q_j (j = 1 to 4) represents the recycling flow rate set by LICOSEP depending on the recycling pump location (in Section I, II, III, or IV). Each of the 12 columns was assumed to be homogeneously packed. The characteristics of the columns packed are used in the simulations and are presented in the Table 2.19.

2.7.4 Mathematical Model

As stated earlier, the SMBR modeling strategy is more precise than the TMBR model since it performs as the actual physical equipment operation. It allows the visualization of the axial movement of concentration profiles and the variations in extract and raffinate concentrations within a period. However, it demands considerably higher computational effort than the TMBR strategy, especially when a large number of columns is involved. Cyclic behavior of the SMBR can be predicted from the steady-state model of the TMBR model with good accuracy. Hence, the mathematical model used to observe the SMBR performance is on the basis of TMBR strategy considering axial dispersion flow for the bulk fluid phase, plug flow for solid phase, linear driving force (LDF) for the particle mass-transfer rate, and multicomponent adsorption equilibria. Length of the column (packed bed length) and porosity of the packed bed are assumed to be constant.[80]

2.7.4.1 Bulk Fluid Mass Balance to Component i and in Section j

$$\varepsilon\frac{\partial C_{ij}}{\partial t} + \varepsilon u_j\frac{\partial C_{ij}}{\partial z} + (1-\varepsilon)\frac{3}{r_p}K_{L,i}(C_{ij} - \bar{C}_{p,ij}) = \varepsilon D_{ax,j}\frac{\partial^2 C_{ij}}{\partial z^2} \quad (2.39)$$

where C_{ij} and $\bar{C}_{p,ij}$ are the bulk and average particle concentrations in the fluid phase of species i in Section j of the TMBR, respectively, $K_{L,i}$ is the global mass transfer coefficient of the component i, ε is the bulk porosity (bed porosity), t is the time variable, z is the axial coordinate, $D_{ax,j}$ and u_j are the axial dispersion coefficient and the interstitial velocity in Section j, respectively, and r_p is the particle radius.

2.7.4.2 Pellet Mass Balance to Component i and in Section j

$$U_s\left[\varepsilon_p\frac{\partial \bar{C}_{p,ij}}{\partial z}+(1-\varepsilon_p)\frac{\partial q_{ij}}{\partial z}\right]+\frac{3}{r_p}K_{L,i}(C_{ij}-\bar{C}_{p,ij})$$
$$=\varepsilon_p\frac{\partial \bar{C}_{p,ij}}{\partial t}+(1-\varepsilon_p)\frac{\partial q_{ij}}{\partial t}-v_i\rho_p\eta\Re \quad (2.40)$$

where q_{ij} is the average adsorbed phase concentration of species i in Section j in equilibrium with $\bar{C}_{p,ij}$, U_s is the solid velocity, and ε_p is the particle porosity, ν_i is the stoichiometric coefficient of component i, ρ_p is the particle density, η is the effectiveness factor of the catalyst, and $\Re$ is the chemical reaction rate relative to the bulk liquid phase, which was given as

$$\eta=\frac{\Re(\bar{C}_p)}{\Re(C_b)} \quad (2.41)$$

The values of the effectiveness factors for the DME and DEE systems are 0.42 at 20°C [81] and 0.34 at 10°C.[38]

Initial and Danckwerts boundary conditions:

$$\text{at } t=0 \qquad C_{ij}=\bar{C}_{p,ij}=C_{ij,0} \quad (2.42)$$

$$\text{at } z=0 \qquad u_jC_{ij}-D_{ax,j}\left.\frac{\partial C_{ij}}{\partial z}\right|_{z=0}=u_jC_{ij,F} \quad (2.43)$$

$$\text{at } z=L \qquad \left.\frac{\partial C_{ij}}{\partial z}\right|_{z=L}=0 \quad \text{and} \quad \bar{C}_{ij,L}=\bar{C}_{ij+1,0} \quad (2.44)$$

where F and 0 refer to the feed and initial states, respectively.

Multicomponent adsorption equilibrium isotherm:

$$q_{ij} = \frac{Q_{ads,i} K_i \bar{C}_{p,kj}}{1 + \sum_{k=1}^{n} K_k \bar{C}_{p,kj}} \tag{2.45}$$

where $Q_{ads,i}$ and K_i represent the total molar capacity per unit volume of resin and the equilibrium constant for component i, respectively.

Mass balances at the nodes of the inlet and outlet lines of the TMBR are:

Eluent Node:

$$C_{ij+1,0} = \frac{u_4}{u_1} C_{ij,L_j} \tag{2.46}$$

Extract Node:

$$C_{ij+1,0} = C_{ij,L_j} \tag{2.47}$$

Feed Node:

$$C_{ij+1,0} = \frac{u_2}{u_3} C_{ij,L_j} + \frac{u_F}{u_3} C_{i,F} \tag{2.48}$$

Raffinate Node:

$$C_{ij+1,0} = C_{ij} \tag{2.49}$$

The relationships between fluid velocities in the four zones (Sections I, II, III, and IV) of the TMBR are:

$$u_1 = u_4 + u_D \text{ , } u_2 = u_1 - u_X \text{ , } u_2 = u_3 - u_F \text{ , } u_4 = u_3 - u_R \tag{2.50}$$

where u_1, u_2, u_3, and u_4 are the fluid velocities in Sections I, II, III, and IV, respectively. The subscripts D, F, R, and X refer to the desorbent, feed, raffinate, and extract streams, respectively.

2.7.4.3 SMBR Performance Criteria

The SMBR process performance can be evaluated according to the following criteria:

Raffinate Purity (%):

$$PUR = \frac{C_{C,R}}{C_{B,R} + C_{C,R} + C_{D,R}} \times 100 \tag{2.51}$$

Extract Purity (%):

$$PUX = \frac{C_{D,R}}{C_{B,R} + C_{C,R} + C_{D,R}} \times 100 \tag{2.52}$$

Acetaldehyde Conversion (%):

$$X = \left(1 - \frac{Q_X C_{B,X} + Q_R C_{B,R}}{Q_F C_{B,F}}\right) \times 100 \tag{2.53}$$

Raffinate Productivity ($kg_{\text{acetal}}/L_{\text{adsorbent/day}}$):

$$PR = \frac{Q_R C_{C,R}}{(1-\varepsilon) V_{unit}} \tag{2.54}$$

Desorbent Consumption ($L_{\text{methanol}}/kg_{\text{acetal}}$):

$$DC = \frac{Q_D C_{A,D} + Q_F \left(C_{A,F} - 2\,X\,C_{B,F}\right)}{Q_R C_{B,R}} V_{ml,A} \tag{2.55}$$

Just for the raffinate stream, the productivity is defined here as acetal (desired product) produced from the raffinate stream. Desorbent consumption as considered here is the amount of methanol consumed as a desorbent and not the methanol consumed for reaction. Above model equations were solved numerically by using gPROMS, the general PROcess Modelling System.[82] gPROMS is a software package for modeling and simulation of processes with both discrete and continuous as well as lumped and distributed characteristics. The mathematical model involves a system of partial and algebraic equations (PDAEs). Third-order orthogonal

TABLE 2.20
Operating Conditions and Experimental/Simulated Performance Parameters of the SMBR Unit for Diethylacetal from Acetaldehyde and Ethanol System

Experiment Conditions and Data	
Switching time (t^*)	3.75 min
Desorbent flow-rate (Q_D)	22.7 mL/min
Section IV recycling flow rate (Q_4)	21.3 mL/min
PUR	84.1 (90.2)%
PUX	95.2 (97.4)%
Conversion of acetaldehyde (X)	98.1 (98.7)%
Productivity	4.65 (4.61) kg L^{-1} day^{-1}
Desorbent consumption	7.45 (7.50) L kg^{-1}

collocation in finite elements method (OCFEM) was used in the discretization of axial domain. Twenty equal elements per section with two collocation points in each element were used. The system of ordinary differential and algebraic equation (ODAEs) was integrated over time using the DASOLV integrator implementation in gPROMS. A tolerance equal to 10^{-5} was fixed for all simulations.

2.7.5 Experimental Results and Discussion

The SMBR experiments were performed under the conditions of completely adsorbent regeneration in Section I. The recycle pump was operated at the minimum flow rate (25 mL/min). For the desorbent, the maximum flow rate of the pump (30 mL/min) was not used as the system was not stable at higher flow rates.

Experiments of DME and DEE[80,83] synthesis were performed in the SMBR unit with the configuration of three columns (3-3-3-3) per section at 20°C and 10°C, respectively. The mixture alcohol (ethanol or methanol) and acetaldehyde was fed to the system with feed composition of 40% acetaldehyde molar fraction. The flow rates of feed, raffinate, and extract streams were Q_F = 3.0 mL/min, Q_R = 8.0 mL/min, and Q_X = 20.0 mL/min. The other experimental conditions and the SMBR experimental and simulated (inside brackets) performance criteria are presented in Table 2.20 and Table 2.21 for DME and DEE synthesis, respectively. The steady-state concentration profiles obtained experimentally at the middle of the switching time after a certain number of cycles are compared with the steady-state profiles obtained from the simulated results by the TMBR model in Figure 2.12 and Figure 2.13 for DME and DEE synthesis, respectively. The stationary steady states predicted with the TMBR model are compared with the experimental results; lines in Figure 2.12 and Figure 2.13 represent the concentration profiles from TMBR model, and points are experimental profiles in a SMBR at the middle of a switching time at cyclic steady state.

TABLE 2.21
Operating Conditions and Experimental/Simulated Performance Parameters of the SMBR Unit for Dimethylacetal from Acetaldehyde and Methanol System

Experiment Conditions and Data	
Switching time (t^*)	3.0 min
Desorbent flow rate (Q_D)	25.0 mL/min
Section IV recycling flow rate (Q_4)	25.0 mL/min
PUR	91.41 (95.33)%
PUX	99.95 (98.74)%
Conversion of acetaldehyde (X)	97.61 (99.08)%
Productivity	8.62 (5.79)
Desorbent consumption	4.52 (6.72)

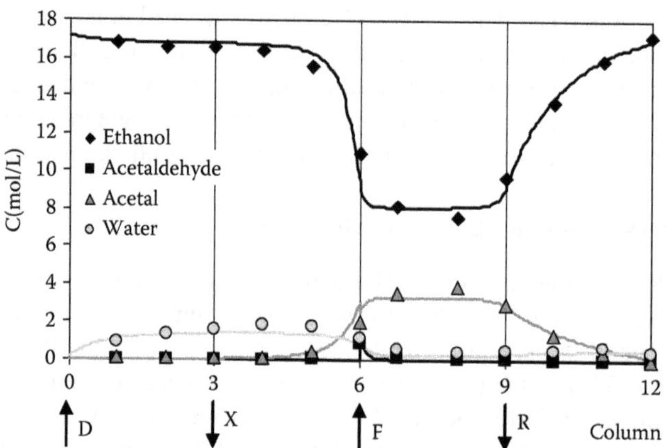

FIGURE 2.10 Experimental and simulated concentration profiles in a SMBR at the middle of a switching time at cyclic steady state (10th cycle).

2.8 CONCLUSIONS

Thermodynamic and kinetic experimental studies for DME and DEE synthesis in a liquid-phase reaction that was catalyzed by Amberlyst 15 resin were performed in a laboratory-scale experimental setup. The thermodynamic equilibrium constant was measured in the temperature range of 293.15 to 333.15 K. The kinetic law expressed in activities

$$\Re = k_c \frac{a_A a_B - \dfrac{a_C a_D}{K_{eq} a_A}}{\left(1 + K_{s,D} a_D\right)^2}$$

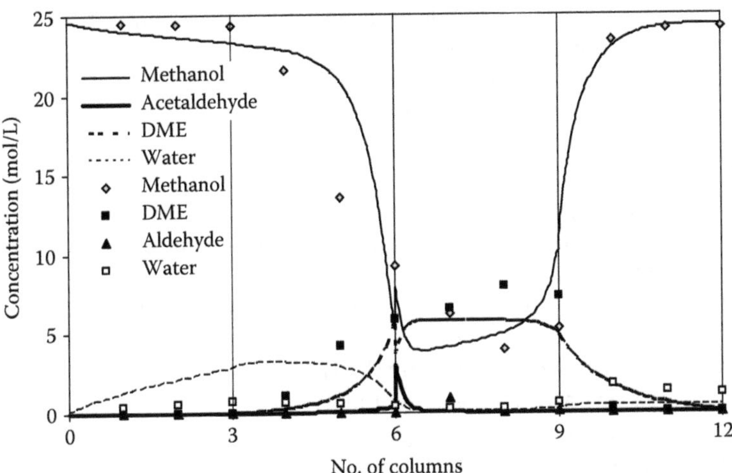

FIGURE 2.11 Experimental and simulated concentration profiles in a SMBR at the middle of a switching time at cyclic steady state (8th cycle).

was proposed for both the systems and was evaluated for the reaction experiments in batch reactor.

The primary experiments were performed to collect the desired adsorption data by performing dynamic binary adsorption experiments in the absence of reaction at 293.15 K, in a laboratory-scale column. Minimizing the difference between the experimental and calculated stoichiometric times of breakthrough curves optimized the adsorption parameters. The mathematical model for the adsorptive reactor was developed in which axial dispersion, external and internal mass-transfer resistances, constant temperature, and multicomponent Langmuir adsorption isotherms were measured on a laboratory-scale experimental setup. The model proposed was compared with the experimental results obtained for reaction and regeneration experiments for both the acetaldehyde–ethanol system and the acetaldehyde–methanol system in a pilot-scale fixed adsorptive reactor.

SMBR experiments were performed in a pilot SMBR unit (Novasep, France) for the synthesis of acetals (DME and DEE) from acetaldehyde and alcohol (ethanol or methanol) as reactants and Amberlyst 15 resin as the catalyst as well as selective adsorbent. The feed was the mixture of acetaldehyde and ethanol or methanol, whereas the desorbent used in the SMBR experiment was ethanol or methanol. The experimental results obtained from physical SMBR unit were given in terms of desired product (acetal) purity.

2.9 NOMENCLATURE

a	liquid-phase activity
A_p	external exchange area between the bulk fluid and the particles, cm^2
C_i	concentration of component i, mol L^{-1}

C_{A_s} methanol concentration at the catalyst surface, mol L^{-1}
C_b bulk concentration, mol L^{-1}
$C_{p,j}$ concentration inside the particle, mol L^{-1}
C_p molar heat capacity at 298 K, J mol^{-1} K^{-1}
d_p mean pellet diameter, m
Da Damköhler number
D_{ax} axial dispersion coefficient (m^2/s)
$D^o_{j,m}$ diffusion coefficient for a dilute solute j in mixture of n components, cm^2 min^{-1}
D_{A_s} effective diffusivity of methanol at catalyst particle surface, cm^2 min^{-1}
DC desorbent consumption (m^3/mol)
$E_{a,C}$ reaction activation energy, J mol^{-1}
ΔG reaction Gibbs free energy, J mol^{-1}
ΔG^o_f standard free energy of formation, J mol^{-1}
ΔH reaction enthalpy, J mol^{-1}
ΔH^o_f standard enthalpy of formation, J mol^{-1}
ΔH_s enthalpy of adsorption, J mol^{-1}
k_c kinetic constant, mol kg^{-1} min^{-1}
K_{eq} equilibrium reaction constant
K_p correction factor
K_s equilibrium adsorption constant
K_L global mass transfer coefficient (m/s)
K^*_L number of mass transfer units
L bed length (m)
K_x constant based on molar fractions
k_e external mass transfer coefficient (m/s)
k_i internal mass transfer coefficient (m/s)
K Langmuir equilibrium parameter (m^3/mol)
K_γ constant based on activity coefficients
m mass, kg
M molecular weight, mol g^{-1}
n number of moles in the liquid phase, mol
P pressure, atm
Pe Peclet number
PR raffinate productivity (mol/min/L)
PUR raffinate purity (%)
PUX extract purity (%)
q solid phase concentration in equilibrium with the fluid concentration inside the particle (mol/m$^3_{res}$)
Q adsorption capacity (mol/m$^3_{res}$)
r particle radial distance, cm
R gas constant, J mol^{-1} K^{-1}
r_p particle radius, μm
$\Re$ rate of reaction, mol kg^{-1} min^{-1}

$r_{A/B}$	initial molar ratio of reactants
Re_P	Reynolds number relative to particle
Sh_P	Sherwood number relative to particle
Sc	Schmidt number
ΔS	reaction entropy, J mol^{-1} K^{-1}
ΔS_f^o	standard entropy of formation, J mol^{-1} K^{-1}
T	temperature, K
t	time, s
t^*	switching time (min)
t_{st}	stoichiometric time defined in equation 22 (s)
u	superficial velocity of fluid in the column (m/s)
V_{ml}	molar volume in liquid phase (m^3/mol)
V	solution volume, m^3
w_{cat}	mass of catalyst, kg
x	molar fraction
X	conversion of the limiting reactant
z	axial coordinate (m)

Greek Letters

ν	stoichiometric coefficient
γ	activity coefficient
ρ	normalized radial variable, dimensionless
ρ_p	particle density, g L^{-1}
ρ_{solid}	true solid density of catalyst, g L^{-1}
ϕ	association factor in Equation (2.29)
ε	porosity
ε_p	particle porosity
σ_A	standard deviation
τ	tortuosity
ζ	dimensionless axial coordinate
θ	dimensionless time coordinate
η	effectiveness factor
μ	fluid viscosity (kg/m^{-1}s^{-1})

Subscripts

0	initial value
a	methanol
B	acetaldehyde
C	acetal (DME)
D	water
eq	equilibrium
exp	experimental data
gas	vapor phase

i	relative to component i
l	relative to limiting reactant
liq	liquid phase
s	surface condition
$theo$	theoretical data
D	relative to the desorbent
i	relative to component i ($i = A, B, C, D$)
j	relative to the section in SMB
p	relative to particle
R	relative to raffinate
SMB	relative to a SMB process
X	relative to extract
0	relative to initial conditions
TMB	relative to a TMB process

REFERENCES

1. Kohlpaintner, C., Schulte, M., Falbe, J., Lappe, P., and Weber, J., Aldehydes, aliphatic and araliphatic, in *Ullmann's Encyclopedia of Industrial Chemistry*, Wiley-VCH, Weinheim, 1999, electronic release.
2. Kelly, J., Chapman, S., Brereton, P., Bertrand, A., Guillou, C., and Wittkowski, R., Gas chromatographic determination of volatile congeners in spirit drinks: interlaboratory study, *J. AOAC Int.*, 82, 1375, 1999.
3. Iwasaki, H., Kitayama, M., and Onishi, T., Process for Producing Acetals, EP Patent No. 0,771,779 A1, 1996.
4. Hoffmann-LaRoche, *Manufacture of Unsaturated Aldehydes,* Great Britain Patent No. 797,200, 1958.
5. Iwai, H. and Fujigaski, J., Antifungal Fragrance Composition, European Patent No. 1,214,879 A2, 2002.
6. Johnson, J., Improvements in the Treatment of Cellulosic Fabrics, Great Britain Patent No. 1,047,694, 1966.
7. Maletra, B., Process for Obtaining Acetals Having a High Degree of Purity, Great Britain Patent No. 607,130, 1948.
8. Handlin, D.L., Stewart, D.R., and Wilkey, J.D., Increased Throughput in the Manufacture of Block Copolymers by Reduction in Polymer Cement Viscosity through the Addition of Polar Solvents, U.S. Patent No. 6,187,873 B1, 2001.
9. Merten, R. and Wagner, K., Process for the Production of Organic Compounds of High Molecular Weight from an Organic Compound Containing at Least One Linear or Cyclic Acetal Group, GB Patent No. 867,461, 1961.
10. Aizawa, T., Nakamura, H., Wakabayashi, K., Kudo, T., and Hasegawa, H., Process for Producing Acetaldehyde Dimethylacetal, U.S. Patent No. 5,362,918, 1994.
11. Sextro, G., Polyoxymethylenes, in *Ullmann's Encyclopedia of Industrial Chemistry,* Wiley-VCH, Weinheim, 1999, electronic release.
12. Thüner et al. 1999.
13. Golubkov, A. and Golubkov, I., Motor Fuel for Diesel, Gas-Turbine and Turbojet Engines, U.S. Patent No. 2002/0026744 A1, 2002.
14. Lü, X., Yang, J., Zhang, W., and Huang, Z., Improving the combustion and emissions of direct injection compression ignition engines using oxygenated fuel additives combined with a cetane number improver, *Energy Fuels*, 19, 1879, 2005.

15. Bauer, K., Garbe, D., and Surburg, H., Flavors and fragrances, in *Ullmann's Encyclopaedia of Industrial Chemistry,* Wiley-VCH, Weinheim, 1999, electronic release.
16. Sturm, W. and Peters, K., Perfumes, in *Ullmann's Encyclopaedia of Industrial Chemistry,* Wiley-VCH, Weinheim, 1999, electronic release.
17. Kaufhold, M.M. and El-Chabawi, M.T., Process for Preparing Acetaldehyde Diethyl Acetal, U.S. Patent No. 5,527,969, 1996.
18. Medina, M., Typical sherry wines, Cordoba University Research Group on Viticulture and Enology, http://www.uco.es/investiga/grupos/vitenol, 2002.
19. SCCNFP, The 1st first update of the inventory of ingredients employed in cosmetic products. 2. Perfume and aromatic raw materials, *SCCNFP/0389/00 Final,* Scientific Committee on Cosmetic Products and Non-Food Products, 1, 2000.
20. Miller, D., Scherl, X., Skrypzak, W., and Wiener, E.A., Aqueous Liquids Comprising a Surfactant and an Acetal, U.S. Patent No. 2002/0061827 A1, 2002.
21. Perron, B., Restle, S., Giroud, F., and Samain, H., Washing Composition Comprising Particles of Aluminium Oxide, at Least One Conditioning Agent and at Least One Detergent Surfactant, U.S. Patent No. 2002/0054862, 2002.
22. Garnier, N. and Samain, H., Hair Styling Composition Comprising Encapsulated Adhesives, U.S. Patent No. 2002/0059941, 2002.
23. Lee, S., *Methane and Its Derivatives,* Marcel Dekker, New York, 1997.
24. Ancillotti, F. and Fattore, V., Oxygenate fuels: market expansion and catalytic aspect of synthesis, *Fuel Process. Technol.,* 57, 163, 1998.
25. Golubkov, A., Motor Fuel for Diesel Engines, World Intellectual Property Organization (WO) Patent No. 01/181,154 A1, 2001.
26. Bramwyche, P.L., Mudgan, M., and Stanley, H.M., Manufacture of Diethyl Acetal, U.S. Patent No. 2,519,540, 1950.
27. Petersen, M.L., Process for the Production of Liquid Acetals, U.S. Patent No. 4,024,159, 1977.
28. Morrison, R. and Boyd, R., *Organic Chemistry,* Allyn and Bacon, London, 1983.
29. Mazzoti, M., Neri, B., Gelosa, D., Kruglov, A., and Morbidelli, M., Kinetics of liquid-phase esterification catalyzed by acidic resins, *Ind. Eng. Chem. Res.,* 36, 3, 1997.
30. Mazzotti, M., Kruglov, A., Neri, B., Gelosa, D., and Morbidelli, M., A continuous chromatographic reactor: SMBR, *Chem. Eng. Sci.,* 51, 1827, 1996.
31. Martin, D.L. and Raynolds, P.W., Process for the Purification of Diethoxymethane from a Mixture with Ethanol and Water, U.S. Patent No. 4,740,273, 1988.
32. Backaus, A.A., Continuous Process for the Manufacture of Esters, U.S. Patent No. 1,400,849, 1921.
33. Stankiewicz, A.I. and Moulijn, J.A., Process intensification: transforming chemical engineering, *Chem. Eng. Prog.,* January, 22, 2000.
34. Reid, R.C., Prausnitz, J.M., and Poling, B.E., *The Properties of Gases and Liquids,* McGraw-Hill, New York, 1987.
35. Fredeslund, A., Gmehling, J., and Rasmussen, P., *Vapor-Liquid Equilibria Using UNIFAC,* Elsevier, Amsterdam, 1977.
36. Dogu, T., Aydin, E., Boz, N., Murtezaoglu, K., and Dogu, G., Diffusion resistances and contribution of surface diffusion in TAME and TAEE production using Amberlyst-15, *Int. J. Chem. Reac. Eng.,* 1, A6, 2003.
37. Sundmacher, K., Zhang, R.S., and Hoffmann, U., Mass-transfer effects on kinetics of nonideal liquid-phase ethyl *tert*-butyl ether formation, *Chem. Eng. Technol.,*18, 269, 1995.

38. Silva, V.M.T.M. and Rodrigues, A.E., Kinetic studies in a batch reactor using ion-exchange resin catalysts for oxygenates production: the role of mass transfer mechanisms, *Chem. Eng. Sci.,* 61, 316, 2006.
39. Sundmacher, K. and Hoffmann, U., Development of a new catalytic distillation process for fuel ethers via a detailed nonequilibrium model, *Chem. Eng. Sci.,* 51, 2359, 1996.
40. Higler, A., Krishna, R., and Taylor, R., Nonequilibrium modelling of reactive distillation: a dusty fluid model for heterogeneously catalyzed processes, *Ind. Eng. Chem. Res.,* 39, 1596, 2000.
41. Sardin, M., Schweich, D., and Villermaux, J., Preparative fixed-bed chromatographic reactor, in *Preparative and Production Scale Chromatography,* Ganetsos, G. and Barker, P.E., Eds., Marcel Dekker, New York, 1993.
42. Carta, G., Simultaneous reaction and chromatography, in *Chromatographic and Membrane Processes in Biotechnology,* Costa, C.A. and Cabral, J.S., Eds., Kluwer Academic Publishers, Dordrecht, 1991; Catalysis, *Catal. J.,* 18, 212, 1970.
43. Kawase, M., Suzuki, T.B., Inoue, K., Yoshimoto, K., and Hashimoto, K., Increased esterification conversion by application of the simulated moving-bed reactor, *Chem. Eng. Sci.,* 51, 2971, 1996.
44. Azevedo, D.C.S. and Rodrigues, A.E., Design methodology and operation of a simulated moving bed reactor for the inversion of sucrose and glucose-fructose separation, *Chem. Eng. J.,* 3766, 1, 2001.
45. Barker, P.E., Ganetsos, G., Ajongwen, J., and Akintoye, A., Bioreaction-Separation on continuous chromatographic systems, *Chem. Eng. J.,* 50, B23, 1992.
46. Kawase, M., Inoue, Y., Araki, T., and Hashimoto, K., The simulated moving-bed reactor for production of bisphenol *a*, *Catal. Today,* 48, 1, 1999.
47. Lode, F., Houmard, M., Migliorini, C., Mazzotti, M., and Morbidelli, M., Continuous reactive chromatography, *Chem. Eng. Sci.,* 561, 269, 2001.
48. Ferreira, R.M.Q., Almeida-Costa, C.A., and Rodrigues, A.E., Heterogeneous models of tubular reactors packed with ion-exchange resins: simulation of the MTBE synthesis, *Ind. Eng. Chem. Res.,* 35, 3827, 1996.
49. Quinta-Ferreira and Rodrigues, 1993.
50. Caetano, N.S., Loureiro, J.M., and Rodrigues, A.E., MTBE synthesis catalysed by acid ion exchange resins: kinetic studies and modeling of multiphase batch reactors, *Chem. Eng. Sci.,* 49, 4589, 1994.
51. Ihm, S.K., Ahn, J.H., and Jo, Y.D., Interaction of reaction and mass transfer in ion-exchange resin catalysts, *Ind. Eng. Chem. Res.,* 35, 2946, 1996.
52. Oktar, N., Murtezaoglu, K., Dogu, T., and Dogu, G., Dynamic analysis of adsorption equilibrium and rate parameters of reactants and products in MTBE, ETBE and TAME production, *Can. J. Chem. Eng.,* 77, 406, 1999.
53. Ruckenstein, E., Vaidyanathan, A.S., and Youngquist, G.R., Sorption by solids with bidisperse pore structures, *Chem. Eng. Sci.,* 26, 1305, 1971.
54. Leitão, A., Dias, M., and Rodrigues, A.E., Effectiveness of bidisperse catalysts with convective flow in the macropores, *Chem. Eng. J.,* 55, 81, 1994.
55. Taqvi, S.M., Vishnoi, A., and Levan, M.D., Effect of macropore convection on mass transfer in a bidisperse adsorbent particle, *Adsorption,* 3, 127, 1997.
56. Ruthven, D.M. and Loughlin, K.F., The diffusion resistance of molecular sieve pellets, *Can. J. Chem. Eng.,* 50, 550, 1972.
57. Turner, G.A., The flow structure in packed beds, *Chem. Eng. Sci.,* 7, 156, 1958.

58. Villermaux, J., Schweich, D., and Authelin, J.R., Le peigne du diable: un modèle d'interface fractale bidimensionnelle, *C. R. Acad. Sci., Ser. II,* t. 304, 307, 1987.
59. Tartarelli, R., Conti, S., and Capovani, M., On the second-order reactions in heterogeneous catalysis, *J. Catal.*, 18, 212, 1970.
60. Silva, V.M.T.M. and Rodrigues, A.E., Adsorption and diffusion in bidisperse pore structures, *Ind. Eng. Chem. Res.*, 38, 4023, 1999.
61. Silva, V.M.T.M. and Rodrigues, A.E., Diethylacetal synthesis with acid resin catalysis: Dynamics of a fixed bed adsorptive reactor, *AIChE J.*, 48, 625, 2002.
62. Gandi, G.K., Silva, V.M.T.M., and Rodrigues, A.E., Synthesis of 1,1-Dimethoxyethane in a fixed bed adsorptive reactor, *Ind. Eng. Chem. Res.*, 45, 2032, 2006.
63. Hashimoto, K., Adachi, S., Noujima, H., and Ueda, Y., A new process combining adsorption and enzyme reaction for producing higher-fructose syrup, *Biotechnol. Bioeng.*, 25, 2371, 1983.
64. Fish, B.B. and Carr, R.W., An experimental study of the countercurrent moving-bed chromatographic reactor, *Chem. Eng. Sci.*, 44, 1773, 1989.
65. Petroulas, T., Aris, R., and Carr, Jr., R.W., Analysis and performance of a countercurrent moving-bed chromatographic reactor, *Chem. Eng. Sci.*, 40, 2233, 1985.
66. Sardimi, M.R. and Barker, P.E., Simultaneous biochemical reaction and separation in a rotating annular chromatograph, *Chem. Eng. Sci.*, 48, 2615, 1993.
67. Ganetsos, G., Barker, P.E., and Ajongwen, J.N., Batch and continuous chromatographic systems as combined bioreactor-separators, in *Preparative and Production Scale Chromatography,* Ganetsos, G. and Barker, P.E., Eds., Marcel Dekker, New York, 1993, pp. 375–394.
68. Storti, G., Masi, M., Morbidelli, M., and Carrá, S., Adsorption separation processes: countercurrent and simulated countercurrent operations, *Comp. Chem. Eng.*, 12, 475, 1988.
69. Pais, L.S., Loureiro, J.M., and Rodrigues, A.E., Modeling strategies for enantiomers separation by SMB chromatography, *AIChE J.*, 44, 561, 1998.
70. Ray, A.K., Carr, R.W., and Aris, R., The simulated countercurrent moving bed chromatographic reactor: a novel reactor-separator, *Chem. Eng. Sci.*, 49, 469, 1994.
71. Tonkovich, A.L.Y. and Carr, R.W., A simulated countercurrent moving-bed chromatographic reactor for the oxidative coupling of methane: experimental results, *Chem. Eng. Sci.*, 49, 4647, 1994.
72. Meurer, M., Altenhöner, U., Strube, J., Untiedt, A., and Schmidt-Traub, H., Dynamic simulation of a simulated-moving-bed chromatographic reactor for the inversion of sucrose, *Starch*, 48, 452, 1996.
73. Kawase, M., Pilgrim, A., Araki, T., and Hashimoto, K., Lactosucrose production using simulated moving-bed reactor, *Chem. Eng. Sci.*, 51, 2971, 2001.
74. Migliorini, C., Fillinger, M., Mazzotti, M., and Morbidelli, M., Analysis of simulated moving-bed reactors, *Chem. Eng. Sci.*, 54, 2475, 1999.
75. Zhang, Z., Hidajat, K., and Ray, A.K., Application of simulated countercurrent moving-bed chromatographic reactor for MTBE synthesis, *Ind. Eng. Chem. Res.*, 40, 5305, 2001.
76. Hotier, G., Dulot, H., Bailly, M., and Ragil, K., Simultaneous Process for Simulated Moving Bed Dismutation and Separation of Toluene into Benzene and Xylenes, U.S. Patent No. 6,429,346 B2, 2002.
77. Fricke, J., Meurer, M., and Schmidt-Traub, H., Design and layout of simulated-moving-bed chromatographic reactors, *Chem. Eng. Technol.*, 22, 835, 1999b.

78. Dünnebier, G., Fricke, J., and Klatt, K.-U., Optimal design and operation of simulated moving bed chromatographic reactors, *Ind. Eng. Chem. Res.*, 39, 2000.
79. Zhang, Z., Hidajat, K., and Ray, A.K., Multiobjective optimization of simulated countercurrent moving bed chromatographic reactor (SCMCR) for MTBE synthesis, *Ind. Eng. Chem. Res.*, 41, 3213, 2002.
80. Silva, V.M.T.M. and Rodrigues, A.E., A novel process for diethylacetal synthesis, *AIChE J.*, 51, 2752, 2005.
81. Gandi, G.K., Silva, V.M.T.M., and Rodrigues, A.E., Process development for the dimethylacetal synthesis: thermodynamics and reaction kinetics, *Ind. Eng. Chem. Res.*, 44, 7287, 2005.
82. gPROMS, *gPROMS v2.2.3 User Guide*, Process System Enterprise Ltd., London, 2003.
83. Silva, V.M.T.M. and Rodrigues, A.E., Industrial Process for Acetals Production in a Simulated Moving Bed Reactor, WO Patent 2005/113476 A1, 2005b.
84. Guinot, H.M., Process for the Manufacture of Acetal, U.S. Patent No. 1,850,836, 1932.
85. British Celanese, Improvements in the Manufacture of Acetals, Great Britain Patent No. 664,956, 1952.
86. Korff, J., Fremery, M., and Zimmermann, J., Process for the Production of Acetaldehyde Dimethyl Acetal, U.S. Patent No. 4,278,819, 1981.
87. Wegman, R., Hydroformylation Process, U.S. Patent No. 4,429,165, 1984.
88. Iwasaki, H., Kitayama, M., and Onishi, T., Process for Producing Acetals, U.S. Patent No. 5,792,876, 1998.
89. Smith, Jr., Lawrence A., and Arganbright, Robert P., Process for Making Acetals, U.S. Patent No. 6,015,875,2000.
90. Therre, J., Kaibel, G., Aquila, W., Wegner, G., and Fuchs, H., Continuous Process for the Preparation of Acetals, U.S. Patent No. 6,518,464, 2003.
91. Boesch, V. and Herguijuela, J.R., Process and Manufacturing Equipment for Preparing Acetals and Ketals, U.S. Patent No. 6,806,392, 2004.
92. Rehfinger, A. and Hoffmann, U., Kinetics of methyl tertiary butyl ether liquid phase synthesis catalyzed by ion exchange resin. I. Intrinsic rate expression in liquid phase activities, *Chem. Eng. Sci.*, 45, 1605, 1990.
93. Zhang, T. and Datta, R., Integral analysis of methyl *tert*-butyl ether synthesis kinetics, *Ind. Eng. Chem. Res.*, 34, 730, 1995.
94. Ali, A. and Bhatia, S., Methyl tertiary butyl ether formation in a catalytic bed reactor — kinetic and modelling study, *Chem. Eng. J.*, 44, 97, 1990.
95. Zhang, T., Jensen, K., Kitchaiya, P., Phillips, C., and Datta, R., Liquid-phase synthesis of ethanol-derived mixed tertiary alkyl ethyl ethers in an isothermal integrated packed-bed reactor, *Ind. Eng. Chem. Res.*, 36, 4586, 1997.
96. Syed, F.H., Egleston, C., and Datta, R., *Tert*-amyl methyl ether (TAME): thermodynamics analysis of reaction equilibria in the liquid phase, *J. Chem. Eng. Data*, 45, 319, 2000.
97. Sircar, S. and Rao, M.B., Liquid-phase sorption-enhanced reaction process, *AIChE J.*, 45, 2326, 1999.
98. Pöpken, T., Götze, L., and Gmelhing, J., Reaction kinetics and chemical equilibrium of homogeneously and heterogeneously catalyzed acetic acid esterification with methanol and methyl acetate hydrolysis, *Ind. Eng. Chem. Res.*, 39, 2601, 2000.
99. Kiviranta-Pääkkönen, P., Struckmann, L., Linnekoski, J., and Krause, A., Dehydration of the alcohol in the etherification of isoamylenes with methanol and ethanol, *Ind. Eng. Chem. Res.*, 37, 18, 1998.

100. Linnekoski, J.A., Krause, A.O.I., and Struckmann, L.K., Etherification and hydration of isoamylenes with ion exchange resin, *Appl. Catal. A,* 170, 117, 1998.
101. Prior, J., Síntese de ETBE, Ph.D. thesis, University of Porto, 2001.
102. Oost, C. and Hoffmann, U., The synthesis of tertiary amyl methyl ether (TAME): microkinetics of the reactions, *Chem. Eng. Sci.,* 51, 329, 1996.
103. Fite, C., Tejero, J., Iborra, M., Cunill, F., Izquierdo, J.F., and Parra, D., The effect of the reaction medium on the kinetics of the liquid-phase addition of the methanol to isobutene, *Appl. Catal.*, 169, 165, 1998.
104. Vila, M., Cunill, F., Izquierdo, J.F., Tejero, J., and Iborra, M., Equilibrium constants for ethyl tert-butyl: the liquid-phase synthesis, *Chem. Eng. Commun.*, 124, 223, 1993.
105. Calderón, A., Tejero, J., Izquierdo, J.F., Iborra, M., and Cunill, F., Equilibrium constants for the liquid-phase synthesis of isopropyl *tert*-butyl ether from 2-propanol and isobutene, *Ind. Eng. Chem. Res.,* 36, 896, 1997.
106. Oudshoorn, O.L., Janissen, M., van Kooten, W.E.J., van Bekkum, H., van den Bleek, C.M., and Calis, H.P.A., A novel structured catalyst packing for catalytic distillation of ETBE, *Chem. Eng. Sci.,* 54, 1413, 1999.
107. Broughton, D.B. and Gerhold, C.G., Continuous Sorption Process Employing Fixed Bed of Sorbent and Moving Inlets and Outlets, U.S. Patent No. 2,985,589, 1961.

3 Ion Exchange Resins in Drug Delivery

Sunil K. Bajpai, Manjula Bajpai, and Sutanjay Saxena

CONTENTS

3.1 Ion Exchange Resins (IERs) in Pharmaceutical Applications 104
3.1.1 IERs in Oral Drug Delivery 104
3.1.1.1 Mechanism of Gastric Drug Delivery 107
3.1.1.2 Selection of Ion Exchange Resin 108
3.1.1.3 Superiority of Drug-Resin Complex to Drug Alone 109
3.1.1.4 Characterization of Resin 111
3.1.1.5 Can Any Drug Be Loaded? 113
3.1.1.6 Preparation of Resinate 113
3.1.1.7 Drug Release Studies 115
3.1.1.8 Kinetic of Drug Release through Ion Exchange 116
3.1.1.9 Factor Affecting Drug Loading into Resin and Release from Resinate 119
3.1.1.10 Use of IER to Modify Release Profiles 128
3.1.1.11 Polysaccharides as Ion Exchange 131
3.1.2 IER as Gastric Retentive Devices 132
3.1.2.1 Preparation of Floating Beads 132
3.1.2.2 Drug Release from Floating and Nonfloating IER Beads 133
3.1.2.3 Drug Release Mechanism 134
3.1.2.4 Evaluation of Gastroretentive Formulations 136
3.1.3 IER for Cancer Treatment 138
3.1.4 Miscellaneous Applications 140
3.1.4.1 Chewable or Dispersible Tablets of Bitter or Nauseous Medications 140
3.1.4.2 Chewing Gum for Buccal Absorption 141
3.1.4.3 Drug Stabilization 141
3.1.4.4 Sigmoidal Release Systems 142
3.1.4.5 Nasal or Ophthalmic Drug Delivery 142
3.1.4.6 Improvement in Tablet Disintegration 142
3.1.4.7 Use of IER as Therapeutics 143

3.1.4.8 Implantation Devices for Water–Soluble/Charged Drugs 143
3.1.4.9 Transdermal Drug Delivery 144
3.2 Concluding Remarks 144
Acknowledgment 145
References 145

3.1 ION EXCHANGE RESINS (IERS) IN PHARMACEUTICAL APPLICATIONS

Until the middle of the 19th century, ion exchange resins (IERs) were frequently used for the purpose of removing toxic metal ions from the domestic waters as well as industrial effluents. However, in 1956, Saunders and Choudhary[1] studied the uptake and release of alkaloids from IER and suggested that these resins might act as a suitable chemical carrier for the development of sustained release formulations. IERs have since been extensively explored in the pharmaceutical field, leading to the some important patents. Extensive research over the past few years has revealed that IERs are equally suitable in a variety of pharmaceutical formulations such as chewable or dispersible tablets, chewing gum for buccal absorption,[2] sustained release preparations[3] such as capsules,[4] liquid orals,[5] bioadhesive systems, transdermal and iontophoretically assisted transdermal systems,[6] ophthalmic delivery systems,[7] and nasal, topical, and taste-masked systems,[8] and so forth.

3.1.1 IERs in Oral Drug Delivery

Ion exchange resins have been used for many years in pharmaceutical formulations. One of the most important properties of ion exchange resins is that they contain functional groups attached to the backbone of the polymer that can exchange ions with ions in the solution. Figure 3.1 shows the equilibrium reaction between an anion exchange resin and a drug molecule. It is important to note that this is a reversible reaction and its equilibrium position will depend upon the environment in which drug and ion exchange resin are present.

In practice, drug in an ionic form (usually in solution) is mixed with the appropriate IER to form a complex, known as a "resinate." The performance of resinate is governed by several factors such as:

- pH and temperature of the drug solution
- Molecular weight and charge intensity of the drug and IER
- Geometry
- Mixing speed
- Ionic strength of the drug solution
- Degree of cross linking and particle size of the IER
- The nature of the solvent
- Contact time between the drug species and the IER[9,10]

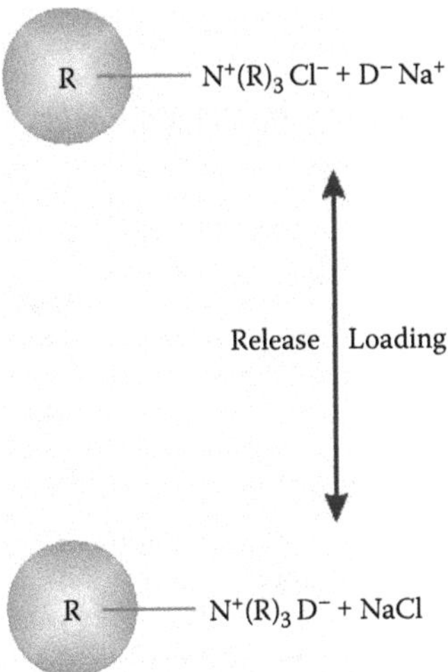

FIGURE 3.1 Equilibrium reaction between an anion exchange resin and a drug molecule.

The chemistry of the resinate is such that the drug retains its characteristics but is immobilized on a solid support.[7] The interactions between the IER and drug, although primarily chemical in nature, are also partially a result of physical adsorption. These interactions are commonly referred to as "adsorption on IER" rather than complexations on most occasions. The ion exchange process is therefore a double-decomposition process, in which the IERs used are able to provide the type of ion required to replace the one that is adsorbed from the solution. The ion of the IER, which can be exchanged for a drug counterpart, is called a "counterion." The affinity of counterions and drug ions towards the IER is competitive. When resinate from the delivery system reaches the site of delivery, the exchange process is reverted, resulting in liberation of free drug ions. Therefore, the ionic strength and pH at the site of delivery play a key role in the liberation of the immobilized drug from the resinate. Drug delivery at the desired target via the ion exchange process occurs because of the presence of highly activated counterions at the site, resulting in the exchange of ions with subsequent drug release. The ion exchange resin devoid of drug is eliminated or biodegraded from or at the site of delivery. From the above discussion, it appears that presence of ions in the release medium is the basic requirement for using ion exchange resins as a drug delivery device. That is the reason that controlled and sustained-release drug delivery have been frequently studied using ion exchange resins, as there is sufficient ionic strength in the gastrointestinal tract for the exchange process to occur. The ion exchange (IE) process might not be optimally applicable to the skin, external canals (e.g., nasal and ear), or other areas with limited concentrations of eluting ions. By contrast, the subcutaneous and intramuscular routes where the pool of ions is more controlled would appear suited for this approach. Some ion exchange resins used in recent past for drug delivery applications have been listed in Table 3.1.

TABLE 3.1
Some Ion Exchange Resins Used in Drug Delivery

Ion Exchange Resin	Drug	Type of System	Remarks	Ref.
Amberlite IRA 69 Amberlite IRA 88	Diclofenac and Propranolol	Resinate encapsulated in HPMC tablets	The drug release from HPMC tablets containing drug-resin complex was slower than HPMC tablets containing drug only	40
Indion 234	Ciprofloxacin HCl	Resinate coated with polyethylene Glycol	The rapid ionic equilibrium drug exchange at gastric pH from DRC was retarded on treatment with PEG	37
Dowex 50W-X4 (200–400 mesh)	Terbutaline hemisulfate	Microencapsulated resinate	The nature of the phase used in the microencapsulation method has crucial role in determining the behavior of resin-containing acrylic microcapsules	8
Dowex Cl-X2 (200–400 mesh)	Diclofenac sodium	Microencapsulated resinate	Prolonged drug release was observed in the microcapsules prepared with Eudragit RS30D	44
Amberlite CG 50 R Amberlite CG 120 R	Diphenhydramine HCL Pseudoephedrine	Cellulose acetate butyrate coated resinates	Drug release was observed to vary from drug to drug and also from resin to resin	86
Dowex 50WX8	Pseudoephedrine	Coated resinates	The coated drug-resin complex particle showed fracturing of the coat, thus necessitating impregnation	91
Indion 244	Bromhexine	Microencapsulated resinate	Controlled release oral liquid suspension was formulated	92
Amberlite IL 120	Metoclopramide	Resinate	Method for determining diffusion-controlled drug release was presented	29
Smopex® 101 (–SO_3 ion exchange groups)	Propranolol HCl and nadolol	Fibers	This study shows that the release of ionic drugs can be controlled by modifying either the fiber type or the external solutions	90
Dowex 1-X2, 1-X4, 1-X8	Theophylline	Microencapsulated resinate	The release profile was affected by the degree of cross-linking and coating process	93
Amberlite IRA 900	Diclofenac	Floating beads coated with hydrophobic polymers	Drug release from both coated and uncoated beads occurs via particle diffusion. The release is prolonged	52

TABLE 3.1 (continued)
Some Ion Exchange Resins Used in Drug Delivery

Ion Exchange Resin	Drug	Type of System	Remarks	Ref.
Sulfopropyl dextran	Doxorubicin	Microspheres	The rate and extent of drug release was affected by the presence of divalent Ca2+ ions and salt concentration	38

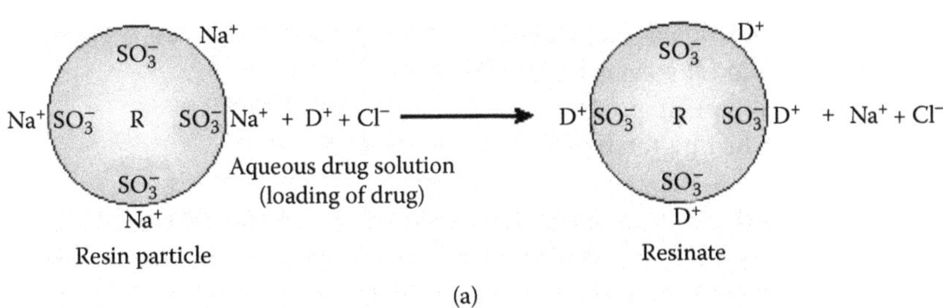

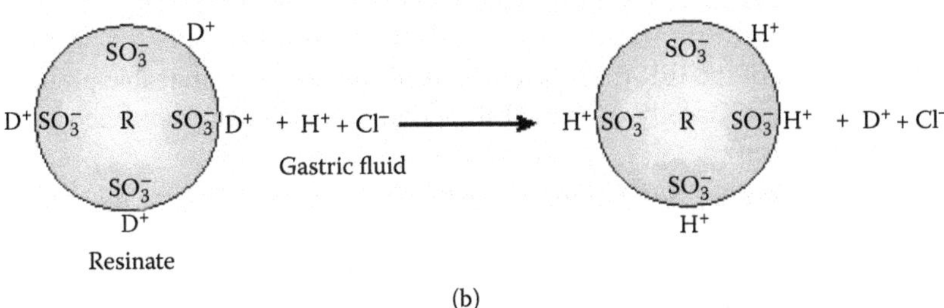

FIGURE 3.2 Complete mechanism of loading and delivery of a cationic drug (D^+, Cl^-) using an ion exchange resin polymer SO_3^-, Na^+.

3.1.1.1 Mechanism of Gastric Drug Delivery

Figure 3.2 shows the complete mechanism of loading and delivery of a cationic drug (D^+, Cl^-) using an ion exchange resin polymer, SO_3^-, Na^+. In order to load the drug, a specific amount of resin is put in a drug solution of known concentration and is constantly stirred. As a result of ion exchange between Na^+ ions from the resin D^+ ions from the drug solution, the D^+ ions are absorbed into the resin matrix, replacing Na^+ ions (see Figure 3.2a). In this way, the drug is loaded into the resin, forming the drug-resin complex (or "resinate").

When the drug-loaded resin particles are taken orally, they reach the highly acidic environment of the stomach. Here, H^+ ions, present in the gastric fluid, undergo ion exchange process with D^+ ions present within the resin matrix, thus resulting in the release of drug ions into the gastric fluid (see Figure 3.2b).

Finally, these drug ions are absorbed through gastric mucosa and the resin particles are either eliminated through biodegradation or passed on to the large intestine and removed along with feces.

3.1.1.2 Selection of Ion Exchange Resin

Typical properties of pharmaceutical-grade ion exchange resins that are pertinent to their use in pharmaceutical formulations are:

- Particle size of 25 to 150 microns
- Contain functional groups capable of exchanging ions and ionic groups
- Insoluble in all solvents at all pH levels
- Fine, free-flowing powders
- Not absorbed by the body

The fact that these materials are totally insoluble in all solvents and at all pH levels, combined with their particle size, means that they are not absorbed by the body, and so have proven to be nontoxic and very safe.

The selection of ion exchange resin for the purpose of drug delivery is mainly governed by the functional group properties of the resin.[11] However, the following points should be considered during the selection:

1. Capacity of the IER (i.e., the concentration of the exchangeable groups in the resin, usually expressed in milliequivalents per gram [meq g^{-1}] of dry resin).
2. Degree of cross-linking in the resin.
3. Particle size of the resin.
4. Nature of the drug and site of drug delivery. It is also important to evaluate the resin in the pH and ionic strength environment, simulating the *in vivo* conditions.
5. Swelling ratio.
6. Biocompatibility and biodegradability.
7. Regulatory status of the IER.

3.1.1.2.1 Illustration

If a resin in a region with a low degree of cross-linking is selected, it will have a more porous structure, thus facilitating the exchange of larger ions. However, it will also cause volume change in the resin upon conversion from one form to another. Similarly, the use of a strong ion exchange resin will provide a rapid rate of exchange, but it could also cause hydrolysis of the labile drugs because strong IERs are effective acid-base catalysts. Therefore, a fine balance of all the parameters mentioned above needs to be made to achieve optimal performance.

3.1.1.3 Superiority of Drug-Resin Complex to Drug Alone

When a suitable resin is put in the drug solution for a specific time, the drug molecules are loaded into the resin matrix through the ion exchange process, thus resulting in the formation of the drug-resin complex (resinate). When the resinate is put in the physiological fluid, it releases the loaded active pharmaceutical agent, again by ion exchange process, as was discussed in the previous section. Now there arises a question. *What are the advantages of getting the drug released from the resinate? Why cannot the drug be taken directly?* The answer to these or similar questions can be obtained by considering the various problems that are created during the process of drug delivery and also their solutions provided by delivering the drug through ion exchange process.

3.1.1.3.1 Stability

The drug resinate is frequently more stable than the original drug. This tendency is exemplified by the stabilization of vitamin B_{12} in the oldest pharmaceutical resinate exemplified applications. Vitamin B_{12} has a shelf life of only a few months, but the resinate is found to be stable for more than 2 years. This technology is still used commercially today, more than 50 years after it was first introduced. Another example is nicotine. Nicotine discolors quickly when exposed to air and light but the resinate (used in nicotine chewing gums and lozenges) has been found to be much more stable.

3.1.1.3.2 Taste

The test of pharmaceutical preparations is an important parameter governing patient compliance and commercial success in the market. Since resinates are insoluble in water, they have no test. This makes them excellent candidates for masking foul-tasting drugs. At salivary pH (6.8), resinate remains in an intact form, making the drug unavailable for the test sensation. As the formulation enters the upper segments of gastrointestinal tract (GIT), the environment changes to acidic and drug release takes place.[12] Polystyrene-matrix-based cation exchange resin (CER) have been used to mask the bitter taste of chlorpheniramine maleate, ephedrine hydrochloride, and diphenylhydramine hydrochloride.[13] The ionic binding of drugs to polymeric materials such as Carbopol® is emerging as an important mechanism of taste masking. Erythromycin and clorithromycin have been taste masked by binding to Carbopol.[12] Chloroquin phosphate and dicyclomine hydrochloride[14] have been successfully taste masked with IERs recently. However, as IERs could also retard the release of drugs, a proper and careful selection of the IER is essential to yield optimal taste masking without affecting the bioavailability. Generally, less cross-linked IERs are helpful in taste masking. This technique is applicable to liquid formulations (suspensions) and mouth-dissolving tablets. It is particularly effective in liquid formulations because the resinate represents the thermodynamically stable form so that leaching of the drug into the aqueous phase will not occur.

TABLE 3.2
Dissolution of the Drug Indomethacin

Time (min)	% Release (22°C)
0	0
10	61
20	78
45	97
120	100

USP: Not less than 80% in 20 min at 37°C.

Data adapted from Dr. Lyn Hughes, *New Uses of Ion Exchange Resins in Pharmaceutical Applications*, www.Rohmhass.com/ionexchange/pharmaceuticals/Formulations_doc/new_uses.pdf.

3.1.1.3.3 Poor Dissolution

The problem of dissolution of poorly soluble drugs is well known in the pharmaceutical field. It has been observed that in the case of poorly soluble ionizable drugs, the release of a drug from a resinate can be faster than the rate of dissolution of the solid form of the drug. Hence, one can increase the rate at which poorly soluble drugs "dissolve." This is demonstrated by the data given in Table 3.2.

It shows the results of a United States Pharmacopeia (USP) constant volume dissolution test on an indomethacin resinate. The footnote refers to a formulation where micronization of the drug has also been used to enhance the dissolute rate. Note that the resinate provides nearly 78% drug release in 20 min, whereas 80% is released in the same period when micronization has been used. The minor difference could be attributed to the fact that test was done at ambient temperature, not the required 37°C. Increasing the temperatures will obviously increase the release rate. This indicates that resinates are able to provide the drug release at the same rate as obtained by micronization. It is also worth mentioning here that using micronization to increase the rate of dissolution can be problematic, frequently requiring specialized equipment and causing problems with agglomerations of the fine particles after grinding. The grinding can also result in melting and conversion to other crystal forms. These problems are completely eliminated using an ion exchange approach.

3.1.1.3.4 Deliquescence

Deliquescence is the property of a solid whereby it absorbs so much water that it dissolves in the water it absorbs. While this is not a common problem, it has been very difficult to solve. It requires the use of specialized equipment or careful scheduling of production in dry seasons. However, it has been found that resinates of deliquescent and highly hygroscopic drugs retain the properties of the resin and are not deliquescent and remain free-flowing powders. Their water absorption characteristics are similar to those of unloaded resins, so that any formulation equipment that can handle the resins can handle the resinate of the deliquescent drug without need for special manufacturing conditions.

TABLE 3.3
Deliquescent Behavior of Valproate Resinates (Conditions: 24°C/55% RH)

Resin Used	30 min	60 min
Cholestyramine USP	Free flowing	Free flowing
Amberlite IRA 458	Free flowing	Free flowing
Amberlite IRA 67	Free flowing	Free flowing
Colestipol USP	Free flowing	Free flowing
Sodium valproate	Sticky	Sticky

Data adapted from Dr. Lyn Hughes, *New Uses of Ion Exchange Resins in Pharmaceutical Applications*, www.Rohmhass.com/ionexchange/pharmaceuticals/Formulations_doc/new_uses.pdf.

For example, sodium valproate is a drug that is well known to be highly deliquescent. However, it has been found that valproate resinates remains free-flowing even after exposure to ambient air. Table 3.3 summarizes results from several resinates using different anion exchange resins.

Sodium valproates became liquid within 1 h. On the other hand, valproate resinates remained free-flowing even after 1 h. This suggests that dosage forms of deliquescent or highly hygroscopic drugs could be manufactured with no special equipment or atmosphere controls, and deliquescence during storage is eliminated, thus simplifying the packaging requirements. *In vitro* release tests on these resinates have confirmed that the drug is released on exposure to GI fluids.

Very similar results have also been obtained with other deliquescent drugs, such as rivastigmine bitartrate, which uses cation exchange resins to make the resinate, showing that this technique is highly significant.

3.1.1.3.5 Polymorphism

Polymorphism has been a very common problem in the pharmaceutical industry. Much funding has been spent trying to identify polymorphs and make stable suitably soluble forms. Failure to resolve such problems can result in significant stability problem for the final dosage form. Ion exchange resins present a unique way of dealing with this problem. A drug resinate is an amorphous solid and cannot crystallize or even form hydrates. In addition, the release of the drug from the resinate is independent of the crystal form that was used to make it. Consequently, the use of resinates can eliminate any problems with polymorphism. Figure 3.3 shows release and dissolution data on lansoprazole and its resinates. The data clearly demonstrates that, although the original crystal forms of the drug had very different dissolution rates, the release rates from the resinates were all the same.

3.1.1.4 Characterization of Resin

As the performance of a drug delivery system depends upon the quality of its IERs, it is important to evaluate the IERs at each stage in the preparation of

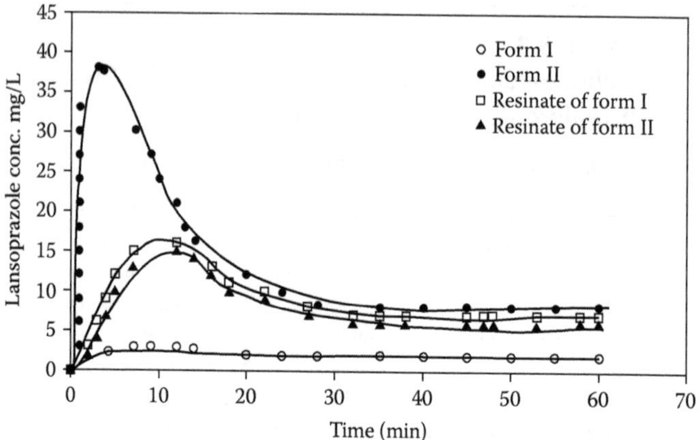

FIGURE 3.3 Release/dissolution data on lansoprazole and its resinates.

resinates. This is one of the most mystifying parts of working with ion exchange resins. When characterizing polymers, it is normal to consider the molecular weight of the polymer. Unfortunately, this has no useful meaning in ion exchange resin because the cross-linking agents in the polymerization process have rendered them insoluble. This results in the polymer being a three-dimensional network for which a single particle is effectively a single molecule, and so molecular weight (MW) is defined by particle size. This has been estimated to be in the range of 10^{17} to 10^{20} Da. For all intents and purposes, it is infinite. All resins will contain some level of low MW material (up to ~ 5×10^6 Da), but this is considered to be an extractable impurity and contributes no beneficial effects during use.

How does one characterize a material that has no sensible MW and is completely insoluble in all known solvents? Typical techniques such as gas chromatograph (GC), high-performance liquid chromatography (HPLC), mass spectrometry (MS), and melting point cannot be used. There are two nondestructive techniques that can be used: infrared (IR) and solid-state nuclear magnetic resonance (NMR), but even these are seriously limited. IR cannot be used quantitatively except in exceptional cases; NMR is useful and can be used quantitatively, but the sensitivity can be poor.

The method used for assay is drug displacement. In this technique, the drug is displaced from the resinate into a solution and the amount displaced is quantified by the usual solution methods. However, there are two problems with this technique: how to create a standard and how to displace the drug.

Creating a standard is best done by careful analysis of the loadings steps. If the amount of drug initially used and the amount left after loading are known accurately, then it is possible to calculate the amount of drug loaded; knowing the amount of resin used, one can calculate the loading.

In order to get reasonable assay, it is essential to displace > 95% of the drug, and this can be problematic if the drug has a high affinity for the resin. Usually aqueous acids, bases, or salt solutions are used as displacement fluids. Optimization of the displacement conditions can be a reference to the guidelines provided.

A common method for characterizing polymer is the glass transition temperature (T_g). While this is of some use in characterizing an ion exchange resin, it contributes little, if any, useful information in characterizing a resinate.

As with the electrolytes, ion exchange resins do not have a single pK_a value. As the groups ionize, they create a charge on the polymer that tends to resist further ionization. For example, consider an ion exchange resin with carboxylic acids groups. The first groups to ionize will have a particular pK_a (probably around 3), but as further groups ionize the charge buildup will make the remaining groups behave as even weaker acids, and so their effective pK_a increases. The greater the density of groups along the polymer, the more severe the effect will be.

Finally, the following significant parameters are also evaluated.

3.1.1.4.1 *Particle Size*

This is measured directly with a set of microsieves by screening.[15] The particle size of IERs can also be determined by microscopy,[16] Coulter counter,[17] and other available techniques.

3.1.1.4.2 *Porosity*

The porosity of dry IERs can be determined through nitrogen adsorption at –195°C, and by measuring the true density (mercury displacement).[18] Scanning electron microscopy reveals the internal pore structure. The use of an air-compression pyknometer for determination of porosity has also been reported.[19]

3.1.1.4.3 *Moisture Content*

This is the amount of water retained by the resin when completely saturated with water. Within any specific type of polymer, it is an indicator of the degree of cross-linking. Typical values can be in the range of 40 to 70% by weight. It is usually determined by Kart Fischer titrimetry. Excess water can be removed by drying in a vacuum desiccator.[20]

3.1.1.4.4 *Ion Exchange Capacity*

The IER capacity of strong CER is determined as meq g^{-1} by evaluating the number of moles of Na^+ that are adsorbed by 1 g of dry resin in the hydrogen form.[21] Similarly, the IE capacity of a strong basic anion exchange resin is evaluated by measuring the amount of Cl^- taken up by 1 g of dry resin in the hydroxide form.

3.1.1.5 Can Any Drug Be Loaded?

The basic condition for loading of the drug into an ion exchange resin is that the drug, to be loaded, should be ionizable. Even very weakly acidic drugs can be loaded. Table 3.4 shows the list of some drugs that have been loaded onto ion exchange resins.

3.1.1.6 Preparation of Resinate

Once the selection of resin and suitable drug is made, the next step involves preparation of drug-resin complex before designing a suitable drug delivery

TABLE 3.4
p*K* Values of Various Drugs

Basic Drugs	pK_b	Acidic Drug	pK_a
Acycloguanosine	1.86	Nicotinic acid	2.17
Tinidazole	2.34	Mefenamic acid	3.69
Deferiprone	3.04	Indomethacin	4.17
Cimetidine	6.73	Diclofenac	4.18
Oxycodone	7.53	Repaglinide	4.19
Remacemide	7.76	Ketoprofen	4.23
Nicotine	8.00	Ibuprofen	4.41
Morphine	8.14	Valproic acid	4.82
Hydrocodone	8.48	Ambroxol	8.69
Rivastigmine	8.62	Omeprazole	9.08
Propranolol	9.14	Acetaminophen	9.86
4-Aminopyridine	9.25	Topiramate	12.37
		Carbamazepine	13.94

Note: The list is a combination of published data and Rohm and Haas in-house studies. The p*K* values have been obtained from Chemical Abstract Services.

system. The main hurdle is to optimize the conditions of preparation in order to obtain the desired drug loading in the resinates. Generally, the following steps are involved in the preparation of resinates:

1. Prior to use in pharmaceutical applications, the resin is purified.[22] For example, anion exchange resin, Dowex® may be purified by the column method. Approximately, 350 g of the IER is allowed to swell in 500 ml of purified water for 12 h. The IER swollen particles are now decanted to remove some floating particles, washed again with purified water until the supernatant becomes transparent, and slurried in 500 ml of purified water. The slurry is then poured into a glass column (55 mm inside diameter × 860 mm) equipped with a cotton plug at the bottom. Three and a half liters of methanol is passed through the column of resin, followed by 17.5 l of 2 *N* NaOH for replacing the Cl^- ions with the OH^- ions. Therefore, the purified water is passed through the column to wash out NaOH until the pH value of eluent becomes neutral. The purified IER is now recovered by vacuum filtration and dried in an oven at 50°C for 72 h. The batch method is also used for purification of resins.[23]
2. Changing the ionic form of the IER might occasionally be required to convert a resin from one form to another if it does not have the desired counterions. Strongly acidic cation exchange resins are usually marketed

in Na^+ form and strongly basic anion exchange resins in Cl^- form. They are generally converted into hydrogen and hydroxide forms, respectively. The conversion can be achieved by soaking the resin with acid or alkali, respectively. After changing the ionic form, the resin is subjected to washing with distilled water until elute becomes neutral in reaction, and finally is dried at 50°C.

3. Preparation of resinate is usually done by two techniques:
   (a) Batch technique: After suitable pretreatment, a specific quantity of the granular IER is agitated with the drug solution until the equilibrium is attained.[24]
   (b) Column technique: Resinate is formed by passing a concentrated solution of drug through the ion exchange resin packed column until the effluent concentration is the same as the eluent concentration.

3.1.1.7 Drug Release Studies

Basket and paddle apparatus (according to the European Pharmacopoeia) or apparatus 1 and 2 (according to the United States Pharmacopoeia) are usually prescribed in different pharmacopeias as conventional testers for characterizing the drug release of active ingredients from different types of formulations and resinates. Dissolution media with different ionic strengths and pH in the range of 1.0 to 4.0 (citrate buffer), 5.8 to 8.0 (phosphate buffers), and 8.0 to 10.0 (alkaline borate buffers) are prepared for drug release studies.

Technical data[25] for the drug release experiments using basket or paddle apparatus are shown in Table 3.5.

However, the release studies may also be performed according to the paddle method as described in Japanese Pharmacopoeia (JP XIII).

TABLE 3.5
Technical Data for the Drug Release Experiments Using Basket or Paddle Apparatus

Number of replicants	6 Experiments/resinate tablets. The results are expressed as the mean of six units
Dissolution medium	Citrate buffer solutions (pH 1–4), phosphate buffer solutions (pH 5.8–8.0) and alkaline borate buffer solutions (pH 8.0–10.0)
Volume of dissolution medium	900 ml
Temperature of dissolution medium	37 ± 0.5°C
Stirring elements	Basket (apparatus 1) and paddles (apparatus 2)
Rotation speed of the stirring elements	20 rpm, 50 rpm, 100 rpm, 150 rpm, and 200 rpm
Sampling times	24 h (Usually at the interval of 1 h)

Source: Adapted from Kinel M., et al., *Acta Chem. Solv.*, 51, 409, 2004.

3.1.1.8 Kinetic of Drug Release through Ion Exchange

Quantitative studies of the ion exchange process already have been reported.[26] On the basis of these studies, various mathematical expressions describing the release of drug from ion exchange resins were formulated. In an ion exchange process, the rate-limiting step is shown to be the diffusion either in the resin particle itself (the so-called particle diffusion) or in an adherent stagnant film (film diffusion). Since particle diffusion and film diffusion are sequential steps, the slower of the two is rate controlling. Under conditions where particle diffusion is the rate-limiting step, the fraction of drug released, F, from spherical resin particles with a uniform diameter in a solution of infinite volume as a function of time is given by the following:

$$F = \frac{Q_t}{Q_\infty} = 1 - \frac{6}{\pi^2} \sum_{n=1}^{\infty} \frac{e^{-n^2 Bt}}{n^2} \tag{3.1}$$

where Q_t and Q_∞ are the amounts of drug released after time t and after infinite time, respectively, and n is the summation variable, B is the rate constant defined as $4\ \pi^2 D/d^2$, where D represents the effective diffusion coefficient of the exchanging ions (drugs) in the resin particle and d is the mean diameter of resin particles. Depending upon the magnitude of F, Reichenberg[26] obtained the following two equations:

$$Bt = 2\pi - \frac{\pi^2 F}{3} - 2\pi\left(1 - \frac{\pi F}{3}\right)^{1/2} \tag{3.2}$$

$$Bt = -\log_e \frac{\pi^2}{6} - 2\pi(1 - F) \tag{3.3}$$

Equation (3.2) is the result of Fourier transformation and integration of Equation (3.1) and is used for F values lower than 0.85. However, Equation (3.3) is applicable for F values larger than 0.85. If a plot of Bt values corresponding to the F values against time t gives a straight line with a slope equal to B (see Figure 3.4), it can be assumed that drug diffusion within the resin particles is the rate-controlling step in the diffusion process.

The slope of the line yields the rate constant B, and the effective diffusion coefficient, D, of the drug can be calculated from this B value. However, this theory cannot be used in the case of coated ion exchange resin unless the IER and the coating material are treated as a homogeneous ion exchange matrix. However, it has also been suggested[27] that in order to establish whether the diffusion through the particle is the rate-limiting step, it is enough to check the direct proportionality between log (1 – released fraction) and time$^{0.65}$. For example, the ion exchange

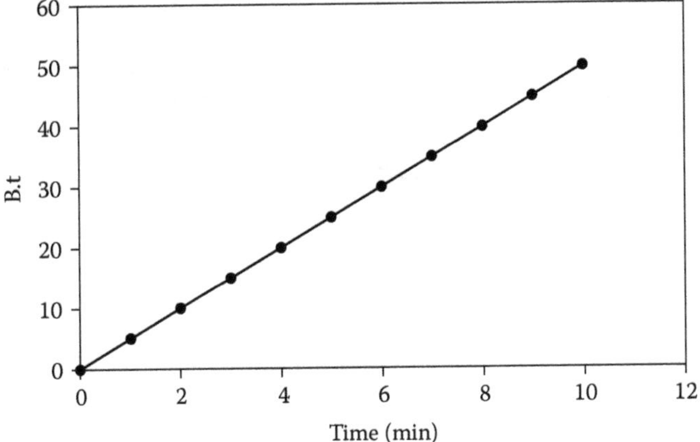

FIGURE 3.4 *Bt* values corresponding to the *F* values against time *t*.

between the diclofenac anion of antiinflammatory diclofenac sodium (DIK–Na) and the Cl^- ion of layered Mg-Al hydrotalcite chloride (HTIC–Cl) results in the formation of intercalation compound HTIC–DIK,[28] which releases the drug in the buffer medium of pH 7.5. The plot between log (1 – released fraction) and time$^{0.5}$ (see Figure 3.5A) is observed to be almost linear, thus confirming that the rate-limiting step is the diffusion through the particle. Moreover, a good linearity between percent release and time$^{0.5}$ is also obtained (see Figure 3.5B) and shows fair agreement with the well-known Higuchi[30] matrix-diffusion-controlled model.

These results confirm the importance of the diffusion through the particle in controlling the release rate. Here it is also worth mentioning that size of the drug molecule and its affinity toward the resin also affect the release kinetics. For example, from the results obtained for the release of another antiinflammatory drug, ibuprofen from hydrotalcite,[29] and its comparison with the results obtained for release of diclofenac sodium from hydrotalcite[28] reveal some interesting facts. First, diclofenac is released more slowly; second, its release depends upon the diffusion through the particle (as discussed above), thus indicating diffusion-controlled kinetics, while the drug ibuprofen demonstrates faster release and the release depends upon the drug concentration, thus indicating first-order kinetics (see Figure 3.6). The different behaviors can be explained by both the greater affinity of diclofenac for hydrotalcite[30] and its bigger molecular size in comparison to ibuprofen. The exchange of anions begins from the external part of the resinate particle and proceeds towards the inside, with the consequent formation of an external phase with smaller distance. Diclofenac release, because of its bigger size, may be slowed down more than ibuprofen by the reduction of the interlayer distance in the external part of the layers. Moreover, the greater the distance that the drug has to travel until the exit (and, as a consequence, the number of exchange sites), the greater the influence of the affinity of the drug for the matrix on the release rate.

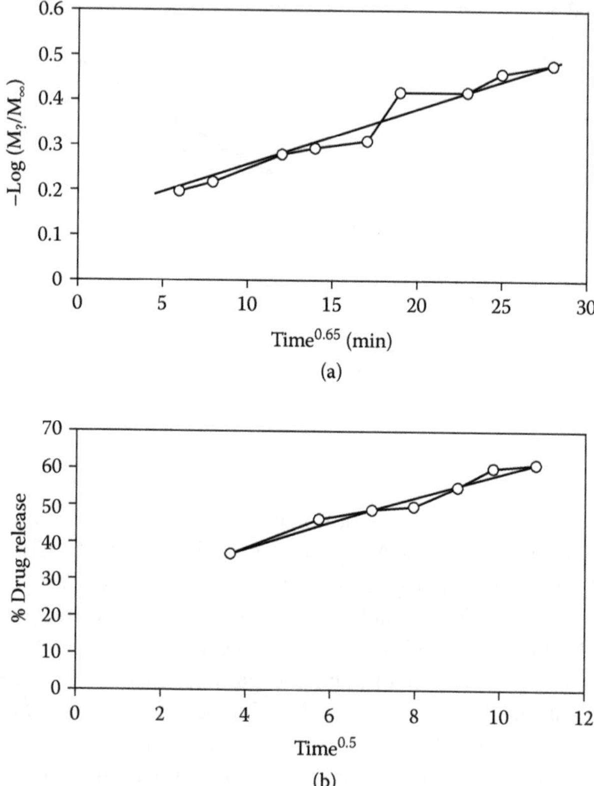

FIGURE 3.5 Log (1 – released fraction) and time$^{0.5}$.

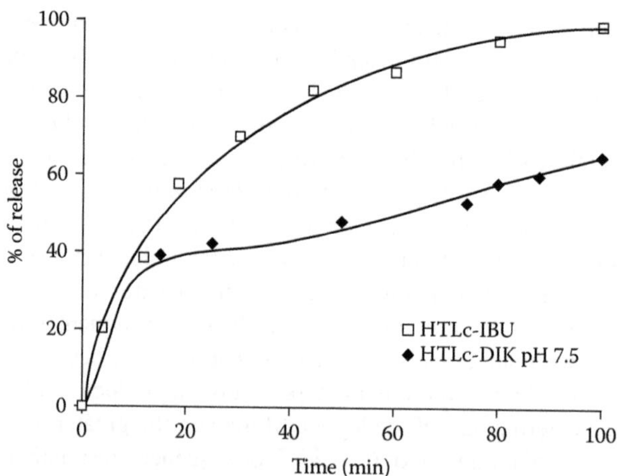

FIGURE 3.6 Ibuprofen diffusion demonstrating faster release dependent on drug concentration.

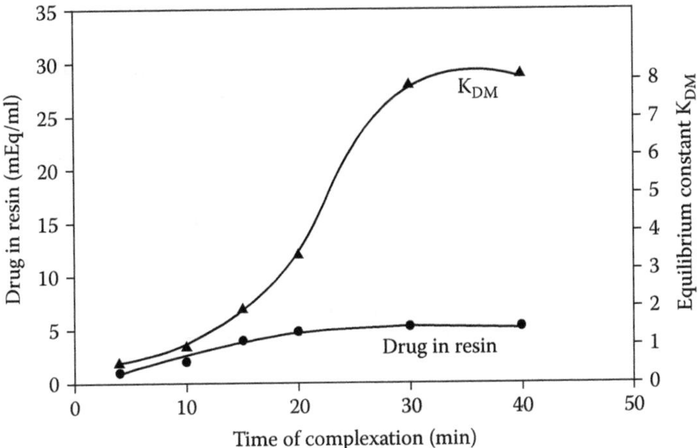

FIGURE 3.7 K_{DM} (equilibrium constant) value after 20 to 30 min.

3.1.1.9 Factor Affecting Drug Loading into Resin and Release from Resinate

The loading of a drug into resin particles and the drug's release from resinate depend upon a number of factors.

3.1.1.9.1 Effect of Swelling

The swelling and hydrating properties of an ion exchange resin affect the rate of ion exchange, which, in turn, affects the percentage of drug loading. In an unswollen resin matrix, the exchangeable groups are latent and coiled towards the backbone, thus causing less drug-loading efficiency.[31] However, a higher degree of swelling of resinate particles causes an increase in the release rates.[32]

3.1.1.9.2 Effect of Stirring Time on Complexation

The equilibrium ion exchange in solution occurs stoichiometrically and is affected by stirring time or time of complexation. For example, let us consider the complexation studies performed for loading of ciprofloxacin into Indion 234 complexes.[33] The equilibrium constant K_{DM} is given as

$$K_{DM} = \frac{[D]^r [M]^s}{[D]^s [M]^r}$$

where $[D]^r$, $[M]^r$, $[D]^s$, and $[M]^s$ are drug and metal concentrations of resin and solution, respectively. The study revealed that as the time increases, the K_{DM} (equilibrium constant) value also increases between 20 and 30 min (see Figure 3.7).

This finding may indicate the significant involvement of van der Waals forces or that chemisorption is taking place along with drug exchange during

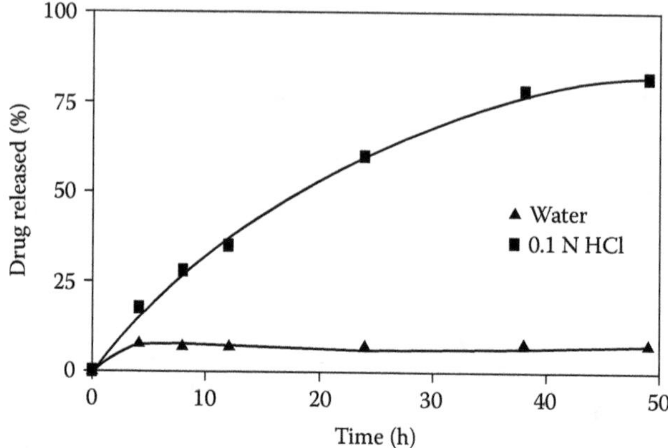

FIGURE 3.8 Propranolol–Amberlite IRP 69 resinate particles a hydroxypropyl methyl cellulose (HPMC) matrix tablets and the release of drug in water and in 0.1 *N* HCL.

complexation.[34] It is also clear that an increase in stirring time beyond 30 min does not further increase the value of K_{DM}. Moreover, the amount of drug loaded into the resin also approaches optimum value.

3.1.1.9.3 Effect of particle size

Particle size does not have effect on drug loading; it affects the rate of exchange of ionic species. The rate of exchange decreases with bead diameter due to the reduction in diffusion path length. Hence, larger particle size affords a slower release pattern.[35] Let us consider the release of drug from propranolol to Amberlite IRP 69 complex particles of different sizes.[40] It has been shown that the resinate particles with size < 45 μm demonstrate faster release as compared with the particles with diameters > 45 μm. This is due to the fact that for a given weight of resinate particles, the particles with the smaller size shall possess greater surface area for the ion exchange process to occur, thus causing a faster release (although ion exchange from resin is not purely a surface phenomenon).

3.1.1.9.4 Presence of Ions in the Release Medium

The release of a drug — say, cationic in nature — from the resinate may be represented as

$$\text{resin}^- - \text{drug}^+ + X^+ \rightarrow \text{resin}^- - \text{X}^+ + \text{drug}^+$$

where X^+ represents the exchangeable cation present in the release medium. Therefore, the release of drug from a resinate depends upon the presence of an exchangeable ion in the release medium. For example, in the study mentioned above, propranolol–Amberlite IRP 69 resinate particles were incorporated in hydroxypropyl methyl cellulose (HPMC) matrix tablets and the release of the drug was studied in water and in 0.1 *N* HCL (Figure 3.8).

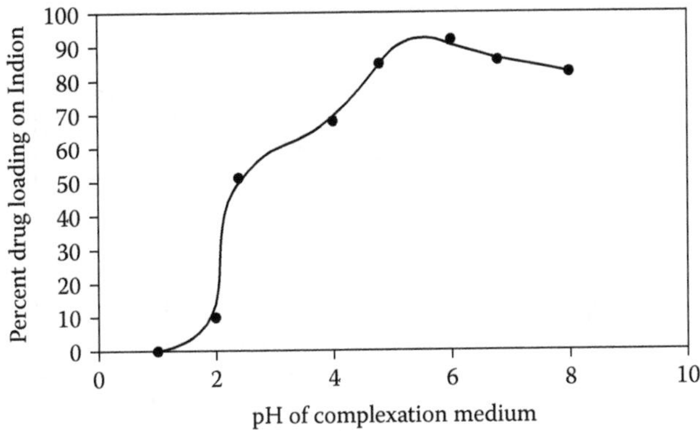

FIGURE 3.9 Percent drug loading as a function of pH of the complexation medium for the complexation between antibiotic drug ciprofloxalin and resin Indion 234.[33]

The drug was not released in water since there were no counterions present in the medium to replace the drug ions from the ion exchange resin particles. The drug was released in 0.1 *N* HCL, indicating that the drug release was initiated by ion exchange process. The counterions present in the dissolution medium diffuse through the gel layer to replace the drug that is then released by diffusion through this gel layer.

3.1.1.9.5 Effect of pH on Complexation or Drug Loading

The formation of an ion exchange resin-drug complex involves the exchange of ionizable drug and metal ions in the resin, which in turn depends upon the pK_a of the drug and resin. Therefore, the pH of the complexation medium plays an effective role in the drug-metal ion exchange process. When the pH of the medium approaches the pK_a of the drug, then optimum complexation (and hence loading) is expected to occur. For example, Figure 3.9 depicts the percent of drug loading as a function of pH of the complexation medium for the complexation between the antibiotic drug ciprofloxacin and resin Indion 234.[33]

It is clear that the complexation (and therefore drug loading) is enhanced with the increase of pH from 1.2 to 6.0. A maximum of 94.3% (wt/wt) drug loading is observed at pH 6.0 (i.e., at pK_a of the drug ciprofloxacin hydrochloride). However, when the pH is increased further beyond 6.0, the percent loading begins to decrease. The pH of the solution affects both the solubility and the degree of ionization of the drug. The above results can also be attributed to the fact that ciprofloxacin hydrochloride has a pK_a between 5.61 and 6.18 and hence will have maximum solubility and complete ionization in this stage. The decreased complexation at lower pH is due to an excess of H^+ ions in the solution, which have more binding affinity to the $-COO^-$ groups of resin and compete with the drug for binding. Such a trend has also been reported in the complexation of chloroquine phosphate with polymethacrylic acid ion exchange.[31] From the above discussion, it is clear that the maximum complexation (and hence loading)

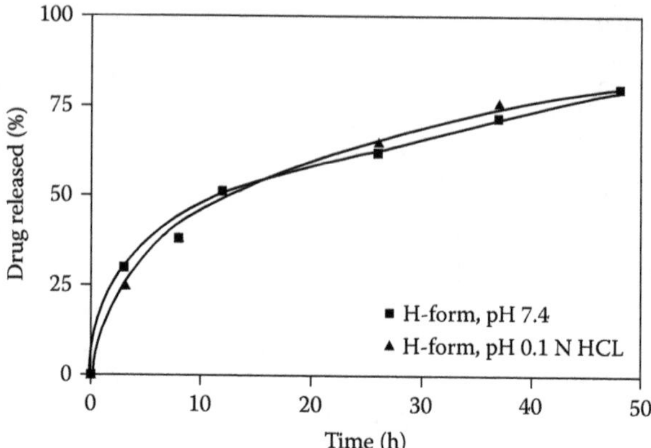

FIGURE 3.10 Release profiles of the amount of drug released at different time intervals in two media of different pH.

occurs when the pH of the complexation of medium approaches the pK_a value of the drug being loaded.

3.1.1.9.6 Effect of pH on the Drug Release

The effect of variation in pH of the dissolution medium on the drug release from the resinates depends on the degree of ionization of the resin. For example, if the release media at two different pH levels contain the same exchangeable counterions, then the amount of drug released from a resinate in these media shall be governed by the degree of ionization of the resin in the release media. Let us consider the previous example[36] where resinate particles formed between the drug propranolol and the strong cation resin Amberlite IRP 69 were incorporated into HPMC matrix tablets and placed in a media of pH 7.4 and in 0.1 *N* HCL, both containing exchangeable H^+ ions. The release profiles, as depicted in Figure 3.10, clearly indicate that the amount of drug released at different time intervals is nearly the same in two media of different pH.

This may be attributed to the fact that Amberlite IRP 69, being a strong resin, is dissociated at both pH values, thus allowing the drug binding. However, in the same study when the propranolol release was carried out with a weak ion exchange resin, Amberlite IRP 88, the results obtained were quite different (see Figure 3.11).

The release is faster in 0.1 *N* HCL, which may be attributed to the nonionization of carboxylic groups at low pH (Amberlite IRP 88 is a weak resin and it contains carboxylic groups) and which results in a weaker binding of drug$^+$ ions within the resinate. Therefore, drug$^+$ ions are easily replaced by H^+ ions from the release medium, thus finally resulting in faster release. On the other hand, the carboxylic groups undergo ionization in the medium of pH 7.4, thus producing strong interactions with drug$^+$ ions within the resinate particles. Therefore, the exchange of drug$^+$ ions with H^+ ions of release medium becomes relatively slower, thus finally resulting in slower drug release.

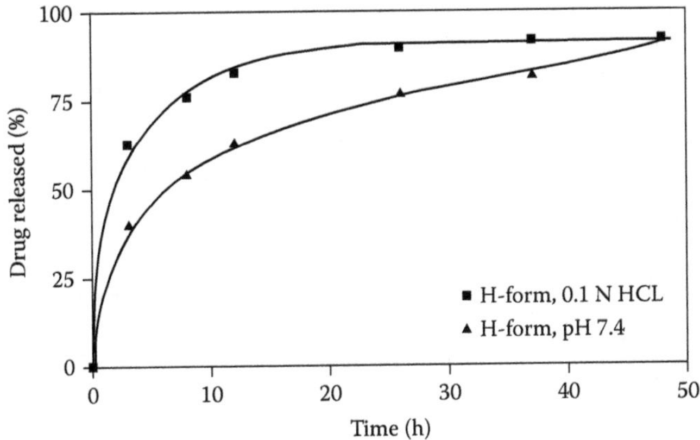

FIGURE 3.11 Propranolol release with Amberlite IRP 88.

Therefore, drug release behaviors of strong and weak cation resins are quite different when exposed to media of pH 7.4 and 0.1 *N* HCL (pH 1.0) solutions.

3.1.1.9.7 Effect of Ionic Strength of Salts on Drug Release

The release of a drug loaded into a resin is due to an ion exchange process between the drug ions present within the resinate and exchangeable ions present in the release medium. Therefore, the concentration of ions in the release medium (i.e., ionic strength of the salt) plays a significant role in governing the release kinetics. In other words, electrostatic interactions govern the equilibrium distribution of the drug species between the resin and solution phases.[37,38]

Let us consider the effect of ionic strength of $CaCl_2$ in the release media on the dynamic release of drug from ciprofloxacin hydrochloride–Indion 234 complex.[33] The data displayed in the Figure 3.12 show that the drug release from the complex increases as the concentration of electrolyte is increased in the release media.

This may be attributed to the fact that the increase in electrolyte or salt concentration is accompanied by a decrease in the Donnan potential, and hence the electrostatic affinity between the drug and the ion exchanger also decreases, thus tending to enhance the drug release. The influence of salt concentration on the release rate can also be explained on the basis of solute diffusion. Since the drug release is due to the ion exchange process, the exchange rate is dominated by the rate at which the competing ions diffuse from the media into the resin. Solute diffusion is driven by concentration gradient. At high salt concentrations, the concentration gradient is greater, thus resulting in a faster diffusion of ions and thereby a higher release rate.

3.1.1.9.8 Effect of Valency of Ions in the Release Medium

The valency of the ions being exchanged with the drug ions present in the resinate seems to play a significant role in governing the drug release rate. A close look at the Figure 3.13, showing release of ciprofloxacin hydrochloride from a ciprofloxacin–Indion 234 complex in the medium of sodium and calcium

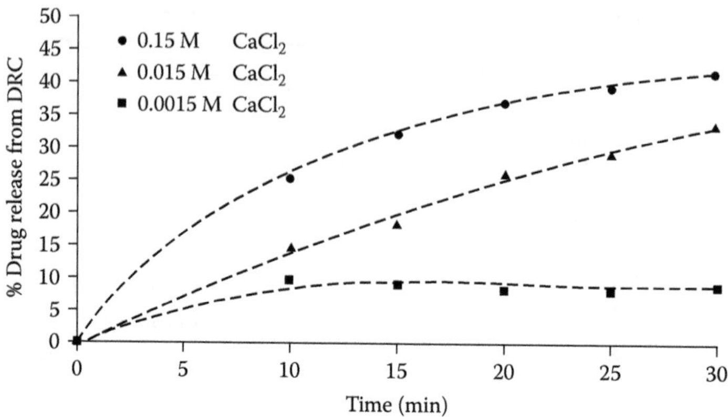

FIGURE 3.12 Drug release from the complex as the concentration of electrolyte is increased in the release media.

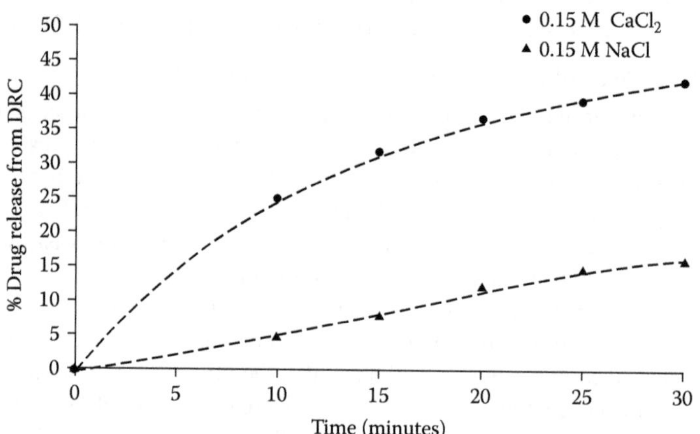

FIGURE 3.13 Release of ciprofloxacin hydrochloride from ciprofloxacin-234 complex in the medium of sodium and calcium ions.

ions, reveals that the drug is released at a faster rate in the medium containing Ca^{2+} ions as compared to monovalent sodium ions.

This is due to the fact that the rate of sorption of the divalent calcium ions is much faster as compared with sodium ions. Also, the selectivity of carboxylic acid resins is higher for divalent calcium ions. Even though larger-sized Ca^{2+} ions are expected to diffuse slowly, their valency enhances the drug release. Calcium ions also reduce the Donnan potential to a greater extent, thus reducing the affinity between the drug ions and the ion exchanger.

3.1.1.9.9 Effect of Coating on Drug Release

The use of ion exchange resins for oral sustained-release formulations occupies an important place. It has several advantages, such as less variability of rate

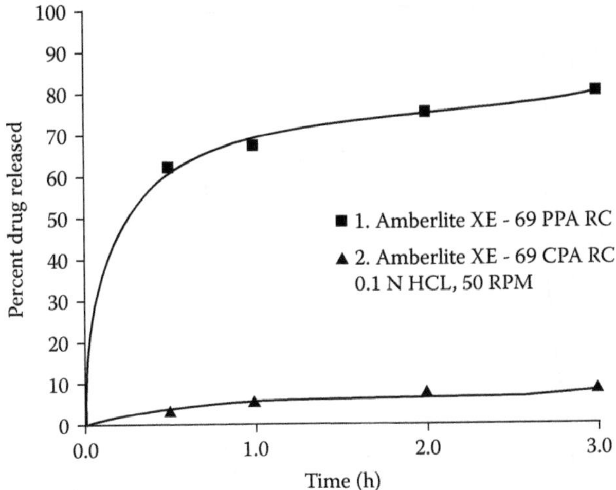

FIGURE 3.14 Release of chloropheneramine from CPARC vs. release of phenylpropanolamine from PPARC.

constant,[39] a readily controllable particle shape, a lack of toxicity, the spherical nature of polymer coating, and so forth. Although it has been a well-established fact that the release of bioactive materials can be retarded by forming drug-resin complexes, in some cases the need to further retard the release is experienced. For example,[40] the drug chloropheniramine is released from the resin complex (CPARC) quite slowly while the release of phenylpropanolamine from its resin complex (PPARC) is much more rapid, with nearly 70% of the drug released within 30 min (see Figure 3.14).

Therefore, in order to prepare sustained-release formulation of phenylpropanolamine, it becomes necessary that something other than simple formation of the drug-resin complex should be done. In this situation, the drug complex is usually coated with some suitable film-forming material so that the prolonged release can be obtained.

Out of different methods employed to coat IER particles, the air-suspension coating process, often referred to as the Wurster process, has been characterized as a mechanical microencapsulation method.[41] In this process, particulates circulating in the coating chamber are encapsulated by a wet spraying. Because of such a simple principle, it provides a unique method for preparing microcapsules with multilayered and composite structures that have potential as functional particulate materials in a variety of industrial fields, including pharmaceuticals.

The major drawback in this process is the difficulty in processing fine particles. For usual pharmaceutical applications, the lower critical size of particles that can be individually coated without agglomerations appears to be approximately 20 μm in diameter.[42] However, the coating process is often hampered by severe agglomerations, even for particles in the 20- to 100-μm size range, depending on the physiochemical properties of the particles, for example, hygroscopicity,

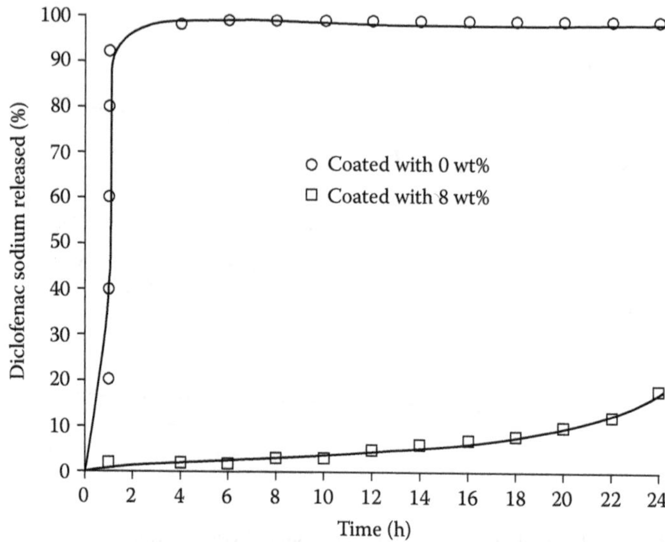

FIGURE 3.15 Release of diclofenac sodium from Eudragit RS30D coated (8 wt%) ion exchange Dowex 1 –X2 resinate particles.

electrostatic charging, and solubility of spray solvent. In this context, it was recently found that aqueous colloidal polymer dispersions can exhibit an extremely low agglomeration tendency even in such fine particles of less than 100-μm size.[43] Another problem encountered in spray coating such fine particles (20 to 100 μm) comes from the difficulty in decreasing the coat thickness. Usually, a coating thickness of 10 μm is required to act as diffusion barrier for the prolonged release of a drug from dosage forms. For prolonged release, coating of the fine particles requires large quantities of coating material due to large surface area. Moreover, the process is also time consuming.

The effect of coating the resinate on the release rate can be seen in the Figure 3.15, which displays release of diclofenac sodium from Eudragit® RS30D coated (8 wt%) ion exchange Dowex 1–X2 resinate particles.[44]

It is clear that uncoated particles release nearly 80% drug in 1 h while coating of these particles with Eudragit RS30D makes the drug release process extremely slow, extended over a duration of 48 h. Therefore, coating of ion exchange resin-drug complex particles with colloidal polymer dispersions seems to be very effective for prolonged release.

Although Eudragit-coated resin particles prove to be quite effective in retarding the drug release rate, the rupture of the coated particles sometimes creates problems. For example,[45] the morphological analysis of Eudragit RS/RL-coated terbutaline–Dowex 50 W-X4 resinate particles show that they are broken upon the first day of storage (see Figure 3.16a). The rupture of the Eudragit-coated resinate particles may be attributed to the swelling shown by the resin particles after contact with the suspending aqueous vehicle. A microscopic observation of

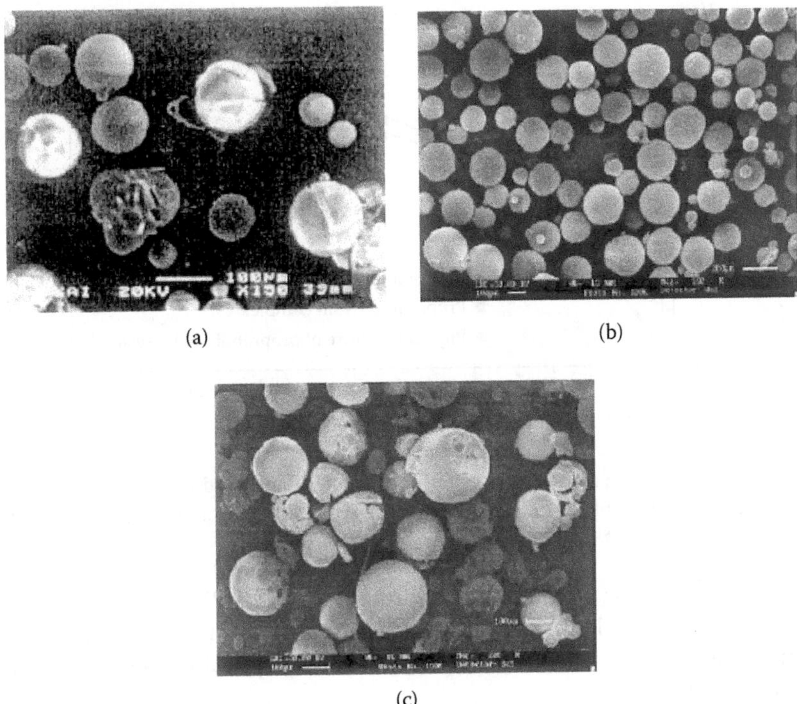

FIGURE 3.16 (a) Eudragit RS/RL coated Terbutaline-Dowex 50 W-X4 resinate particles upon the first day of storage; (b) same microcapsules prepared without resin, after 30 days of storage; (c) resin particles pretreated with PEG 4000 after 1 day of storage.

the same microcapsules prepared without resin, after 30 days of storage, shows that they maintain their integrity (see Figure 3.16b). A possible alternative to overcome the problem of fracture in the coating could be the pretreatment of the resin particles with polyethylene glycol (PEG 4000). This excipient acts as an impregnating agent and has an essential role in retaining the geometry of the particles when coated by air-suspension technique. However, this approach is also ineffective in this particular case because the microcapsules containing resin particles pretreated with PEG 4000 appeared identically fractured after 1 day of storage (Figure 3.16c). This indicates that that treatment with polyethylene glycol is not suitable for in-liquid drying microencapsulation methods but only for the air-suspension coating procedure.

3.1.1.9.10 Effect of Physical Mixing of Drug and Resin on Release

It may be very interesting to investigate whether a physical mixture of drug and resin particles could result in modified release. In an experiment, the release of propranolol HCL from HPMC tablets containing a drug without resin, propranolol–Amberlite IRP 69 complex, and a physical mixture of propanolol HCL and resin was studied in 0.1 *N* HCL. The results, as depicted in Figure 3.17,

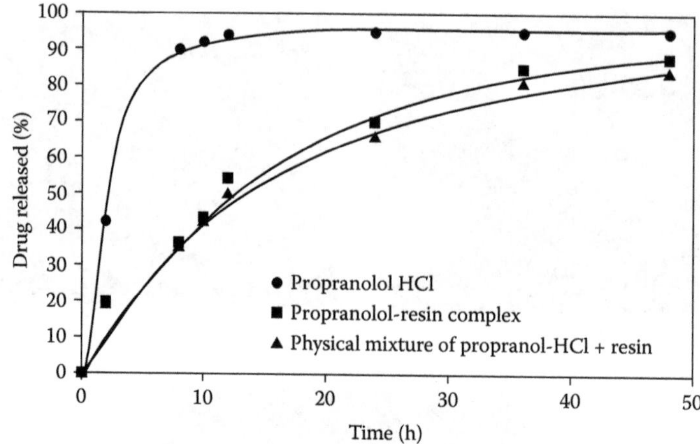

FIGURE 3.17 Release of drug from resin-free HPMC tablets vs. drug released at a slower rate from the HPMC tablets containing drug-resin complex and from the tablets containing the physical mixture of drug and resin particles.

indicate that the drug is released faster from resin-free HPMC tablets, while the drug is released at a slower rate from the HPMC tablets containing the drug-resin complex as well as from the tablets containing the physical mixture of drug and resin particles.

Interestingly, the release profiles from the drug-resin complex and a physical mixture of drug and resin almost coincide with each other. In the latter case, what happens is that upon contact with the dissolution medium, a gel layer of HPMC is formed rapidly around the solid tablet core. The complex between the drug and the resin is formed *in situ* in the gelled regions. The drug is then replaced by the counterions of the dissolution medium and released via diffusion through the gel layer. This *in situ* method is advantageous with regard to simplifying the manufacturing process when compared to the use of performed drug-resin complexes. The steps involved in the complex formation — such as loading, washing, and drying of the resin — can be eliminated through the *in situ* formation of the drug-resin complex. However, this method cannot be adopted when oral suspension is to be used or when the bitter taste of any particular drug is to be masked. This approach is only adoptable for using ion exchange resins as release modifiers in matrix formulations containing oppositely charged drugs.

3.1.1.10 Use of IER to Modify Release Profiles

As mentioned in the above section, the ion exchange resin can be used to modify the drug release profile in very interesting ways. Let us consider Figure 3.18, which shows the release of diclofenac loaded onto a strongly basic ion exchange resin. It is clear that a very significant extended release is achieved, with about 70% of the drug released over an 8 h period. The curve is typified by the rapid release at the start with logarithmic decay in the release rate. However, this profile

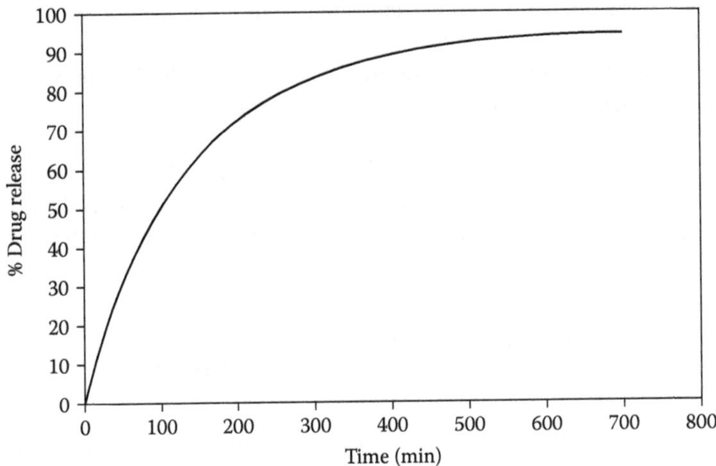

FIGURE 3.18 Release of diclofenac loaded onto a strongly basic ion exchange resin.

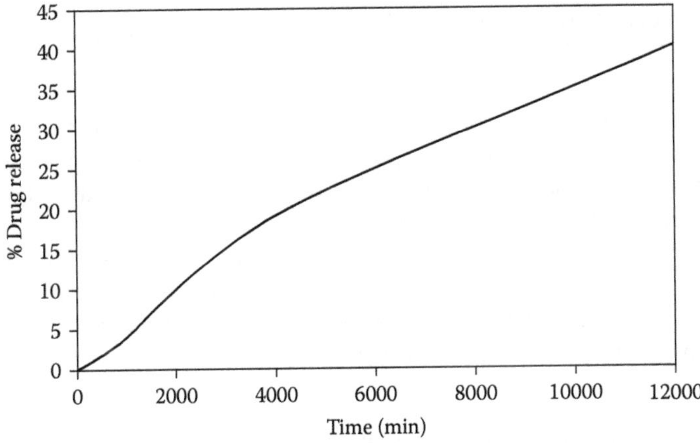

FIGURE 3.19 Same drug-resin complex coadministered with unloaded resin.

is limited in the use of ion exchange resins in controlled release applications because, although the overall release rate may be changed by varying loading, coating, and resin type, the shape of the curve is always the same.

Interestingly, when the same drug-resin complex is coadministered with unloaded resin, an almost different release profile, shown in Figure 3.19, is obtained. As is clear, the release rate is essentially constant over the period tested. By changing variables such as loading, particle size, and the ratio of loaded to unloaded resin particles, it is possible to achieve release rate curves between the two extremes represented in Figure 3.18 and Figure 3.19, and even go beyond Figure 3.18 into a curve that has a gradually increasing release rate during the first part of the profile.

In this particular example, the unloaded resin is the same type as used for the loaded resin.

The nearly "zero-order" release profile, depicted in the Figure 3.19, can be explained on the basis of the following mechanism:

1. When a mixture of unloaded and loaded resin particles comes in contact with the gastric fluid, the drug is released from the drug-loaded particles due to the ion exchange process between the drug ions and the H^+ ions of gastric fluid.

$$\underset{\text{in gastric fluid}}{\text{drug loaded resin}} \xrightleftharpoons{-\text{drug}^+} \text{drug}^+ \text{ in gastric fluid}$$

2. The drug released is partly absorbed in the body through gastric mucosa. At the same time, unloaded resin also begins to absorb the drug present in the gastric fluid through the ion exchange.

$$\underset{\text{in gastric fluid}}{\text{Unloaded resin particle}} \xrightleftharpoons{-\text{drug}^+} \text{loaded particle in gastric fluid}$$

The result is a decrease in drug concentration in the gastric fluid, thus shifting Equilibrium (I) towards the right, that is, more drug is released from the resinates and some part of this is always absorbed by the resin particles that were initially unloaded.

3. This situation is continuous until the concentration of the drug in both the resins is the same. Since the drug is continuously being removed from gastric fluid by the solution through absorption into body, the newly formed resinate will also start to release the drug that it absorbed earlier.

$$\underset{\text{in gastric fluid}}{\text{newly loaded resin}} \xrightleftharpoons{-\text{drug}^+} \text{drug}^+ \text{ in gastric medium}$$

In this way, the concentration of drug in the gastric fluid remains almost constant with time, thus yielding a nearly zero-order profile. It should also be noted that the shape of the release profile can be changed. The shape is affected by the ratio of resinate to resin, particle size of resinate and resin, and loading on the resinate.

An extension of the use of resinate and unloaded resin, as mentioned above, is to use the drug and an unloaded resin. The unloaded resin absorbs the drug from the medium in which the drug is dissolved, regardless of how the drug got into the solutions. Not creating the resinate in the first place and simply coadministrating the drug and unloaded resin has some advantages in that the resinate need not to be manufactured prior to the formulation. In this case, the drug and the resin

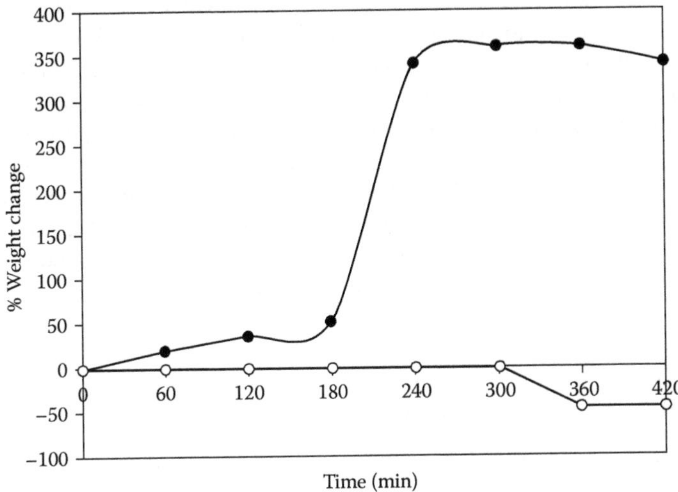

FIGURE 3.20 Beads in distilled water.

would simply be coformulated. This is also likely to be easier in terms of regulatory requirements as the resinate may have required some characterization, whereas using this approach, the drug is used in its original form and ion exchange resin is added as part of the formulation.

3.1.1.11 Polysaccharides as Ion Exchange

From the above discussion, it appears that ion exchanger resins used in drug delivery are composed of synthetic polymers and they can only deliver those drugs that are either acidic or basic in nature, so that these drugs can be loaded into the resin via ion exchange process.

However, recently we explored possibilities of using ionically cross-linked calcium alginate beads as a drug delivery vehicle, operating through ion exchange mechanism. Alginate is a common term used for a family of unbranched polymers composed of 1,4-β-D-mannuronic and –L-guluronic acid residues in varying proportions, sequence, and molecular weight. Alginate gelation takes place when divalent cations (usually Ca^{2+} ions) interact ionically with blocks of guluronic acid reduces, resulting in the formation of a three-dimensional network that is usually described by an "egg-box" model.[46]

Although alginate are hydrophilic and water-soluble polysaccharides, the Ca^{2+} ion induced cross-linked beads are quite stable in aqueous media and take up water and swell only when Ca^{2+} ions are exchanged with Na^+ ions present in the external solution. This causes swelling followed by degradation. In order to confirm this, calcium alginate beads, prepared in 4% $CaCl_2$ solution through ionotropic gelation, were put in pure distilled water and in a 3% solution of sodium chloride.[47] The results, as depicted in the Figure 3.20, show that the beads put in distilled water do not show any tendency to take up water and swell.

Moreover, they remain stable for more than 6 h. On the other hand, the beads put in 3% saline begin to take up water and demonstrate nearly 3.6 times swelling in 6 h. This finding may be attributed to the fact that beads that are put in water do not show any tendency to absorb water and swell because no sodium ions are available in the external solution that could undergo ion exchange with calcium ions present within the beads, while for the beads present in 3% saline, the situation is quite different. Here, there starts an ion exchange process between Na^+ and Ca^{2+} ions. Due to their small size and monovalency, Na^+ ions are not able to bind the carboxylate groups of poly mannuronic and guluronic residues. Therefore, the bead structure becomes rather loose and begins to absorb water. In this way, these beads can be used to deliver calcium ions through the ion exchange mechanism.

3.1.2 IER as Gastric Retentive Devices

Gastric employment of dosage forms is an extremely variable process, and the ability to prolong and control the emptying time is a valuable asset for dosage forms that reside in the stomach for a longer period of time than conventional dosage forms. The gastroretentive systems can remain in the gastric region for several hours and hence significantly prolong the gastric residence time of drugs. Prolonged gastric retention improves bioavailability, reduces drug waste, and improves the solubility of drugs that are less soluble in a high-pH environment. It has applications for local drug delivery to the stomach and proximal small intestine. Gastric retention helps to provide better availability of new products with new therapeutic possibilities and substantial benefits for patients. Gastric retention of oral dosage forms can be achieved in many ways (see Figure 3.21).

Out of the various approaches employed for gastric retention of oral-dosage forms,[48–51] ion exchange resins based on floating dosage forms are the ones designed to prolong gastric residence of drugs. Such a dosage form is attractive in that it theoretically permits control over the time and site of drug release. This would be particularly valuable for drugs exhibiting an absorption window in the small intestine or for drugs such as weak bases that dissolve better in the acid environment of the stomach. In addition, the devices may be useful for local treatment of the stomach or, if formulated in a particular manner, prevent or limit damage limit of gastroesophageal reflux.

Recently, a novel gastric retentive system based on ion exchange resins[52] was described. Resin particles are loaded with bicarbonate and coated with a semipermeable membrane. On exposure to gastric media, exchange of bicarbonate and chloride takes place, thus releasing carbon dioxide gas. The gas is entrapped within the membrane, causing the particles to float. *In vivo* studies in human volunteers have also confirmed the potential of such ion exchange resin–based systems.[53]

3.1.2.1 Preparation of Floating Beads

The ion exchange resin (Amberlite IRA 400 or Dowex 2 × 10) particles are first loaded with bicarbonate by mixing the resin beads with 1 *M* $NaHCO_3$ solution

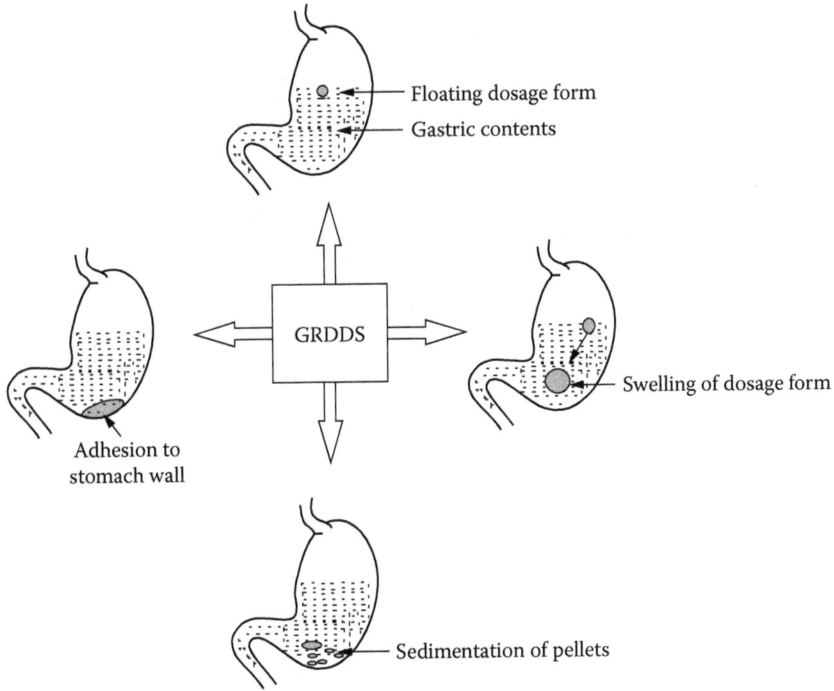

FIGURE 3.21 Gastric retention of oral dosage forms.

for 15 min followed by decanting and then further mixing for nearly 15 min with a fresh solution. The beads are then filtered, washed with deionized water, and dried overnight at 40°C.

The bicarbonate-loaded beads are now loaded with cationic drug by adding an aqueous solution of a drug of known concentration for a specific period of time. The drug-resin complex is then washed with deionized water and dried at room temperature. Finally, the beads are coated by an emulsification-solvent evaporation method for coating material, such as ethyl cellulose (EC), and a coacervation method[54] using a nonsolvent addition technique for film-forming agents such as Eudragit RS 100.

3.1.2.2 Drug Release from Floating and Nonfloating IER Beads

The advantages of using floating beads for prolonged gastric delivery can be seen from the profiles obtained for the release of drug theophylline from a coated and uncoated drug-resin (Dowex 2 × 10, 50 to 60 mesh) complex[55] in simulated gastric fluid at 37°C (See Figure 3.22).

For an uncoated drug-resin complex in simulated gastric fluid (pH 1.5), the elution half-life ($t_{50\%}$) is nearly 80 min and 90% of the drug loaded is released in 2 h. On the other hand, the drug-resin complexes coated with 10 and 20% Eudragit (w/v) demonstrate prolonged release. The rate-controlling step in ion

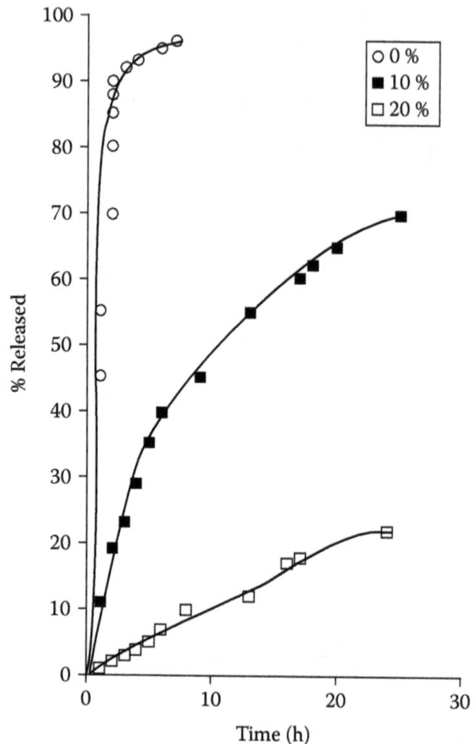

FIGURE 3.22 Advantages of using floating beads for prolonged gastric delivery.

exchange may be film diffusion or particle diffusion. Film diffusion is the process of bringing dissolved ions up to and away from the surface of ion exchange resin. Particle diffusion is the diffusion of ions within the resin particles. In order to assess which process is the rate-controlling step, the amount of exchange after time t is expressed as a function of the amount of exchange at infinite time and related to a rate constant, B, as described previously in Section 3.1.1.8.

It is clear from the Figure 3.22 that the rate of drug release from the resin beads is decreased by coating, but the release takes place for a much longer period. For the membrane-coated resin beads, release of the drug will be zero-order provided diffusion through the membrane is the rate-limiting step and a saturated solution of drug is maintained in the core. This appears to be the case for 20% coating in this particular example.

In this way, these systems not only provide a prolonged drug release by floating in the gastric fluid, but they may also demonstrate zero-order release of the drug.

3.1.2.3 Drug Release Mechanism

The overall loading of drug into resin particles with subsequent drug release from the floating devices can be described as follows (see Figure 3.23A,B,C,D,E,F):

Here it is worth mentioning that this process also takes place within the matrix, but for the sake of convenience, reactions only at the surface are shown.

Some IER, especially anion exchange resins (AER) such as cholestyramine, possess bio/mucoadhesive properties that might be caused by their electrostatic interaction with the mucin and epithelial cell surface. The use of such bioadhesive IERs is another attractive approach in the development of targeted formulation for the gastrointestinal tract. This approach would enhance the localized delivery of antibiotics, such as tetracycline, to the sites of *Helicobacter pylori* colonization (fundus), which conventional dosage forms fail to reach.[56] *H. pylori* is a highly virulent pathogen of the stomach and duodenum that infects up to one in four people in the adult population. It has been shown to cause both acute and chronic gastritis, and it has also been implicated in the occurrence and relapse of peptic ulcer disease and possibly the development of gastric cancer.[57] The bacterium colonizes the gastric mucus throughout the stomach and has proven extremely difficult to eradicate. The difficulty of eradicating *H. pylori* can be directly linked to the absence of a suitable therapeutic procedure for attaining an appropriate concentration of antibiotics at the gastric mucosa over a suitably extended period together with increasing bacterial resistance to the antibiotics used. Current treatment includes a triple therapy consisting of colloidal bismuth, which has a direct effect on the pathogen, together with two antibiotics such as amoxicillin and metronidazole; more recently, clarithromycin has been preferred. The tablets and capsules used to deliver these drugs are, however, distributed unevenly throughout the stomach, with the majority of the dose being delivered to the base of the stomach and relatively little drug reaching the fundus. However, it has been shown that targeted delivery to the gastric mucosa can be achieved using small doses of finely powdered ion exchange resins. When administered in small volume to the fasted subjects, these demonstrate prolonged gastric residence and uniform distribution over the gastric mucosa.[58] The mechanism by which the resins become mucoadherent is not clear. Previously it was thought to be due to ionic interactions between the charged surfaces of the resin and mucus. In order to investigate whether administering larger doses of resin by feeding subjects with dosing resins or masking surface charges on the resin particle using an inert polymer had any effect on coating and distribution properties of the resin in the stomach, Thairs and coworkers[59] coated cationic ion exchange resin cholestyramine with inert polymer ethyl cellulose and determined the gastric residence and distribution on mucus of uncoated and coated ion exchange resins. It was calculated that prolonged gastric residence and uniform distribution of ionic resin was not influenced by the dose size and that the binding of the dose to the mucosa was sufficiently strong. In addition, the resin coated with inert polymer also displayed extended gastric retention. This study suggests that mechanism of mucoadhesion is unlikely to be due to charge-based attraction. Similar types of studies have been reported by others with ion exchange resins.[60,61] However, it appears that gastric mucoadhesion does not tend to be strong enough to impart to dosage forms the ability to resist the strong propulsion of mucous by the gastric mucosa to replace the mucous that is lost through peristaltic contractions, and the dilution

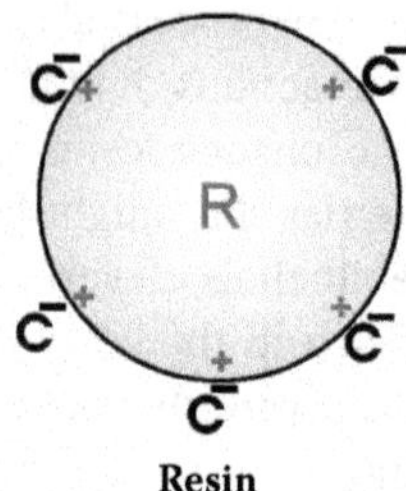

FIGURE 3.23A Let us consider a resin particle *R* with positively charged integral ions and negatively charged C^- counterions.

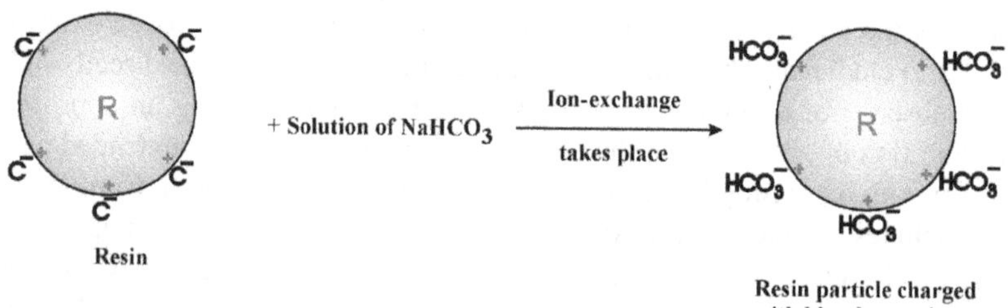

FIGURE 3.23B When resin is placed in sodium bicarbonate solution, the HCO_3^- ions enter into the resin matrix and replace some of the C^- counterions.

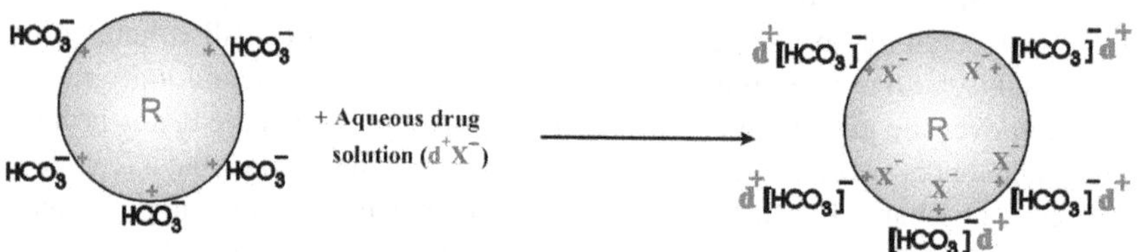

FIGURE 3.23C Now these bicarbonate-loaded resin particles are placed in an aqueous solution of cationic drug (d^+X^-). As a result, d^+ ions of the drug are bound electrostatically to the HCO_3^- ions present in the resin matrix.

of the stomach content also seems to limit the potential of mucoadhesion as a gastroretentive force.

3.1.2.4 Evaluation of Gastroretentive Formulations

Evaluation for gastroretention of coated IER is carried out by means of x-rays or gamma scientigraphic monitoring of the dosage-form transit in the GI tract. The modern technique of gamma scientigraphic now makes it possible to follow the transit behavior of dosage forms in human volunteers in a noninvasive manner.

First of all, bicarbonate-loaded ion exchange resin beads are soaked in deionized water for a specific time, say 15 min. A small amount of sodium pertechnetate

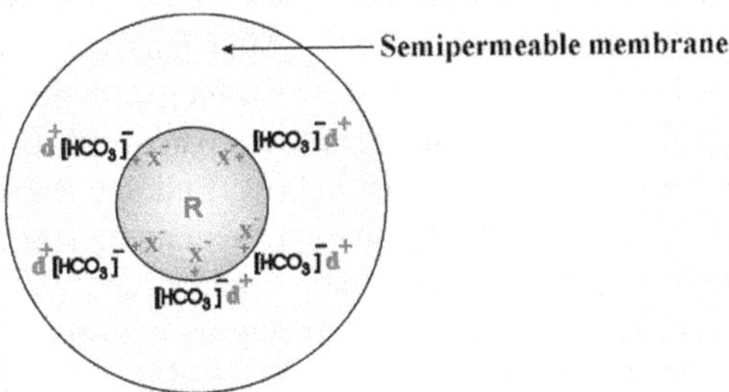

FIGURE 3.23D Now, the drug-loaded particles are coated with a semipermeable membrane of some coating material such as Eudragit RS/ethyl cellulose. This results in the formation of coated resinate particles.

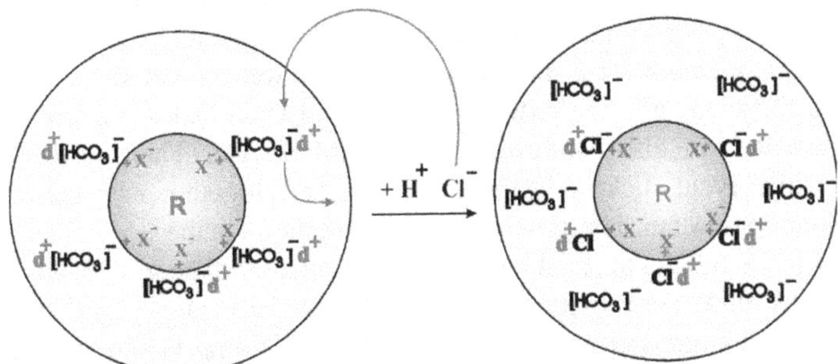

FIGURE 3.23E When the coated resin-drug complex particles are placed in the gastric medium, the Cl^- ions present in gastric fluid enter into the resin matrix and replace the HCO_3^- ions. At the same time, to maintain the electronegativity, H^+ ions from the gastric fluid also enter into the matrix and react with HCO_3^- ions of the resin matrix to produce CO_2, according to the reaction H+ $HCO_3^- \rightarrow H_2O + CO_2\uparrow$. The CO_2 produced remains entrapped inside the membrane, thus making the beads float in the gastric fluid.

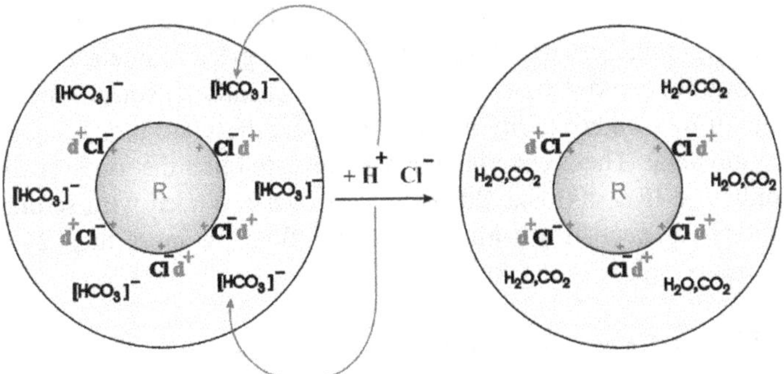

FIGURE 3.23F Finally, the H^+ ions of gastric fluid enter into the resin matrix and undergo ion exchange with the d^+ ions, thus causing them to diffuse out of the matrix.

solution in a Sterlite vial is added. The suspension is now mixed intermittently using a vortex mixer and then left to settle. After removing the supernatant, the beads are removed by filtration, followed by washing with deionized water. After drying for some time, the damp resin beads are coated with Eudragit RS using a coacervation phase separation technique. In brief, Eudragit RS (0.1 g) is dissolved in 10 g of a 3% (w/v) solution of polyisobutylene in methylene chloride/*n*-hexane (60/40 v/v). Now one gram of damp resin is added and 15 ml of *n*-hexane is dropped in at 1 ml/min with continuous stirring to form the coating. The beads are then removed, washed with *n*-hexane, and stored in a desiccator.

The behavior of coated resin beads is monitored using a single-channel analyzing study in healthy human volunteers by coadministrating a radio-labeled meal with the formulation. The test meal is usually a light breakfast, consisting of 200 ml milk, 40 g cornflakes, and 6 g sugar. A small amount of $^{99m}T_c$–DTPA solution is added to the test meal to give an activity of 1.5 MBq at the time of administration. To standardize conditions of GI motility, the subjects are required to fast for 12 h prior to the commencement of experiment.

After an overnight fast, the subjects are allowed to eat the radio-labeled breakfast. Subsequently, the radio-labeled coated resin beads, in zero-size hard gelatin capsules, are also swallowed with 200 ml of water. This volume is almost sufficient to prevent sticking in the esophagus and, together with the breakfast, allows floating to occur. The position of the formulation in the stomach is followed by allowing the subject to stand before a gamma camera. The images are recorded on a magnetic disk. Three external marks containing less than 1 MBq $^{99m}T_c$ sodium pertechnetate are attached to the skin (with lead shielding facing the body): one overlying the liver to the right of the stomach and two on either side of the lower abdomen. These marks are used as reference points during the image analysis. During the experiments, the subjects may carry out their normal activities but they are not allowed to take food or drink milk until the formulation has left the stomach completely.

Finally, each image is analyzed for radioactivity remaining in the stomach by creating two regions of interest, one around the upper half of the stomach and a second around the lower half. An additional region of interest is created around the whole stomach from the images taken before administration of the formulation to assess background activity. The counts from the regions of interest are corrected for the background and decay. The values subsequently calculated represent the percentage of total radioactivity administered and these are plotted against time for each individual. These experiments have revealed that the ion exchange resin beads Amberlite IRA 400 and Dowex 1 × 10, when loaded with sodium bicarbonate followed by coating with Eudragit RS float for more than 24 h, while uncoated beads sink within 1 to 3 h.[53]

3.1.3 IER for Cancer Treatment

Entrapment of anticancer drugs within the particulate carriers (microcapsules, nanoparticles) is the most popular approach for the development of delivery systems for cancer treatment. However, drugs such as doxorubicin, which are ionic in nature, can be complexed with ion exchange resins. Doxorubincin (Dox), an anthracycline, is an effective anticancer drug used for the treatment of a number of carcinomas, such as breast, bladder, and gastric cancers. Like most anticancer drugs, Dox can cause severe toxicity to the body when administered at high doses systemically. Its acute toxicities include myelosuppression, loss of hair, nausea, vomiting, mucositis, and local tissue necrosis when it leaks into the extra vascular space at the site of injection. Moreover, repeated administration of Dox is often limited by irreversible cardiac toxicity.[62] Hence, many efforts have been made to deliver Dox to tumor tissues by either targeting moieties or by locoregional delivery, while reducing systemic exposure of the drug. A number of microsphere systems, including albumin microspheres, polysaccharide microspheres, and polystyrene microspheres, have been studied for locoregional delivery of Dox to solid tumors.[63] Of these various types of microspheres, ion exchange microspheres exhibit the highest loading for ionic drugs such as Dox, whereas those prepared by chemical cross-linking or physical entrapment display far lower drug loading levels. This is because ion exchange microspheres contain charged functional groups that have a high affinity for oppositely charged counterions. Thereby, ionic drugs can be readily loaded onto the microspheres by a simple absorption method.[64] The advantage of high drug-loading capacity, ease of loading process, and the avoidance of chemical reactions make ion exchange microspheres more attractive than other microsphere systems.[65] However, recently it was proved that, as compared to polystyrene-based ion exchange microspheres, natural polysaccharides such as dextran-based ion exchange microspheres have proved to be more advantageous because of additional advantages such as biocompatibility and biodegradibility.[66] The ionic dextran microspheres for simultaneous delivery of ionic anticancer drugs such as vinblastine and Dox, and chemosensitizing agents such as verapamil, to multi-drug-resistant (MRD) cells and tumors have also been used.[67] It has been proved that tumor growth and systemic toxicity are reduced to a greater extent when Dox is delivered by microspheres via intratumoral injection to the EMT6 murine breast tumor model rather than injection of free Dox solution.

The loading of doxorubicin into sulfopropyl groups containing dextran microspheres (MS) and its release in the medium containing sodium/calcium ions is governed by the ion exchange process as shown in the following two equations:

$$\text{MS–}SO_3^-\ Na^+ + Dox^+\ Cl^- \,\Delta\, \text{MS–}SO_3^-\ Dox^+ + Na^+ + Cl^-$$

$$\text{MS–}(SO_3^-)_2Ca^{2+} + 2Dox^+\ Cl^- \,\Delta\, 2\ \text{MS–}SO_3^-\ Dox^+ + Ca^{2+} + 2Cl^-$$

The relative amount of the drug and the competing ions in the MS and in the solution determines the maximum amount of the drug that can be released from the MS. At equilibrium, their relationship can be expressed by the selectivity coefficients of the reaction[68]:

$$K_1 = \frac{[Dox]_{MS}[Na]_s}{[Dox]_s[Na]_{MS}}$$

and

$$K_2 = \frac{[Dox]_{MS}^2[Ca]_s}{[Dox]_S^2[Ca]_{MS}}$$

where K_1 and K_2 are, respectively, the selectivity coefficients for the drug and competing ions, Na^+ and Ca^{2+}; the subscripts *MS* and *S* denote drug-ion concentration in the microspheres and in the solution, respectively. This equation indicates that for a fixed value of selectivity coefficient, an increase in salt concentration in the solution leads to a higher release of Dox from microspheres at equilibrium.

Recently it was found[69] that sulfopropyl dextran MSs have a high loading capacity and high loading efficiency for the drug doxorubicin. The high loading efficiency reduces wastage of expensive drugs and the high drug loading capacity makes it possible to use a small quantity of carrier material, which is desirable for repeated injections. These two properties give the dextran MS an advantage over the albumin MS. Additionally, the dextran MS with various drug loadings can be prepared readily by varying the MS:drug ratio without a noticeable compromise of loadings efficiency. The high loading capacity and efficiency of the dextran MS can be attributed to the large pores in the MS because they enable easy access of the ionic binding sites in the MS and, thus, high drug loading, as also reported by others.[64] Finally, it can be said that microspheres, functioning through an ion exchange mechanism, can be treated as the most efficient and suitable device for the delivery of ionic anticancer drugs in cancer treatment.

3.1.4 Miscellaneous Applications

In the preceding sections, we have discussed some of the major applications of ion exchange resins in oral drug delivery. In addition to these applications, they are also used in a variety of other pharmaceutical applications such as chewable or dispersible tablets, chewing gum for buccal absorption, in drug stabilization, bioadhesive systems, taste masking of bitter drugs, and in nasal or ophthalmic drug delivery, and so forth.

3.1.4.1 Chewable or Dispersible Tablets of Bitter or Nauseous Medications

The taste of pharmaceutical preparations is an important parameter governing patient compliance and commercial success in the market. The scope of IERs for masking the undesirable taste of pharmaceuticals is unlimited. Since the drug-resin complex is insoluble, it has virtually no taste, so that even bitter drugs lose their taste when converted into a drug resinate with proper selection of ion exchange resins. The drug resinate can be made sufficiently stable so that it does not break down in the mouth and so that the patient does not taste the drug when it is swallowed. However, when the drug resinates come in contact with gastrointestinal fluid, usually the acid of the stomach, the complex is broken down quickly and completely. The drug is released and then absorbed in the usual way. The resin passes through the GI tract without being absorbed. Weak acid cation exchange resins with carboxylic acid functionality can be used to formulate chewable or dispersible tablets of bitter drugs, for example, the Rodec® decongestant tablet containing pseudoephedrine.[70] The complex of cationic drug and the weak cation exchange resin does not break at a pH of 6 to 7 of saliva with cation concentration of 40 meq/lit. But at higher cation concentrations in the stomach and a pH of 1 to 3, the free drug is immediately released. The taste of antibacterials belonging to the quinolone category, such as ciprofloxacin, when loaded on the cation exchanger and administered to animals was improved as judged by animals accepting the material more readily.[71] Binding to a cation exchange resin Amberlite IRP 6910 masked the taste of peripheral vasodilator buflomedil. Polystyrene matrix CER have been used to mask the bitter taste of chlorpheniramine maleate, ephedrine hydrochloride, and diphenhydramine hydrochloride.[72] The ionic binding of the drugs to polymeric materials such as Carbopol is emerging as an important mechanism of taste masking. Erythromycin and clarithromycin have been taste masked by binding to Carbopol.[12] Similarly, chloroquine phosphate and dicyclomine hydrochloride have also been successfully taste masked with IER recently. However, since IER could also retard the release of drugs, a proper and careful selection of IERs is essential to yield optimal taste masking without affecting the bioavailability. Generally, less cross-linked IERs are helpful in taste masking.

3.1.4.2 Chewing Gum for Buccal Absorption

Nicorette® is a widely patented product for smoking cessation programs. It contains nicotine adsorbed on an ion exchange resin with carboxylic group acid functionality and formulated in a flavored chewing-gum base to provide gradual drug release through buccal mucosa as the gum is chewed, offering fresh saliva as solvent for elution.[73] In this way, nicotine is released only during chewing, thus providing the minimal supply to facilitate smoking cessation.

3.1.4.3 Drug Stabilization

Complexing active ingredients with ion exchange resin prevents harmful interaction with other components (e.g., complex between vitamin B_{12} and carboxylic acid), and this is as effective as free drugs.[74] Ion exchange resins can also be used as carriers for immobilized enzymes to provide extended activity at localized sites. IERs may have inherent bioadhesive properties similar to those of highly charged polyanions.[68] Hence, they may be useful for prolonging the gastric residence of antibiotic drugs like amoxicillin and cimetidine.[75]

3.1.4.4 Sigmoidal Release Systems

The drug release should be controlled in accordance with the therapeutic purpose and the pharmacological properties of the active substances. Accordingly, the maintenance of a constant drug to blood level is not always required, as in the case of nitrates, antibiotics, and contraceptive steroids. To avoid the development of tolerance, the systemic variations of blood concentration must be maintained for such medications.

A sigmoidal release system rapidly releases the drug from a multiple unit device after a predetermined lag time, and can achieve both time-controlled and rhythmic release. IERs have been studied in the development of sigmoidal release systems. Eudragit RS (Rohm, Darmstadt, Germany), an AER with limited quaternary ammonium groups, is coated over beads with a sugar core surrounded by an organic acid and drug mixture. The ionic environment induced by the addition of an organic acid to the system is responsible for pulsatile release.[76] A hydration study of Eudragit RS films suggested that the increase in drug release was attributable to structural changes of the film induced by polymer acid interactions.

3.1.4.5 Nasal or Ophthalmic Drug Delivery

Attempts have been made to deliver therapeutic peptides or synthetic drugs via nasal mucosa with the IER complexation approach. A composition has been developed to deliver nicotine in pulsatile fashion to the systemic circulation via the nasal route.[77] An excess amount of nicotine, as an immediate dose, is either dispersed in a non-IE material or overloaded in IER. The excess uncomplexed nicotine is thus available for immediate absorption. The prerequisite for nasal delivery by the IER approach is a high ion exchange capacity of the resin.

The IER have also been used for ophthalmic drug delivery. The drug ciprofloxacin in complexation with polystyrene sulfonate has been successfully used for the treatment of eye infections.[78] Similarly, a novel suspension containing 0.25% betaxolol bound to a cationic ion exchange resin (5 μm in diameter), incorporated in a well-structured polymeric vehicle, has been claimed to provide uniform dosing of betaxolol for 4 weeks.[6]

3.1.4.6 Improvement in Tablet Disintegration

Many tablet disintegrants owe their action to their capacity to absorb water and swell up. Fine-particle-size ion exchange resins have shown superiority as disintegrating agents due to their considerable swelling pressure upon hydration. The advantages of ion exchange resins over conventional disintegrating agents include:

1. The rate of permeation of water and subsequent swelling is very fast and cuts down the disintegrating time.
2. Ion exchange resins do not have adhesive tendency on hydration; hence, tablets disintegrate evenly without lumps.

3. Ion exchange resin is effective in low concentration as a disintegrant.
4. After incorporation of ion exchange resins, the hardness of the tablets increases.

Ion exchange resins work equally effectively with hydrophilic as well as hydrophobic formulations, especially the latter, where most of the conventional disintegrants fail. Because of their usually large swelling capacities, poly-methacrylic carboxylic acid ion exchange resins have found usage in pharmacy as tablet disintegrants, for example, in polacrillin, a potassium salt of a weakly acidic cation exchange resin with methacrylic acid–divinylbenzene matrix.[79] It has been reported that interference of cation exchanger disintegrants with drug availability is not affected *in vivo*.[80] However, it is questionable that the use of an ion exchanger in the tablet would cause any delay in the release.

3.1.4.7 Use of IER as Therapeutics

In addition to the utility of ion exchange resins as excipients, these materials are explored for their therapeutic potential as well. Cholestyramine was the first polymeric resin-based drug approved for the treatment of high cholesterol. The cationic resin in this case serves as a sequestrant to bind bile acids in the gastro-intestinal tract. The binding and effective removal of bile acids forces the liver to consume cholesterol and synthesize more bile acids. This leads to an indirect reduction in cholesterol level. The advantage of this therapy is that it does not use conventional drugs and hence shows fewer side effects as compared with conventional therapies. However, a drawback of this therapy is the higher dose requirement for the first-generation resins (4 tablespoons to be administered via fruit juice). Subsequent generation therapies have utilized molecular modeling strategies to impart specificity to the resin, thereby decreasing the dose to capsule size (e.g., Welchol® developed by Genzyme). Specificity for these resins has been achieved by the use of polymers that bind not only via electrostatic interaction but also by other forces such as hydrophobic interactions. Ion exchange resins have also been used for hemoperfusion and management of drug overdoses (poisoning). At present, cholestipol (anion exchange resin) is used in the treatment of type II hyperlipoproteinemia and familial hyperlipoproteinemia in children and young adults.[81] The resin is mixed with fluids and administered as slurry.

3.1.4.8 Implantation Devices for Water–Soluble/Charged Drugs

Because of minimal body tissue response and relatively high tissue diffusivity of hydrophobic solutes, polydimethylsilicone is widely used in several implantation devices. Limited permeability of charged or water soluble drugs has restricted the use of silicons in drug delivery. Electrostatic repulsion between the resin and cations can increase their diffusion rate through polydimethylsilicone membranes. Trimethylbenzyl ammonium chloride polymer, an anion exchange resin, along with cations in the silicone tubes has been studied and an increase in the diffusion of cations was observed in the order: iron > calcium > sodium > potassium.[82] Resin introduces an additional force, which reduces the activation energy required for each ion to permeate through the membrane. This approach could be employed in the delivery of cationic drugs from silicon devices.

3.1.4.9 Transdermal Drug Delivery

The transdermal delivery of drugs offers substantial advantages over traditional oral and parenteral routes. The transdermal delivery system essentially consists of a drug reservoir from which a supply of active constituent is maintained and a device to control the rate of drug diffusion across the surface of the skin. This device is usually a suitable membrane or membrane laminate acting as rate-controlling barrier.

The drugs such as theophylline, hydralazine, and propranolol–HCl have been loaded into suitable ion exchange resins and these drug-loaded resins when coated with polymer coating or semipermeable membrane demonstrate diffusion-controlled[83,84] release. The release of drugs also depends upon the chemical nature of the resin used. For example, the rate of release of propranolol–HCL from Amberlite IRC 50 is reported to be faster than that from the strong resin Amberlite IR 120.[85] Similarly, chlorpheniramine is released more quickly from a carboxylic acid resin than from one containing sulfonic acid.[86] Transdermal systems are also developed by incorporating a drug-loaded resin into a polymeric hydrogel.[87] Recently, the iontophoretically assisted transport of nicotine across artificial and human skin membranes from heterogeneous gel vehicle comprising a mixture of nicotine-loaded ion exchange resins and agar hydrogel has been reported.[5] The heterogeneous vehicles show more advantages over simple hydrogel vehicles in their versatility, and in their capacities to store the drug and to control both its delivery rate and the pH of the vehicle during iontophoresis. *In vitro* studies with tacrine nadolol, and sodium salicylate have revealed that IERs are more suitable as delivery vehicles in iontophoretic drug delivery.[88] It is well known that biological factors of transdermal drug delivery, such as intra- and intersubject variability, regional blood flow, and skin pH are less easily controlled.[89] Better control of biological variability in transdermal drug absorption has been attempted with ion exchange systems. Selection of external conditions (e.g., ionic strength, pH, and the choice of the salt in the release medium), drug properties (charge, lipophilicity, and molecular weight), and resin quality (ion exchange groups, capacity, grafting) may be utilized in better ways to control the release kinetics of a drug from ion exchange resin.[90]

3.2 CONCLUDING REMARKS

In the recent past, a number of ion exchange resin–based drug formulations have been patented, which indicates that the use of IERs in drug delivery is gaining importance and commercial success. In addition to oral drug delivery, IERs are being explored for site-specific or targeted, transdermal, nasal, and ophthalmic routes. Moreover, several novel concepts, such as sigmoidal release, floating, and pH and ionic strength responsive systems have shown potential uses of IERs in drug delivery.

However, it appears that there is need to carry out more and more *in vitro* and *in vivo* drug-release studies with IER and establishment of *in vitro–in vivo* correlation so that they can be used in a better way for the service of society and humanity.

ACKNOWLEDGMENT

The authors are very thankful to Dr. Y. Murli Mohan, research scientist, Department of Material Science & Engineering, Gwangju Institute of Science & Technology, South Korea, for providing relevant literature for manuscript preparation.

REFERENCES

1. Sounders, L. and Choudhary, N.C., Sustained release of drug from ion-exchange resin, *J. Pharm. Pharmacol.*, 8, 875, 1956.
2. Borodkin, S.S., Ion exchange resins and sustained release, in *Encyclopedia of Pharmaceutical Technology*, Vol. 8, Swarbrick, J. and Boylan, J.C., eds., Marcel Dekker, New York, 1960, p. 410.
3. Deeb, G. and Becker, B., *Toxicol. Appl. Pharmacol.*, 2, 410, 1960.
4. Cuna, M., Jato, Vila, J.L., and Torres, D., Controlled release liquid suspension based on ion-exchange particles entrapped within acrylic microspheres, *Int. J. Pharm.*, 199, 151, 2001.
5. Conaghey, O.M., Corish, J., and Corrigan, O.I., Iontophoretically assisted *in vitro* membrane transport of nicotine from a hydrogel containing ion exchange resin, *J. Cont. Rel.*, 170, 225, 1998.
6. Jani, R., Gan, O., Ali, Y., Rodstrom, R., and Hancock, S., Ion exchange resins for ophthalmic delivery, *J. Ocul. Pharmacol.*, 10, 57, 1994.
7. Jenke, D.R., Drug delivery via ion exchange across a fiber membrane, *Pharm. Res.*, 6, 96, 1989.
8. Borodkin, P. and Sundberg, P.D., Polycarboxylic acid ion-exchange resin absorbate for taste coverage of chewable tablets, *J. Pharm. Sci.*, 60, 1523, 1971.
9. Irwin, W.J. and Belaid, K.A., Drug delivery by ion exchange. I. Ester prodrugs of propranolol, *Drug Dev. Ind. Pharm.*, 13, 2017, 1987.
10. Plaizier-Vercammen, J.A., Investigation of the bioavailability of codeine from a cation ion exchange sulphonic acid. I. Effect of parameters, *Int. J. Pharm.*, 85, 45, 1992.
11. Saunders, L., Ion exchange resins in organic analysis, *J. Pharm. Pharmacol.*, 5, 569, 1953.
12. Glen, M., Taste masking in oral pharmaceuticals, *Pharm. Technol.*, 18(4), 84, 1994.
13. Manek, S.P. and Kumar, V.S., Evaluation of Indion CRP 244 and CRP 254 as sustained release and taste masking agents, *Indian J. Pharm. Sci.*, 26, 773, 1981.
14. Nanda, A., Development and Evaluation of Novel Taste Masked Suspension and Mouth Dissolving Dosage Forms of Dicyclomine Hydrochloride, Ph.D. thesis, National Institute of Pharmaceutical Education and Research (NIPER), Mohali, India, 1989.
15. Raghunathan, Y., Amsel, L., Hinsbark, O., and Bryant, W., Sustained release drug delivery system. I. Coated ion exchange resin system for phenylpropanolamine and other drugs, *J. Pharm. Sci.*, 70, 379, 1981.

16. Burk, G.M., Mendes, R., and Jambhekar, S., Investigation of the applicability of the ion exchange resins as a sustained release drug delivery system for propranolol hydrochloride, *Drug Dev. Ind. Pharm.,* 12, 713, 1986.
17. Torres, D., Boado, L., Blanco, D., and Vila-Jato, J.L., Comparison between aqueous and non-aqueous solvent evaporation methods for microencapsulation drug resin complexes, *Int. J. Pharm.*, 173, 171, 1998.
18. Liu, Z., Wu, X.Y., Ballinger, J., and Bendayan, R., Synthesis and characterization of surface-hydrophobic ion exchange microspheres and the effect of coating on drug release rate, *J. Pharm. Sci.*, 89, 807, 2000.
19. Whitechead, L., Fell, J.T., Collett, J.H., Sharma, H.L., and Smith, A.M., Floating dosage forms and *in vivo* study demonstrating prolonged gastric retention, *J. Cont. Release*, 55, 3, 1998.
20. Farag, Y. and Nairn, J.G., Rate of release of organic carboxylic acids from ion exchange resins, *J. Pharm. Sci.*, 77, 872, 1988.
21. Sawaya, A., Fickat, R., Benoit, J.P., Puisieux, F., and Benita, S., Ion exchange albumin microcapsules of doxorubicin and *in vitro* kinetic evaluation, *J. Microencapsulation*, 5, 255, 1988.
22. Ichikawa, H., Fukumori, Y., and Adeyeye, C.M., Design of prolonged-release microcapsules containing diclofenac sodium for oral suspensions and their preparation by the Wurster process, *Int. J. Pharm.*, 156, 39, 1997.
23. Irwin, W.J., Belaid, K.A., and Alpar, H.O., Drug delivery by ion exchange. III. Interaction of ester pro-drugs of propranolol with cation exchange resins, *Drug Dev. Ind. Pharm.*, 13, 2047, 1987.
24. Sandhavi, N.M., Kamath, P.R., and Amin, D.S., Ion exchange resins as matrices for controlled drug release, *Indian Drugs*, 26, 27, 1988.
25. Kincl, M., Meleh, M., Veber, M., and Vreier, F., Study of physiological parameters affecting the release of diclofenac sodium from lipophilic matrix tablets, *Acta. Chem. Slov.*, 51, 409, 2004.
26. Reichenberg, D., Properties of ion-exchange resins in relation to their structure. III. Kinetics of exchange, *J. Am. Chem. Soc.*, 75, 589, 1953.
27. Bhasker, R., Murthy, S.R.S., Miglani, B.D., and Vishwanathan, K., Novel method to evaluate diffusion controlled release from resinate, *Int. J. Pharm.*, 28, 59, 1986.
28. Ambrogi, V., Fardella, G., Grandolini, G., Perioli, L., and Tiralti, M.C., Intercalation compounds of hydrotalcite like anionic clays with anti-inflammatory agent. II. Uptake of diclofenac for a controlled release formulation, *AAPS Pharm Sci. Tech*, 3(3), 1, 2002.
29. Ambrogi, V., Fardella, G., Grandoline, G., and Perioli, L., Interaction compounds of hydrotalcite-like anionic clays with anti-inflammatory agents. I. Interaction and in vitro release of ibuprofen, *Int. J. Pharm.*, 220, 23, 2001.
30. Costantino, U. and Nocchetti, M., Layered double hydroxides and their intercalation compounds in photochemistry and medicinal chemistry. In *Layered Double Hydroxides: Present and Future,* Rives, V., Ed., Nova Science Publishers, New York, 2001, pp. 383–411.
31. Agrawal, R. and Mittal, R., Studies on ion exchange of chloroquine phosphate, *Drug Dev. Ind. Pharm.*, 27, 359, 2001.
32. Burke, G.M., Mendes, R.W., and Jambheker, S.S., Investigation of the applicability of ion-exchange resins as a sustained release drug delivery system for propranolol hydrochloride, *Drug Dev. Ind. Pharm.*, 12, 713, 1986.

33. Pisal, S., Zainnuddin, R., Nalawade, P., Mahadik, K., and Kadam, S., Drug release properties of polyethylene glycol-treated Ciprofloxacin–Indion 234 complexes, *AAPS Pharm. Sci. Tech.*, 5(4), 1, 2004.
34. Glasstone, S. and Lewis, D., *Elements of Physical Chemistry,* Macmillan & Co. Ltd., London, 1960.
35. Irawin, W.J. and Belaid, K.A., Drug delivery by ion exchange. I. Ester pro–drugs of propranolol, *Drug Dev. Ind. Pharm.*, 13, 2017, 1987.
36. Sriwongjanya, M. and Bodmeier, R., Effect of ion-exchange resins on the drug release from matrix tablets, *Europ. J. Pharm. Biopharm.*, 46, 321, 1998.
37. Herrera, T., Matejka, Z., and Eliasek, J., Cation exchange kinetics on carboxylic acid resins in a multicomponent system $H–Me^{+}–Me^{2+}$, *Desalination*, 48, 161, 1983.
38. Liu, Z., Cheung, R., Yu, W., and James, R., A study of doxorubicin loading onto and release from sufoproyal dextran ion-exchange microspheres, *J. Cont. Rel.*, 77, 213, 2001.
39. Hinsvark, O.N., Truant, A.P., Jenden, D.J., and Steinborn, J.A., The oral bioavailability and pharmacokinetics of soluble and resin-bound forms of amphetamine and phentermine in man, *J. Pharm. Biopharm.*, 1, 319, 1973.
40. Hall, H.S., Sustained Release from Coated Ion-Exchange Resins, paper presented at the Sixth International Symposium on Controlled Release of Bioactive Materials, August 1979, New Orleans, LA.
41. Thies, C., A survey of microencapsulation process, in *Microencapsulation,* Benita, S., Ed., Marcel Dekker, New York, 1996, pp. 1–19.
42. Fukumori, Y., Ichikawa, H., Yamaoka, Y., Akaho, E., Takeuchi, T., Fukuda, T., Kanamori, R., and Osako, Y., Microgranulation and encapsulation of pulverized pharmaceutical powders with ethyl cellulose by the Wurster process, *Chem. Pharm. Bull.*, 39, 1806, 1991.
43. Ichikwa, H., Jono, K., Tokimitsu, H., Fukuda, T., and Fukumori, Y., Coating of pharmaceutical powders by fluidized bed process. V. Agglomeration and coating efficiency in the coating with aqueous lattices of copoly(ethylacrylate-methyl methacrylate-2-hydroxyethyl methacrylate), *Chem. Pharm. Bull.*, 41, 1132, 1993.
44. Ichikawa, H., Fujioka, K., Adeyeye, M.C., and Fukumori, Y., Use of ion-exchange resins to prepare 100 µm-sized microcapsules with prolonged release by the Wurster process, *Int. J. Pharm.*, 216, 67, 2001.
45. Cuna, M., Vila Jato, J.L., and Torres, D., Controlled release liquid suspension based on ion-exchange particles entrapped within acrylic microspheres, *Int. J. Pharm.*, 199, 151, 2000.
46. Gombotz, W.R. and Wee, S.F., Protein release from alginate matrices, *Adv. Drug Deliv. Rev.*, 31, 267, 1998.
47. Bajpai, S.K. and Sharma, S., Investigation of swelling/degradation behaviour of alginate beads cross-linked with Ca^{2+} and Ba^{2+} ions, *React. Funct. Polym.*, 59, 129, 2004.
48. Singh, B.M. and Kim, K.H., Floating drug delivery systems: an approach to controlled drug delivery gastric retention, *J. Cont. Release*, 63, 235, 2000.
49. Desai, S. and Botton, S.A., Floating controlled release drug delivery system: *in vitro–in vivo* evaluation, *Pharm. Res.*, 10(9), 321, 1993.
50. Iannuccelli, V., Coppi, G., Sansone, R., and Ferolla, G., Air component multiple-unit system for prolonged gastric residence. II. *In vivo* evaluation, *Int. J. Pharm.* 174(1–2), 55, 1998.

51. Whitehead, L., Fell, J.T., and Collett, J.H., Development of gastroretentive dosage form, *Europ. J. Pharm. Sci.*, 4, 5182, 1996.
52. Kouchak, M. and Atyabi, F., Ion exchange: an approach to prepare an oral floating drug delivery system for diclofenac, *Iran J. Pharm. Res.*, 2, 93, 2004.
53. Atyabi, F., Sharma, H.L., Mohammad, H.A.H., and Fell, J.T., *In vivo* evaluation of a novel gastric retentive formulation based on ion exchange resins, *J. Cont. Release*, 42, 105, 1996.
54. Atyabi, F., Sharma, H.L., Mohammad, H.A.H., and Fell, J.T., A novel floating system using ion exchange resins, *Proc. Int. Symp. Control Release Bioact. Mater.*, 21, 806, 1994.
55. Atyabi, F., Sharma, H.L., Mohammad, H.A.H., and Fell, J.T., Controlled drug release from coated floating ion-exchange resin beads, *J. Cont. Release*, 42, 25, 1996.
56. Jackson, S.J., Bush, D., and Perkins, A.C., Comparative scientigraphic assessment of the intragastric distribution and residence of cholestyramine, Carbopol 934 P and Sucralfate, *Int. J. Pharm.*, 212, 55, 2001.
57. Barreto-Zuniga, R., Maruyama, M., Kato, Y., Aizu, K., Ohta, H., Takekoshi, T., and Bernal, S.F., Significance of *Helicobacter Pylori* infection as a risk factor in gastric cancer: serological and histological studies, *J. Gastroenterol*, 32, 289, 1997.
58. Washington, N., Wilson, C.G., Greaves, J.L., Norman, S., Peach, J.M., and Pugh, K., A gamma scientigraphic study of gastric coating by micronized ion-exchange resin delivered is expidet tablet and liquid formulation in healthy volunteers, *Int. J. Pharm.*, 57, 17, 1989.
59. Thairs, S., Ruck, S., Jackson, S.J., Steele, R.J.C., Feely, L.C., Washington, C., and Washington, N., Effect of dose size, food and surface coating on the gastric residence and distribution of ion-exchange resin, *Int. J. Pharm.*, 176, 47, 1998.
60. Irwin W.J., MacHale, R., and Watts, P.J., Drug delivery by ion exchange. VII. Release of acidic drugs from anionic exchange resinate complexes, *Drug Dev. Ind. Pharm.*, 16, 883, 1990.
61. Ko, H. and Royer, M.E., *In vitro* binding of drugs to Colestipol hydrochloride, *J. Pharm. Sci.*, 63, 1914, 1974.
62. Chen, Y., Burton, M.A., Coddle, J.P., Napoli, S., Martins, I.J., and Gray, B.N., Evaluation of ion-exchange microspheres as carrier for anticancer drug Dox: in vitro studies, *J. Pharm. Pharmacol.*, 44, 211, 1992.
63. Lu, J.Y., Lowe, D.A., Kennedy, M.D., and Low, P.S., Folate-targeted enzyme pro-drug cancer therapy utilizing Penicillin V amidase and a doxorubicin prodrug, *J. Drug Target.*, 7, 43, 1999.
64. Range, V.V. and Hollinger, M.A., in *Drug Delivery Systems,* CRC Press, Boca Raton, 1996, pp. 73–76.
65. Codde, J.P., Lumsden, A.J., Napoli, S., Burton, M.A., and Gray, B.N., A comparative study of the anticancer efficiency of Dox carrying microspheres and liposomes using a rat liver tumor model, *Anticancer Research*, 13, 539, 1993.
66. Kamath, K.R. and Park, K., Biodegradable hydrogels in drug delivery, *Adv. Drug Deliv. Reviews*, 11, 59, 1993.
67. Wu, X.Y., Liu, Z., and Bendayan, R., Development of a particulate delivery system with surface-immobilized Chemosensitizer, *Pharm. Res.*, 13(9), 9303, 1996.
68. Borodkin, S., Ion-exchange resin delivery system, in *Polymers for Controlled Drug Delivery,* Tarcha, P.J., Ed., CRC Press, Boca Raton, FL, 1991, pp. 215–230.

69. Liu, Z., Cheung, R., Wu, X.Y., Ballinger, J.R., Bendayan, R., and Rauth, A.M., A study of doxorubicin loading onto and release from sulfopropyl dextran ion-exchange microspheres, *J. Cont. Release*, 77, 213, 2001.
70. Borodkin, S. and Sundeberg, D.P., Chewable Tablets Including Coated Particles of Psuedo-Ephedine Weak Cation Exchange Resin, U.S. Patent No. 3,594,4700, 1971.
71. Bice, W.O. and Koble, R.A., Filtering Process and Apparatus, U.S. Patent No. 3,152,986, 1960.
72. Manek, S.P. and Kamat, V.S., Evaluation of Indion CRP 244 and CRP 254 as sustained release and taste masking agent, Indion, *J. Pharm. Sci.*, 26, 773, 1981.
73. Litchenckert, S., Lundgren, C., and Ferno, O., Chewable Smoking Substitute Composition, U.S. Patent No. 3,901,248, 1975.
74. Siegel, S., Reiner, R.H., Zelinskie, J.A., and Hanus, E.J., Tablets of pyrilamine resin adsorbate with aspirin and vitamin C, *J. Pharm. Sci.*, 51, 1068, 1962.
75. Burton, S., Washington, N., Steele, R.J.C., Musson, R., and Feely, L.C., Intragastric distribution of ion-exchange resins: a drug delivery system for the topical treatment of the gastric mucosa, *J. Pharm. Pharmacol.*, 47, 901, 1995.
76. Narisawa, S., Nagata, M., Danyoshi, C., Yoshino, H., Murata, K., Hirakawa, Y., and Noda, K., An organic acid induced sigmoidal release system for oral controlled release preparations, *Pharm. Res.*, 11, 111, 1994.
77. Illum, L., Nasal Drug Delivery Compositions Containing Nicotine, U.S. Patent No. 5,935,604, 1999.
78. Moreau, J.M., Green, L.C., Engel, L.S., Hill, J.M., and O'Callaghan, R.J., Effectiveness of ciprofloxacin-polystyrene sulfonate, ciprofloxacin ofloxacin in a staphylococcus keratitis model, *Curr. Eye Res.*, 17(8), 808, 1998.
79. Van Abbe, N.J. and Rees, J.T., Amberlite resin XE–88 as a tablet disintegrate, *J. Amer. Pharm. Assoc. Sci.*, 47, 477, 1998.
80. Saul, B. and Yunker, M.H., Interaction of amine drugs with a polycarboxylic acid acid ion–exchange resin, *J. Pharm. Sci.*, 59, 227, 1970.
81. Witzmar, J.L., Drug used in the treatment of hyperlipoproteinemias, in *The Pharmacological Basis of Therapeutics,* Hardman, J.G. and Limbird, L.E., Eds., McGraw Hill, New York, 1996, pp. 875–897.
82. Christy, D.P., et al., Effect of temperature and ion-exchange resin on cation diffusion through silicon polymer tubings, *J. Pharm. Sci.*, 68, 1102, 1979.
83. Moldenhauer, M.G. and Nairn, J.G., Formulation parameters affecting the preparation and properties of microencapsulated ion-exchange resins combining theophylline, *J. Pharm. Sci.*, 79, 659, 1990.
84. Woodworth, J.R., Ludder, T.M., Ludder, L.K., Sheperd, A.M.M., and Rotenberg, K.S., Comparative bioavailability of sustained release ion-exchange hydralazine product with a potassium challenge, *J. Pharm. Sci.*, 81, 541, 1992.
85. Jaiswal, W.J. and Bedi, G.S., Studies on sustained release formulations of propanolol–HCL with ion-exchange resins, *Indian Drugs*, 17, 102, 1980.
86. Sprockel, O.L. and Price, J.C., Evaluation of sustained release aqueous suspension containing microencapsulated drug-resin complexes, *Drug. Dev. Ind. Pharm.*, 15, 1275, 1989.
87. Conaghey, O.M., Corish, J., and Corrigan, O.I., The release of nicotine from a hydrogel containing ion-exchange resins, *Int. J. Pharm.*, 170, 215, 1998.
88. Jaskari, T., Controlled transdermal iontophoresis by ion exchange fibers, *J. Control. Rel.*, 67, 179, 2000.

89. Guy, R.H. and Hadgraft, J., Rate control in transdermal drug delivery, *Int. J. Pharm.*, 82, 121, 1992.
90. Jaskari, T., Vuorio, M., Kontturi, K.K., Manzanares, J.H., and Hirvonen, J., Ion exchange fibers and drugs: an equilibrium study, *J. Cont. Releases*, 70, 219, 2001.
91. Imtiaz, C., Patricia, K., Edward, R., and Joel, S., Sustained Release and Suspensions, U.S. Patent No. 4,999,189, 1991.
92. Sayed, U.G. and Bajaj, A.N., Oral controlled release bromphexine ion-exchange resinate suspension formulation, *Indian Drugs*, 37, 27, 2000.
93. Motycka, S., Newth, C.J.L., and Nairn, J.G., Preparation and evaluation of microencapsulated and coated ion-exchange resin beads containing theophylline, *J. Pharm. Sci.*, 74, 643, 1985.

4 Biopolymers as Supports for Heterogeneous Catalysis: Focus on Chitosan, a Promising Aminopolysaccharide

Eric Guibal, Thierry Vincent, and Francisco Peirano Blondet

CONTENTS

4.1 Introduction 153
4.2 General Properties of Selected Biopolymers for Supported Catalysis 155
4.2.1 Structure of Biopolymers and Functional Groups 155
4.2.1.1 Chitosan 155
4.2.1.2 Alginate 158
4.2.1.3 Cellulose and Starch 160
4.2.1.4 Gelatin 161
4.2.1.5 Wool and Silk 165
4.2.1.6 Casein 166
4.2.1.7 Miscellaneous Biopolymers 168
4.2.2 Biopolymer–Metal Ion Interactions 170
4.2.2.1 Ion Exchange Mechanisms 170
4.2.2.2 Chelation Mechanisms 177
4.2.2.3 Reduction and Precipitation Mechanisms 181
4.2.2.4 Encapsulation Mechanism 182
4.2.3 Conditioning of Biopolymer Support 183
4.2.3.1 Dissolved State 183
4.2.3.2 Composite Materials 185
4.2.3.3 Gel Beads 186

4.2.3.4 Fiber and Hollow Fiber ... 187
4.2.3.5 Flat Membranes ... 188
4.3 Biopolymer-Based Heterogeneous Catalysts ... 189
4.3.1 Synthesis ... 189
4.3.1.1 Conditioning ... 189
4.3.1.2 Metal and Catalyst Immobilization ... 191
4.3.1.3 Activation of Catalytic Metal ... 197
4.3.2 Critical Parameters for the Design of Biopolymer-Supported Catalysts ... 198
4.3.2.1 Size of Crystallites ... 199
4.3.2.2 Metal Content on the Solid and Metal Oxidation State ... 204
4.3.2.3 Diffusion Properties ... 207
4.3.2.4 Molecular Structure of Biopolymer–Metal Ion Assemblies ... 212
4.3.2.5 Stability ... 215
4.4 Examples of Reactions Catalyzed by Biopolymer-Supported Catalysts ... 215
4.4.1 Hydrogenation and Hydroformylation Reactions ... 215
4.4.1.1 Hydrogenation of Alcohols ... 215
4.4.1.2 Hydrogenation of Ketones ... 216
4.4.1.3 Hydrogenation of Aromatic Nitro-Compounds ... 221
4.4.1.4 Hydrogenation of Olefins ... 222
4.4.1.5 Transfer Hydrogenation ... 225
4.4.1.6 Miscellaneous Reactions of Hydrogenation ... 226
4.4.2 Oxidation Reactions ... 227
4.4.3 Polymerization Reactions ... 235
4.4.4 Cyclopropanation of Olefins ... 237
4.4.5 Hydration Reactions ... 237
4.4.6 Hydroxylation Reactions ... 239
4.4.7 Carbonylation Reactions ... 239
4.4.8 Cross-Coupling Reactions: Heck, Sonogashira, Suzuki, and Trost-Tsuji Reactions ... 241
4.4.9 Miscellaneous ... 245
4.5 Chitosan-Supported Pd Catalyst for Hydrogenation Reactions, from Flakes to Catalytic Hollow Fiber ... 247
4.5.1 Chitosan Flakes for Hydrogenation Transfer ... 247
4.5.1.1 Synthesis of Chitosan-Supported Pd Catalyst ... 247
4.5.1.2 Characterization of Catalytic Performance ... 249
4.5.2 Chitosan Hollow Fiber for Catalytic Hydrogenation ... 256
4.5.2.1 Manufacturing of Catalytic Hollow Fibers and Characterization ... 256
4.5.2.2 Nitrophenol Hydrogenation ... 263
4.5.2.3 Nitrotoluene Hydrogenation ... 266
4.6 Conclusion ... 269
References ... 270

4.1 INTRODUCTION

Catalytic phenomena affect virtually all aspects of our lives: enzymatic catalysis, for example, constitutes the basis for most of the biochemistry associated with life, and about 80% of industrial chemical processes are concerned with at least one catalytic reaction. In 2000, the worldwide market for catalysts was valued at about $10 billion, with applications in areas as different as food processing, fine chemical synthesis, petroleum refining, manufacture of fibers and plastics, and environmental protection.[1]

The increasing demand of industry for catalytic materials and processes, the limitation of available resources (especially for precious and strategic metals), and the cost of a number of catalytic formulations (including metal complexes) may explain the growing interest of research community for the development of alternative heterogeneous catalysts. Indeed, industry is requiring the development of chiral catalysts with enhanced stereo-, chemo-, and regioselectivity.[2–6] These materials are generally more expensive than conventional catalysts — up to $500,000 kg^{-1} for chiral phosphine, for example — and their recycling is an important feature in the design of the catalytic process, and a real challenge when the catalyst-to-substrate ratios are greater than 1:10,000.[7]

Homogeneous catalysis was first developed with the objective of simulating enzymatic and biological catalytic systems for improving reaction kinetics and selectively orientating reaction pathways. Nanostructured metal clusters stabilized by surrounding polymers can act as catalysts, mimetic to enzyme systems. The surrounding polymer of metal clusters can be assigned as a polymeric field and has functions similar to those of the protein surrounding the active site of the enzyme.[8] However, the necessity to recover the catalytic system (especially for expensive formulations such as chiral catalysts or for hazardous materials) requires recycling the material, and a number of techniques have been developed for recovering homogeneous catalysts: membrane filtration, precipitation, two-phase systems, and, of course, heterogeneization on an insoluble inorganic or organic support. The immobilization of catalytic metals at the surface of inert materials facilitates the recovery of active substances at the end of the process, and its recycling contributes to a more efficient use of the resource and the enhancement of economic balance. However, depending on the mode of insertion of the catalytic metal (an active vs. a passive mechanism), the environment of ligand-metal center may be affected changing its reactivity.[9] Additionally, the immobilization process may also contribute to the limitation of kinetics due to diffusion restrictions. This significant drawback can be counterbalanced by the stereoselectivity the material can bring to the reaction due to the conformation of the support. The selection of the support requires considering different parameters that may have a significant impact on catalytic performance:

- Affinity for the metal (binding capacity, metal content on the support)
- Stability of the metal on the catalyst (degradation, poisoning)
- Porosity (accessibility to reactive sites, mass transport)
- Physical versatility (conditioning of the support, physical form of the support)

- Conformational effect (orientation of the reaction, stereoselectivity)
- Crystallite size (preventing aggregation of crystallites)

Additionally, at the end of catalyst life cycle, the possibility of recovering the metal may have an important impact on the cost effectiveness of the catalyst.

Although a number of reactions continue to process with homogeneous catalysis, an increasing number of processes use supported catalysis. The first processes designed in heterogeneous catalysis involved mineral materials, such as alumina,[10–14] silica,[15–20] modified clay,[21,22] and activated carbon.[23–29] The support is usually impregnated with the metal-containing solution and the immobilization proceeds by solvent evaporation and metal entrapment into the porous network of the support. These materials are widely available and cost effective, and it is possible to design especially tailored porous forms; however, they usually do not bring any stereoselectivity to the reaction and, in some cases, the recovery of the metal requires drastic operations to make it accessible and completely recoverable (grinding, leaching, etc.). Additionally, the binding of the metal to the mineral supports is not strong enough to limit its leachability, which in turn induces a significant loss of active material and a progressive decrease of catalytic activity.

For the past few decades, a number of studies focused on the design of supported catalysts based on polymer materials.[30–51] Ion exchange resins have been investigated for their use in immobilizing a wide range of metals, including conventional catalytic metals.[52–63] In this case, the metal is bound to the resin through stronger interactions, preventing (or at least limiting) the leaching of catalytic metals and maintaining the catalytic activity for a longer time. Additionally, the polymeric network can affect significantly the activity, stability, and selectivity of chiral catalysts; an appropriate selection of the auxiliaries and the immobilization protocol can afford catalysts that are more active and even more selective than the homogeneous analogues.[30,64–68] Polymers can also be used in the dissolved state for the protection of metal nanoparticles in order to increase the reactivity of the catalyst.[69,70] This is the basis of colloid-supported catalysis.[33,35,69,71] In this case, the binding of the metal to the polymer in solution decreases the tendency of metal nanoparticles to aggregate.[72]

More recently, many studies have focused on the use of biological materials and, more specifically, biopolymers for the immobilization of catalytic metals for the preparation of heterogeneous catalysts. Many reasons may explain this increasing interest:

- Availability of the resource
- High binding ability of these materials for selected metals
- Physical and chemical versatility of these materials (which can be easily modified)
- Easy degradation of the organic material at the end of the life cycle (less toxic degradation products than conventional resins)
- Possible conformational effects (in relation with the physical structure of these materials)

Many of these biopolymers have been used for the elaboration of supports for affinity chromatography with highly selective properties for the separation of optical isomers.[73–81]

This review focuses on the use of materials of biological origin, including polysaccharides (chitin, chitosan, alginate, starch, and cellulose-based materials), and proteins (gelatin, casein, wool, etc.) for the elaboration of supported catalysts. Special attention will be paid to chitosan, an aminopolysaccharide with high binding capacities for metals ions and high chemical and physical versatility. This biopolymer was widely carried out in applications in heterogeneous catalysis. Binding mechanisms will be considered, with attention paid to the influence of metal speciation. Indeed, changing the speciation of the metal (with pH, presence of ligands, etc.) may change the types of interactions established between functional groups of the biopolymer and metal ions.[82]

4.2 GENERAL PROPERTIES OF SELECTED BIOPOLYMERS FOR SUPPORTED CATALYSIS

Biopolymers offer a wide resource in terms of both the quantities annually produced and the diversity of functional groups available for interactions with solutes (metals, organic compounds, organometallics, etc.). Their limited application in the industrial field for wastewater treatment may be explained in some cases by: (a) the variability in their composition depending on the production and extraction processes; the default in standardization may cause differences in the binding properties of materials and difficulties in the design of treatment plants; and (b) the difficulty in getting appropriate conditioning of these materials for large scale applications. For this last point, a great effort has been devoted over the past few decades to the elaboration of new physical forms of biopolymers.

This section successively describes the structure of the biopolymers carried out as support for catalytic metals, the types of interactions involved in metal binding, and the techniques developed for the modification of biopolymer conditioning for catalytic applications.

4.2.1 Structure of Biopolymers and Functional Groups

Polysaccharide-based materials (including chitosan, alginate, and cellulose and starch derivatives) and protein-based materials (including gelatin, wool, casein, and other derivatives) are the biopolymers most frequently cited for the preparation of heterogeneous catalysts. They can be used as encapsulating materials (gels) or through physicochemical interactions. This section describes the principal properties of these biopolymers with respect to their chemical and physical structures.

4.2.1.1 Chitosan

Chitosan is an aminopolysaccharide [poly-β-,(1→4),2,amino-2-deoxy-D-glucan, poly-D-glucosamine] obtained by the alkaline deacetylation of chitin [poly-β-,(1→4),2,acetamido-2-deoxy-D-glucan, poly-D-acetylglucosamine]. Chitin is one

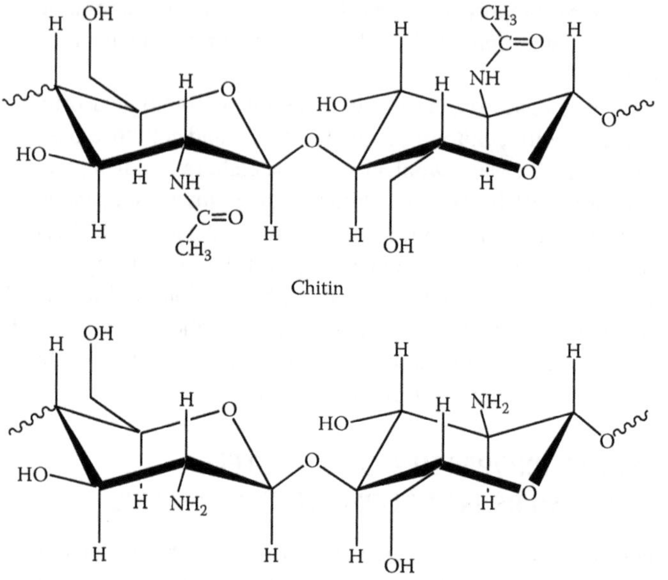

Chitin

Chitosan

FIGURE 4.1 Structure of chitin and chitosan.

of the most abundant polysaccharides in nature, being present in the cuticle of insects, the cell wall of fungi, the shell of crustaceans, and the pen of squids. Crustacean shells represent the most representative source of chitin for commercial production. Chitin is produced from washed and ground shells by a series of acidic and alkaline treatments to successively remove carbonates, proteins, and natural colorants.[83] The deacetylation procedure consists of the alkaline treatment at boiling temperature of chitin powder. Actually, in most cases, chitosan is not fully deacetylated and the biopolymer should be more accurately described as a copolymer formed of β, (1→4)-linked acetylglucosamine and glucosamine units, whose distribution defines the appropriate name (chitin versus chitosan). This definition is still debatable in the chitin scientific community. However, it is generally considered that chitosan is a chitin derivative where more than 50% of acetamido groups have been deacetylated, and which is soluble in 1 M acetic acid solutions. Figure 4.1 shows the chemical structure of these biopolymers.

The properties of chitosan strongly depend on two critical parameters: the deacetylation degree and the molecular weight of the polymer. The presence of numerous hydroxyl groups confers to the biopolymer interesting hydrophilic properties that are very helpful for sorption properties in aqueous solutions and for some catalytic reactions. However, the most important reactive group for the design of supported catalysts is the amino group present on chitosan. These amino groups are very reactive for metal binding; their action may proceed by chelation on the free electronic doublet (at near-neutral pH) or by electrostatic attraction and ion exchange on protonated amine groups (in acidic solutions).

The presence of amine groups leads to important acid-base properties responsible of biopolymer dissolving, in addition to sorption properties. The polyelectrolyte behavior of the polymer obeys the Katchalsky's equation:

$$pK_a = \text{pH} + \log\left(\frac{1-\alpha}{\alpha}\right) = pK_0 - \frac{\varepsilon \Delta\Psi(\alpha)}{KT} \tag{4.1}$$

where α is the neutralization degree, ε is the dielectric constant of the solvent, KT is the Boltzmann parameter, pK_0 is the intrinsic pK of the amine site supposed isolated and undissociated, (ΔΨ(α)) is the potential difference between the surface of the polyelectrolyte ion and the reference.

The pK_a of chitosan amino groups strongly depends on the deacetylation degree and the degree of neutralization of these functional groups; Sorlier et al. found that the pK_a varied between 6.4 and 7.2 at complete neutralization of protonated amine groups when decreasing the degree of deacetylation.[84] In acidic solutions, the protonation of amine groups causes the dissolving of chitosan in most mineral and organic acids, with the remarkable exception of sulfuric acid. This solubility is a critical parameter for polymer stability (necessary to cross-link the material in order to maintain the sorbent in a solid sate), and also for the conditioning of the biopolymer (allowing the preparation of gels). These amine groups are also very important for the chemical modification of chitosan; a number of chitosan derivatives have been prepared by grafting new functional groups on amine groups, in addition to the modifications operated on hydroxyl groups.[85–90] Amine functions have also been used for the cross-linking of chitosan to prevent its dissolving in acidic solutions through the formation of imine linkages, for example, in the case of glutaraldehyde reactions. Amine groups of chitosan react with aldehyde functions of the cross-linking agent through the so-called Schiff base reaction. These supplementary chemical linkages limit biopolymer dissolving.

The physical structure — that is the conformation of the polymer — is of critical importance regarding its ability to match steric access to metal ions (the stiffness of the chain influences the flexibility of the polymer and its rearrangement for the interaction of functional groups with metal ions) and to substrate molecules (this is especially important for enantioselective reactions). This forms the base of the properties of chitin and chitosan materials for enantioseparation. Chitin may exist in three different forms, depending on the origin of the material. These forms correspond to the arrangement of polymer chains: α-chitin (typical of shrimp material) corresponds to parallel arrangements of chains, β-chitin (typical of squid material) is found in antiparallel arrangements, while γ-chitin (typical of the presence of chitin in internal organs of some crustacean) is characteristic of an alternate arrangement of chains.

The most frequent forms are α-chitin (crystallized under orthorhombic structure) and β-chitin (arranged according a monoclinic cell unit). For β-chitin, the antiparallel structure reduces the possibility of forming hydrogen bonds between

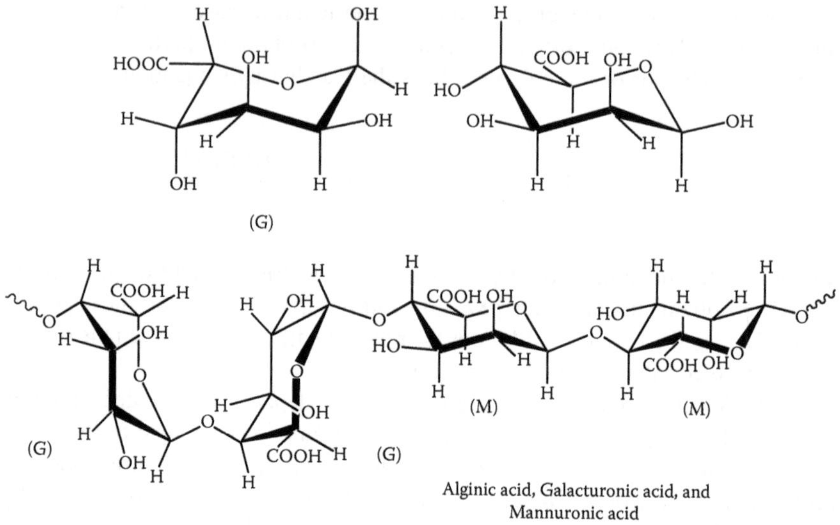

Alginic acid, Galacturonic acid, and Mannuronic acid

FIGURE 4.2 Structure of alginic acid: Galacturonic acid and mannuronic acid.

the chains, which in turn results in better reactivity, swelling, and solubility. However, α-chitin is much more abundant, cheaper, and therefore more frequently used than β-chitin, which is reserved for highly specific applications such as biomedical. For chitosan, the chain segments are antiparallel inside a sheet of polymer chains, as they are for α-chitin. The orthorhombic cell typical of chitosan is strongly affected by the water content; hence, the unit cell dimensions (Å) varied from (8.28, 10.43, 8.62) in the annealed form to (8.95, 10.34, 16.97) in the standard form (structural water content, 10%).[85–90] The presence of water has an important impact on the stability of the structure (formation of hydrogen bonds with water molecules) and the reactivity of the polymer.

4.2.1.2 Alginate

Alginate is another example of a polysaccharide mainly produced from marine resources. Alginate and alginic acid are produced by the alkaline extraction ($NaCO_3$ treatment) from brown algae (*Laminaria*, *Ascophyllum*, and *Fucus*, for example). Present in the form of insoluble salts of calcium in the biomass, the displacement of the counterion with sodium makes possible biopolymer extraction; the reprecipitation with calcium salt allows obtaining purified calcium alginate, while a treatment with acid solutions produces a gel of alginic acid. This is a hetero polymer constituted of α-L-guluronate units (*G*) and β-D-mannuronate units (*M*), whose fractions (*M*/*G*) depend on the origin of the biomass, for example, close to 1.6 for *Macrocystis pyrifera* and close to 0.45 for *Laminaria hyperborea* (Figure 4.2). Actually, it is a linear polysaccharide arranged in an irregular blockwise pattern consisting of three types of blocks: *GG*, *MG*, and *MM*.

The key functional groups are therefore the carboxylic functions. When alginic acid or the alkaline form of alginate (i.e., in the form of ammonium, potassium, or sodium alginate salt) is dissolved in water, all the acidic groups of the biopolymer are ionized. However, when decreasing the pH, the protonation of carboxylate groups results in charge neutralization and progressive precipitation. This behavior is the complete opposite of that observed with chitosan.

The properties of this heteropolymer strongly depend on the distribution of mannuronic and galacturonic units because of their different conformations in aqueous solutions. Hence, the tendency of guluronic units to form intermolecular cross-links may explain that cross-linking or ionotropic gelation occurs in the presence of divalent or trivalent cations and results in the formation of stable gels. This cross-linking will thus be enhanced in the case of alginate rich in guluronic units (high *G*/*M* ratio). This property is very important for the comprehension of metal sorption, the stability of metal bound to the biopolymer, and the structure of gel material (when alginate is only used for the entrapment of catalytic metals).

The physical structure is also influenced by the heterogeneities of the monomer units present on the biopolymer. Indeed, guluronic units in homopolymer series are in the conformation ${}^{1}C_{4}$, while mannuronic units are in a ${}^{4}C_{1}$ conformation. Therefore, four different glycosidic links may exist between the different units: di-equatorial (for *MM* association), di-axial (for *GG* association), equatorial-axial (for *MG* association), and axial-equatorial (for *GM* association), as shown on Figure 4.2. The steric hindrance near the di-axial glycosidic linkage for *GG* blocks results in very rigid and extended structures for cross-linked gels. Poly β-(1→4)-linked D-mannuronate units preferentially forms a threefold left-handed helix with intramolecular hydrogen bonding between the hydroxyl group in the 3 position and the following ring oxygen. On the opposite hand, poly α-(1→4)-linked L-guluronate units form stiffer twofold screw helical chains, due to intramolecular hydrogen bonding between the carboxyl group and the 2-OH group of the prior residues, and the 3-OH of the following residues (weaker bonds). The diaxal links cause less flexibility to the molecule. In the case of alternate linkages between *G* and *M* units, the alternation of equatorial-axial and axial-equatorial links causes dissimilar conformations. This induces a greater flexibility for these alternate structures compared to mannuronate structures, due to the higher degrees of freedom offered by hydrogen linkages between carboxyl groups of the mannuronate units and 2-OH and 3-OH of vicinal guluronate units. Hydrogen bonding on carboxylate groups may be replaced by Ca^{2+} ions, causing the formation of guluronate zipped ribbon. Calcium ions arrange in an egg-box-like conformation (six oxygen ligands supplied by two parallel chains come from 2-OH, 3-OH, and carboxylate oxygen of the following residue). It is noteworthy to observe that mannuronate units cannot contribute to this egg-box mechanism.

The pK_a of carboxylic groups depends on the type of unit; hence, the pK_a of mannuronic acid is close to 3.38, while it tends to 3.65 for guluronic acid units. This leads to alginate precipitation when the pH is below 3.5. However, this solubility depends on biopolymer molecular weight, ionic strength, and the nature of ions present.

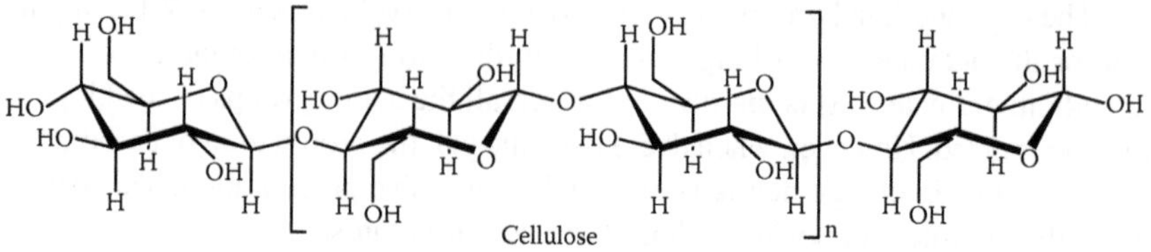

FIGURE 4.3 Structure of cellulose.

4.2.1.3 Cellulose and Starch

Cellulose and starch are very similar materials in terms of structural units; each of them is constituted by glucopyranose units. The main differences consist in the linkages involved in the polymer structure (β-linkage for cellulose, α-linkage for the constituents of starch). These carbohydrates are part of the main constituents of vegetal biomass. Cellulose constitutes the structurally strong framework of cell walls in plants (being present in the form of microfibrils), while starch is the major carbohydrate reserve in plants.

Cellulose is a linear polymer of β-(1→4)-D-glucopyranose units in $^4C_1$ conformation. The fully equatorial conformation of this polymer stabilizes the chair structure, limiting its flexibility (compared to other α-glucopyranose polymers) (Figure 4.3). This is a homogeneous biopolymer constituted of 2,000 to 14,000 units (although it is possible to produce shorter chains). The formation of intramolecular and intrachain hydrogen bonds gives cellulose crystal-like structure that may explain its insolubility in aqueous solutions.[91] The natural form of cellulose (called cellulose I) is characterized by parallel ribbons without intersheet hydrogen bonding formed of two kinds of cellulose crystallized, respectively, in triclinic (identical chains with two alternating glucose conformers) and monoclinic phases (two conformationally distinct alternating staggered sheets, made of identical crystallographic glucose conformers). The proportions of these distinct phases depend on the origin of the cellulose biopolymer and its treatment: the annealing converts triclinic cellulose into monoclinic cellulose. Other forms of cellulose (i.e., cellulose II and III) correspond to recrystallization of cellulose I and chemical treatments.[91] The chemical structure, principally made of OH groups, makes the polymer hydrophilic (except in the case of crystalline forms of cellulose). Therefore, the material has less reactive sites than chitosan and alginate biopolymers.[92] For these reasons, a series of cellulose derivatives has been developed, such as methylcellulose, hydroxypropylmethylcellulose, and carboxymethylcellulose.[91] These modifications contribute to the increased solubility of cellulose material in water. Despite their poor reactivity, cellulose-based materials can be developed using the ability of these materials to form gel for the entrapment or the encapsulation of target molecules, metals, and so forth.

Starch is typically made of two different types of molecules, amylose and amylopectin, which usually represent around 20 to 30% and 70 to 80%, respectively (Figure 4.4). Both amylose and amylopectin are formed of α-D-glucopyranose (in the $^4C_1$ conformation) linked by α(1→4) links (ring oxygen atoms being on the same

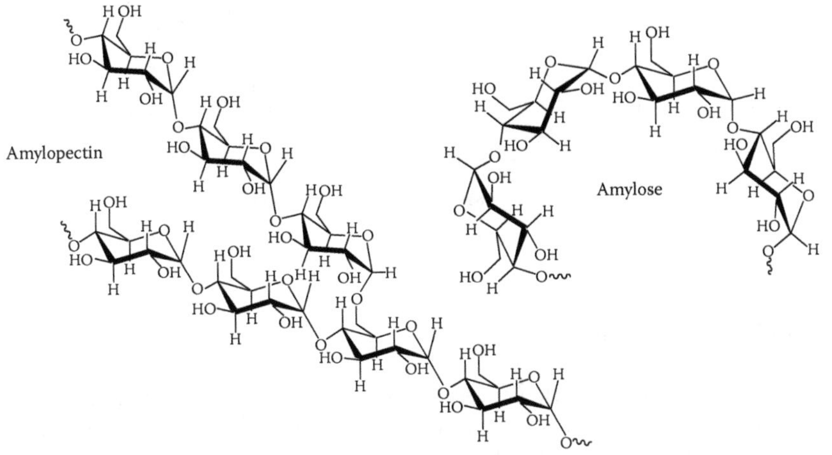

FIGURE 4.4 Structure of starch: Amylose and amylopectin.

side of the chain). In the case of amylopectin, additional linkages are established at regular intervals (every 20 to 30 units) between *C*1 and *C*6, forming branch points.

Amylose tends to form a stiff, left-handed, single helix or to form stiffer parallel, left-handed, double-helical junction zones. The number of inter- and intrachain hydrogen bonds may explain the hydrophobic structure of amylose and its weak solubility in water. By analogy to cyclodextrins, the relatively hydrophobic inner surface holds water molecules that can be easily replaced by other hydrophobic molecules.

Amylopectin contains up to two million glycopyranose units, arranged in a compact structure (hydrodynamic radius 21 to 75 nm, much greater than that of amylose materials). Actually, two forms of amylopectin have been identified: the double-helical chain can be transformed, depending on polymer hydration, into hexagonal crystallites or staggered dense monoclinic packing.

Again, in the case of starch constituents, the functional groups (i.e., –OH) are not very reactive for metal binding. However, these –OH groups can be used for the preparation of starch derivatives, obtained, for example, by the quaternarization procedure (involving the grafting of new functional groups, bearing positive charge).[93–95] Additionally, starch derivatives, like cellulose, can be used for the entrapment or encapsulation of active functional groups (immobilized on a suitable support).

4.2.1.4 Gelatin

Gelatin is prepared by the thermal degeneration of collagen, which is usually isolated from animal skin and bones and fish skin by weak acidic treatment. The composition of collagen from different sources differs somewhat in amino acid sequence. However, most contain about 27% glycine (Gly), 11% alanine (Ala), 15% proline (Pro), and 13% 4-hydroxyproline (4Hyp), or 12% glutamic acid

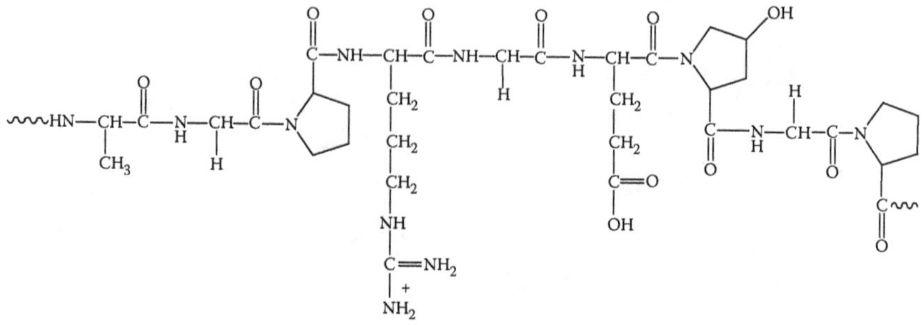

FIGURE 4.5 Structure of gelatin.

(Glu) or 9% arginine (Arg), 7% aspartic acid (Asp), in addition to a number of less represented amino acids. A typical series of amino acids for the representation of gelatin could be schematized as: -Ala-Gly-Pro-Arg-Gly-Glu-4Hyp-Gly-Pro-. Figure 4.5 shows an example of structure for gelatin.

The physical structure of the polymer corresponds to the heterogeneous mixture of single and multistranded polypeptides, each characterized by extended left-handed proline helix conformations and containing between 300 and 4000 amino acids. The triple helix of collagen is composed of two $\alpha 1$(I) and one $\alpha 2$(I) chains, with molecular mass close to 95 kD, (width: ≈ 1.5 nm, length: 0.3 μm). Actually, gelatin consists of a mixture of these strands, with their oligomers and their breakdown polypeptides resulting from thermal degradation. Chemical cross-links can be established through amine groups of different amino acids (and, more specifically, lysine amine acids) using, for example, glutaraldehyde to reinforce the stability of the polymer.

The most important properties for metal binding on gelatin-based materials are related to the presence of carboxylic groups and amine groups (sulfur-amino acids are not present in significant quantities). The coexistence of basic (amine functions) and acid (carboxylic functions) groups confers to the polymer amphiphilic properties and multiple functionalities for metal binding (chelation, ion exchange facilities). Table 4.1 reports the acid-base properties of the most frequent amino acids in proteins such as collagen (for gelatin), keratin (for wool), or caseins. The pK_a of carboxylic functions vary between 1.7 and 2.6, while those of amine functions vary between 9 and 11. These values are important for predicting the charge of functional groups and their ion exchange and electrostatic attraction properties. Hence, in moderate acidic conditions (around a pH of 4), amine groups are protonated while carboxylic groups are dissociated, giving the proteins the possibility of interacting with both cations and anions. It is interesting to observe that the α-carboxyl group of monoamino monocarboxylic acids is stronger than the carboxyl group of comparable aliphatic acid. The increased acid strength of these groups is caused by the presence of the electron withdrawing ammonium group and its positive charge (giving a strong field effect and the tendency for the carboxylic hydrogen to dissociate as a proton). The α-amino group

TABLE 4.1
The p*K* Values for the Ionizing Groups of Some Amino Acids at 25°C.

Amino acid		pK (-COOH)	pK ($-NH_3^+$)	pK (*R*-group)[a]
Glycine	$^{+}H_3N-CH(H)-C(=O)-O^-$	2.34	9.6	
Alanine	$^{+}H_3N-CH(CH_3)-C(=O)-O^-$	2.34	9.69	
Leucine	$^{-}O-C(=O)-CH(NH_3^+)-CH_2-CH(CH_3)-CH_3$	2.36	9.6	
Serine	$^{-}O-C(=O)-CH(NH_3^+)-CH_2-OH$	2.21	9.15	
Threonine	$^{+}H_3N-CH(CH(OH)CH_3)-C(=O)-O^-$	2.63	10.43	
Glutamine	$H_2N-C(=O)-CH_2-CH_2-CH(NH_3^+)-C(=O)-O^-$	2.17	9.13	

(Continued)

TABLE 4.1 (Continued)
The p*K* Values for the Ionizing Groups of Some Amino Acids at 25°C.

Amino acid	pK (-COOH)	pK ($-NH_3^+$)	pK (*R*-group)[a]
Aspartic acid	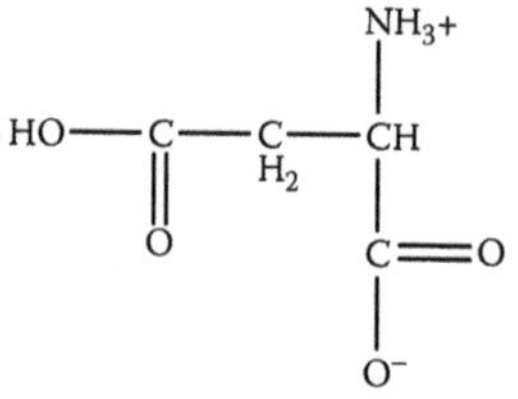 2.09	9.82	3.86
Glutamic acid	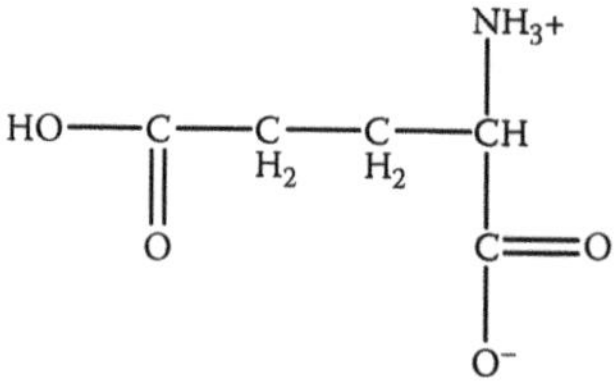 2.19	9.67	4.25
Histidine	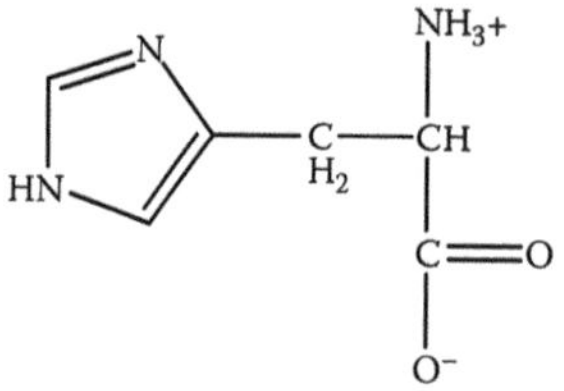 1.82	9.17	6.00
Cysteine	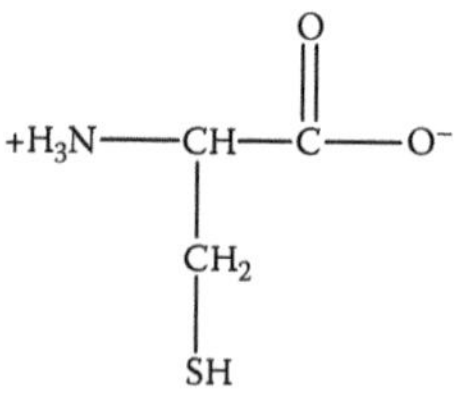 1.71	10.78	8.33
Tyrosine	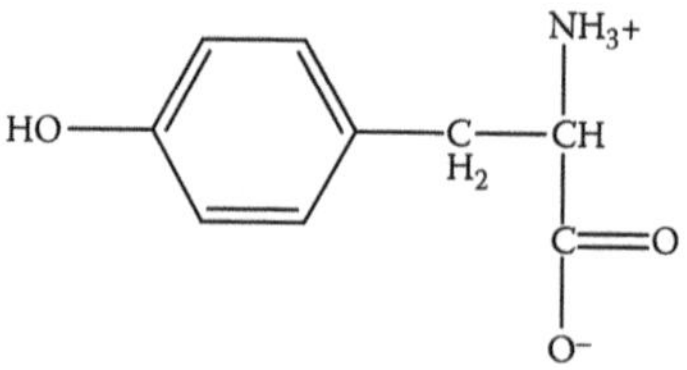 2.2	9.11	10.07

TABLE 4.1 (Continued)
The pK Values for the Ionizing Groups of Some Amino Acids at 25°C.

Amino acid	Structure	pK (-COOH)	pK ($-NH_3^+$)	pK (*R*-group)[a]
Lysine	$H_2N{-}CH_2{-}CH_2{-}CH_2{-}CH_2{-}CH(NH_3^+){-}C(=O){-}O^-$	2.18	8.95	10.53
Arginine	$H_2N{-}C(=NH){-}NH{-}CH_2{-}CH_2{-}CH_2{-}CH(NH_3^+){-}C(=O){-}O^-$	2.17	9.04	12.48

[a] *R* groups (bold symbols)

of monoamino monocarboxylic acids is a stronger acid (and reversibly weaker base) than the amino group of comparable aliphatic amine or amino groups of chitosan. The thiol or sulfhydryl groups (–SH) of cysteine and *p*-hydroxyl group of tyrosine are very weakly acidic, while the ε-amino group of lysine is strongly basic (losing its protons only in very alkaline solutions).[96] However, in the proteins, the association of amino acid with peptidic bonds changes their acid-base properties, which depend on the length of protein chain and electrostatic and inductive interactions of residues in the intermediate positions. Although for short peptides the acid-base properties of terminal groups are close to those of free amino acids, when the number of units increases, the p*K* values of terminal α-carboxyl groups increase and those of α-amino groups decrease (compared to their free groups).

4.2.1.5 Wool and Silk

Wool is another example of protein made of fibrous proteins, with keratin being the most representative protein, somewhat the equivalent of collagen for gelatin. There are two classes of keratin, the so-called α-keratin and β-keratin, differentiated by their composition and their physical structure (x-ray diffraction patterns). While α-keratin is characterized by the presence of a significant amount of cysteine or cystine residues (bearing sulfhydryl groups able to form disulfide cross bridges), β-keratin does not contain sulfur-based amino acids but is characterized by significant proportions of amino acids with small side chains (such as glycine, alanine, and serine). Additionally, α-keratin is characterized by the α helix structure: the backbone is arranged in a helical coil having about 3.6 amino acids per turn, with the *R* groups of the amino acid extending outward from the rather tight helix. Hence, the repeat unit is close to 0.5 to 0.55 nm along its axes. This structure makes possible the formation of intrachain hydrogen bonds between successive coils of the helix. In wool, three right-handed α helixes are coiled around each other to form three-stranded ropes linked together by disulfide cross-links, while in hair α-keratin seven α helixes are involved in the formation of supercoiled structures. The α helix structure is characterized by a strong dextrorotatory power; this is especially important for the optical properties of the protein and also for anticipating possible stereoselectivity for catalytic applications. It is interesting to observe that the optical rotation is very sensitive to pH due to the possible rearrangement of the structure when changing the acidity of the solution, passing from an α helix in acid media to a random conformation in alkaline media (and reciprocally depending on the type of polyamino acid). Submitted to moist heat, α-keratin (in hair, for example) fibrous protein stretches (to almost double, and the repeat unit increases to 0.65 to 0.7 nm) and contracts to its normal length on cooling. This is caused by the thermal breakage of the intrachain hydrogen bonds that normally stabilize the α helix. This is another significant difference from β-keratin, which does not change in length when submitted to a similar heating treatment.

In the case of β-keratin (present in silk, for example), the polypeptide chains are arranged in a zigzag conformation: pleated sheets are cross-linked by interchain hydrogen bonds. The *R* groups (small functional groups such as –H or $-CH_3$ in glycine and alanine, respectively) lie above or below the zigzagging planes of the pleated sheet. These differences can be also explained by the relative arrangement of polypeptide chains: in α-keratin the chains are arranged in a parallel mode, while in β-keratin the chains are antiparallel. Glycine, alanine, serine, and tyrosine constitute more than 90% mol of the whole volume of silk fibroin.[97]

4.2.1.6 Casein

Casein is a protein produced from milk. Thus, this is a cheap and abundant material. Casein is formed of various amino acids and a small amount of phosphoric acid (phosphate esterified to serine residues, less than 4%). The composition of casein constitutes about 200 residues (Table 4.2) with nonpolar, polar (but uncharged), negatively charged, and positively charged units (classified by function of amino acid charge at pH 6).

This chemical composition shows that casein, like gelatin and other proteins, is characterized by a large diversity of functional groups bearing different ionic charges, which in turn may affect the affinity of the biopolymer for some metal ions. Casein forms stable micelles in the presence of Ca^{2+} ions. These micelles are roughly spherical particles with a mean radius of 80 nm. They are principally composed of four different casein molecules, called α_{s1}-, α_{s2}-, β-, and κ-casein, in

TABLE 4.2
Amino Acid Composition of Casein (Classified by Function of Overall Charge of the Amino Acid at pH 6)

Nonpolar	Polar (Uncharged)	Negatively Charged	Positively Charged
Alanine [9]	Glycine [9]	Aspartic acid [7]	Lysine [14]
Valine [11]	Serine [16]	Glutamic acid [25]	Arginine [6]
Leucine [17]	Threonine [5]		Histidine [5]
Isoleucine [11]	Tyrosine [10]		
Proline [17]	Asparagine [8]		
Methionine [5]	Glutamine [14]		
Phenylalanine [8]			
Tryptophane [2]			

Note: Brackets indicate the number of amino acid residues per molecule of casein.

the approximate ratio of 4:1:4:1, respectively; other molecules have been also identified but with lower proportions (including the γ series).[98] In addition to enzymatic systems, various processes can be used for casein precipitation and coagulation, including thermal treatment and reaction with neutral salts (at slightly acidic pH, i.e., pH 4.7). The precipitation of casein in the presence of Ca^{2+} depends on the type of protein: α_{s1}-, α_{s2}-, and β-casein precipitate even with low concentrations of calcium ions (at millimole levels), while κ-casein remains stable. These differences can be explained by a phosphorylation ratio much lower for κ-casein (κ-casein < β-casein < α_{s1}-, α_{s2}-casein). Another difference consists of the presence of a small fraction of glucides in κ-casein (less than 5% in the form of galactose, *N*-acetylgalactosamine, and *N*-acetylneuraminic acid). The mixture of the different proteins ensures that the casein does not precipitate in milk and forms micelles. Dauphas et al. discuss the supramolecular organization of β-casein[99]: they show that the casein may exist in solution under different forms depending on the temperature and the presence of calcium ions. At low temperature (below 15°C), β-casein exists in a molecular state in the absence of Ca^{2+} and in a polymeric form at 100 m*M* Ca^{2+} concentration; increasing the temperature above 35°C causes the formation of free or aggregated micelles (in the absence and presence of Ca^{2+}, respectively).

The structure of casein strongly depends on the type of protein and its environment[100]; this could explain the discrepancies observed in the description of the secondary structure of these proteins. Horne describes a series of self-assembly of casein depending on the type of casein: for β-casein, the interaction of hydrophobic regions on neighboring molecules leads to the formation of detergent-like micelles while α_{s1}-casein forms worm-like chains.[100] Guo et al. model the precipitation of caseins as a function of the composition of the precipitating solution and the composition of casein.[101] In the presence of Ca^{2+} ions, the neutralization of β-casein promotes its precipitation while other types of casein remain in solution; the presence of phosphate causes the coprecipitation of protein

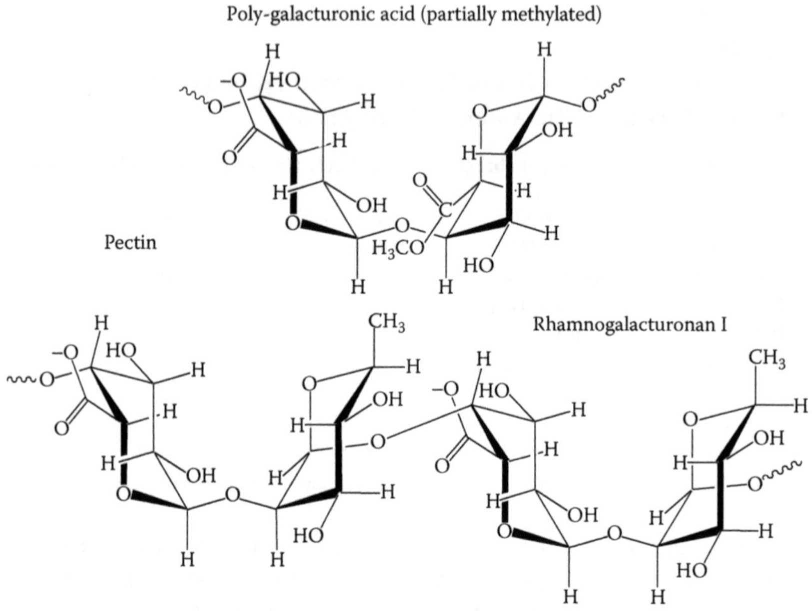

FIGURE 4.6 Structure of pectin.

(regardless of the type of casein) and inorganic microcrystals (calcium phosphate acting as germ for precipitation). The effect of phosphate is enhanced at pH levels close to neutrality and by cyclic phosphate.

4.2.1.7 Miscellaneous Biopolymers

A number of other biopolymers could be effective at supporting catalytic metals, such as pectins, carrageenan, or other proteins. Although these biopolymers have not been extensively carried out for the synthesis of supported catalysts at the moment, their ability to bind metal ions and their structure similar to that of previously cited materials may, in the near future, open the route to new catalytic supports.

For example, pectin is a heterogeneous grouping of acidic structural polysaccharides found in fruit and vegetables, obtained at the commercial level from waste citrus peel and apple pomace. The structure is quite complex and depends on both the material source and the extraction procedure. The commercial extraction of pectin frequently causes extensive degradation of neutral sugar-containing side chains. Typically, pectin can be considered being constituted of partially methylated poly-α-(1→4)-D-galacturonic acid residues, complete with "hairy" nongelling areas made of alternating α-(1→2)L-rhamnosyl-α-(1→4)-D-galacturonosyl residues, with branched units bearing mostly neutral side chains of L-arabinose and D-galactose (rhamnogalacturosan I). Pectin may also contain rhamnogalacturosan II (a form containing much complex and variable sugar residues such as D-xylose, L-fucose, and so forth). Figure 4.6 shows representation of these main constituents.

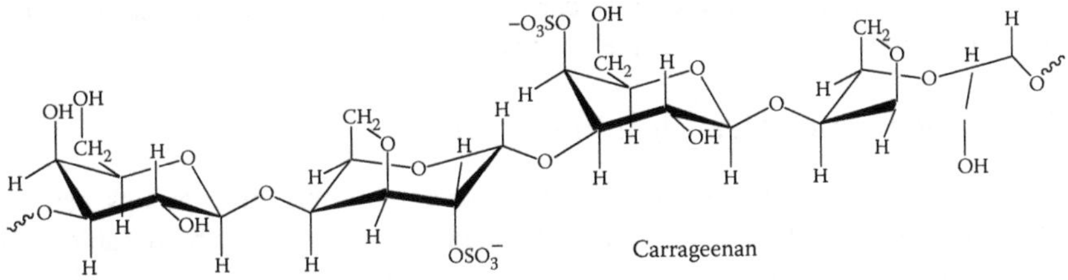

FIGURE 4.7 Structure of carrageenan.

The most important properties of pectin (gelation, metal binding) are related to the presence of carboxylic groups, whose pK_a are close to 2.9. The structure of pectin remains debatable; the molecule does not adopt a straight conformation in solution but is extended and curved, wormlike, with great flexibility, increased in the regions enriched in rhamnogalaturosan-based compounds. The properties of pectin strongly depend on the degree of methylation; hence, the methylation of carboxylic acids forms methyl esters with increased hydrophobicity. With low methylation degree (about 35%), pectins gel in the presence of calcium ions by bridging between adjacent twofold helical chains; this is similar to the egg-box gelling systems observed with alginate. When increasing the degree of methylation (above 50%), pectins do not gel, despite their interactions with calcium ions; the only way to reach pectin gelling requires controlling the pH close to 3 to decrease electrostatic repulsions (in order to form hydrogen bonds and improve hydrophobic interactions) or by addition of sugars (to reduce polymer–water interactions). Actually, pectins behave similarly to alginate: poly-α-(1→4)-D-galacturonic replaces poly-α-(1→4)-L-guluronic; the only significant difference is the presence of the 3-hydroxyl group in an axial position in pectins.

Carrageenan is another example of a carbohydrate with potential applications in supporting catalytic metals. The terminology of carrageenan covers a wide range of polysaccharides obtained by alkaline extraction from red seaweed. The composition of the biopolymer strongly depends on its source and the extraction procedure. It can be considered as a biopolymer constituted of alternating 3-linked-β-D-galactopyranose and 4-linked-α-D-galactopyranose units bearing sulfate groups (Figure 4.7). Actually, three different forms of carrageenan polymers have been identified:

1. κ-*Carrageenan*(-(1→3)-β-D-galactopyranose-4-sulfate-(1→4)-3,6-anhydro-α-D-galactopyranose-(1→3)-)
2. ι-*Carrageenan*(-(1→3)-β-D-galactopyranose-4-sulfate-(1→4)-3,6-anhydro-α-D-galactopyranose-2-sulfate-(1→3)-)
3. λ-*Carrageenan*(-(1→3)-β-D-galactopyranose-2-sulfate-(1→4)-α-D-galactopyranose-2,6-disulfate-(1→3)-) (less abundant than the other forms)

These are flexible molecules forming double-helical zones with potential for thermoreversible gelling at cooling warm polymer solution in the presence of K^+

and Ca^{2+} ions, especially for κ- and ι-carrageenan. The key functions for reactivity with metal ions are anionic sulfate groups.

4.2.2 Biopolymer–Metal Ion Interactions

The preceding section has shown that a number of functional groups can be identified on these biopolymers. These materials are also readily modifiable; this makes the potential for using biopolymers as support for heterogeneous catalysis very large, and this extended choice is further increased by the diversity of metals that can be used for the preparation of supported catalysts. The interaction of metal ions with these biopolymers is a key criterion for the development of these materials, together with the conditioning of the catalyst (changing the diffusion properties of the support, for example, or the activation of the metal). The objective of the supported catalysis is the ready recovery of catalytic metals at the end of the reaction step. The immobilization of the catalytic metals may proceed through different mechanisms, including active mechanisms (based on chemical or physicochemical interactions) or passive phenomena (*in situ* precipitation of catalytic metals inside the porous structure of the support or inclusion of catalytic metal into a gel-like material), depending on the couple (metal and biopolymer). The mechanisms involved in metal binding are mainly illustrated using alginate and chitosan as generic examples of amine-bearing materials and carboxylic acid-bearing materials; more complex materials bearing both carboxylic acid and amine groups, such as proteins, may react with metal ions through similar mechanisms.

4.2.2.1 Ion Exchange Mechanisms

A number of metal ions can be bound to biomass and especially biopolymers by ion exchange mechanisms. Some examples are cited based on the diversity of functional groups present on selected biopolymers: (a) carboxylate and alginate, (b) protonated amine groups and chitosan.

4.2.2.1.1 Alginate-Based Materials

For example, the binding of divalent or trivalent metal cations to alginate materials is frequently attributed to ion exchange mechanisms. Na-Alginate is dissolved in water, dropped into calcium chloride solutions, the cation is exchanged ($Na^+ \leftrightarrow Ca^{2+}$), and the biopolymer forms stable gel spheres.[102–105] As pointed out above, alginate is a heteropolymer constituted of both guluronic and mannuronic units. The guluronic units are the only residues involved in the gelation of alginic acid. The electrostatic interactions between Ca^{2+} and guluronic acid units contribute to form a three-dimensional network, conventionally described as the egg-box model (Figure 4.8). The gelation mechanism can also occur directly with other divalent or trivalent cations (Cu^{2+}, Al^{3+}); in this case, the Na-alginate solution is directly dropped into the metal ion solution.[106–109] However, in most cases, metal ion binding occurs on Ca^{2+} preformed hydrogels.[110–117] Several studies have established that metal binding occurs by ion exchange with carboxylic groups; the protons on carboxylic functions are exchanged with metal cations in a pH range

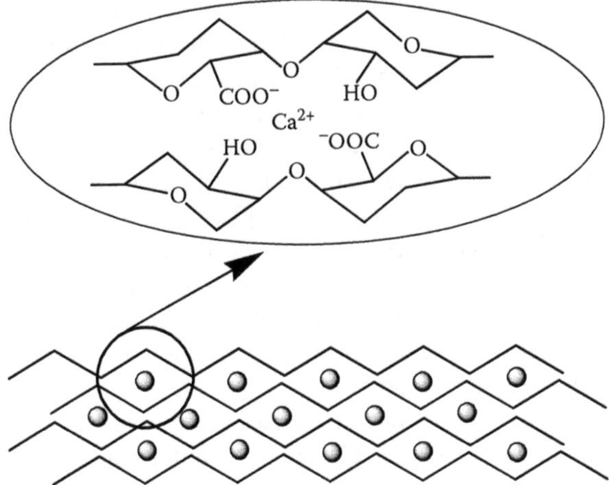

FIGURE 4.8 Gelation of alginate with Ca^{2+}: Egg-box model.

between 2 and 4.[103,113–115] This is consistent with the pK_a values cited above for mannuronic and guluronic acids (3.38 and 3.65, respectively).

In the case of alginic acid beads prepared by gelation of alginic acid into a HCl solution (dropwise addition of the viscous alginic acid solution in a 0.1 *M* HCl solution), Konishi et al. also observed the exchange of three protons for each rare earth [La(III), Nd(III), Sm(III), Dy(III) and Yb(III)] bound to alginic acid beads.[118]

In the presence of lactic acid, which is a complexing agent forming different complexes with rare-earth cations (with significantly different equilibrium constants), the ion exchange mechanism involves both the trivalent cations and the divalent lactate complexes. It is noteworthy to observe that the presence of lactate increased significantly the possibility of selectively separating the different rare-earth cations. Indeed, in the absence of lactic acid, the distribution coefficients were very similar for the five members of the series, while in the presence of an excess of lactic acid, the formation of metal–lactate complexes with significantly different equilibrium constants facilitates the separation of the metals. The complexation of target metal with an appropriate ligand is a powerful tool for enhancing sorption performance and improving sorption selectivity.

A similar impact of metal speciation was observed on the ion exchange properties of protonated alginate beads for Cr^{3+} binding; depending on the pH [and the formation of hydroxo-complexes: $CrOH^{2+}$, $Cr(OH)_2^{+}$] metal binding causes the release of three, two, or one proton.[114] Calcium alginate beads were carried out for the removal of gold and silver from aqueous solutions.[115] Fourier transform infrared (FT-IR) analysis has been used for the identification of functional groups involved in metal uptake; although most of binding occurs through interactions with carboxylic groups, hydroxyl groups also contributed to metal binding. However, x-ray diffraction analysis (together with scanning electron microscopy, SEM) shows that metal nanoparticles were also formed, indicating that a reduction mechanism was also involved in metal removal.[115] At higher pH, ion exchange properties may be also involved in metal binding through an exchange of light metal cations (Na^+ on alginate flakes, Ca^{2+} or Ba^{2+} for alginate beads). The oxidation of alginic acid using potassium permanganate (for the substitution of carboxylic groups onto OH groups at C2 and C3 positions) contributes to its carboxylation. This treatment allows increasing the sorption capacity for lead in relation with the decrease of the pK_a of carboxylic groups on modified alginate: the pH for zero proton condition (pH_{ZPC}) (corresponding to the pH at which the total surface charge becomes zero) decreased from 2.83 (for the studied alginic acid sample) to 2.31 (for the modified biopolymer).[119–121] Comparing pH variations (i.e., released protons) and lead binding, it was possible to conclude that about 2 mol of protons were displaced when 1 mol of lead ions was adsorbed, consistently with the egg-box model.

Chen et al. and Chen and Wang observed an increase in metal recovery (Pb^{2+}, Cu^{2+}, and Zn^{2+}) with increasing the pH and the simultaneous release of Ca^{2+}.[122,123] Chen et al. used spectrometric analysis and molecular modeling to propose a binding mechanism of Pb^{2+} and Cu^{2+}.[122] They concluded that lead ions are removed by pure ion exchange mechanism with carboxylic groups (exchange of calcium ions) while for Cu^{2+} there is a complementary mechanism to ion exchange consisting of coordination with other groups present on the biopolymer.

The carboxylic groups present on alginate (mannuronate and guluronate units), and more specifically their relative proportions, control the ability of the biopolymer for gelling and also its affinity for metal ions and the selectivity properties.[124] While mannuronic acid residues do not develop any selectivity for the separation of metals from binary mixtures such as Ca–Mg, Ca–Sr, Sr–Mg, and Co–Ca, the guluronic acid units show a more marked selectivity (from 150 for the couple Sr–Mg to less than 0.2 for Co–Ca). The selectivity is brought by guluronic acid, probably due the different arrangements of polymer chains. Indeed, several binding groups are involved in the uptake of metal cations, and there is an appropriate distance required between the reactive groups for optimized interactions and this distance depends on the size of metal ions.[125] However, in the case of copper binding, the sorption was less affected by the relative proportions of mannuronic acid and guluronic acid; but Haug and Smidsrød observed that when the structure of the polyuronides allows a preferential binding of Ca^{2+} compared to Mg^{2+}, Cu^{2+} is also bound preferentially, probably by the same mechanism.

There are other methods to improve the binding selectivity for given metals, including template formation.[126] A new alginate gel (Cu-alginate) was prepared by ion imprinting technology in which the copper ion was used as a template ion. The templated gel has a high selective adsorption capability with copper ion in the presence of non-heavy-metal ions such as K^+, Na^+, and Ca^{2+}, and heavy metal ions such as Ni^{2+} and Cd^{2+}. The selectivity and adsorption capacity of Cu-alginate gel were superior to that of Ca-alginate gel in which the calcium ion was used as a templating ion.

Among the important parameters for metal binding are pH (as cited above), ionic strength,[117] and temperature.[121] Increasing the ionic strength of the solution generally decreases the removal efficiency as well as sorption kinetics.[117]

Mass-transfer resistance into the polymer material is generally the controlling step in the binding kinetics. This important parameter should be taken into account not only for metal binding but also for supported catalysis. The high water content of the beads means that the volumetric sorption capacity is somewhat much lower than that of conventional resins; drying the material causes the irreversible collapse of the structure, which in turn drastically reduces mass-transfer properties of the original material. This is one of the reasons that motivated the development of different polymer conditioning (this will be discussed later), including deposition on inorganic materials characterized by a high specific surface area. Several studies have also focused on the preparation of composite materials, that is, in association with chitosan[102,127] or with cellulose.[128]

4.2.2.1.2 Chitosan-Based Materials

The high reactivity of chitosan amino groups for metal ions is well documented.[129,130] This reactivity may proceed through different mechanisms depending on the metal and the pH of the solution. While metal cations are generally sorbed on chitosan in a near-neutral solution by chelation on the free electronic doublet of nitrogen, the protonation of amine functions in acid solutions makes possible the electrostatic attraction of anionic species (metal anions,[131] anionic dyes[132,133]).

Because chitosan is soluble in most acids, it requires a stabilization treatment prior to metal binding, using, for example, cross-linking[134–137] or controlling the pH with sulfuric acid when this acid is compatible with metal chemistry and the formation of adsorbable species (see below). The cross-linking treatment may operate on amine functions (using dialdehydes such as glutaraldehyde[135,138,139] or cyclodextrin,[140–142] by Schiff base reaction), or on hydroxyl groups (epichlorohydrine or di-glycidyl ether derivatives, and similar group) after protecting amine groups or directly on amine functions.[143,144] The reaction of the cross-linking agent with amine functions may immobilize amine functions. Actually, in the case of Pd sorption, increasing the concentration of glutaraldehyde in the cross-linking bath (aldehyde:amine ratio) had a limited impact on Pd uptake. This means that the ion exchange mechanism is not very affected by steric hindrance around the reactive group and that the availability of amine groups is not the controlling step, contrary to the case of chelation mechanism (see below).

The pK_a of amine groups on chitosan polymer strongly depends on both the degree of dissociation and the degree of deacetylation (Figure 4.9).[84] Sorlier et al. investigated a series of chitosan samples partially reacetylated and show that when the degree of deacetylation (DD) is higher than 80%, the charge density increased with decreasing the DD and the pK_a tends to a constant value (pK_0) with increasing the degree of neutralization close to 6.4. For a DD close to 20%, the pK_a is almost constant and just slightly affected by the degree of dissociation; the system behaves as a simple electrolyte.[84] For 25% < DD < 75%, more significant changes are observed: at low dissociation, the pK_a strongly increased with decreasing the DD, but decreased monotonously with increasing the DD. At full neutralization, the pK_a varies between 6.6 and 7.3 (gradually increasing with decreasing the DD). For most of commercial chitosan samples, whose DD is

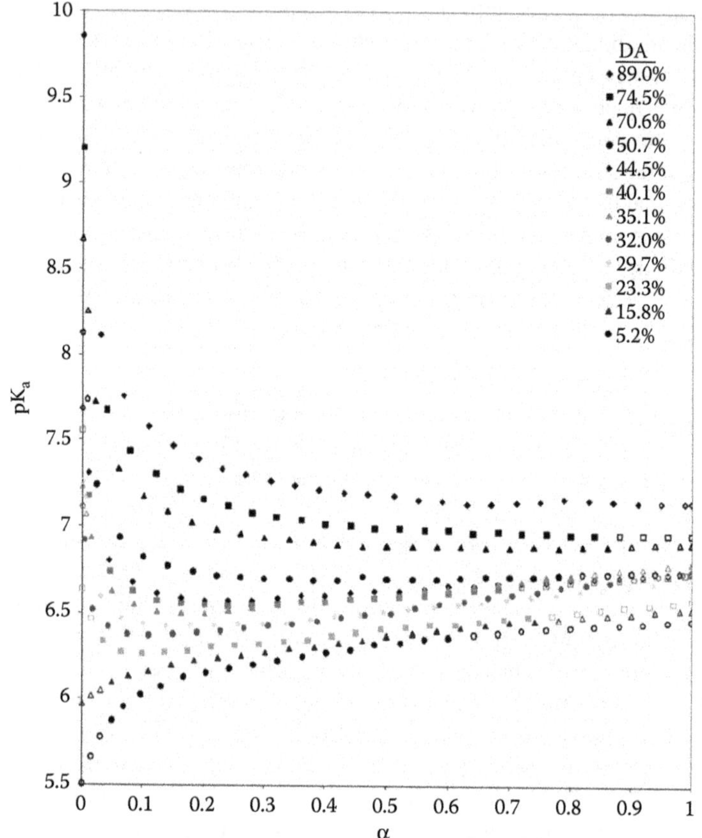

FIGURE 4.9 pK_a of amine groups of chitosan in function of the degree of dissociation and the degree of deacetylation. (Reprinted from P. Sorlier, A. Denuzière, C. Viton, and A. Domard, *Biomacromolecules*, 2, 765–772. With permission. Copyright 2001, American Chemical Society.)

close to 80 to 90%, the pK_a at full neutralization can be considered varying between 6.3 and 6.5; at low dissociation degree, the variations are more marked (between 5.8 and 6.3). This means that when the pH is decreased below 4, a large majority of amine groups are protonated and available for the binding of anionic species: dyes[145–150] and metal anions.[134,151–163]

Actually, in most cases, with raw chitosan, the binding efficiency reaches a maximum before it decreases at low pH. This decrease may be explained by a strong competitor effect of the anions brought about by the dissociation of the acid used for pH control, or present in the matrix of the solution. The optimum pH is frequently found around pH 2 to 4.[82,131,134,152,157,158,163–167] Below this limit value, usually a large excess of competitor anions limits sorption efficiency. A number of chitosan derivatives have been developed to limit the impact of these competitor anions, including the grafting of new functional groups such as sulfur

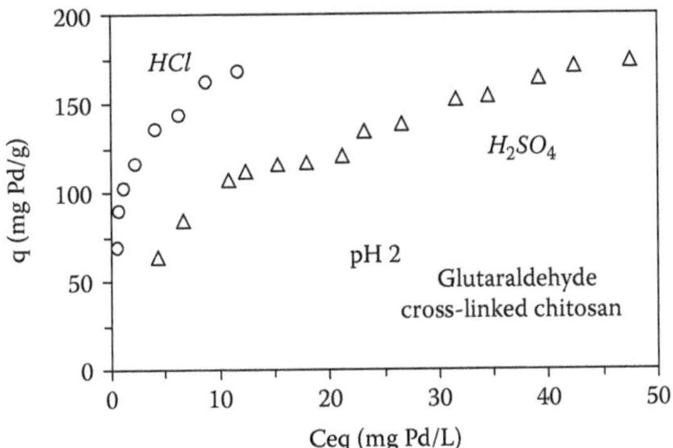

FIGURE 4.10 Influence of the acid used for pH control (pH 2) on Pd sorption using glutaraldehyde cross-linked chitosan. (Reprinted from M. Ruiz, A.M. Sastre, and E. Guibal, *React. Funct. Polym.*, 45, 155–173, 2000. With permission. Copyright 2000, Elsevier.)

functions.[159,160,168,169] This electrostatic attraction may occur by direct interaction with free metal anions, but that mechanism may be also involved in the sorption of metal complexes, as a result of the interaction of metal cations with ligands in the solution.[143,144,170,171] Actually, the formation of complexes, with ligands or OH, influences the speciation of metals ions,[172] and thus the sorption efficiency and uptake mechanism. In the case of vanadium and molybdenum uptake, the study of the influence of pH and total metal concentration has shown that the key parameter for increasing metal sorption capacities is the formation of decavanadate and polynuclear molybdate species.[155,164] Below a limit value of concentration, the sorption capacity remains negligible; above this limit value, the sorption capacity strongly increases. The limit value depends on the pH and corresponds to the beginning of the formation of polynuclear hydrolyzed species. In the case of palladium and platinum, the shape of sorption isotherm is affected by the type of acid used for pH control (close to pH 2). With sulfuric acid, the speciation of metals is displaced: the concentration of chloride ions is not sufficient to form adsorbable chloro-anionic species and the sorption isotherms show a weakly favorable profile (Figure 4.10). On the other hand, when the pH of the solution is controlled with hydrochloric acid, the tetrachloropalladate species and the hexachloroplatinate species are preferentially formed; the sorption is very favorable (it should be considered as a quasi-irreversible sorption isotherm, characterized by a sharp initial curve followed by the formation of a plateau at low residual metal concentration).[134] In platinum or palladium solutions, whose pH was controlled to 2 with sulfuric acid, the sorption capacity was increased by adding small amount of sodium chloride (up to 0.05 *M*); above, the sorption capacity tended to significantly decrease (Figure 4.11). At low chloride concentration, the speciation of platinum and palladium is displaced toward the formation of chloro-anionic species, which are favorably adsorbed by protonated amine groups. At high chloride concentration, there is a strong competition of chloride anions that reduces the binding of chloro-anionic metal species.[168]

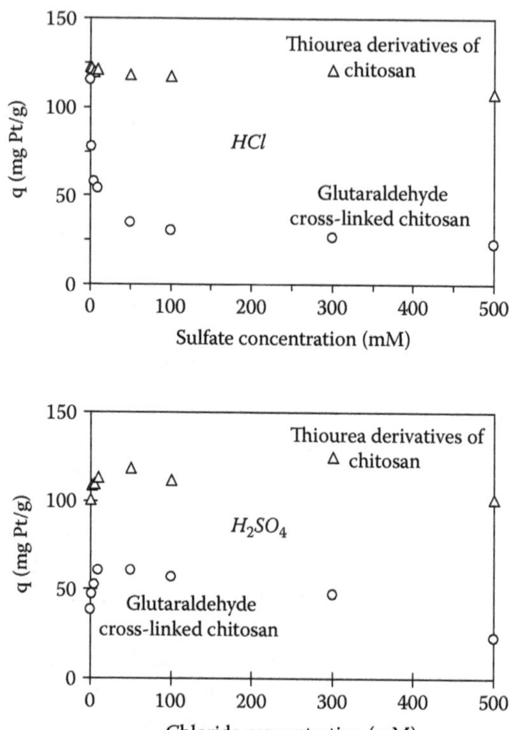

FIGURE 4.11 Influence of the acid and salt addition (NaCl, Na_2SO_4) on Pt sorption using glutaraldehyde cross-linked chitosan and chitosan modified by thiourea grafting. (Reprinted from E. Guibal, M. Ruiz, T. Vincent, A. Sastre, and R. Navarro Mendoza, *Sep. Sci. Technol.*, 36, 1017–1040, 2001. Copyright 2001, Taylor & Francis Group.)

This property of electrostatic interaction between protonated amine groups and anions has been used for the gelation of chitosan.[154,173–175] For example, in the case of molybdate, polynuclear species may interact with several amine groups from the same chain or different chains, strengthening the structure of the polymer and preventing its dissolving in moderate acidic solutions. Polyoxoanions and polyphosphate anions are very efficient for the formation of these multiple bonds.[154,175,176]

The ionic strength and the presence of large concentrations of competitor anions contribute to decreasing the binding capacity of metal anions on chitosan-based materials.[134,152,168] This should be taken into account in the preparation of supported catalysts.

4.2.2.1.3 Miscellaneous Materials

Cellulose and starch have been frequently carried out for the binding of metal ions. However, the poor reactivity of functional groups present on the biopolymers made the sorption capacities considerably lower than those of conventional materials, and it is generally necessary to modify the raw material by grafting reactive groups.[93–95,177–180] Kabay et al. investigated the sorption of a number of metal ions, including Cd(II), Cu(II), Co(II), Zn(II), Pb(II), Cr(III), V(IV), and V(V), using two cellulose derivatives obtained by grafting of (a) phosphate groups (ester-linked orthophosphoric acid cellulose, cellulose–O–$PO_3H^-.NH_4^+$), and (b) diethylaminoethyl groups (DEAE-cellulose, or cellulose–O–$(CH_2)_2N(C_2H_5)_2$,HCl form).[179] The kind of grafted group strongly influences metal sorption and optimum pH. With DEAE-cellulose, the sorption of Cd(II), Cu(II), Co(II), Zn(II), Pb(II), and Cr(III) ions is very low (below 10% for a sorbent dosage of 1 g L^{-1} and metal concentration: 50 mg L^{-1}) and weakly affected by the pH. Metal sorption is strongly increased when increasing the pH and sorption efficiency is greater than 90% at pH higher than 3 and 5 for V(V) and V(IV), respectively. On the other hand, the phosphate-cellulose material is very efficient at removing Cd(II), Cu(II), Co(II), Zn(II), and Pb(II) at pH greater than 3; for Cr(III) sorption, a sharp optimum is found close to 3, V(IV) recovery is optimum between pH 2 and 4, and V(V) binding reaches a maximum above pH 2. The differences are related to the formation of anionic species for vanadium at pH greater than 3 [for V(V)] and 4 [for V(IV)], highly adsorbable on DEAE-cellulose, but poorly adsorbable on phosphate derivative of cellulose.

Simkovic et al. prepared a number of basic ion exchanger derivatives of starch by grafting ammonium groups (through reaction with epichlorohydrin in the presence of NH_4OH)[93–95,181] and imidazolium derivatives.[180] Matsumoto et al. described the synthesis of a series of starch derivatives bearing cation exchange or anion exchange groups.[182] These modifications significantly increase the ion exchange capacities instead of raw materials.

4.2.2.2 Chelation Mechanisms

The theory of hard and soft acids and bases (HSAB) has been defined by Pearson.[183] It describes the ability of ions to interact or enter into coordinate bonding with other ions or with ligands. This ability depends on the availability of their outmost electrons and empty molecular orbitals. This must be considered on top of any electrostatic effects due to ion-ion, ion-dipole, and ion-higher multipole interactions. This last type of effect is governed primarily by the charge and size of the ion. The first type of effect can be described by means of the softness parameters and the Lewis acid/base parameters of the ions.[184] The HSAB concept describes the capacities of ions to prefer ligands of the same kind (soft-soft and hard-hard) to those of the different kinds when forming coordinative bonds. Softness of ions generally correlates to their polarizability, and hardness with their electrostatic field strength. This concept can be helpful for predicting the functional groups the most appropriate for efficient binding of given metal ions (Table 4.3).

4.2.2.2.1 Alginate-Based Materials

Although most of the studies on the interactions established between alginate and metal ions refer to ion exchange mechanisms, especially in acidic solutions (proton

TABLE 4.3
HSAB Theory: Affinity Series

Hard Acids	Borderline	Soft Acids
Li^+, Na^+, K^+, Be^{2+}, Mg^{2+}, Ca^{2+}, Sr^{2+}, UO_2^{2+}, VO^{2+}, Al^{3+}, Sc^{3+}, La^{3+}, Cr^{3+}, Mn^{3+}, Si^{4+}, Ti^{4+}, Zr^{4+}, Th^{4+}, …	V^{2+}, Cr^{2+}, Mn^{2+}, Fe^{2+}, Co^{2+}, Ni^{2+}, Cu^{2+}, Zn^{2+}, Pb^{2+}, Bi^{3+}, …	Cu^+, Ag^+, Au^+, Tl^+, Ga^+, Cd^{2+}, Hg^{2+}, Sn^{2+}, Tl^{3+}, Au^{3+}, In^{3+}, Pd^{2+}, Pt^{2+}, …
Preference for ligand atom:		
N >> P		P >> N
O >> S		S >> O
F >> Cl		I >> F
Ligand classification:		
F > O > N _ Cl > Br > I > S		S > I > Br > Cl ≈ > O > F
$OH^- > RO^- > RCO_2^-$		
$CO_3^{2-} >> NO_3^-$		
$PO_4^{3-} >> SO_4^{2-} >> ClO_4^-$		

exchange or calcium exchange with divalent metals), chelation is also cited as a possible binding mechanism, generally at less acidic pH.[106,185–188] DeRamos et al.[186] compared the binding of a series of alkaline-earth metals (Mg^{2+}, Ca^{2+}, Sr^{2+}, Ba^{2+}) with that of a series of lanthanide metals (La^{3+}, Pr^{3+}, Nd^{3+}, Eu^{3+}, Tb^{3+}); using nuclear magnetic resonance (NMR) analysis and molecular modeling, they observed significant differences in the affinity of alginic acid for these metals and the mechanism of binding. Although alkaline-earth metal ions exclusively bind to guluronic acid residues on alginate (with an affinity that increases with the ionic radius of the metal), in the case of lanthanide metal ions, the metal ions also bind to mannuronic acid residues despite a marked preference for guluronic acid groups (and the affinity increases with charge density).

For alkaline-earth metals, the key parameter seems to be the secondary structure of alginic acid. Indeed, the diaxial linkage pattern of guluronic acid residues results in buckled structure with deep cavities for metal ion, while mannuronic acid residues that are linked diequatorially result in a ribbon-like structure with shallow cavities.[186] Additionally, the type of oxygen atom involved in metal uptake changes: one ring oxygen on each dimmer for mannuronic dimer, and hydroxyl oxygen on guluronic dimer. Hydroxyl oxygen is a stronger Brønsted-Lowry base than ether oxygen; this may explain the greater affinity of guluronic acid for Ca^{2+} ions (and related metals). The smaller distance between negatively charged carboxylate groups in the buckle-shaped guluronate dimmers results in a higher charge density and a stronger interaction with positively charged counterions. Because of the flatter structure of mannuronic acid dimmers, the block charge density is considerably lower than for guluronic units, limiting the affinity of the material for alkaline-earth metals. The interaction increases at increasing the ionic radius the distance between alginate oxygen and metal.[186] The additional water molecules of hydration in the inner sphere of the lanthanides affect their packing into guluronic acid and mannuronic acid blocks, especially compared to dehydrated alkaline-earth metal ions. Binding results are much more influenced by the charge density of the metal than by the geometry of alginate chain. The strength and selectivity of cooperative binding are determined by two parameters: the size of metal ions (and its hydration sphere) and the ease of packing of the alginate chains around the metal ion (depending on the distribution of guluronic and mannuronic acid units).[186]

To prevent the aggregation of alginate chains, Lamelas et al. worked with low polymer concentrations and investigated the influence of pH, ionic strength, and the metal:alginate (or metal:carboxylic group) ratio on Pb and Cd binding, with the final objective of separating the effects of intrinsic chemical binding from the electrostatic and macromolecular (i.e., conformational) contributions.[187] Metal complexation by alginate is influenced by pH through competition due to the protonation of binding sites (pK_a close to 3.4) and the modification of the overall molecular charge. While Pb sorption is not significantly affected by pH change between 4 and 8, Cd sorption increases with pH between 4 and 5.5, before stabilizing. The ionic strength may have a screening effect on molecular charges of polyelectrolytes, which in turn affects their binding affinity; even with a 100-fold increase of ionic strength, Pb binding was not drastically reduced while Cd uptake suffered a threefold decrease. This can be attributed to a greater effect of electrostatic mechanism for Cd sorption than for Pb uptake. High metal:ligand ratios affect macromolecular conformations because of the formation of metal bridges and aggregation phenomena (superstructure), corresponding to the egg-box model. Lamelas et al. used the nonideal competitive adsorption isotherm combined with the Donnan approach (NICA-Donnan); they concluded that the Coulombic contribution to metal binding was quite higher for Cd binding on alginate (close to 50%) than for Pb uptake (about 15%). Metals can be found in three forms in alginate gels (with different relative proportions): dissolved in the water-phase of the gel, bound within the Donnan gel, and specifically bound to carboxylate groups.[187]

Ferreira and Gschaider compared the binding of Pb^{2+} and Hg^{2+} using pectic acid. They used molecular orbital calculations for the modeling of the interactions between carboxylic groups of galacturonic acid residues and metal ions.[189] They concluded that intermolecular chelation, involving the same $-COO^-$ groups, is preferred over intramolecular chelation. The sorption is controlled by hydrolysis mechanisms and by water coordination.

4.2.2.2.2 Chitosan-Based Materials

The amino groups present on chitosan have been used in acid solutions for the binding of metal anions; however, in near-neutral solutions (where a number of amine groups are free), the free electronic doublet of nitrogen is responsible for the binding of a number of metal cations, and more specifically transition metal cations.[130]

The key parameter for the binding of metal cations on chitosan is the pH of the solution: the competition of protons at acidic pH may considerably limit sorption performance. Another important parameter is the degree of acetylation and the availability of amine groups. Indeed, the cross-linking of chitosan with glutaraldehyde in order to increase its stability in acidic solutions causes a drastic decrease of sorption properties; this is a confirmation of the involvement of free amine groups in metal chelation.[136,190,191]

The mechanisms involved in metal binding remains a controversial subject. For example, the structure of the very simple complex between copper and chitosan opposes two concepts: (a) the "bridge" model, and (b) the "pendant" model. In the bridge model, metal ions are bound with several amine groups from the same chain or from different chains, via inter- or intramolecular complexation.[192–195] In the pendant model, the metal ion is bound to an amino group in a pendant fashion[196–198]: the coordination sphere of copper is completed at the fourth site by either a water molecule or the –OH group at the C3 position.

These studies have shown that the sorption mechanisms and the species adsorbed can be significantly modified by the experimental conditions (pH, metal concentration, metal:ligand ratio). The coordination number (ligand:metal molar ratio) varied from 1 at pH 5.3 to 2 at pH 5. The change in the conformation of the polymer (dissolved state vs. solid state, stiffness of polymer chain controlled by the molecular weight, the degree of deacetylation, the distribution of acetyl groups, etc.) may change the coordination mechanism and the kind of complex formed between copper and amine groups.[193] Most of the studies performed with chitosan in solutions used oligomers or polymers; indeed, it is generally accepted that the monomer is not efficient at complexing copper. Shahgholi et al. comment that copper forms a strong chelate with tetrasaccharide,[199] while Rhazi et al. show that a higher stability of Cu-chitosan complex is obtained with a higher degree of polymerization (DP); they propose a DP of 6.[193] This conclusion confirms that several glucosamine units are involved in the binding mechanism, probably through the contribution of hydroxyl groups of vicinal units together with amine group of a given monomer. Ferreira and Gschaider confirmed, using molecular modeling, that Pb^{2+} and Hg^{2+} bind to amino groups from different chains (intermolecular chelation is preferred to intramolecular chelation).[189] It is noteworthy that chitosan does not bind alkaline and alkaline-earth metals due to the absence of *d* and *f* unsaturated orbitals (unlike transition metals).[136] This means that chitosan will be selective for transition metals over common nontransition metals. However, the binding of alkaline and alkaline-earth metals can be achieved using (a) phosphorylated derivatives of chitosan,[200–204] or (b) by formation of a ternary complexes.[205,206] Calcium was bound to chitosan after ion pair formation with two carboxylate groups of undecylenate before the ion pair complexes with the amine groups of chitosan[205]; in the case of strontium, the formation of an ion pair with carbonate allowed binding the metal through formation of ternary complex.[206]

A number of derivatives have been developed for improving metal sorption properties by grafting functional groups such as carboxylic functions, phosphonic functions, sulfur groups, pyridyl groups, and so forth. These modifications may contribute to (a) increased sorption capacities (density of sorption sites), (b) increased pH range for efficient metal sorption (new functional groups for which optimum pH range is shifted), or (c) reduced impact of environmental parameters on metal binding (changing the speciation of metal in solution or changing the binding mechanism). Comprehensive reviews of the modifications brought to chitosan for the design of chelating resins have been published.[83,129,130,207]

Lasko and Hurst commented on the influence of metal speciation, due to pH changes and the presence of ligands in the solution on the binding of silver.[171] The sorption of copper was modified in terms of sorption capacities and pH range by the presence of ligands such as ethylene diamine tetraacetic acid (EDTA).[143,144] Guzman et al. observed that sorption occurs in the presence of citrate copper through electrostatic attraction of anionic complexes instead of chelation with a shift in optimum pH towards acidic solutions.[82]

4.2.2.3 Reduction and Precipitation Mechanisms

The structure of these biopolymers (the presence of reducing ends on polymer chains) and the redox properties of the metals may cause secondary reactions involving changes in the metal valence. This was observed in the case of chitosan sorbents.[208,209] The reduction mechanism was observed in the case of uranium binding to chitosan; the difference in potential was measured between two compartments containing uranyl solutions and chitosan connected by a conducting agar bridge.[208] The difference in potential became significant when one of the compartments was submitted to ultraviolet irradiation. Bubbling air through the compartment decreased the photoreduction effect due to the reoxidation effect of the oxygen. The reducing effect of chitosan, completed by a photochemical effect, is not very strong but cannot be neglected. Yonozewa et al. observed the tendency of gold deposited onto chitosan membranes to form a gold mirror when submitted to photoirradiation: the photoirradiation, combined to chitosan, contributes to reduce gold ions to metal colloids.[210,211] By surface reflectance analysis, it was also possible to observe the partial reduction of molybdate species adsorbed on chitosan gel beads.[156] X-ray photoelectron spectroscopy (XPS) analysis has been used to determine the oxidation state of metal ions after sorption on chitosan, and more specifically to characterize metal reduction.[208] However, the reducing activity strongly depends on (a) the oxidation potential of the metal (correlated to the normal redox potential scale), and (b) the structure of the polymer (glutaraldehyde cross-linking significantly increases the reducing effect). Chromate is almost completely reduced on the sorbent, while molybdate is only partially reduced, particularly on the external surface of chitosan beads due to a combined effect of the reducing ends of polymer chains and a photochemical effect.

A similar mechanism of metal reduction was observed and used for the recovery of some base metals (including lead and chromium) or precious metals (including gold, palladium, and platinum) on tannin-based polymers.[212–219] Tannins are natural polyphenols widely distributed in roots, barks, and stalks. Actually, their structure is very complex and they can be classified into three groups: hydrolysable tannins (tannic acid, for example), condensed tannins (polymerized products of flavan-3-ols and flavan-3,4-diols, such as black wattle tannin), and complex tannins (such as bayberry tannins).[212] Tannins are very efficient for precipitating proteins and metal ions; this precipitating effect is due to the presence in close vicinal positions of phenolic adjacent groups, which can form stable complexes with metal ions. The solubility of tannins in water requires immobilizing the material onto supports through cross-linking on natural or synthetic polymers.[212] The immobilization of tannins on collagen fibers proceeds in a two-steps procedure: (a) first immobilization of tannins on collagen fibers through hydrophobic and hydrophilic bonds, followed by (b) covalent bonding through cross-linking reaction.[212] These tannin-collagen composites can adsorb uranyl ions in the pH range 5 to 8 by chelation on hydroxyl and galloyl groups of tannin.[216] Using the same sorbent gold recovery is enhanced by acidic solutions (below pH 5).[215] A more complex phenomenon is observed in the case of platinum and palladium sorption with coexistence of chelation and oxido-reduction mechanisms, leading to changes in the coordination compounds along the sorption process.[212] Nakano and coworkers prepared another kind of tannin derivative by polycondensation of tannin in alkaline solutions with formaldehyde.[213,214,217–219] They identify different sorption mechanisms, depending on the type of metal. For Cr(VI) sorption at the optimum pH (i.e., pH 2), four steps have been identified: (1) esterification of chromate with tannin, (2) Cr(VI) reduction, (3) formation of carboxyl groups (by oxidation of tannin molecules), and (4) ion exchange of reduced Cr(III) with carboxyl and hydroxyl groups.[218,219] In the case of lead recovery, three mechanisms are involved: ion exchange, hydrolytic adsorption, and surface precipitation.[217] In the case of precious metals [Au(III), Pd(II)], the simultaneous reduction of the metal [to Au(0) and Pd(0), respectively] and oxidation of hydroxyl groups on tannins are responsible of the deposition of metals at the surface of the sorbent.[213,214] Torres et al. also observed nanoprecipitation of silver and gold in the presence of alginate.[115]

A number of catalysts prepared by metal deposition onto biopolymers have been prepared by a mixed complexation-reduction-precipitation process. The process will be described in detail below. The metal is mixed with the support and a solvent (generally ethanol) before the mixture is refluxed. The *in situ* reduction results in the immobilization of catalytic metal in a reduced chemical state.

4.2.2.4 Encapsulation Mechanism

Encapsulation has been frequently used for the preparation of a number of supported catalysts based on polymers. Köckritz et al. described a new process for the entrapment of transition metal complexes using both synthetic polyelectrolytes and biopolymers (including alginate, carrageenan-cellulose derivatives, and pectinate).[220] The homogeneous catalyst solution is mixed with the polyelectrolyte solution; the homogeneous mixture is thus deposited on a polypropylene or polyethylene foil before being dried under controlled atmosphere (air or argon). The catalyst is conditioned under the form of lens-shaped particles with strong electrostatic interactions between the catalyst and the polyelectrolyte. Indeed, the polyelectrolyte is selected for electrostatic compatibility, depending on the charge of catalyst complex; metal leaching is significantly decreased due to both physical entrapment and electrostatic interactions.

The encapsulation of the catalytic metal can be also obtained by a coprecipitation process consisting of the preparation of a mixture of polyelectrolyte and catalytic metal solutions, which is dropped into a coagulating bath. The coagulating agent obviously depends on the biopolymer: the gelling of metal-chitosan acid solutions can be obtained by neutralization in alkaline solutions.[221,222]

4.2.3 Conditioning of Biopolymer Support

In addition to chemical versatility, which allows (a) increasing sorption capacities, (b) enhancing binding selectivity, or (c) improving polymer stability, it is possible to physically modify these biopolymers for preparing new conditioning.[129] These physical modifications tend to (i) facilitate their use, (ii) diversify their mode of application, or (iii) enhance the diffusion properties of these materials. Indeed, raw materials are generally characterized by poor porosity, which strongly limits mass transfer and gives weak kinetic performance. Weak diffusion properties may result from different causes, including hydrophilic or hydrophobic characteristics,[223] pore size, and crystallinity.[158,167,224,225] These diffusion restrictions may cause limitations in the binding of catalytic metals (catalyst preparation) and also on the diffusion of reaction substrates and products. The improvement of mass-transfer characteristics of the biopolymer supports is thus a key parameter for the development of competitive systems. The biopolymers can be used in dissolved state, solid, or gel forms. The process usually consists of two steps: (a) biopolymer dissolving in appropriate solvent, followed by (b) polymer gelling (casting, ionotropic gelation, etc.) in an appropriate coagulation or neutralization batch. A third step may be involved as a post-treatment for the control of drying procedure (freeze-drying, drying under CO_2 supercritical conditions).[226,227] An alternative for increasing the specific surface area may consist in depositing the biopolymer on a high specific surface-area porous material.

4.2.3.1 Dissolved State

Many of these biopolymers have been tested for metal binding in the dissolved state. The interaction of the biopolymer with metal ions may induce its gelation and metal recovery as a precipitate or a gel. However, the gelation is controlled by the pH of the solution and the concentrations of biopolymer and metal (and, more specifically, their molar ratio); in some cases, the experimental conditions are not favorable enough for complex recovery in the solid state. This forms the basis of metal recovery by polymer-enhanced ultrafiltration (PEUF).[228–231] The process counts on the chelation or ion exchange properties of the biopolymer for the binding of metal ions and the separation effect of an ultrafiltration membrane that retains the loaded macromolecules. Figure 4.12 shows the ultraviolet (UV) spectra of Pd solutions (at 12.5 mg L^{-1} concentration), chitosan and PEI (polyethyleneimine, synthetic polymer) (at 100 mg L^{-1} concentration), and

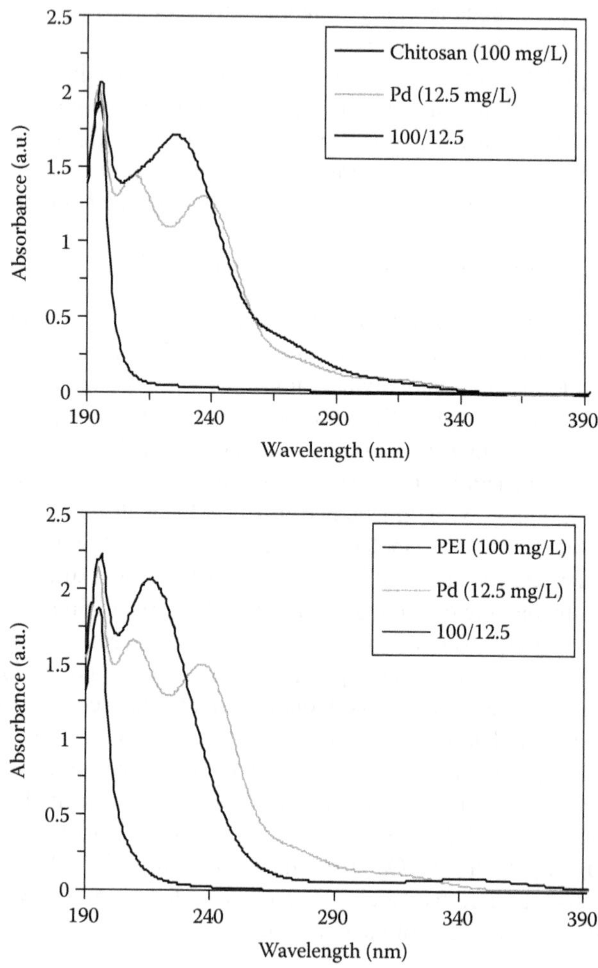

FIGURE 4.12 Pd interactions of PEI and chitosan dissolved in HCl solution (pH 2) characterized by UV spectra (polymers alone, Pd salt alone, and mixed [bold spectra]).

their mixtures at pH 2 (in HCl solutions). This figure shows that the interaction of Pd with the polymers induces a significant change in the spectrum of metal ions (chloropalladate species). Two peaks at 209 nm and 236.5 nm wavelengths characterize chloropalladate species. The intensity of these peaks considerably decreases with the presence of polymer, and a new peak, representative of the interaction of the polymer with metal, appears at 224.5 nm for chitosan and 216.5 nm for PEI. The retentate, resulting from the ultrafiltration of loaded macromolecules, gives the same UV-spectra polymer-Pd complex (not shown). The "complex" (in a broad sense, regardless of the kind of interaction established between the polymer and metal ions, i.e., ion exchange or chelation) can be directly used for catalysis or for the preparation of the precursor of the catalyst (when required, a supplementary reduction step is added). In some cases, the presence of the polymer allows stabilizing metal ions in solution preventing their precipitation, or the agglomeration of metal particles.

It is interesting to observe that the molar ratio between solute and amine groups of chitosan, for example, may be significantly increased when using chitosan in a dissolved state, indicating that the reactive groups are much more available and accessible for interaction with solute molecules than when the polymer is used in a solid state.[232,233] This may be explained by the "opening" of the polymer structure: in the solid state, the supramolecular structure of the polymer results from the hydrogen bonds between polymer chains decreasing the availability of reactive groups. Polymer dissolution breaks these hydrogen bonds, and reactive groups are available for interacting with solute molecules. Additionally, polymer dissolution reduces its residual crystallinity and increases the accessibility, which can be limited by resistance to intraparticle mass resistance in solid-state adsorbent.

Nanoparticles are attracting growing interest due to (i) large surface area to mass ratios, (ii) high surface energy, and (iii) many surface defects involving high catalytic activity. The tendency of these nanoparticles to aggregate or precipitate requires using protective colloid or polymer coating during the synthesis of nanosized metal catalysts.[36,234–240] The polymer solution is mixed with metal precursors under heating until the color of the solution changes; after cooling, nanoparticles (size smaller than 30 nm) can be recovered by centrifugation.

The main drawback of dissolved-state biopolymers as support for heterogeneous catalysis (colloid-supported catalysis) is due to the difficult separation of colloids at the end of the reaction, requiring, for example, microfiltration units.

4.2.3.2 Composite Materials (Biopolymer Deposition on Organic and Inorganic Surfaces)

A number of processes have been developed for increasing the accessibility to reactive groups by depositing biopolymer thin films at the surface of materials of high specific surface area. Basically, the procedure consists of the impregnation of the porous material with the biopolymer solution or the amino acid, followed by a post-treatment consisting of a drying step,[222,241–244] a coagulation and neutralization process,[245,246] or a cross-linking treatment.[247,248]

The case of the coagulation and neutralization procedure can be illustrated by the case of chitosan. Wei et al. prepared an acetic acid solution of chitosan mixed with silica particles[246]; the dropwise addition of NaOH concentrated solution (to reach pH 13) causes the deposition of chitosan at the surface of silica. After rinsing (to pH 8), the support can be dried and sieved to an appropriate size for metal binding or deposition. Zhou et al. proceeds the same way for the immobilization of chitosan on magnesium oxide.[249] The treatment can be completed by a cross-linking step using, for example, glutaraldehyde.[250] Tang et al. used the solvent precipitation procedure for depositing carboxymethyl cellulose onto silica for the preparation of a Pt-supported catalyst.[251] Wei et al. also developed a precipitation procedure for the deposition of alginic acid at the surface of silica: a volume of 1 *M* HCl solution was added to a mixture of alginic acid (completed with amino acids) and silica.[246] The addition of acid causes alginate precipitation at the surface of silica (entrapping amino acid moieties).

Polymer precipitation at the surface of the composite material can also be obtained by dropping the mixture (silica-chitosan or a silica-chitosan-synthetic polymer, such as polyacrylic, polymethacrylic, polyethyeneimine) into a solvent (acetone, for example).[252] Huang et al. also used the solvent precipitation procedure for the precipitation of chitosan-silica materials for the preparation of nanosized Pd catalysts.[253] Kucherov et al. immobilized a copper-chitosan complex onto silica gel using the precipitation procedure[222]: a homogeneous copper-chitosan solution is mixed with silica gel prior to being transferred into a 0.05 *M* NaOH solution for precipitation of the complex at the surface of the matrix.

Huang et al. grafted starch onto silica by activation of silica with ethyl silicate in ethanol-water-sulfuric acid solution (15-200-15 v/v).[247] Activated silica was used for the immobilization of starch (in toluene solvent). This support was used for Pt immobilization to prepare hydrogenation catalysts. In the case of nonporous materials (glass beads, for example), a preactivation of the support is required, attaching reactive groups at its surface.[244,250,254]

The drying procedure was essentially used for the immobilization of alginate,[241,245,246] starch,[241,255] methylcellulose,[241,243] protein (gelatin,[241,242,244,256] and casein[244,257]) materials on silica,[241,243–245,255,257] and other mineral materials (such as zeolite[242,256]) or synthetic polymers (such as polysulfostyrene[258]). The biopolymer solution in water, which can be completed by addition of an amino acid, is mixed with silica under heating in the range 50 to 60°C. The slurry is dried, ground, and sieved.

4.2.3.3 Gel Beads

Gel-bead conditioning has been mainly developed on chitosan-based and alginate-based materials. Typically, the procedure consists of dissolving a biopolymer in an appropriate solvent (acidic solution for chitosan, water for alginate) before the viscous solution is distributed dropwise in the coagulating bath. The coagulation may consist of (a) a neutralization step using concentrated alkaline solution (generally NaOH) for chitosan,[259–261] or (b) the ionotropic gelation of biopolymer drops, using solutions of calcium chloride solution,[185,262–264] or hydrochloric acid for sodium alginate,[118] or metal salts such as copper chloride for sodium alginate,[265–267] and molybdate salts[154,174,268] or polyphosphate for chitosan.[175,176,269]

The physical properties of the beads are controlled by the concentration of the biopolymer solution and its characteristics [molecular weight, acetylation degree for chitosan, proportion of guluronic (*G*) and mannuronic (*M*) acid residues], the concentration, and the type of coagulating bath. Hence, due to the higher affinity of guluronic acid residues for binding metal ions, with calcium alginate samples characterized by higher *G*:*M* ratio produce much stronger gels in the ionotropic gelation procedure. The gelation with the acidic procedure tends to the same conclusion: stronger gels are obtained with alginate samples of higher *G:M* ratios. The viscosity of the solution has an important impact on the control of the shape of the beads. These beads generally contain a very high percentage of water (generally greater than 90%). This may significantly affect the volumetric density of the sorbent. This means a low volumetric density of sorption sites and weak catalytic activity. The drying generally contributes to significantly decreasing diffusion properties, due to the irreversible collapse of the porous network of gel structure. This irreversible degradation can be limited using appropriate drying procedure. Freeze-drying allows reducing structure modification but, generally, this is not sufficient to maintain the original structure. Drying using CO_2 under supercritical conditions shows much better respect for the raw structure of the material; the shrinkage induced by capillary evaporation is prevented and high surface areas are maintained: up to 300 $m^2\ g^{-1}$ for an aerogel produced by supercritical CO_2 conditions vs. a few $m^2\ g^{-1}$ for conventionally dried materials.[226,227,270,271] Beads are dehydrated with a series of successive ethanol-water baths of increasing alcohol concentration. Then the beads are dried under supercritical CO_2 conditions [(i.e., $P(CO_2)$: 74 bar; T: 31.5°C]. Another procedure has been successfully applied in the controlled drying of chitosan gels beads: this consists of the impregnation of chitosan gel beads with sucrose prior to the drying step.[272] The presence of sucrose inside the porous network avoids, at least partially, the collapse of the structure; after rehydration, the gels almost regain their initial volume, and diffusion properties are almost totally restored.

4.2.3.4 Fiber and Hollow Fiber

Hollow fibers made of cellulose derivatives are commercially available and they are extensively used for the preparation of dialysis modules. Kurokawa and Hanaya described the synthesis of fibers made of cellulose acetate functionalized with metal alkoxide for the immobilization of enzymes.[273] Cellulose acetate spontaneously gelled in the presence of metal alkoxide, due to coordination bonding between the hydroxyl groups on pyranose rings and the polyvalent metal. The strength of the fiber strongly depends on alkoxide content. Fibers and hollow fibers made of chitosan and alginate materials are much rarer, especially in the field of supported catalysis.

The preparation of fibers and hollow fibers obeys the same two-step procedure: (a) dissolving the biopolymer, followed by (b) the extrusion of the solution into a coagulation bath, as described for gel-bead preparation.[274–276] The fiber is extruded through a thin nozzle into a bath falling from a spinneret directly into the coagulation bath or alternatively in air to stretch the extruded fiber and reduce its diameter (before it enters into the coagulating bath).[148,277,278] In the case of solid fibers, the coagulated fiber can be rinsed and dried to increase the strength of the fiber, at the expense of a loss of diffusion properties.

In the case of hollow fibers, the manufacturing procedure consists of coextruding the viscous solution of chitosan and a core liquid (Figure 4.13). The core liquid is pumped concentrically to the chitosan solution using a double spinneret.[275,279]

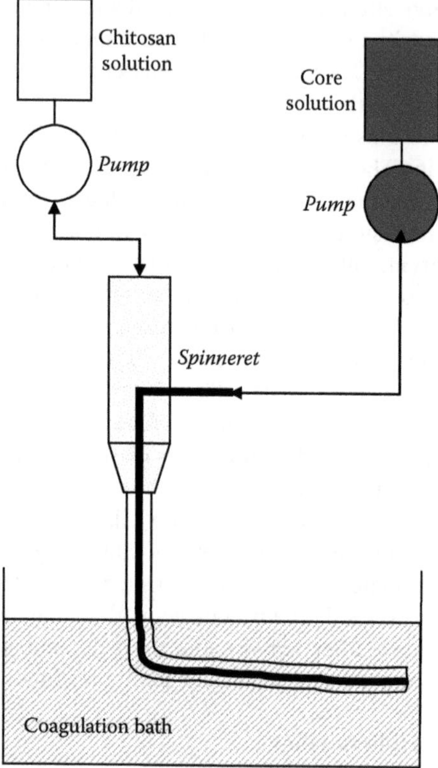

FIGURE 4.13 System for hollow fiber manufacturing.

A second technique has been described for the preparation of chitosan hollow fibers; it consists of the coagulation of the chitosan fiber in an ionotropic gelation bath followed by forced extrusion of the noncoagulated core.[280,281]

These conditions have not been widely investigated for catalytic applications and only a few papers focus on, for example, the testing of catalytic hollow fibers made of biopolymers.[282,283] The use of hollow fibers offers interesting perspectives in catalysis: the solution to be treated is circulating through the lumen of the fiber, while the reagent can be provided in a liquid or gas state at the outer side of the fiber. These techniques can be developed using the same experimental procedures (especially in the choice of coagulating agent) described for the preparation of gel beads to manufacture a larger portfolio of supported catalysts and to extend the offer of new materials, with potential applications in fuel cells, in addition to catalytic areas.

4.2.3.5 Flat Membranes

Flat membranes made of cellulose-based materials are extensively used in separation processes, especially filtration. These materials can be used for the deposition of other biopolymers (composite membranes) by impregnation followed by precipitation or coagulation. In this case, the cellulose membrane brings the mechanical strength while the other biopolymer brings additional reactive groups and increases site density.

Flat membranes have been also prepared using chitosan and alginate alone,[284–295] or combined with other polymers (natural or synthetic)[296] or inorganic compounds.[273,297,298] The procedure for manufacturing alginate or chitosan flat membranes consists of three steps: (a) biopolymer dissolution (see above), followed by (b) casting on a glass, polystyrene, polyethylene, or polycarbonate surface and solvent evaporation, and (c) neutralization of the acid (using NaOH for chitosan) or coagulation (using HCl or calcium chloride for sodium alginate).[288,290,295,298] The drying step is a key parameter for maintaining the porous structure of the membrane. To prevent the irreversible collapse of the polymer structure, it may be interesting to limit the drying of the membrane to 50%; this allows maintaining the rigid structure of the membrane prior to its neutralization or coagulation. An insufficient drying may cause disturbances at the surface of the membrane at the moment of its contact with the neutralization or coagulation bath. It is difficult to control the porosity properties of the materials produced using this procedure. Zheng and Ruckenstein controlled the size of the pore of chitosan membranes using a porogen (silica nanoparticles), directly introduced in the chitosan viscous solution.[299] After membrane casting and drying, the membrane is maintained in contact with NaOH solution; this contributes to both the neutralization of the solution and the dissolving of silica particles, which, in turn, maintain a highly porous structure.

4.3 BIOPOLYMER-BASED HETEROGENEOUS CATALYSTS

This section describes (a) the principal procedures developed over the past 10 years for the preparation of biopolymer-based supported catalysts, and (b) the main characteristics of these materials (together with the techniques that can be used for their determination).

4.3.1 Synthesis

The synthesis of biopolymer-supported catalysts may involve three successive steps: (a) the conditioning of the biopolymer (physical or chemical modification), (b) the metal binding or immobilization (active uptake or entrapment), and (c) the activation (the reduction of the metal, when required).

4.3.1.1 Conditioning

The previous section described a number of processes that can be used for the physical modification of biopolymers. In most cases biopolymers are used as (a) raw particles, (b) in the form of gel beads, or (c) immobilized on high-surface-area supports. The identification of limiting steps, especially the resistance to intraparticle mass transfer, may explain why small particles are preferred for catalyst immobilization and why alternative conditionings are developed. These criteria are critical for both metal uptake and diffusion of substrate and products during the catalytic reaction.[55,300,301]

Using small particles increases the external surface area, but it does not improve intraparticle diffusion. Biopolymer gels offer enhanced diffusion properties; however, the volumetric density of reactive groups is decreased (due to high water content). Drying the gels may cause irreversible collapse of the structure (as pointed out above). Controlled drying, in the presence of a spacer (such as sucrose) or with freeze-drying and preferentially with drying in supercritical CO_2 conditions, offers interesting perspectives.[226,227,302]

However, the technique most frequently used for improving mass transfer consisted of depositing the biopolymer at the surface of a support with high surface area, typically silica. Supporting the biopolymer on silica increases the contact surface, if the biopolymer is deposited as a thin layer, and also improves mechanical properties, compared to dried or wet gels. The immobilization procedure was described in the previous section. It is noteworthy that this immobilization can also be used for the coimmobilization of amino acids on silica through the encapsulating effect of biopolymer deposition. Immobilization of amino acid brings some interesting properties for the attachment of metals and for the enhancement of catalytic activity of polymer-supported catalysts.[67,303,304] A number of papers cite the combination of amino acids with encapsulating biopolymers, especially for polymers that have poor intrinsic properties for target metal binding or when it is required to increase their enantioselective properties.[245,255,304,305] The amino acid solution is usually mixed with the polymer solution prior to silica introduction, and the slurry is then dried to physically entrap the amino acid in the thin biopolymer deposited at the surface of mineral support. Acetate cellulose membranes have been functionalized by grafting polyamino acid (poly-L-glutamic acid); the acetate cellulose membrane is regenerated by an alkaline treatment ($NaOH/NH_4OH$), and the polymer surface is activated using sodium periodate in a phosphate buffer, giving aldehyde functionalized cellulose that can react with poly-L-glutamic acid. The functionalized membrane is treated with sodium borohydride and finally with acidic solution (pH in the range 3 to 4) for metal binding, with increased sorption properties.

Liu et al. prepared silica-supported oxalic acid materials for the immobilization of carboxymethylcellulose, alginic acid, casein, and gelatin prior to platinum binding for the synthesis of hydrogenation catalysts.[241] Silica is mixed with oxalic acid solution; after being dried, the powder is mixed with biopolymer solution; the impregnation phase is followed by a new drying step. A similar procedure was used by Liu et al. for the preparation of Pt catalysts supported on silica-methylcellulose material.[243] These treatments contribute to activate the support for biopolymer and metal immobilization.

In addition to physical modifications of the biopolymers, the grafting of functional groups onto a polymer backbone can increase their sorption properties and their selectivity for target metal, and changes the reactivity of immobilized metals by modification of their environment. Comprehensive reviews on chitin and chitosan modifications have been extensively documented by Roberts,[83] Kurita,[207] and Sashiwa and Aiba.[306] The chemical modification of chitosan by grafting new functional groups has been much more frequently used than for other biopolymers in the preparation of supported catalysts,[307] including Schiff base derivatives[308–311] or Salen complexes.[312–314] *Bis*(salicylidene-ethylenediamine)-Co complexes (tetra-coordinated complex, CoSalen) has been used for reversible binding of molecular oxygen; however, its tendency to form dimmers and peroxy-bonded adducts may make the complex inactive toward reversible oxygen binding. The immobilization of the complex on a support prevents this inactivation mechanism; Finashina et al. successfully used CoSalen immobilized on chitosan and chitosan-silica composite supports for oxidation of catecholamines.[312] Macquarrie and Hardy recently reviewed the functionalization of chitosan for catalytic applications, including the description of Schiff base formation, reductive amination, amide formation, alkylation, and Michael addition.[307]

As pointed out above, alternative conditionings (such as hollow fiber systems) may contribute to develop new way to manage these catalytic reactions with a separation of the substrate and the oxidizing or reducing compartments.[282,283,315,316]

4.3.1.2 Metal and Catalyst Immobilization

Catalytic groups (either free metal ions or complexes) can be immobilized by at least three different methods: (a) adsorption (regardless of the mechanism involved, i.e., chelation or ion exchange), (b) precipitation or coprecipitation, and (c) encapsulation. The general properties of selected biopolymers for metal binding, gel forming, and precipitation have been detailed above; in this section, information focuses on the processes employed in the synthesis of heterogeneous catalysts. The mechanisms involved depend on the metal-biopolymer and the type of solvent used for metal binding: water or alcoholic solutions.

4.3.1.2.1 Chelation and Ion Exchange

Most of the literature reports the use of ethanol media for metal binding (regardless of the molecular mechanism, that is, coordination, ion exchange, and so on) for the preparation of heterogeneous catalysts based on biopolymers. However, some supported catalysts have been prepared by metal sorption from aqueous solutions.

Copper complexes formed with chitosan in aqueous solutions have been used for the preparation of a series of catalysts.[221,222,317] The complex formed in acidic solution can be (a) adsorbed on preformed chitosan gel beads, or (b) impregnated onto macroporous silica or mesoporous mobile crystalline material (MCM-41, in particular), or zeolite support before being precipitated (in the case of silica support) or cross-linked (in the case of zeolite). The Cu-chitosan complex immobilized on macroporous silica exhibits a much greater activity for oxidation of di-hydroxybenzene than homogeneous complex: the inhibiting effect of copper-hydroquinone binding is reduced. Additionally, the matrix of heterogenized chitosan stabilizes isolated copper ions in a coordinatively unsaturated state favorable for catalytic activity (unsaturated square-planar geometry of the complex).

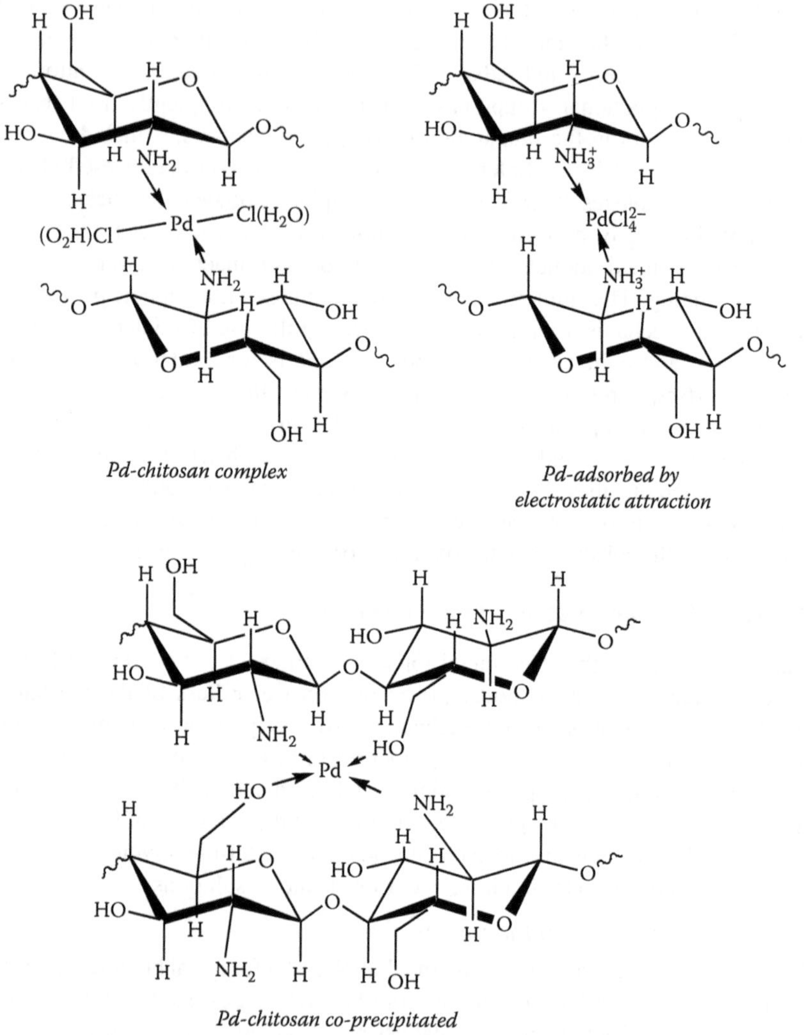

Pd-chitosan complex

Pd-adsorbed by electrostatic attraction

Pd-chitosan co-precipitated

FIGURE 4.14 Structure of Pd complexes with chitosan prepared by different methods. (Reprinted from N.V. Kramareva, A.E. Koklin, E.D. Finashina, N.S. Telegina, A. Yu. Stakheev, and L.M. Kustov, *Kinetics Catal.*, 45, 751–753, 2004. Copyright 2004, MAIK "Nauka/Interperiodica," Springer.)

Similar procedures have been developed for the preparation of heterogeneous Pd catalysts for terminal olefin oxidation.[318,319] Chitosan cross-linking was performed by contact of an acidic solution of chitosan with hexane solution (completed with an emulsifier) under strong agitation, followed by the addition of a glutaraldehyde cross-linking agent. Kramareva et al.[318,319] discuss the binding mechanism of Pd as a function of the conditioning of the catalyst (free chitosan vs. glutaraldehyde cross-linked) and the contact procedure (adsorption vs. coprecipitation) (Figure 4.14). In the case of the adsorption method, the complex is formed by ligand exchange and has the hypothetical structure $[Pd(RNH_2)_2Cl_2]$, while in the case of the coprecipitation method, chelate complexes are the most probable forms with coordination of the amino and hydroxyl groups of neighboring chitosan units (or from adjacent polymer chains).

Different interpretations of the interaction of chitosan materials with palladium and platinum have been presented.[134,152,163,320,321] These differences may be attributed to different experimental conditions, especially relative to pH. In acidic solutions, the protonation of amine groups limits the possibility for amine groups to coordinate with palladium or platinum species and an anion exchange mechanism is preferred.[320] In less acidic solutions, Brack et al. also commented on chelation mechanism for explaining the gelation of chitosan in slightly neutral solutions containing platinum chloride species.[173]

The preparation of a cobalt-chitosan catalyst is described by Guan and Cheng.[322] They soaked a preformed chitosan film in a $CoCl_2$ aqueous solution for 24 h. Washed membranes were dehydrated before being dried at 60°C under vacuum. The authors commented on the structure of coordinated cobalt; they described the formation of a high spin complex with a coordination number of Co(II) equal to 4. The catalyst is used for the polymerization of vinyl acetate.

More numerous are the examples of catalysts prepared by metal binding from alcoholic solutions. The process consists of the contact of the metal salt with the polymer in an alcoholic solvent (for example, ethanol, methanol, or butanol) maintained under reflux for 6 to 16 h, until there is a complete change of sorbent color and the metal is removed from the solution. A simultaneous reduction of the metal is frequently observed, especially in the case of platinum- and palladium-based catalysts. The process has been applied with raw, modified, and conditioned biopolymers.

This technique has been used for manufacturing silica-supported chitosan catalyst with palladium,[323–329] with platinum,[247,330] with Pt-Fe,[331] with osmium,[332] with rhodium,[249] and nonnoble metals.[305,333,334] It was also used for preparing supported catalysts based on modified chitosan (chemical derivatives obtained by Schiff base reaction, by interaction with other synthetic polymers, and by other methods).[252,308,310,311,335,336] For example, in the case of Pd binding, XPS analyses have shown that after metal uptake the binding energy of nitrogen (N_{1s}) is increased, confirming that amino groups are involved in coordination bonds. The simultaneous decrease of the binding energy of Cl_{2p} shows that nitrogen in the amino group of chitosan supplied part of an electron to Pd after coordination.

The technique of deposition and simultaneous reduction of catalytic metals has been also used for the synthesis of catalysts based on other polysaccharides (including starch, alginic acid, methylcellulose, and carboxymethylcellulose) and proteins (including gelatin, casein, wool, or silk fibroin) deposited on silica or other mineral oxides, or self-supported (in the case of silk fibroin and wool). To emphasize the above, when the biopolymer is not reactive against the target metal (this is the case for alginate, starch, and methylcellulose for Pd or Pt binding), the biopolymer action is completed by the introduction of amino acid residues or treatment with oxalic acid. XPS and FTIR analyses confirm that in this case

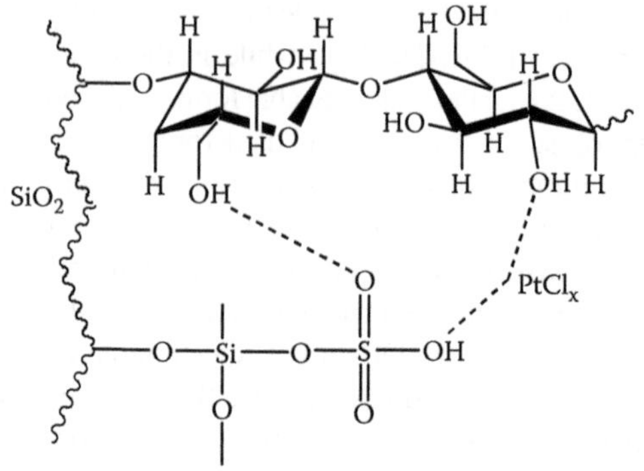

SiO_2-Cellulose/Starch-Polysulfoxane-Pt Complex

FIGURE 4.15 Structure of Pt complex with composite SiO_2-cellulose/starch-polysulfoxane. (Reprinted from K. Huang, L. Xue, Y.-C. Hu, M.-Y. Huang, and Y.-Y. Jiang, *React. Funct. Polym.*, 50, 199–203, 2002. With permission. Copyright 2002, Elsevier.)

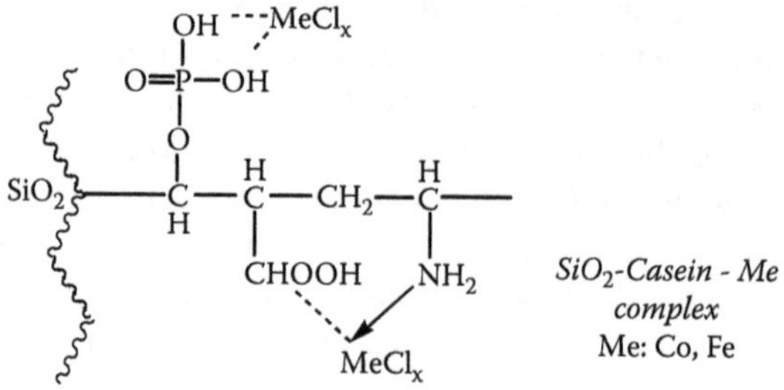

SiO_2-Casein - Me complex
Me: Co, Fe

FIGURE 4.16 Structure of Co complex with composite SiO_2-casein. (Reprinted from X. Zhang, B. Han, Y.-N. Hua, M.-Y. Huang, and Y.-Y. Jiang, *Polym. Adv. Technol.*, 13, 216–219, 2002. With permission. Copyright 2002, John Wiley & Sons Ltd.; (Reprinted from L. Shen, J.-L. Ye, M.-Y. Huang, and Y.-Y. Jiang, *Polym. Adv. Technol.*, 13, 173–177, 2002. With permission. Copyright 2002, John Wiley & Sons Ltd.)

the coordination occurred through either the carboxylic groups of oxalic acid,[241,243] or the amino groups and carboxylic groups of amino acid.[246,255,305] In the case of Pt catalysts immobilized on starch-polysulfoxane, starch was reacted with silica-supported polysulfosiloxane (produced by the reaction of silica with ethyl silicate); platinum mainly reacts with –OH groups on both the biopolymer and sulfosiloxane (Figure 4.15).[247] These catalysts have been used for hydrogenation reactions. For proteins containing amino groups and carboxylic acid functions, many different interactions may coexist. In the case of casein deposited on silica, Shen et al. and Zhang et al. suggest that iron and cobalt are immobilized on the support through combined interactions with –COOH and $–NH_2$ groups, and through interactions with OH groups on phosphate functional groups (Figure 4.16).[257,337] In the case of gelatin-based supports (immobilized on zeolite), Zhang et al. immobilized iron and Co/Ru through combined effects of amine groups and carboxylic functions.[242,256]

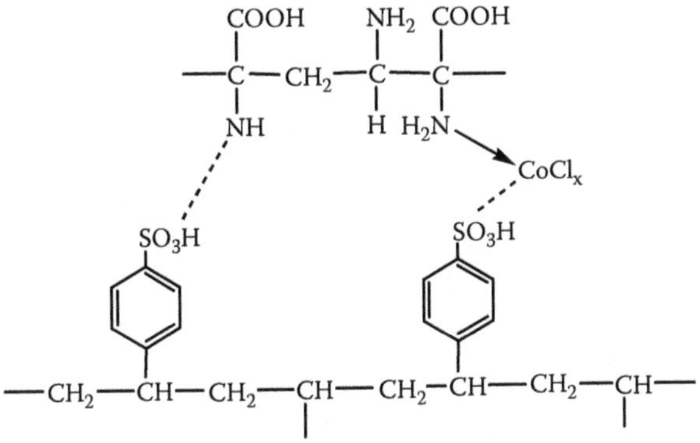

Polysulfostyrene-Gelatin-Co Complex

FIGURE 4.17 Structure of Co complex with composite polysulfostyrene-gelatin. (Reprinted from X. Zhang, Y.-J. Li, M.-Y. Huang, and Y.-Y. Jiang, *Polym. Adv. Technol.*, 13, 305–309, 2002. With permission. Copyright 2002, John Wiley & Sons Ltd.)

In the case of gelatin immobilized on polysulfostyrene carrier, Zhang et al. observed that cobalt binding affects not only the amine groups and carboxylic functions but also sulfonic groups on the support (Figure 4.17).[258] The catalysts prepared with gelatin and casein have been tested for asymmetric reactions (epoxidation of alcohols, hydrogenation of ketones, hydroformylation, and hydration). The structure of wool is very complex, containing a wide range of functional groups, from carboxylic groups to amine and amide groups, including sulfur containing groups ($-SO_3H$, –SH, and –S–S). The most significant differences in the binding energies following Pd or Pt sorption are obtained for N_{1s} (–NH–CO– and $-NH_2$ groups), for S_{2p} (–SH and –S–S groups, the shift of binding energy for $-SO_3H$ groups is negligible) (Figure 4.18)[338]; carboxylic functions do not appear to contribute significantly to Pd binding.[326] Wool has been carried out for the immobilization of rhodium,[339] platinum,[338] palladium,[326] and Pd/Fe.[340] These catalysts have been tested for asymmetric hydrogenation of ketones and for hydration of alkenes.

4.3.1.2.2 Precipitation

Kurokawa and Hanaya described the impregnation of cellulose membranes with metal ions by a counterdiffusion method.[273] Cellulose acetate membranes, obtained by gelation of cellulose acetate in an acetone-formamide mixture (40% acetyl content of cellulose acetate-acetone-formamide: 25-30-45) are disposed at the interface between two compartments, one being filled with metal ions solution ($AgNO_3$, $RuCl_3$, $PdCl_2$, or $RhCl_3$), the other filled with a 0.1 *M* NaOH solution. The metal-impregnated membrane is finally immersed in 0.1 *M* $NaBH_4$ solution for activation of catalytic metal by reduction.

Several supported catalysts have been prepared by a multistep procedure involving the binding of the metal to the biopolymer in solution, followed by its coprecipitation to obtain a solid-state or gel-form material. Zeng et al. prepared

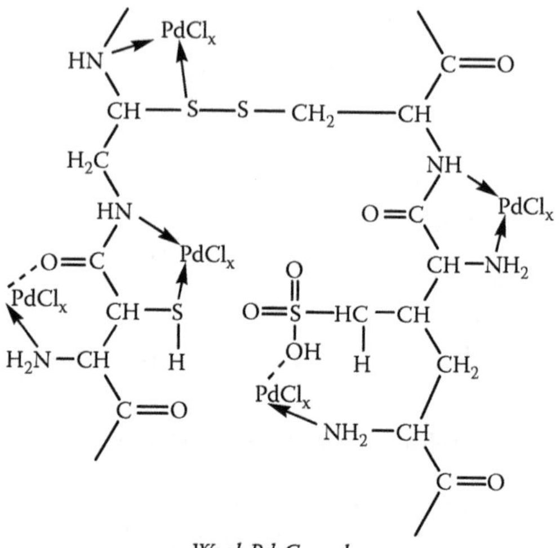

Wool-Pd-Complex

FIGURE 4.18 Structure of Pt complex with wool. (Reprinted from M.-Y. Yin, G.-L. Yuan, M.-Y. Huang, and Y.-Y. Jiang, *J. Mol. Catal. a: Chem.*, 147, 89–92, 1999. With permission. Copyright 1999, Elsevier.)

a chitosan-based catalyst for the ring-opening polymerization of propylene oxide by immobilization of rare-earth metal.[341] Metal oxide is dissolved in a concentrated HCl solution, before being added to a chitosan solution. The pH of the mixture is controlled to 6.7 using ammonium hydroxide. The resulting "complex" is precipitated with acetone:ethanol (1:1).

Isaeva et al. prepared a series of catalysts supported on chitosan with Pd, Rh, Zn, and Pb metals for hydrogenation reactions (cyclopentadiene and 1,4-butynediol hydrogenation).[248] The acidic complex metal-chitosan is dropped into an alkaline precipitating bath.

Kramareva et al. prepare copper-chitosan gel beads by direct precipitation of chitosan-Cu complexes (formed by copper binding to chitosan in acidic solution) into 0.5 *M* NaOH solution.[317] The procedure is also used for the preparation of gel beads of chitosan-Pd complex: the Pd-chitosan complex is coprecipitated into an alkaline NaOH solution, producing a spherical heterogeneous catalyst for the oxidation of olefin.[318,319]

4.3.1.2.3 Impregnation and Encapsulation

Shim et al. mix cellulose acetate membranes with copper salts (both acetate and chloride salts) in tetrahydrofuran at boiling temperature for 15 min.[236] The mixture is partially evaporated before being cast on glass plates. Copper can be washed up from the membranes in water, with more than 70% of the copper being removed, indicating that there is no significant direct interaction between metal and cellulose support. The metal can be activated with hydrogen gas at high temperatures (160°C).

In addition to the precipitation procedure described above, Isaeva et al. prepared a series of catalysts supported on chitosan (free form, succinyl derivative, or modified with glutaraldehyde or pyridinealdehyde, and eventually deposited on silica) with Pd, Rh, Zn, and Pb metals for hydrogenation reactions (cyclopentadiene and 1,4-butynediol hydrogenation).[248] After impregnation of the solid-state materials with a metal ion solution, the sorbent is precipitated into a neutralization bath.

Köckritz et al. obtain very stable catalysts (in terms of both metal stability and catalytic activity) for the preparation of noble metal catalysts.[220] The technique consists of mixing the polyelectrolyte solution with a homogeneous catalyst solution, followed by the distribution of the mixture as droplets on a suitable surface (polypropylene, polyethylene) and drying under air or argon flux. The advantage of this process is that the catalyst (metal complex) can be immobilized in its optimized form (with limited effect on the interaction of the metal complex with the encapsulation medium). The diffusion properties are probably affected by the drying step. They investigated several synthetic polycationic polyelectrolytes and several anionic biopolymers (alginate, cellulose derivatives, carrageenan, pectinate). In the case of biopolymers, they prepared Pt-colloids by reaction of $PtCl_4$ with protonated 10,11-dihydrocinchonidine in formic acid. The stabilized colloid particles are then mixed with biopolymer solutions and sprayed on the plate for their encapsulation.

4.3.1.3 Activation of Catalytic Metal

In the case where Pd or Pt was immobilized onto the support via the alcoholic mediated reaction, the metal is generally reduced to its metal form. These catalysts are usually synthesized for hydrogenation reactions; the simultaneous binding and reduction allows using the heterogeneous catalyst as produced. When the catalyst is designed for reductive hydrogenation reactions, it may be necessary to proceed to the reduction of the metal for an immobilization process operated in aqueous solutions, where the binding process does not cause the simultaneous change of the metal oxidation state. Several processes can be used by utilizing reducing solutions, hydrogen gas at room temperature, or with heating.

This post-treatment of activation has been used for the preparation of chitosan-supported Pd-catalysts synthesized in aqueous solutions,[282,283,342–346] and for the preparation of colloid-supported materials and nanoparticles.[208,210,239,240,347,348] Three different routes have been described for the preparation of colloids and nanoparticles based on biopolymers: (a) chemical reduction, (b) radiolytic treatment, and (c) UV irradiation.

Belapurkar et al. prepared Pt colloids stabilized by gelatin using gamma irradiation.[235] Gelatin and tetrachloroplatinic acid solutions are mixed with methanol (OH radical scavenger). The mixed solution is purged with N_2 before being gamma irradiated with a ^{60}Co source (dose rate 20 Gy min^{-1}). The colloids are very stable, although the reactivity of free colloids was lower than that obtained with glass-deposited nanoparticles.

In the case of photochemical activation, Kundu et al. mixed a warm gelatin solution with silver nitrate; some drops of NaOH allow removal of the turbidity that appears in the solution.[349] After pH control to 8, the solution is cooled and silica particles can be added for nanoparticle desposition. The paste after being dried is spread on a surface submitted to UV irradiation (wavelength: 365 nm). The size of nanoparticles is in the 10 to 20 nm range. The catalytic activity of deposited catalyst is higher than that of free nanoparticles. Pal et al. used a similar procedure for the preparation of gold nanoparticles, in this case using sodium alginate and a stronger UV irradiation system (i.e., low pressure Hg lamp; wavelength: 254 nm).[350] However, the most frequently cited process uses the chemical reduction of catalytic metal in solution through reaction with reducing agents such as hydrazine,[234,351] formaldehyde,[234] pentaerythritol,[234] sodium borohydride,[240,348,351,352] tannin,[353] sodium citrate,[347] or methanol.[351] In some cases, the biopolymer supports simultaneously contribute to both the stabilization and the reduction of metal. The process allows synthesizing multi-layers of reactive metals.[347] The addition of tripolyphosphate in the reactive media changes the size of nanoparticles and their distribution; more surprisingly, it also changes their shape: in the presence of tripolyphosphate, nanoparticles adopt a polygonal shape instead of the conventional spherical form.[237]

4.3.2 Critical Parameters for the Design of Biopolymer-Supported Catalysts

The catalytic activity of a supported catalyst depends on several intrinsic characteristics of the supported system, including parameters related to (a) metal state (size of metal crystallites or the oxidation state of the metal), (b) support structure (size, porosity, hydrophilic character), (c) support-metal interactions (stereoselectivity, stability). The kinetics of heterogeneous catalysis is generally separated in five steps[354]:

1. Mass transfer to the solid surface, including film diffusion and intraparticle mass transport
2. Adsorption of the substrate (or the substrates) on the surface of the catalyst, and more specifically on the catalytic site (or in its neighborhood)
3. Chemical reaction at the surface of the catalyst involving at least one of the adsorbed species
4. Desorption of reaction products from the reactive surface
5. Diffusion of desorbed products away from the catalyst (through film and intraparticle diffusion)

Steps 1 and 5, are due to diffusion constraints controlled by the characteristics of catalyst particles (physical properties, affected by particle size, porosity, and so forth). Steps 2 and 4 are controlled by the affinity of the substrates and products for the reactive surface (i.e., their affinity for crystallite, but also for the support; crystallite size, hydrophilic or hydrophobic character, etc.). The chemical reaction may be controlled by the conformational effects of the support, the oxidation state (and, more generally, the chemical state of the metal: free, complex, etc.), and the size of metal crystallites.

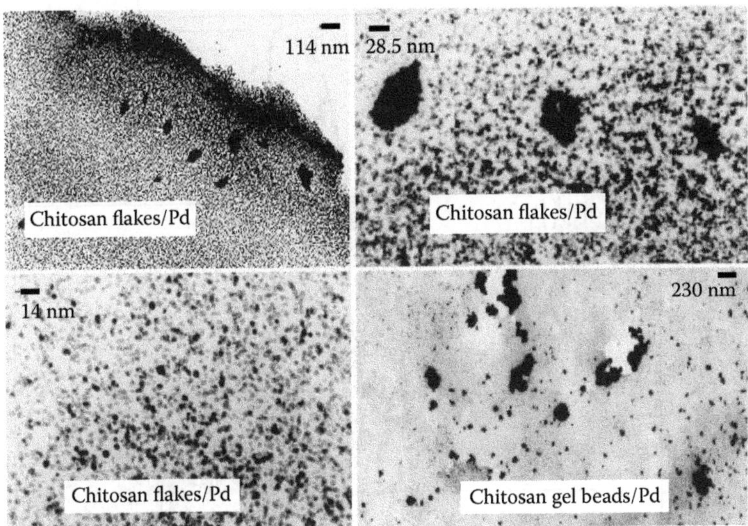

FIGURE 4.19 TEM microphotographs of Pd nanoparticles on different chitosan-based supports (flakes, gel beads).

4.3.2.1 Size of Crystallites

The size of the crystallite is a critical parameter in the design of a heterogeneous catalyst; increasing the size of metal crystal may have either a detrimental or an enhancing effect, depending on the material. Several studies have shown the correlation between the size of metal crystallites and their catalytic activity for colloids,[71,234,351] nanoparticles,[348,351] or immobilized on inorganic supports[10,14,23,355] or organic supports.[253]

Transmission electron microscopy (TEM) and HR-TEM (high-resolution TEM) are the techniques most frequently cited for the determination of the cross-sectional average particle size. Figure 4.19 shows the TEM microphotograph of Pd nanocrystals deposited on glutaraldehyde cross-linked chitosan flakes (after chemical reduction using *in situ* generated hydrogen gas). The size of crystals is in the range of 3 to 5 nm. In the case of gel beads, nanoparticles tend to aggregate, forming agglomerates of much larger sizes. Stevens et al. commented that the technique does not detect heterogeneities at the surface of the catalyst; this may induce discrepancies in the determination of particle sizes.[24]

Neri et al. compared the size of Pd crystallites (formed on carbon support) measured by CO chemisorption and TEM analysis.[355] They concluded that the techniques fairly give the same order of magnitude for crystals, despite some discrepancies (especially for large particles forming aggregates; in this case, CO chemisorption involves a Pd:CO ratio of 2 instead of 1 in the case of small supported crystals).

Wide-angle x-ray scattering analysis (WAXS) is also frequently used; the Scherrer equation allows calculating the crystal diameter (d, nm) as a function

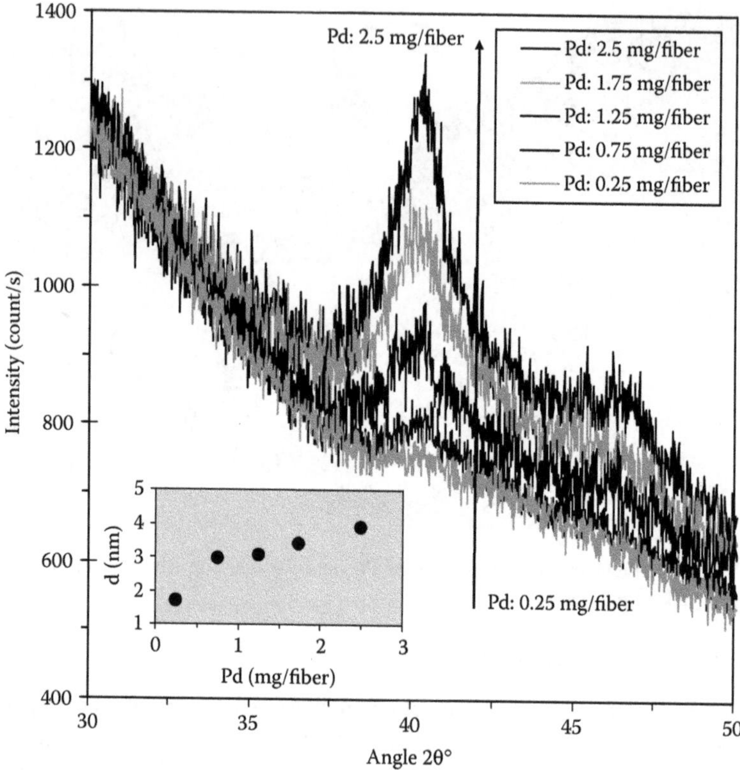

FIGURE 4.20 X-ray diffraction patterns of Pd crystallites on chitosan hollow fibers with different metal loadings (0.25–2.5 mg Pd/fiber, i.e., 1.1–10% w/w).

of the x-ray wavelength (nm), the full width at half maximum [$\beta(\theta)$, rad] of the identification peak, the diffraction angle and (θ, rad), and k a constant typical of the equipment:

$$d = \frac{k\lambda}{\beta(\theta)\ \cos\theta} \tag{4.2}$$

This method is efficient for medium-size nanoparticles; however, in the case of very small crystals (below 3 nm), the poorly resolved peaks do not allow a precise determination of crystal size. Stevens et al. suggested using small-angle x-ray scattering analysis.[24] Figure 4.20 shows an example of x-ray diffraction patterns obtained with Pd deposited on chitosan hollow fibers. The size of crystallites, deduced from the Scherrer equation, increases with Pd loading on the fiber.

The dispersion of the crystallites in the polymer matrix is also an important parameter in the prediction of catalytic properties. Douidah et al. compared the catalytic activity for oxidation of Pt immobilized on different materials (alumina, carbon, and silica).[356] They observed that the interactions of Pt with the support decreases according $Al_2O_3 > C > SiO_2$. Some of these supports have heterogeneous distribution of Pt particle size, and they show that reaction occurs first on small particles and that at the end of the reaction the only remaining particles of the catalyst have sizes ranging between 90 and 150 Å. The size of the most stable particles is defined as the compromise between a decrease of the stability when the dispersion increases and an increase of the metal-support interaction when the particle size decreases.

The effect of crystal size and metal dispersion on catalytic activity is mainly due to an increase in the external surface area available for reaction. Decreasing the size of metal crystallites generally enhances the catalytic activity,[36,71,357–359] although, in some cases, the catalytic activity passes by an optimum value of particle size. Neri et al. observed that for Pd immobilized on carbon, the catalytic hydrogenation of 2,4-dinitrotoluene is increased by large Pd crystallites.[355] The increase of catalytic activity with particle size is attributed to (a) formation of a palladium β-hydride phase (i.e., hydrogen reservoir), (b) mass-transfer limitations, or (c) changes in the interaction between the substrate and metal crystals (conformation of the complex).

The beneficial effect of decreasing metal particle size is not limited to catalytic efficiency since it may also increase the selectivity of the reaction. This can be explained by different modes of interaction for the substrate on the crystals (due to different adsorption geometry)[355] or by different electron-deficient properties of immobilized metals (a change in the surface coordination number).[14,360] Actually, the size of metal crystallites depends on several experimental parameters. In the case of synthetic polymers, Hirai et al. and Chen et al. showed that the molecular weight of the polymer and the molar ratio of monomeric unit:metal strongly influence the size of metal crystals and their distribution.[36,358] Palladium and platinum ions are deposited on the colloid by the reduction procedure in an alcohol solution. The size of the metal crystals increases with increasing the molecular weight of the polymer, and with decreasing the monomeric unit:metal ratio. They also show that the catalytic activity depends on specific surface area of Pd nanoparticles much more strongly than the thickness of the adsorbed layer of the polymer.[36] In the case of supported catalysts prepared with biopolymers, the influence of molecular weight has not retained a great attention, contrary to the effect of metal:monomer concentration ratio. Huang and Yang reported the preparation of chitosan-stabilized gold nanoparticles[237]; they observed that the size and the distribution of gold nanocrystals slightly changes when changing the molecular weight of the biopolymer and its concentration. The size of crystallites increases with decreasing the concentration of chitosan (below 0.01%; above, the concentration has a limited effect, i.e., in the range 0.05% to 0.25%), and with increasing its molecular weight (especially at low chitosan concentration) (Figure 4.21). The shape of the metal crystals also changes with experimental conditions: from spherical shape with low-molecular-weight chitosan to polygonal shape when increasing biopolymer molecular weight (Figure 4.21). The films made by slow evaporation of the solvent of these gold-chitosan nanocomposites

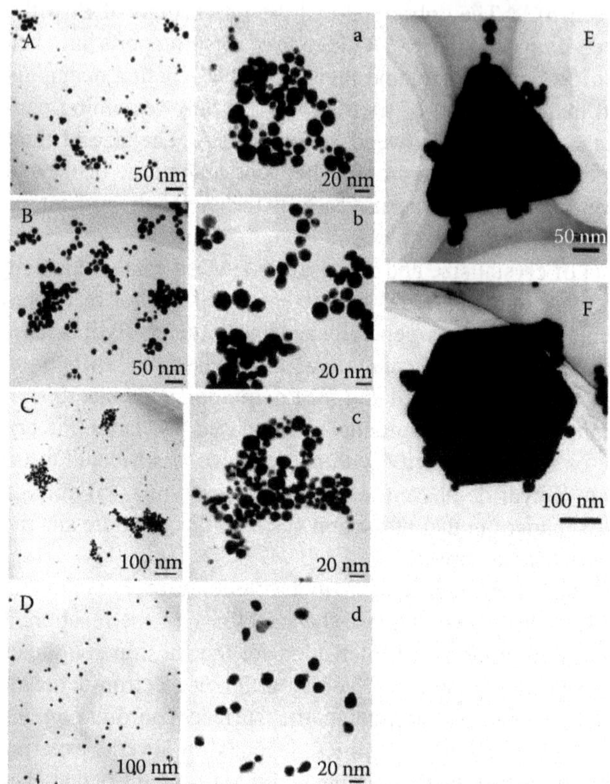

FIGURE 4.21 TEM images of gold nanoparticles prepared by 0.05% (A,a) and 0.25% (B,b) medium molecular weight chitosan and 0.05% (C,c) and 0.25% lower molecular weight chitosan; and after addition of tripolyphosphate (E,F). (Reprinted from H. Huang and X. Yang, *Biomacromolecules,* 5, 2340–2346, 2004. With permission. Copyright 2004, American Chemical Society.)

show a branched-like structure or a cross-linked needlelike structure when observed under cross-polarized light.[240] The addition of tripolyphosphate induces the ionotropic gelation of chitosan in the form of nanoparticles; in gold solution, it causes the formation of gold nanoparticles with bimodal size distribution and polygonal structure.[237] Ishizuki et al. prepared a series of Au, Pt, Pd, and Rh (alone or in bimetallic form) catalysts supported on chitosan for the hydrogenation of methylvinylketone and the decomposition of hydrogen peroxide.[234] Regardless of the metal, they observe that the size of the crystals decreases with increasing the chitosan:metal ratio, while the dispersion stability increases.

With bimetallic compositions, the effect of chitosan concentration is less marked. When increasing the molar ratio of chitosan to Pd, Huang et al. observed a significant decrease of the size of Pd crystallites.[253] However, they observed that the catalytic activity, measured by hydrogen uptake, is not directly correlated to the size of metal crystallites: the catalytic efficiency increases with decreasing the size of palladium at first before declining after a critical size. Esumi et al. also observed a discontinuous variation of the size of gold nanoparticles (mixing gold hydrochloric acid solution with chitosan, before chemical reduction with sodium borohydride) with the concentration of chitosan.[348] The size of crystallites increases with chitosan concentration up to a critical concentration (obtained for a Au:chitosan ratio close to 32:1), followed by a decrease of crystal size when increasing the amount of chitosan. The catalytic activity also reaches an optimum value at a critical chitosan concentration, which is lower than that determined for the size of metal crystallites. Adlim et al. compared the effect of different reducing agents (methanol, sodium borohydride, hydrazine) on the size of Pd and Pt nanoparticles stabilized with chitosan.[351] The colloids are prepared in a chitosan-acetic acid-methanol solution by addition of a metal salt solution, followed by the appropriate reduction method. They concluded that the smallest particles are obtained with methanol-induced and sodium borohydride methods, with particles ranging between 1.9 and 2.2 nm. The stabilized particles are very homogeneous in size and well dispersed, while hydrazine contributes to form large particles with significant aggregation. The increase of the chitosan:metal ratio has a limited effect on the size of metal crystallites, except for materials produced by the hydrazine method and in the case of Pd supported on chitosan prepared by the methanol method. The size of Pd nanoparticles, reduced with methanol under reflux, decreases with increasing chitosan concentration. In this case, the catalytic activity of these materials for the hydrogenation of octane and cyclooctene decreases with increasing chitosan concentration on the nanoparticles. In addition to x-ray diffractograms, Figure 4.20 shows the decrease in Pd particle size with an increase in the amount of Pd immobilized on chitosan fibers. Palladium ions are adsorbed from aqueous solutions through electrostatic attraction, followed by reduction using *in situ* produced hydrogen. Although the experimental procedure used for the preparation of these chitosan-based catalytic fibers is not the same as cited above, the order of magnitude of the diameter of Pd nanoparticles (i.e., 1.7 to 3.9 nm) and the effect of Pd content (i.e., the detrimental effect of increasing Pd content) are consistent with those observed with other cited systems.

Other biopolymers tested for the heterogeneization of catalytic metals have received less attention regarding the optimization of nanoparticle size. Ascencio et al. prepared Eu-Au nanoparticles by contact of bimetallic solutions with milled alfalfa (high tannin content), and they obtain stable particles with sizes from 2 to 50 nm (corresponding to clusters made of 100 to several thousands Eu atoms).[353] The size of the particles is controlled by the pH: smaller particles are obtained with pH close to 8. At pH 8, all the particles are smaller than 20 nm (and more than 60% are smaller than 10 nm). The mechanism of metal reduction is modeled according the following equations[353]:

$$M(NO_2)_3 + H_2O \rightarrow M^{3+}_{aq} + 3\,NO_3^- \rightarrow [M(H_2O)_n]^{3+} + 3\,NO_3^-$$

The reduction of metal by tannins occurs in two steps:

$$[M(H_2O)_n]^{3+} + R^- + 3\,NO_3^- + H_2O \xrightarrow{H^+} [M(H_2O)_n]^{2+}$$
$$+R\text{-}OH + H_2O + 3\,NO_2\uparrow$$
$$[M(H_2O)_n]^{2+} + 2\,R^- + H_2O \xrightarrow{H^+} M^0 + 2R\text{-}OH + H_2\uparrow$$

where R is the radical $[C_{14}H_4O_8X_5]$, X being H, alkyl groups, or halogens.

Pal et al. used sodium alginate for the stabilization and the reduction of gold and prepared encapsulated material with a particle size smaller than 30 nm, through UV photoirradiation.[350] This stabilizing effect is confirmed by the increase of particle size when decreasing the concentration of alginate: the biopolymer prevents the agglomeration of nanocrystals. Shim et al. prepared copper nanoparticles immobilized on cellulose acetate membranes[236]: cellulose acetate and copper complex are dissolved in tetrahydrofuran cosolvent, and the mixture is boiled until partial evaporation of the solvent (up to a concentration in the solution close to 12 wt%). The solution is cooled and cast on a glass plate. The reduction process of Cu(II) complexes to Cu_2O and copper nanoparticles by reacting with H_2 gas produces particles in the 30 to 120 nm range, depending on reduction conditions and the amount of Cu in the membrane.

4.3.2.2 Metal Content on the Solid and Metal Oxidation State

The oxidation state of the metal may be of critical importance for the efficiency of the catalytic process, especially for reactions involving reduction mechanisms. It is usually necessary to reduce the metal to a zero-valent state. A partial reduction may result in a significant decrease of the catalytic activity, a fraction of the metal being inactive. This reduction can be included in the metal immobilization step; this is the case for a process involving metal binding in alcoholic media (see above). Alternatively, the catalyst requires an additional reduction step. The processes for metal activation have been described above. The evaluation of the fraction of catalytic metal truly reduced is thus an important parameter for predicting the catalytic activity of the material [especially the turnover frequency (TOF), mol product mol^{-1} metal h^{-1}]. X-ray photoelectron spectroscopy (XPS) is the usual analytical tool for the determination of the oxidation state of the metal immobilized on the supported catalyst, and eventually the fraction of metal present under its different oxidation state. The change in the binding energy (BE, eV) of the atom in reactive chemical groups is helpful for identifying the contribution of the different groups and the mechanisms involved in metal binding. However, for some metals, including platinum group metals, it is necessary to take into account the possible modification of the oxidation state of the metal when submitted to prolonged x-ray bombardment.[361] Noble metal particles dispersed on a high-surface-area carrier exhibit a low rate of photoelectron pulses; short pulse counts (less than 1 h) are thus required, at the expense of lowering the resolution of the analysis. Additionally, XPS analysis (except in the case where ablation techniques are applied) can only measure the photoelectrons that originate from within a 3γ distance from the surface (γ being the attenuation length for photoelectrons in the material, depending on the energy of the beam). This analysis is thus only a mirror of the oxidation state of the metal at the external surface of the catalyst, and it does not bring information on the heterogeneities of the material, except when the analysis is simultaneously performed on the solid and its fine-ground form.[208]

XPS analysis has been widely used for the characterization of biopolymer-supported catalysts for hydrogenation reactions (requiring zero-valent metal catalysts). However, in most cases, the full XPS spectra are not presented and the relative fraction of oxidized and reduced forms is not fully given.[246,247,251,313,317,319,336,339,362–365] Jin et al. observed that about 20% of Pd bound to a composite polyacrylic acid-chitosan (deposited on silica) remains in the oxidized form.[252] The alcoholic impregnation and reduction does not achieve the complete reduction of the metal. Yin et al. prepared a wool-supported Pd catalyst and observed that the impregnation and reduction procedure in alcoholic media reduced about 75% of the metal.[326] In the case of Pd immobilization on chitosan flakes using the sorption process in aqueous media, Vincent and Guibal observed that Pd reduction does not exceed 60%.[343]

The content of catalytic metal controls the size of metal nanoparticles and crystallites, which in turn affects the catalytic activity. In addition to this effect, the relative fractions of the metal-biopolymer and metal-substrate may influence both the yield of the reaction and the stereoselectivity of the reaction. Inaki et al. observed an optimum molar ratio Cu:–NH_2 of chitosan derivatives for the polymerization of vinyl monomers[366,367]; an excess of ligand is required for the maximum polymerization of the substrate. The reducing ends on the polymer may contribute to the initiation of the polymerization. Yin et al. varied the amount of Pd immobilized on silica-chitosan composite and carried out the hydrogenation of a series of ketones[363]; the optical yield is very sensitive to the N:Pd mole ratio in the complex and small changes in the structure of ketone. When substituting iron–nickel with palladium, similar trends are obtained, although the conversion yields (global and optical) are slightly lower.[368] For the hydrogenation of α-phenylethanol by Pt supported on a chitin-silica support, Yuan et al. observed that a 100% optical yield can be obtained with intermediary Pd loading, although the total conversion continuously increases with increasing Pt content.[330] Under optimum experimental conditions, the asymmetric hydrogenation of α-phenylethanol to *R*-(+)-1-cyclohexyl ethanol reaches 65% conversion with 100% enantiomeric excess (ee). In the case of asymmetric hydrogenation of diketones by Rh immobilized on composite MgO-chitosan material, Zhou et al. showed the critical effect of metal loading on the optical yield.[249] The conversion yield continuously increases with metal content while the optical yield reaches a maximum at intermediary values of Rh content. In the cyclopropanation of olefins with Cu-immobilized on Schiff base derivatives of chitosan, the conversion yield is hardly affected by metal content, while the highest optical selectivity is obtained with an intermediary copper dosage.[311] Xue et al. immobilized cobalt on a silica-chitosan composite for the hydration of 1-octene[305]; the maximum yield and optical yield are obtained for Co content close to 0.5 mmol Co g^{-1}. The yield of hydrogenation of 4-methyl-2-pentanone with Pt immobilized on an alginic acid-glutamic acid-silica composite increases with metal content, but the enantioselectivity reaches a maximum for an intermediary Pt concentration (close to 0.08 mmol Pt g^{-1}).[245] For the same reaction, Huang et al. found the same optimum Pt loading when using silica-supported starch-polysulfosiloxane-Pt complexes.[247] Wei et al. carried out the grafting of several amino acids on alginic acid-silica composite for the hydrogenation of furfuryl alcohol.[246] Glutamic acid is the most efficient amino acid: the conversion yield reaches 100% and the maximum optical yield is obtained for Pt content close to 0.15 mmol Pt g^{-1}. In the case of the hydrogenation of nitrobenzene, cyclohexanone and octene catalyzed by SiO_2-starch-glutamic acid-Fe composite, the maximum conversion is obtained with a metal content in the range 0.1 to 0.15 mmol Fe g^{-1}.[255] Liu et al. immobilized Pt on a composite SiO_2-starch-oxalic acid for the asymmetric hydrogenation of 2-butanone and itaconic acid.[243] They observed that Pt loading strongly influences both the global yield of the reaction and the optical yield; intermediary Pt content is required for the hydrogenation of 2-butanone (close to 0.1 mmol Pt g^{-1}) and itaconic acid (i.e., 0.15 mmol Pt g^{-1}). Similar investigations were performed with proteins, including wool, gelatin, and casein, immobilized on silica and zeolite materials with Pt, Pd, Fe, Co-Ru, Co, or Rh catalytic metals, by the 4060 group of the Institute of Chemistry, at the Chinese Academy of Sciences.[242,256–258,326,337–339] In the catalytic hydrogenation of ketones, they found that 3-methyl-2-butanone is more easily converted [to (*R*)-methyl-2-butanol] than diacetone alcohol, acetophenone, 4-methyl-2-pentanone, and propiophenone, with higher production and optical yields, regardless of the catalyst carried out.[256,326,338] The optimum selectivity is obtained at intermediary metal loading, at the expense of a little decrease of product yield. In the case of 3-methyl-2-butanone hydrogenation, the product and optical yields can reach values as high as 80% and 100%, respectively, for selected experimental conditions. For cyclohexanone hydrogenation (conversion to cyclohexanol) using Fe supported on a SiO_2-casein carrier, the optimum metal loading is in the range 0.05 mmol to 0.20 mmol Fe g^{-1}.[257] Cobalt immobilized on a gelatin-polysulfostyrene composite reaches very high production and optical yields for the hydration of allyl alcohol, which is selectively converted to (*S*)-(–)-1,2-propanediol at the optimum Co loading (i.e., 0.4 mmol Co g^{-1}).[258] The optimum Co loading on the SiO_2-casein carrier is close to 0.2 mmol Co g^{-1} for the epoxidation of cinnamyl alcohol [converted to (2*R*,3*R*)-(+)-3-phenylglycidol]; the production and optical yields are close to 70% and 92%, respectively.[337]

However, Jiang et al. pointed out the difficulty in reproducing the high enantioselectivity of hydrogenation of 1-phenyl ethanol and ketones obtained with natural biopolymers,[369] possibly due to the variability in the composition of the materials with material source and material pretreatment. Additionally, they underline the analytical limitations of the experimental procedures followed for the determination of optical yields, which may overestimate the enantioselectivity of the transformation. Indeed, the optical rotation of the solution can be influenced by the partial dissolving of the chiral support and by the presence of different molecules with different intrinsic optical rotations. These reasons make mandatory the utilization of complementary analytical techniques such as NMR, chiral liquid chromatography (GLC), or high pressure liquid chromatography (HPLC). Therefore, the cited optical yields should be only taken as indicative of the potential of these materials for orienting the reactions.

These studies have shown that metal content is an important parameter in the design of a supported catalyst, involving significant changes in the conversion efficiency and on the optical yield. However, Vincent and Guibal have observed a limited impact of metal loading on the hydrogenation of nitrophenol by Pd supported on chitosan flakes.[343] In this case, this is probably due to restricted access to Pd crystals inside the material: the hydrogenation reaction is limited to the external surface of chitosan particles due to the poor porosity of raw material. The influence of diffusion properties should be taken into account for a true evaluation of the impact of metal content on catalytic activity. While metal ions can diffuse and saturate internal sorption sites, the access to these sites for substrate molecules may be limited, or at least be controlled, by mass-transfer properties.

4.3.2.3 Diffusion Properties

Nonporous supports involve only the external diffusion of substrate and product molecules across the fluid film surrounding them, while porous supports also entail internal diffusion. Obviously, diffusion restrictions are limited for nonporous materials, but at the expense of low superficial surface area and catalyst loadability (low volumetric density of reactive sites). Small particles address these limitations (Figure 4.22); however, small-size catalysts may induce drawbacks such as pressure drops in fixed bed reactors, low flow rates in fluidized reactors, and complex solid–liquid separation steps in batch reactors. Porous materials offer much greater surface areas and higher volumetric densities for catalytic sites, at the expense of diffusional limitations. The challenge is thus the control of both pore size and pore diameter.

Synthetic resins are generally classified in two groups regarding their morphology: (a) gel-type (microporous) resins without appreciable porosity in the dry state, whose interior is accessible only after swelling, and (b) macroreticular (macroporous) resins with macropores stable even in the dry state, in addition to the micropores generated by the swelling of the polymer skeleton.[301] These diffusional constraints may control metal binding, as pointed out above, and accessibility of substrate and gaseous reactants.[52]

The influence of diffusion restrictions depends on the characteristics of the polymer and the solvent. Duff et al. prepared a number of platinum colloids stabilized with polyvinylpyrrolidone, used as free colloids or immobilized on mineral surfaces for hydrogenation reactions in methanol solutions.[33] The polymer layer is expected to reduce the accessibility to catalytic sites; however, because the polymer was well solvated in methanol, diffusion properties are enhanced by

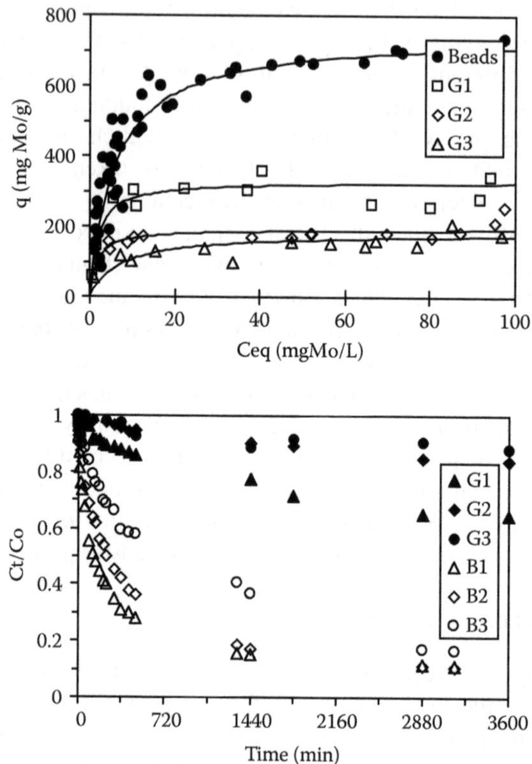

FIGURE 4.22 Influence of particle size of chitosan flakes (G1 < 125 μm < G2 < 250 μm < G3 < 500 μm) and chitosan gel beads (B1: 0.9 mm; B2: 1.6 mm; B3: 2.3 mm) on Mo sorption at pH 3 using glutaraldehyde cross-linking materials: Sorption isotherms (top, equilibrium data are superimposed for B1, B2, B3 series) and sorption kinetics (Co: 100 mg L^{-1}). (Reprinted from E. Guibal, C. Milot, and T. Vincent, *Ind. Eng. Chem. Res.*, 37, 1454–1463, 1998. With permission. Copyright 1998, American Chemical Society.)

the formation of a solvent rich layer at the surface of the catalyst. The diffusion properties may be limited when the polymer-stabilized catalyst is immobilized on a surface; the calcination of the catalyst oxidizes the polymer, and diffusion holes are created, improving the accessibility to reactive sites.

The contribution of diffusion limitations to the control of hydrogenation kinetics has been extensively studied on chitosan-supported Pd catalysts prepared by the adsorption process in aqueous solutions.[342–346] The influence of catalyst particle size has been carried out on the hydrogenation of nitrophenol and nitroaniline using sodium formate as the hydrogen donor.[343,345] Increasing the size of catalyst particles leads to a significant increase in the time required for reaching equilibrium (Figure 4.23). The conversion of substrate was modeled using the Langmuir-Hinshelwood equation; however, in most cases the pseudo-first equation, applied to supported catalysis,[12] fits the experimental data well,

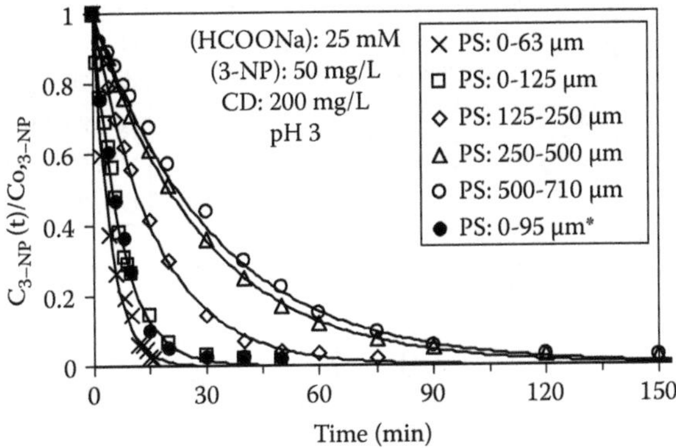

FIGURE 4.23 Influence of particle size of the support on the hydrogenation of nitrophenol using Pd immobilized on chitosan flakes and sodium formate as the hydrogen donor (PS: 0–95 μm corresponds to the grinding and sieving of PS: 500–710 μm size fraction). (Reprinted with permission from T. Vincent and E. Guibal, *Langmuir,* 19, 8475–8483, 2003. With permission. Copyright 2003, American Chemical Society.)

$$\frac{dC(t)}{dt} = \frac{-k_1 C(t)}{1 + k_2 C(t)} \tag{4.3}$$

where k_1 (min^{-1}) and k_2 ($L\ mg^{-1}$) are the kinetic parameters.

The kinetic coefficients vary with the reciprocal of the diameter of the particles. The reaction is located at the external layers of catalyst particles; the intraparticle diffusion is the rate-limiting step. The agitation speed has a limited effect on the hydrogenation of nitrophenol[343]; this is more evidence of the limited impact of external diffusion on kinetic control and an indication that the predominant limiting step is the intraparticle mass-transfer resistance (Figure 4.24).[370] This may explain the number of studies focusing on the deposition of biopolymers on materials with high superficial areas, such as silica and alumina. The immobilization of a thin biopolymer layer on these materials combines the specific properties of the coating molecules with the intrinsic properties of the carrier (i.e., high superficial surface areas and mechanical properties).

Huang et al. prepared SiO_2-chitosan composites for the immobilization of Pd nanoparticles.[253] The catalyst is used for nitrobenzene hydrogenation in ethanol solutions. During preparation, first chitosan is dissolved in acetic acid solution, then silica is added to the solution. The mixture is dipped into acetone to give a chitosan shell and silica core. After drying, the powder is dipped into a $PdCl_2$ solution (pH 1 to 2). After refluxing the mixture, the catalyst is dried and washed with ethanol. The reduction of palladium is performed in ethanol with hydrogen. The size of catalytic clusters increases when decreasing the polymer:metal ratio. However, the catalytic activity (measured by hydrogen uptake rate) is not proportional to the size of clusters. The optimum catalytic

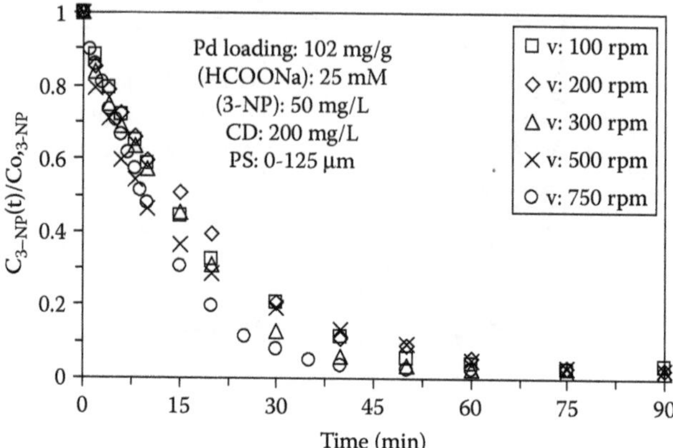

FIGURE 4.24 Influence of agitation speed on the hydrogenation of 3-nitrophenol using Pd immobilized on chitosan flakes and sodium formate as the hydrogen donor. (Reprinted from E. Guibal, T. Vincent, and S. Spinelli, *Sep. Sci. Technol.*, 40, 633–657, 2005. Copyright 2005, Taylor & Francis Group.)

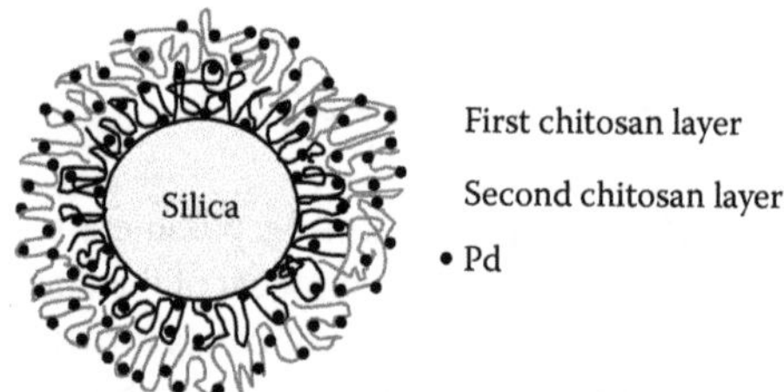

FIGURE 4.25 Modeling of chitosan layers at the surface of SiO_2 surface and interactions with Pd. (Reprinted from A. Huang, Y. Liu, L. Chen, and J. Hua, *J. Appl. Polym. Sci.*, 85, 989–994, 2002. With permission. Copyright 2002, John Wiley & Sons Ltd.)

activity is found for a silica:chitosan ratio around 7:1 to 8:1. The catalyst is modeled as a mineral core coated with a first polymer layer on which other polymer chains can be adsorbed, forming multilayers. This structure causes Pd to be distributed on the particles in three different forms: (a) immobilized on silica surface (internal sites), (b) scattered in the polymer networks, and (c) adsorbed on the outer surface (Figure 4.25). The number of layers immobilized on the silica surface depends on the amount of chitosan mixed with silica particles. Lower accessibility to internal metal sites can reduce the kinetic rate and even the reactivity of metal crystals. Optimum catalytic efficiency is found for monolayer coverage of the silica surface, corresponding to a thickness of 40 to 50 nm (which corresponds, more or less, to double the gyration radius of the polymer). The external layer is directly in contact with the solution and the reaction is thus instantaneous, but the aggregation of particles may induce significant loss of catalytic activity. Within the polymer network, the reaction rate can be controlled by diffusion properties, while the innermost surface of

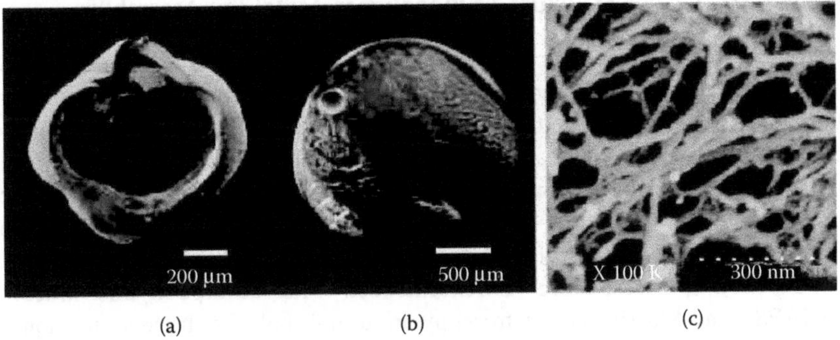

FIGURE 4.26 SEM picture of a cross-section of a G-20 bead: xerogel (a), aerogel (b), and inner part of the aerogel (c) at different magnifications. (Reprinted from R. Valentin, K. Molvinger, C. Viton, A. Domard, and F. Quignard, *Biomacromolecules,* 6, 2785–2792, 2005. With permission. Copyright 2005, American Chemical Society.)

the polymer layer is much less active due to strong limitations in penetrating the polymer layers.

Another possibility for improving the diffusion properties of catalytic materials consists of conditioning the material in the form of thin layers (membranes, fibers, hollow fibers) with diffusion lengths reduced as much as possible. The preparation of gels with high water content (and a large porous network) is a first step in the improvement of diffusion properties. The major drawback remains the difficulty of managing material drying and increasing the mass density of sorption sites. Indeed, the uncontrolled drying results in the irreversible collapse of the porous structure. Freeze-drying slightly improves the reversibility of the drying step, relative to porous structure. However, the most promising technique for the enhancement of mass transfer in biopolymer catalysts with dry materials consists of drying the xerogels in CO_2 supercritical conditions.[226,227,270] Solvent exchange (ethanol exchange with water) followed by drying under supercritical CO_2 conditions, allows the porous structure of the biopolymers to be maintained (Figure 4.26), as also evidenced by the measurement of specific surface areas using nitrogen adsorption/desorption isotherms and the Brunauer-Emmet-Teller (BET isotherm) method. The method requires drying the material before measuring the sorption of nitrogen. In the case of alginate aerogels (obtained with supercritical CO_2 conditions), the BET surface varies between 190 and 400 $m^2\ g^{-1}$ (depending on the metal used for ionotropic gelation), while xerogel and original materials do not exceed a few $m^2\ g^{-1}$.[271] Valentin et al. evaluated the impact of drying methods (freeze-drying vs. drying under supercritical CO_2 conditions) on the percentage of primary amine groups really accessible.[226] This is measured by the degree of substitution of salicylaldehyde on amine groups (by formation of Schiff base derivatives); the drying under supercritical CO_2 conditions strongly increased the percentage of accessible amine groups (from 4 to 54% and from 27 to 73% for cyclohexane and ethanol solvents, respectively). This is clear evidence of the critical importance of the final treatment of catalyst particles on their diffusion properties.

4.3.2.4 Molecular Structure of Biopolymer–Metal Ion Assemblies

Conformational changes to catalytic metals may change their enantioselectivity. This is more likely to occur when the catalyst is immobilized on the support through strong interactions (covalent bonding). The support can form part of the conformational constraint that controls the enantio- or regioselectivity. By attaching a catalytic metal to a chiral support, the substrate may be induced to interact with both the metal and the support. This geometric and morphological control may change the activity of the catalyst and also influence the conversion route.

Electron spin resonance (ESR), extended x-ray absorption fine structure (EXAFS), and Mössbauer spectroscopy are useful tools for the identification of the structure of metal-chitosan complexes.[371,372] The flexibility of biopolymer chains is an important advantage for metal uptake, but it makes the metal loaded on biopolymers stereoselective, which is responsible for chiral reactions (asymmetric hydrogenation for example, with the restrictions cited above).[249,330,371–373]

Inaki et al. clearly showed the impact of chitosan conformation on the reaction rate of vinyl polymerization.[366,367] The neutralization of cationic charges of amino groups by pH change and by the screening effect of salts (ionic strength), as well as temperature, controls the flexibility of the polymer chain, which in turn interferes with the geometry of the metal-biopolymer complex and its reactivity.

Chang et al. showed that the spatial arrangement of ligands (*m*- and *o*-nitrobenzaldehyde) around Mn and Ni affects the oxidation mechanism of hydrocarbons (Figure 4.27).[364] *Ortho*- or metaposition on nitrobenzaldehyde ligand controls the reactivity and the selectivity of the reaction. They investigated a series of benzene derivatives and show that *m*- and *o*-nitrobenzaldehyde do not systematically oxidize the same substrates and that the final product is influenced by the type of ligand. In the case of cyclopropanation of olefins, the enantioselectivity of the reaction is influenced by the chemical state of copper: the conversion is enhanced when copper is immobilized on Schiff base derivatives of chitosan.[311] The enantioselectivity of the reaction is controlled by the substituents on the salicylaldehyde moiety of the catalysts; Wang et al. concluded that the steric effect is more important than the electronic effect.[311] This is consistent with the conclusions reached by Sun et al.[336] Comparing the cyclopropanation of olefins with several copper-based derivatives of chitosan, they observe that Schiff base derivatives of chitosan have much greater activity and higher selectivity than copper immobilized on raw chitosan.

Hu et al. used a CoSalen complex immobilized on chitosan functionalized with pyridylmethylidene grafting for the oxidation of DOPA (3-(3,4-dihydroxyphenyl)alanine).[313] The catalytic efficiency is strongly increased when the complex is immobilized on the polymer. This enhancement of catalytic activity is explained by a site isolation effect caused by the polymer chain. Hence, CoSalen [a four-coordinated Co(II) complex] does not bind oxygen, which strongly limits the formation of inactive dimmers. However, it can bind oxygen in the presence of a suitable monodentate Lewis base [the peroxo-bridged species of CoSalen,

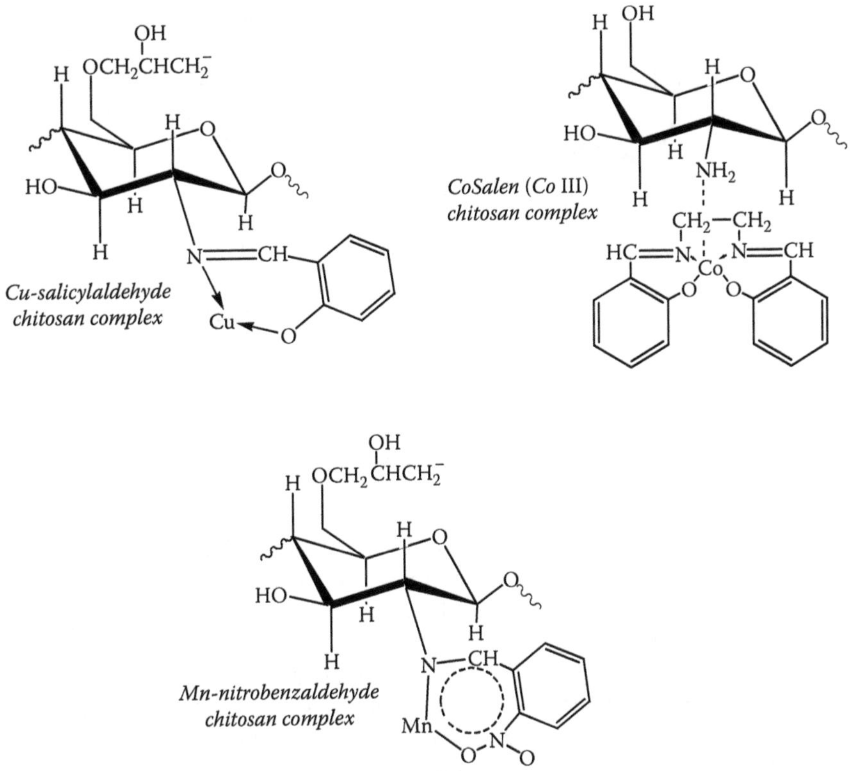

FIGURE 4.27 Structure of Cu(II), Co(II) and Mn(III) complexes with Schiff's base derivatives of chitosan. (Reprinted from Y. Chang, Y. Wang, and Z. Su, *J. Appl. Polym. Sci.*, 83, 2188–2194, 2002. With permission. Copyright 2002, John Wiley & Sons.)

$(CoSalen)_2\ O_2]$, which is unable to directly catalyze oxidation. The site isolation effect brought by pyridylmethylidene chitosan prevents the formation of dimmers and peroxo-bridged dimmers, improving the formation of an active species of superoxo cobalt complex $[CO(III)O_2^-]$.

Paradossi et al. utilized the stiffness of the gel formed by the reaction of oxidized β cyclodextrin to prepare a Cu-catalyst.[372] The host cavities in the chitosan-based network might be modulated according to the geometry of the guest molecule or multivalent ions to be trapped in the hydrogel. This may contribute to higher catalytic selectivity.

Kucherov, Kramareva, and coworkers extensively discuss the interactions between chitosan and copper,[221,222,317] and palladium.[319] They correlate these properties (stereochemistry of the complexes) with their catalytic activity (Figure 4.28). More specifically, they show that the matrix of heterogenized chitosan is able to stabilize and retain copper ions in a highly unsaturated, coordinative state (water molecules penetrating the catalyst do not coordinate with copper as additional ligands),

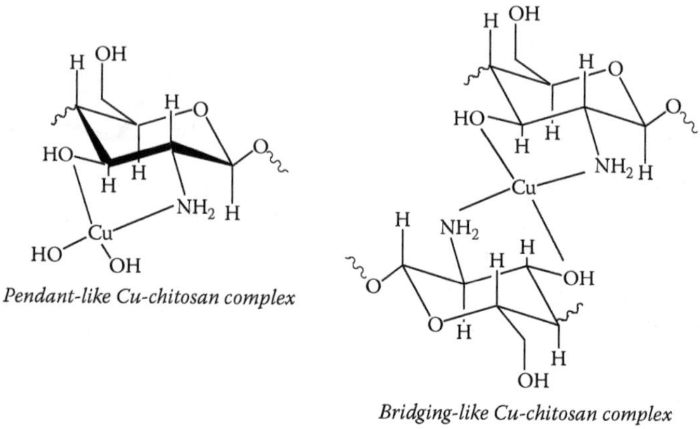

Pendant-like Cu-chitosan complex

Bridging-like Cu-chitosan complex

FIGURE 4.28 Structure of Cu(II) complexes with chitosan: Pendant-like and bridging-like models.

which allows the uptake of molecular oxygen and the reoxidation of copper sites (reduced in the course of the oxidation reaction). The optimum use of metal is obtained for low copper content in the catalyst. In this case, the symmetry of copper ions in chitosan is close to square-planar coordination. They compared the catalytic performance of Cu-catalyst on different supports and correlated the catalytic activity with the hydrophilicity of the support. The most important parameter remains the structural parameter, requiring a high degree of coordinative unsaturation.[222] They prepared a series of Pd catalysts using adsorption and precipitation processes.[319] In the case of the adsorption process, they concluded that metal binding occurs by coordination with two amino groups and the contribution of two ligands (chloride ions), being close to the coordination sphere of a $Pd(NH_2\text{-chit})_2Cl_2$ complex. In the case of the coprecipitation process, the metal forms a chelate complex (palladium coordinated to amino and hydroxyl groups of chitosan) (Figure 4.14). Metal reduction is carried out using sodium borohydride, and the structure of the metal may be strongly changed due to possible aggregation phenomena. For precipitated material, the aggregation degree is greater, with formation of Pd_3 triangular structures, compared to the adsorption process that induces the formation of Pd_2 dimmers.

Guan and Cheng showed that Co is bound to chitosan via both nitrogen and oxygen atoms in the polymer chain and that the coordination polymer possesses the same shake-up peaks as $CoCl_2$,[374] and they concluded that the complex is high spin and the coordination number of Co(II) centers is 4. They used the chitosan-Co complex for the polymerization of vinyl acetate and they correlated its high catalytic activity to the high spin structure of the complex.

Köckritz et al. have shown that the catalytic activity of osmium tetroxide immobilized on chitin is much higher than that obtained by metal immobilization on chitosan.[375] The weaker activity on chitosan-supported catalysts is attributed to a higher binding of the metal on free amine groups that decreases the availability of metal for catalyzing the hydroxylation of olefins. This is further evidence of the impact of a metal-chitosan interaction mechanism (and strength) on the catalytic activity of the heterogeneous material.

Despite the number of studies published on the use of other biopolymers as support for heterogeneous catalysis, most of them only focus on the characterization of metal-biopolymer binding and the determination of catalytic activities, with less attention to the impact of metal-biopolymer structures on the reaction.

4.3.2.5 Stability

The stability of metal complexes is another important concern for the design of a heterogeneous catalyst. This stability may be affected by metal leaching, by catalyst poisoning, or by deactivation. The leaching is controlled by the strength of the interaction of the metal with the support[222,376,377] or by the physical entrapment of the catalyst complex in the polymer matrix.[220]

The deactivation of the catalyst may proceed by progressive change of the oxidation state of the catalytic metal, which may require reactivation (rehydrogenation or reoxidation, depending on the reaction).

In most cases, the immobilization allows maintaining production yield approximately constant; the optical yield is most frequently affected by the recycling of the catalyst over three to five cycles. For the hydrogenation of ethyl pyruvate, Köckritz et al. observed that chirally stabilized Pt-colloids immobilized by encapsulation in alginate maintain a constant enantioselectivity and a good activity during 25 cycles.[220]

Kramareva et al. commented on the irreversible inactivation of a CoSalen complex by the formation of a complex with hydroquinone in the oxidation of dihydroxybenzene compounds, while the heterogenized complex CoSalen-chitosan immobilized on SiO_2 remained active.[317] More specific information on chitosan as a support for heterogeneous catalysis can be found in the comprehensive review published by Guibal.[378]

4.4 EXAMPLES OF REACTIONS CATALYZED BY BIOPOLYMER-SUPPORTED CATALYSTS

4.4.1 Hydrogenation and Hydroformylation Reactions

Hydrogenation is the reaction the most frequently cited for biopolymer-supported catalysis. Various biopolymers (alginic acid, starch, cellulose, chitosan, wool, gelatin, silk, or casein) have been tested for the conversion of nitro-compounds (benzene derivatives), alcohols, ketones, and olefins.

4.4.1.1 Hydrogenation of Alcohols

Yuan et al. described the preparation of SiO_2-chitin composite material by solid mixing of SiO_2 particles with chitin powder, followed by impregnation with

α-phenylethanol (racemate) $\xrightarrow[\text{Cat.}]{H_2}$ (R)-(+)-1-cyclohexyl ethanol

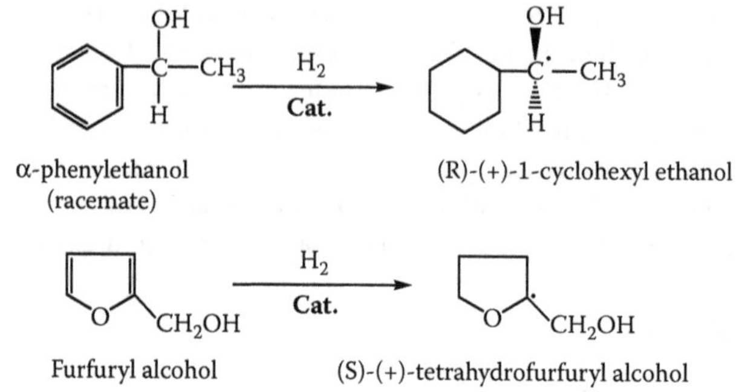

Furfuryl alcohol $\xrightarrow[\text{Cat.}]{H_2}$ (S)-(+)-tetrahydrofurfuryl alcohol

FIGURE 4.29 Examples of asymmetric hydrogenation of alcohols.

hexachloroplatinic acid in ethanol under reflux.[330] However, the procedure is not fully documented, making difficult the understanding of the mechanism allowing chitin to bind to silica. They use SiO_2-chitin-Pt material for the hydrogenation of α-phenyl ethanol dissolved in a solvent (ethanol, methanol, 1,2 dichloroethanol, cyclohexane, 1,4-dioxane) with hydrogen gas (Figure 4.29). Substrate conversion increases with Pt content; however, the selectivity for the production of (*R*)-(+)-1-cyclohexyl ethanol reaches a maximum for Pt content close to 0.1 mmol Pt g^{-1}. They concluded that only a few Pt active centers are chiral. The conversion also changes with the temperature, reaching a maximum at 40°C, with a maximum selectivity around 30°C. At high temperature (60°C and above), the product becomes an optically inactive racemate. Best results are obtained using ethanol as the solvent of the system in terms of both conversion yield and enantiomeric excess; adding HCl at small concentration (i.e., 4 m*M*) improves the hydrogenation rate. The catalyst can be reused for at least six cycles without significant loss of catalytic activity and selectivity. Wei et al. prepared a SiO_2-alginic acid-amino acid support: they mixed sodium alginate with amino acid solution (L-alanine, L-glutamic acid, L-serine or L-lysine) before the addition of SiO_2 and final precipitation of composite material by addition of HCl.[246] Hexachloroplatinic acid is used in ethanol under reflux for preparation of Pt catalyst immobilized on SiO_2-alginic acid-amino acid support. Furfuryl alcohol hydrogenation to tetrahydrofurfuryl alcohol is catalyzed under hydrogen atmosphere in solvent (Figure 4.29). The highest conversion and best selectivity is obtained using L-glutamic acid. With proper reaction conditions, the yield and the *ee* reach 84% and 98%, respectively. The optimum experimental conditions correspond to 0.15 mmol Pt g^{-1} Pt content, 2:1 alginic/L-glutamic acid molar ratio, ethanol used as solvent, temperature close to 30°C, a reaction time longer than 15 h, and a 280:1 substrate:Pt molar ratio.

4.4.1.2 Hydrogenation of Ketones

Since 1999, the 4060 group at the Institute of Chemistry of the Chinese Academy of Sciences (Zhongguancun, China) has developed a number of catalysts deposited on biopolymers for the hydrogenation of ketones.[243,245,247,249,256,257,326,331,338,363,368]

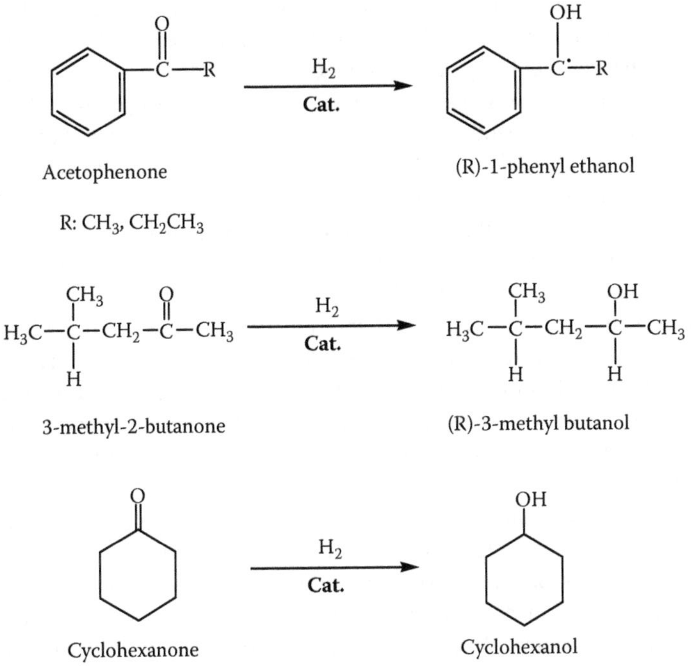

Acetophenone → (R)-1-phenyl ethanol (H_2, Cat.)

R: CH_3, CH_2CH_3

3-methyl-2-butanone → (R)-3-methyl butanol (H_2, Cat.)

Cyclohexanone → Cyclohexanol (H_2, Cat.)

FIGURE 4.30 Examples of asymmetric hydrogenation of ketones.

They systematically use the impregnation-reduction procedure in alcohol under reflux procedure for the immobilization of catalytic metals.

The hydrogenation of propiophenone and acetophenone to (*R*)-(+)-1-phenyl-1-propanol and (*R*)-(+)-1-phenyl ethanol in solvent under H_2 atmosphere is investigated using SiO_2-chitosan-supported catalysts, immobilizing Pd,[363] or Fe-Ni (Figure 4.30).[368] Chitosan dissolved in acetic acid solution is mixed with silica before being precipitated in NaOH solution. The composite material is mixed with $PdCl_2$ or $NiCl_2/FeCl_3$ in ethanol under reflux. The optimum temperature for highest yield and best selectivity is increased from ambient to 110°C when changing the catalytic metal from Pd to Ni/Fe. The nonnoble metal also requires higher H_2 pressure for completing the conversion and reaching appreciable selectivity (5 to 7 bar instead of 1 bar). The SiO_2-chitosan-Pd catalyst is generally more reactive than the nonnoble metal catalysts. The chitosan:metal molar ratio considerably affects the orientation and the yield of the reaction: the optimum is found close to 4 to 5 for Ni/Fe catalysts (at the same molar proportion) and around 2.5 for Pd catalysts. While for nonnoble catalysts, acetophenone and propiophenone follow the same hydrogenation trend, in the case of Pd catalyst the structure of the substituent has a more marked effect: the optical yield is generally lower for propiophenone. The hydrogenation of these ketones was also studied by Yuan et al. using a wool-Pt complex,[338] and by Zhang et al. using a zeolite-supported gelatin-Fe complex.[256] Zeolite is mixed to a gelatin solution (at 60°C), before being dried, to prepare the reactive support. For the immobilization of the metals ($FeCl_3$ or hexachloroplatinic acid), the salts are dissolved in ethanol and mixed with the supports (zeolite-gelatin, or wool). The hydrogenation is operated in ethanol under hydrogen atmosphere. The production and optical yields for Pd-wool catalysts are comparable to those obtained with above-cited heterogeneous catalysts (i.e., 50 to 70% and 40 to 60%, respectively). For a Pd-wool catalyst, the optimum dosage of Pd is close to 0.15 mmol g^{-1}.

Sun et al. carried out the catalytic transfer hydrogenation process (CTH) for the conversion of acetophenone to (*R*)-1-phenyl ethanol.[329] Chitosan is immobilized on silica by the conventional dissolving-precipitation procedure; $PdCl_2$ is reacted with the support in ethanol solution under reflux. Different hydrogen transfer agents are tested; they can be classified according to: $HCOONH_4$ > HCOONa > HCOOH/trietylamine, HCOOK, HCOOH. The enantiomeric excess is comparable to those obtained with other systems, but the conversion yield is substantially reduced. The catalytic performance increases with the temperature (up to 80°C).

Ishizuki et al. prepared a series of precious metal particles (Au, Pd, Pt, Rh) stabilized by chitosan.[234] The metal solutions (single or binary metal) are mixed with polymer solution; the mixture is reduced using hydrazine, formaldehyde (in a KOH solution) or pentaerythritol. Hydrazine is the most powerful reducing agent for tested systems. When increasing the proportion of chitosan, the size of metal colloids decreases, while the dispersion stability increases. In binary-metal colloids (Au-Pd, Au-Pt, and Au-Rh), the size of the nanoparticles is hardly affected by the composition of the solution, while the catalytic rate for the hydrogenation of vinyl ketone to ethyl methyl ketone is significantly reduced when increasing the molar proportion of Au in the binary catalysts.

A number of studies deal with the hydrogenation of 3-methyl-2-butanone (3M2B), and 4-methyl-2-pentanone (4M2P) in solvent medium (ethanol reveals the most efficient) using H_2 atmosphere with different biopolymer-supported systems. Silica is used for the immobilization of chitosan,[363] starch (combined with polysulfosiloxane),[247] alginic acid (combined with L-glutamic acid), according to methods described above (Section 4.3.1.1).[245] Zhang et al. immobilized gelatin on zeolite,[256] while Yin et al. and Yuan et al. directly used wool as a metal carrier.[326,338] For 3M2B hydrogenation to (*R*)-3-methyl-2-butanol using precious metals immobilized on wool at 20 to 30°C under 1 bar H_2, the catalytic efficiency and selectivity depend on metal content: the optimum is found close to 0.12 mmol Pt g^{-1} for Pt,[338] and close to 0.24 to 0.3 mmol Pt g^{-1} for Pd.[326] With Pd immobilized on a SiO_2-chitosan composite, the hydrogenation of 3M2B is hardly affected by Pd content; however, when recycled, the catalyst progressively loses optical yield, except when using high-metal loading.[363] For 3M2B hydrogenation using Fe immobilized on a zeolite-gelatin composite, optimum conversion yield and enantiomeric excess are obtained for a temperature close to 20°C, a substrate:Fe molar ratio close to 150. The optical yield decreases and the production yield increases when increasing Fe content; the best conditions (100% selectivity) with acceptable conversion yield (51%) are obtained when Fe content is close to

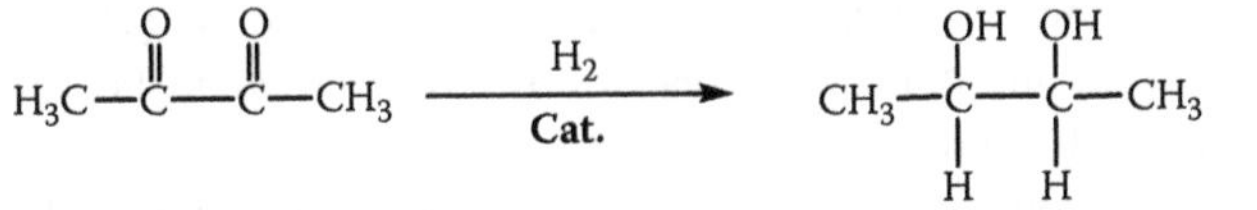

$$H_3C-C(=O)-C(=O)-CH_3 \xrightarrow[\text{Cat.}]{H_2} CH_3-CH(OH)-CH(OH)-CH_3$$

2, 3-butadione (2*S*, 3*S*)-(+)-2, 3-butanediol

FIGURE 4.31 Example of asymmetric hydrogenation of diketone.

0.05 mmol Fe g^{-1}. SiO_2-chitosan-Pd and wool-Pt catalysts have been tested for 4M2P hydrogenation to (*R*)-4-methyl-2-pentanol; the optical yields are generally lower than those observed with the same materials for the hydrogenation of 3M2B. Much better results are cited for 4M2P hydrogenation using starch-polysulfosiloxane-Pt and SiO_2-alginic acid-L glutamic acid-Pt catalysts. For starch-supported Pt catalysts, a good compromise between conversion yield and optical yield is obtained when using ethanol as the solvent, Pt content close to 0.08 mmol Pt g^{-1}, and substrate:Pt molar ratio close to 80:1; conversion and optical yield exceed 73% and 94%, respectively. With alginate-supported material, hydrogenation reaches a maximum efficiency and selectivity for Pt content close to 0.08 mmol Pt g^{-1}, and substrate:Pt molar ratio close to 100:1.

Zhou et al. investigated the hydrogenation of 2,3-butanedione and 2,4-pentanedione, respectively converted to (2S,3S)-(+)-2,3-butanediol and (2S,4S)-(+)-2,4-pentanediol in solvent media (preferentially ethanol) under 1 bar H_2 pressure using MgO-chitosan-Rh catalyst (Figure 4.31).[249] Chitosan is dissolved in acetic acid solution; the pH is thus controlled to 7 with NaOH, before adding MgO. A new addition of NaOH (controlling the pH to 13) allows depositing chitosan at the surface of MgO. The composite material is tested for Rh immobilization by contact of $RhCl_3$ in ethanol under reflux and N_2 atmosphere. The production yield increases with Rh content and temperature, and the optical yield reaches a maximum when Rh content reaches 0.1 mmol Rh g^{-1} and for temperature close to 28°C. The catalyst is successfully reused for up to five cycles without significant loss of conversion efficiency and optical yield.

Cyclohexanone is converted to cyclohexanol in ethanol under hydrogen atmosphere using Fe catalyst immobilized on SiO_2-casein support (Figure 4.30).[257] The support is obtained by mixing casein with silica under heating before being dried. Iron(III) chloride is deposited on the support by the ethanol procedure. For selected experimental conditions, a higher conversion yield is obtained with Fe content close to 0.05 mmol Fe g^{-1} and a contact time greater than 24 h. The substrate:Fe molar ratio does not change the conversion yield but increases significantly the reaction time required for achieving complete conversion of the substrate. Tested for the hydrogenation of other substrates, using comparable experimental conditions, the catalyst shows high conversion rates for propiophenone ≈ benzaldehyde >> cyclohexanone >> cyclohexene >> acetylacetone, and very weak activity for the hydrogenation of styrene, acrylic acid, cyclopentanone, and nitrobenzene. Liu et al. also used an Fe(III) catalyst immobilized on SiO_2-starch-L glutamic acid support for the conversion of cyclohexanone to cyclohexanol in ethanol under

1 bar H_2 atmosphere.[255] At a fixed substrate:Fe molar ratio (75:1), the initial hydrogenation rate and conversion reaches a maximum when Fe content in the catalyst reaches 0.1 mmol Fe g^{-1}.

For the hydrogenation of 2-butanone to (*S*)-(+)-2-hexanol and methylacetoacetate to methyl-(*S*)-(+)-3-hydroxybutyrate, Wei et al. carry out a bimetallic Pd/Fe catalyst deposited on SiO_2-chitosan composite.[331] The ethanol procedure under reflux (nitrogen atmosphere) is used for the immobilization of Pt (hexachloroplatinic acid) and Fe ($FeCl_3$ salt) onto SiO_2-chitosan composite. Hydrogenation is carried out under 1 bar H_2 pressure in ethanol (methanol, water, and propanol reveal inefficient for substrate conversion), at the optimum temperature (30°C). For given amounts of Pd and Fe, the production and optical yields increase with decreasing the amount of chitosan immobilized on silica particles, probably due to the greater dispersion of sorption sites, avoiding the agglomeration of metal crystals. The Pd:Fe molar ratio is fixed to 10, and the optimum metal content for enhanced catalytic activity is found close to 0.2 mmol Pt g^{-1}. The conversion and optical yields continuously decrease with increasing the substrate:Pt ratio for 2-hexanone conversion, while for methylacetoacetate the production decreases with decreasing substrate:Pt ratio and optical yield reaches a maximum for substrate:Pt close to 150:1.

Methylcellulose (combined with oxalic acid) has been deposited on SiO_2 for the immobilization of Pt and the hydrogenation of 2-butanone and itaconic acid.[243] Methylcellulose is mixed with oxalic acid-SiO_2 suspension in water; after drying the slurry, the support serves to immobilize Pt by contact with hexachloroplatinic acid in ethanol under reflux and nitrogen atmosphere. The hydrogenation reaction takes place under hydrogen atmosphere in ethanol. The initial hydrogenation rate for 2-butanone and itaconic acid increases with Pt content, but the product yield (measured at 12 h) decreases (more significantly for 2-butanone than for itaconic acid). This parameter has an important effect on the optical yield, which reaches a maximum at 0.1 mmol Pt g^{-1} for 2-butanone and 0.15 mmol Pt g^{-1} for itaconic acid. The initial hydrogenation rate and conversion yield decrease with increasing the substrate:Pt molar ratio for each substrate, while the optical yield reaches a maximum at 1500:1 and 100:1 for 2-butanone and itaconic acid, respectively.

The hydrogenation of phenol and cresols has been investigated by Tang et al. using SiO_2-chitosan-Pd catalyst.[324,362] They observed that the conversion of phenol is very selective for cyclohexanone, preventing further hydrogenation of the product to cyclohexanol, as it may occur in conventional systems involving Pt catalysts; more specifically, the hydrogenation of cyclohexanone, described by Shen et al. using SiO_2-casein-Fe (see above),[257] is not observed. The conversion reaches a maximum with the chitosan:Pd molar ratio close to 8, at 70°C in cyclohexane as the solvent. A progressive deactivation of the catalyst is observed, though the conversion maintains its selectivity over five cycles. The reactivity is compared for phenol and cresols; the reactivity is controlled by the substituents and their position on the phenolic cycle: phenol > *m*-cresol > *p*-cresol >> *o*-cresol.

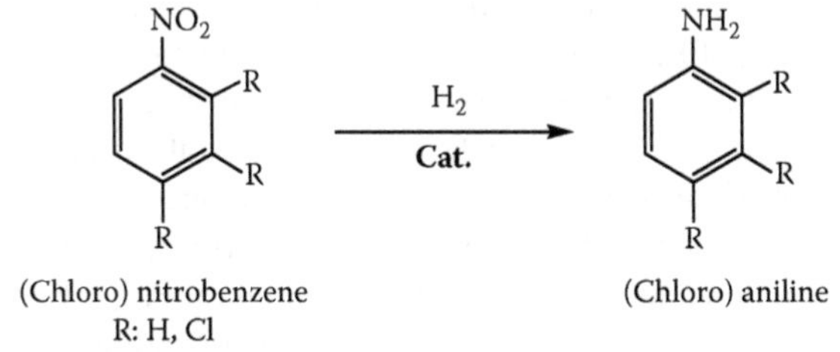

FIGURE 4.32 Example of hydrogenation of nitro-aromatic compound.

4.4.1.3 Hydrogenation of Aromatic Nitro-Compounds

Palladium catalyst supported on SiO_2-chitosan composite has been prepared using the conventional two-step procedure (chitosan precipitation on silica particles, followed by metal binding in ethanol under reflux).[379] The catalyst is tested for the hydrogenation of a series of chloro-nitrobenzenes in solvent under 1 bar H_2 pressure (Figure 4.32). The reaction is specific and the side-reaction is not obtained (i.e., the substrates are not dehalogenated): *p*-, *m*-, *o*-chloronitrobenzene, and 4-chloro-1,3 dinitrobenzene are quantitatively converted to *p*-, *m*-, *o*-chloroaniline and 4-chloro-3-aminoaniline, respectively. The initial hydrogenation rates are found to be comparable for the different substrates, but it depends on the chitosan:Pd molar ratio (based on glucosamine monomer group), reaching a maximum around 50. The catalytic activity is comparable at using methanol, ethanol cyclohexane, or toluene as the solvent, while the hydrogenation rate strongly decreases in tetrahydrofuran. The optimum temperature is close to 45°C. Han et al. used the same SiO_2-chitosan composite for the preparation of a series of nonnoble single metallic [Ni, Cu, Cr, Co, Fe(II), and Fe(III)] or bimetallic (Cu/Zn, Cu/Cr) catalysts; they tested their catalytic activity for the hydrogenation of nitrobenzene to aniline in ethanol under 15 bar pressure.[333] Single-metal catalysts are poorly reactive compared to bimetallic catalysts: the yield of the reaction increases from 10% to 96–100% when combining Cu and Zn or Cr as catalytic active sites. The reaction requires working at high temperature: the conversion yield sharply increases above 120°C. The highest yield is obtained when using chitosan (on monomer unit basis):Cu molar ratio close to 12 (maintaining the Cu:Zn molar ratio constant at 1:1). The solvent influences conversion yield and solvents can be classified according to: ethanol > methanol > water >> tetrahydrofuran (THF) Å 1,4-dioxane. The catalyst has also been successfully tested for the hydrogenation of 2-nitroanisole, 1-chloro-4-nitrobenzene, 2-nitroaniline, and 2-nitrotoluene to 2-anisidine, 4-chloroaniline, 1,4-phenylenediamine, and 2-toluidin, respectively. In this case, again, no by-products are obtained, indicating that a side reaction is not involved and that the reaction is highly selective.

Liu et al. achieved the complete hydrogenation of nitrobenzene to aniline using a Fe(III) catalyst immobilized on SiO_2-starch-L glutamic acid support, at 45°C under 1 bar H_2 pressure.[255] With the substrate/Fe molar ratio fixed to 100:1, the full conversion is obtained for Fe(III) content close to 0.1 mmol Fe g^{-1}. Gelatin, alginic acid, casein, and carboxymethylcellulose have also been immobilized on SiO_2 for the binding of Pt.[241] The biopolymers, dissolved in water, are mixed with silica pretreated with oxalic acid; the supports, after drying, are impregnated with hexachloroplatinic acid in ethanol under reflux. The catalysts have been tested for the hydrogenation of nitrobenzene to aniline, which takes place in ethanol under 1 bar H_2 pressure. The substrate:Pt ratio is fixed to 240:1; the initial hydrogenation rate reaches a maximum at Pt content close to 0.8 mmol Pt g^{-1}. Except in the case of casein-based support, under selected experimental conditions, nitrobenzene is selectively converted to aniline. Based on initial hydrogenation rate, the SiO_2-oxalic acid-biopolymer-Pt catalysts can be classified according the biopolymer series: carboxymethyl cellulose ≈ alginic acid > gelatin >> casein. The catalysts the most stable (for recycling) are catalysts prepared using gelatin and alginic acid biopolymers.

Zhu et al. described the synthesis of different dimethylaminoethyl derivatives of chitosan for the immobilization of Rh cluster, and the reduction of nitrobenzene and benzaldehyde.[380,381] Instead of molecular hydrogen, water is used as the hydrogen supply for this water–gas shift reaction. Benzaldehyde is transformed to benzyl alcohol with a maximum yield obtained at 80°C, and its conversion rate increases with the amount of catalyst. The solvent has a large impact on the conversion: 2-ethoxyethanol is much more efficient than toluene. Under optimum conditions (2-ethoxyethanol, *T*: 80°C), benzaldehyde conversion exceeds 96%. Nitrobenzene is converted to aniline with yield of 70% in 2-ethoxyethanol solvent (which is more appropriate than toluene).

Chitosan dissolved in acetic acid has been added simultaneously with an aqueous solution of polyacrylic acid (PAA) [or polymethacrylic acid (PMAA)] to silica particles to prepare SiO_2-chitosan-PAA (or PMAA) support.[252] Palladium chloride is immobilized on these supports by the ethanol procedure (under reflux). The hydrogenation reaction of nitrobenzene to aniline (acrylic acid to propionic acid) takes place in methanol under 1 bar H_2 pressure, at 30°C. The relative proportions of chitosan and PAA control the initial hydrogenation rate of nitrobenzene and acrylic acid: the optimum chitosan:PAA acid molar ratio (based on monomer molar units) is obtained at 44:56 and 50:50 for nitrobenzene and acrylic acid hydrogenation, respectively. For chitosan/PMAA support, the optimum proportion is 45:55 for both nitrobenzene and acrylic acid hydrogenation.

4.4.1.4 Hydrogenation of Olefins

Isaeva et al. investigated the liquid-phase hydrogenation of cyclopentadiene using different chitosan-supported Pd catalysts (Figure 4.33).[248] Chitosan is conditioned in the form of beads, fibers, deposited on mineral surfaces, or chemically modified with succinate groups or with pyridinealdehyde. Two methods, not fully described, are used for metal binding: (a) impregnation with metals (mono- or bimetallic impregnation) in aqueous or ethanol solutions, and (b) coprecipitation. They test the activity of their catalyst for cyclopentadiene hydrogenation (and 1,4-butynediol hydrogenation) under hydrogen atmosphere, measuring hydrogen

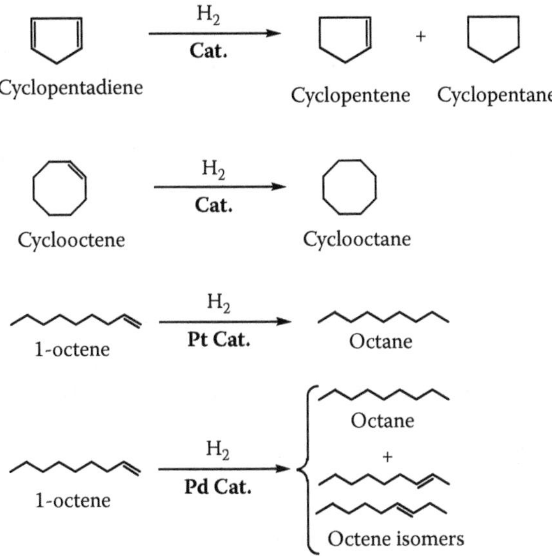

FIGURE 4.33 Examples of hydrogenation of olefins.

consumption. Best results (based on conversion yield, selectivity for cyclopentene, and hydrogenation ratio olefine/diene) are obtained when using pyridine-modified chitosan deposited on SiO_2 for the immobilization of Pd.

Chitosan-stabilized Pt or Pd colloids, prepared by (a) mixing a proper amount of chitosan-acetic acid with palladium chloride or potassium tetrachloroplatinate solutions, and (b) methanol or sodium borohydride reduction, have been tested for the hydrogenation of octene and cyclooctene in methanol under hydrogen atmosphere.[351] The activity of the catalyst, which is generally higher for Pt-stabilized colloids compared to Pd-stabilized materials, decreases with increasing the chitosan:metal molar ratio. Additionally, Pt catalysts provide in some cases a greater selectivity: cyclooctene is converted in cyclooctane (for both Pd and Pt catalysts), while octene is fully and selectively hydrogenated to octane for Pt catalysts and it is converted to octane and isomerized products of 2-octene and 3-octene for chitosan-stabilized Pd colloids.

Sajiki et al. tested the chemoselective hydrogenation of olefins (and azides) in methanol under 1 to 5 bar H_2 atmosphere with Pd immobilized on silk fibroin.[97,382] Palladium acetate dissolved in methanol is reacted under reflux with the biopolymer; palladium is simultaneously reduced to Pd(0) (and production of formaldehyde and acetic acid). They tested the hydrogenation of a number of olefins substituted with ketones, aldehydes, halogens, and benzyl ether (Figure 4.34). It is usually very difficult to achieve the chemoselective hydrogenation of olefin functionalities leaving intact the aromatic carbonyl and halide functions, due to their high reactivity for hydrogenation with conventional catalysts. Under selected experimental conditions, Sajiki et al. obtained a selective hydrogenation of double bonds, preserving the ketone, aldehyde, halogen, and benzyl ester moieties.[97,382]

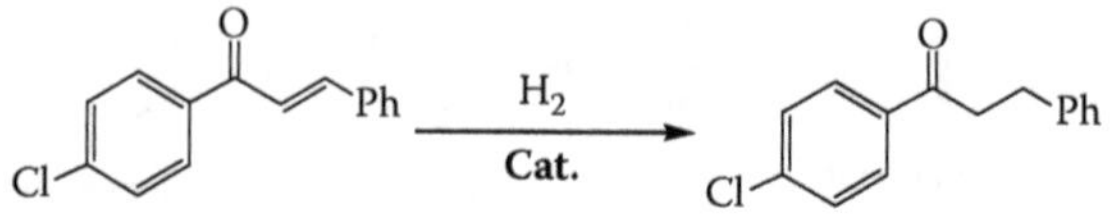

FIGURE 4.34 Example of hydrogenation of olefins with protection of halogen and aromatic moities.

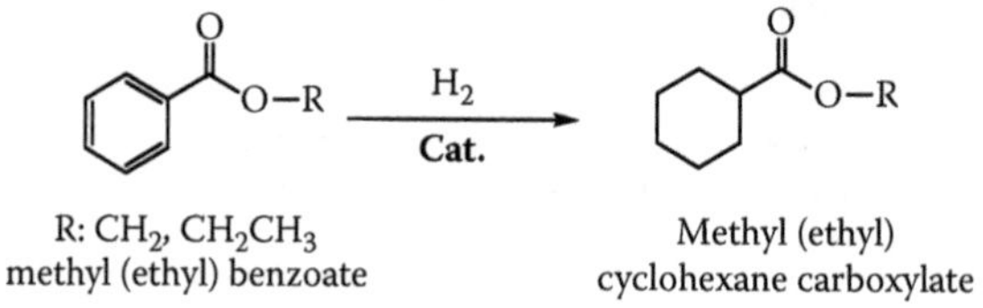

R: CH_2, CH_2CH_3
methyl (ethyl) benzoate

Methyl (ethyl)
cyclohexane carboxylate

FIGURE 4.35 Example of hydrogenation of substituted aromatic ring with protection of substituents.

Tang et al. immobilized carboxymethyl cellulose (CMC) on silica by contact of SiO_2 with a CMC solution, followed by precipitation in acetone. The solid is finally acidified (at pH 2 with acetic acid/HCl solution) and rinsed with acetone.[251] The support is mixed with hexachloroplatinic acid solution in ethanol under reflux. The SiO_2-CMC-Pt catalyst can be used for the hydrogenation of methyl benzoate to methyl cyclohexane carboxylate in 1-propanol under hydrogen atmosphere (Figure 4.35). The optimum substrate:Pt molar ratio is close to 6. The reaction temperature influences the conversion yield and the selectivity of the reaction: an optimum is obtained in the range 60 to 70°C; increasing the temperature to 90°C contributes to the formation of cyclohexane methanol. They also observed that the hydrogenation of aromatic rings without affecting their substituents is difficult due to the easy hydrogenolysis of the functional groups, such as hydroxyl and alkoxy. The SiO_2-CMC-Pt catalyst is also used for the selective hydrogenation of anisol to cyclohexyl methyl ether[383]; best conditions for maintaining high conversion yield, initial rate of hydrogenation, and selectivity are obtained with cyclohexane solvent, temperature close to 30°C, and molar carboxylic group:Pt ratio close to 8. Reused for four cycles, the catalyst maintains its selectivity close to 100% while the conversion yield slightly decreases, but remains higher than 94%.

Liu et al. obtained a conversion yield of 84% for the hydrogenation of octene to octane using a Fe(III) catalyst immobilized on SiO_2-starch-L glutamic acid support.[255] With the substrate:Fe molar ratio fixed to 100:1, the maximum conversion (close to 84%) is obtained for a Fe(III) content close to 0.15 mmol Fe g^{-1}. Gelatin, alginic acid, casein, and carboxymethylcellulose have been also immobilized on SiO_2 for the binding of Pt (see above) and the hydrogenation of 1-heptene to *n*-heptane has been investigated.[241] The hydrogenation takes place in ethanol under 1 bar H_2 pressure. The substrate:Pt ratio is fixed to 160:1; the initial hydrogenation rate reaches a maximum at Pt content close to 0.8 mmol Pt g^{-1}. Except in the case of casein-based support, under selected experimental conditions 1-heptene

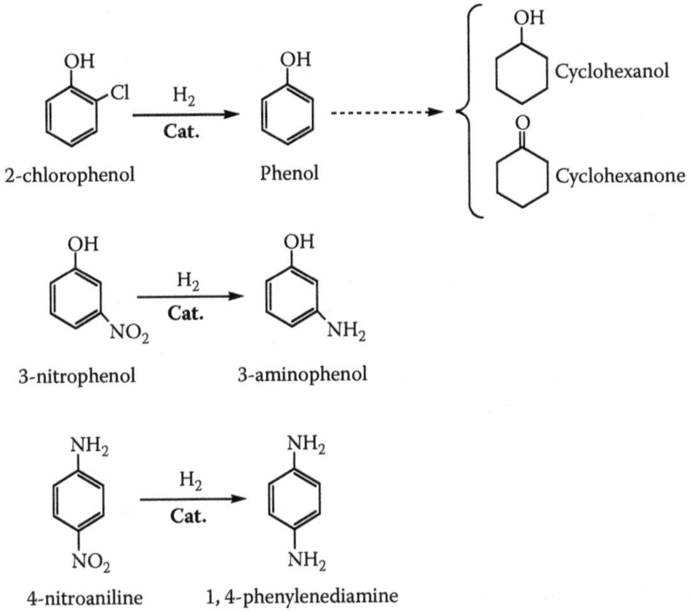

2-chlorophenol → (H_2, Cat.) Phenol → Cyclohexanol, Cyclohexanone

3-nitrophenol → (H_2, Cat.) 3-aminophenol

4-nitroaniline → (H_2, Cat.) 1, 4-phenylenediamine

FIGURE 4.36 Examples of hydrogen transfer reactions.

is converted selectively to 1-heptane. Based on initial hydrogenation rate the SiO_2-oxalic acid-biopolymer-Pt catalysts can be ranked according the series carboxymethyl cellulose > alginic acid >> gelatin >> casein. The initial hydrogenation rates follow the same order as those obtained for nitrobenzene hydrogenation.

4.4.1.5 Transfer Hydrogenation

The ability of chitosan as a Pd support for hydrogenation transfer reaction has been explored by Guibal and coworkers.[342–346] Chitosan flakes are used for the sorption of Pd at pH 2 (the optimum pH for Pd sorption from HCl solutions). To prevent chitosan dissolving in HCl solutions, the sorbent is cross-linked with glutaraldehyde prior to metal binding; alternatively, it is possible to treat the biopolymer with sodium sulfate (for a "soft" cross-linking of amine groups by interchain bonds). Different reducing agents have been tested for the conversion of Pd(II) to Pd(0), including hydrazine, sodium formate, and sodium borohydride; the most efficient treatment is obtained with a combination of sodium borohydride and Zn/sulfuric acid (*in situ* H_2 production) treatments.[344] These materials have been tested for hydrogen transfer reactions using sodium formate as the hydrogen donor in aqueous solutions; a very simple reaction (chromate reduction) was used for screening the synthesis procedures.[344] The optimized catalysts are tested for chlorophenol dehalogenation, leading to phenol and in some cases cyclohexanone and cyclohexanol[346]; for nitrophenol hydrogenation to aminophenol[342,343]; and 4-nitroaniline hydrogenation to 1,4-phenylenediamine (Figure 4.36).[345] The influence of reaction parameters

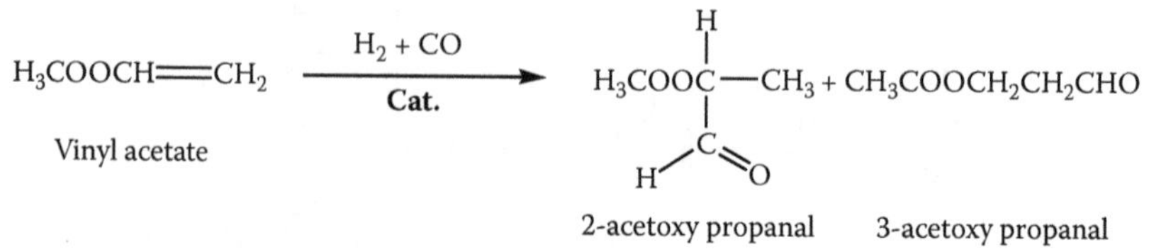

$$H_3COOCH{=}CH_2 \xrightarrow[\textbf{Cat.}]{H_2 + CO} H_3COOC(H)(CHO){-}CH_3 + CH_3COOCH_2CH_2CHO$$

Vinyl acetate — 2-acetoxy propanal — 3-acetoxy propanal

FIGURE 4.37 Example of hydroformylation reaction.

(substrate concentration, hydrogen donor concentration, temperature, pH, etc.) and catalyst parameters (Pd content, support particle size, etc.) is considered in order to identify the limiting steps. The influence of diffusion limitations on kinetic control is underlined, showing the importance of controlling (a) the porosity of the support (appropriate drying procedure), and (b) the thickness of the active layer (depositing on high surface area support), or designing new catalytic systems (using, for example, catalytic hollow fibers). Chitosan hollow fibers are tested as Pd support for the hydrogenation of nitrophenol by hydrogen transfer but also with H_2 gas.[282] This is discussed in detail in Section 4.5.

4.4.1.6 Miscellaneous Reactions of Hydrogenation

Gelatin has been used for coating zeolite material and subsequent immobilization of Co-Ru to prepare a hydroformylation catalyst, which was tested for the conversion of vinyl acetate to (*R*)-2-acetoxy-propanal (Figure 4.37).[242] Zeolite is coated with gelatin by contact of the support with a gelatin solution (at 60°C). The slurry is dried to prepare the support for metal impregnation, which is performed by mixing the support with $CoCl_2$ + $RuCl_3$ solution in ethanol under reflux. The catalyst is used for hydroformylation of vinyl acetate in solvent under H_2 + CO total pressure of 90 bar, at 130°C. The production and optical yields are controlled by Co:Ru molar ratio, substrate:metal molar ratio, solvent, H_2:CO pressure ratio, and reaction time. The best Co:Ru ratio is obtained at equal proportion of metals or in slight excess of Ru. Both production and optical yields reach a maximum when the substrate:metal molar ratio tends to 100. Although the production yield is minimal (close to 19%) when total metal content in the catalyst is 0.2 mmol g^{-1}, the optical yield exceeds 95%. Best results (83% and 91% for production and optical yield, respectively) are obtained for H_2:CO pressure ratio close to 1:1, benzene as the solvent, and 48 h reaction time.

Wool has been used for immobilizing Rh by contact of the support with $RhCl_3$ ethanol solution under reflux.[339] The catalysts is carried out for the hydrogenation of 2-methyl furan to (*S*)-(+)-2-methyl tetrahydrofuran at 28°C under 1 bar H_2 pressure (Figure 4.38). The most efficient solvent is 1-propanol. The highest optical yield (close to 77%, under selected experimental conditions) is obtained after 24 h of reaction time. Although conversion yield increases with Rh content in the catalyst (0.03 to 0.15 mmol Rh g^{-1}), the highest optical yield is found with intermediary Rh content, that is, 0.08 mmol Rh g^{-1}. The production and optical yields vary by less than 5% after four reuses of the catalyst.

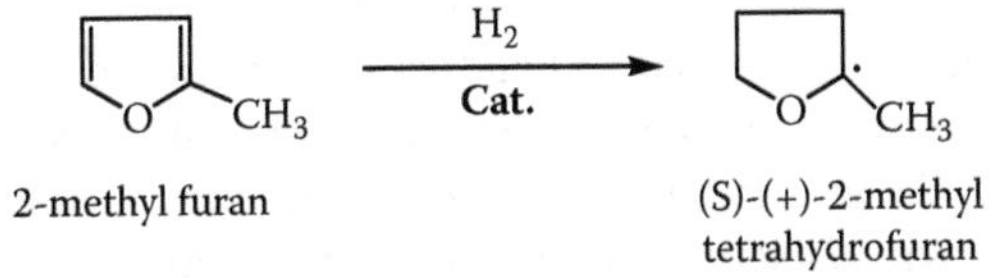

$$\text{2-methyl furan} \xrightarrow[\text{Cat.}]{H_2} \text{(S)-(+)-2-methyl tetrahydrofuran}$$

FIGURE 4.38 Example of hydrogenation of furan-based compound.

Kurita et al. and Nishiyama et al. synthesized chitosan derivatives bearing dihydronicotinamide groups through a complex series of polymer modifications and utilizing amino acids (L-alanine, L- or D-phenylalamine) as spacer agents.[386,385] They use the chitosan derivative for the reduction of ethyl benzoylformate to ethyl mandelate in a mixture of acetonitrile and magnesium perchlorate solution at 40°C. Although the system cannot be strictly considered a catalytic system since the "catalyst" is not self-regenerated at the end of the reaction, the treatment of the exhausted material can be readily regenerated with sodium hydrosulfite, opening the route for the development of new chitosan-based catalytic systems. The spacer contributes to enhancing both the asymmetric selectivity and the chemical conversion yield. The reduction of ethyl benzoylformate with conjugate bearing L- and D-phenylalanine spacer arms, is oriented toward the production of (–)-excess and (+)-excess ethyl mandalate enantiomer, respectively. This appears to be a promising technique for improving the chiral selectivity of chitosan-based catalysts.

4.4.2 Oxidation Reactions

Several catalysts supported on biopolymers (essentially chitosan) have been designed for oxidation reactions. These supported catalysts have been mainly carried out for the oxidation of carbon monoxide,[386,387] of olefins,[235,308,310,318,319,364,388–390] and phenol derivatives.[222,312,313,352,370,371,389]

Qin et al. deposited gold nanoparticles on Al_2O_3-chitosan using the homogeneous deposition-precipitation procedure (with urea as the precipitation agent).[387] Particles as small as 3 nm are obtained and their distribution is homogeneous on the support; the basicity of the support controls the density of gold nanoparticles (as observed when tested different alumina surfaces).[387] Conversion can be efficiently operated at ambient temperature. Kapoor et al. used gelatin for the stabilization of platinum nanoparticles to prepare a CO oxidation catalyst.[386] Gelatin (dissolved in water at 40 to 50°C) is mixed with a tetrachloroplatinic acid solution in the presence of methanol (which acts as a $^{\bullet}OH$ radical scavenger) under N_2 atmosphere. The mixture is gamma irradiated with a ^{60}Co source (at the dose of 1 kGy) before being chemically reduced using dropwise ice-cold sodium borohydride solution. They immobilized these gold nanoparticles (10 to 20 nm in size) on the glass walls of the oxidation reactor. Carbon monoxide mixed with oxygen is oxidized to carbon dioxide. The catalytic activity increases with the temperature and the activation energy is found to be close to 109 kJ mol^{-1}. As expected, increasing the concentration of CO decreases the percentage of conversion; however, the rate of CO_2 formation remains constant for given O_2 concentration.

Kapoor et al. concluded that the mechanism of oxidation occurs via Eley-Rideal mechanism between chemisorbed CO and gaseous-physisorbed O_2. Compared to Pt nanoparticles produced by the chemical reduction of tetrachloroplatinate ions dispersed on SiO_2, the gelatin-based Pt nanoparticles deposited on glass walls give higher catalytic activity. Gelatin preventing Pt agglomeration increases its reactivity.

Guo et el. immobilized iron and cobalt derivatives of porphyrins on chitosan for the catalytic aerobic oxidation of cyclohexane (Figure 4.39).[101,389,390] A chitosan solution (prepared by polymer dissolving in acetic acid solution) is mixed with $NaCO_3$ solution (to form a neutral colloidal solution) and then with chloro[tetra-phenyl-porphinato]iron(III) solution (prepared in benzene).[388] The solid, after washing and drying at room temperature, is used for the aerobic oxidation of cyclohexane at 45°C (the optimum temperature when considering both conversion yield and selectivity) under an air pressure of 8 bar. The turnover frequency and the conversion yield are considerably increased when immobilizing iron(III) tetraprophyrin catalyst on chitosan: the catalytic activity is halved when homogeneous catalyst is used (and the turnover frequency is four times lower). The authors explain this improvement of catalytic performance by the protective effect of chitosan against iron(III) tetraporphyrin oxidation. In the case of chitosan-supported cobalt tetraphenylporphyrin, Huang et al.[389] also observed that the catalytic activity is considerably increased (by 27 times) compared to homogeneous catalyst. The catalyst is prepared by direct adsorption of cobalt tetraphenyl-porphyrin on chitosan particles. The improvement of catalytic activity is explained by both the protecting effect of the biopolymer and the activation of oxygen. Schiff base derivatives of chitosan have been also carried out for the immobilization of catalytic metals for oxidation of cyclohexane,[308] and cyclohexene.[310] Chang et al. described the modification of chitosan with salicylaldehyde[308]: the polymer and the aldehyde are mixed in benzene at room temperature overnight before being cross-linked with epoxy chloropropane in NaOH/dichloroethane solution for 7 h at 40°C. Copper is adsorbed under reflux on Schiff base derivative of chitosan by contact of copper acetate and polymer in ethanol. A similar procedure is used for the preparation of nitrobenzaldehyde (NBA) (in place of salicylaldehyde) derivatives of chitosan (with *o*- and *m*-NBA). Cyclohexene is oxidized under oxygen atmosphere, with formation of 2-cyclohexene-ol, 2-cyclo-hexene-one, and cyclohexene hydroperoxide.[308] The most active catalyst is obtained when using *m*-nitrobenzaldehyde for the preparation of chitosan derivatives: the turnover is found several orders of magnitude greater than that of the salicyladehyde derivative. Adding an acid (acetic acid) or a base (pyridine) significantly affects the conversion yield and the selectivity of the reaction. Chang et al. used nitrobenzaldehyde derivatives of chitosan for binding Ni and Mn (added under the form of acetate salts, following the procedure previously described) and test the catalytic oxidation of ethylbenzene, *n*-propylbenzene, and isopropylbenzene in an oxygen atmosphere.[364] The reactivity of the catalysts depends on the metal, the substituents on chitosan. Hence, with *m*-NBA/chitosan/Mn, none of the alkenes are oxidized, while *o*-NBA/chitosan/Mn oxidize all the tested alkenes, as does *m*-NBA/chitosan/Ni; *o*-NBA/chitosan/Ni oxidizes only *n*-propylbenzene and

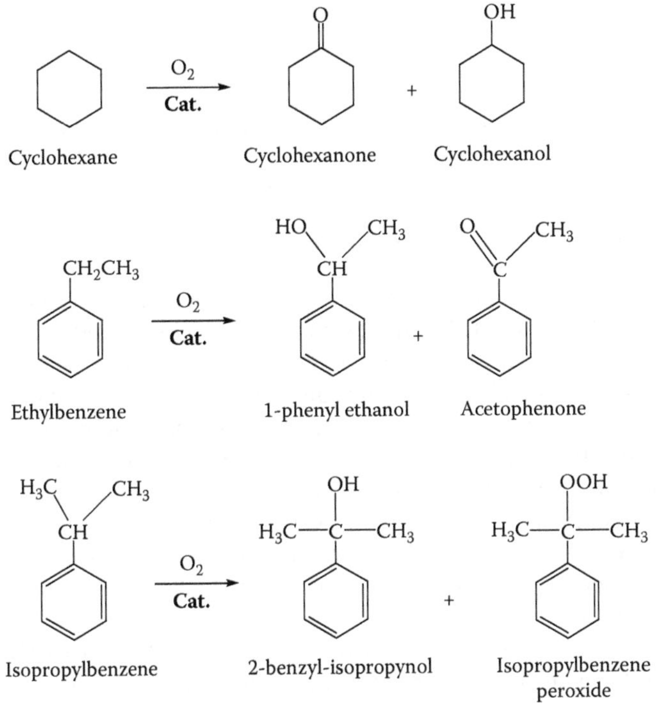

O_2 / Cat.

Cyclohexane → Cyclohexanone + Cyclohexanol

Ethylbenzene → 1-phenyl ethanol + Acetophenone

Isopropylbenzene → 2-benzyl-isopropynol + Isopropylbenzene peroxide

FIGURE 4.39 Examples of oxidation of olefins.

isopropylbenzene (with negligible effect on ethylbenzene oxidation) (Figure 4.39). They show that the oxidation state changes in the course of the reaction and that the nature and the spatial arrangement of ligands around the metal influence the oxidation process. Tong et al. prepared two Schiff base derivatives of chitosan by grafting salicylaldehyde (SA) or pyridine-2-carboxaldehyde (P2CA) on the biopolymer in ethanol (under reflux for 30 h).[310] Chitosan Schiff base derivatives are mixed in ethanol solution under reflux with cobalt acetate (both SA- and P2CA-chitosan), or palladium acetate (P2CA-chitosan). Cyclohexane is oxidized by molecular oxygen to form cyclohexanol and cyclohexanone (and by-products, such as dicyclohexyl, adipic acid, and esters). Regardless of the catalyst, the conversion increases with oxygen pressure, while the selectivity for cyclohexanone reached a maximum around 15 bar. The optimum temperature is found to be around 150°C; the selectivity for cyclohexanone increases with temperature. The catalyst P2CA-chitosan-Cu shows the highest and the most selective catalytic activity for cyclohexanone production; however, the recycling of the catalyst shows that this catalyst is more rapidly deactivated than the others. Under selected experimental conditions, the conversion yield exceeds 10%, several percent more than the values obtained with homogeneous catalysts; similar conclusions on the beneficial effect of supporting the catalytic metals are observed with the turnover number and the selectivity. Manganese has been immobilized on salophen deriv-

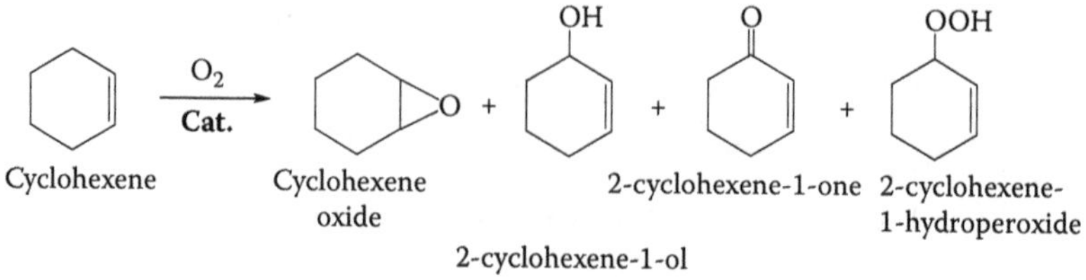

Cyclohexene $\xrightarrow[\text{Cat.}]{O_2}$ Cyclohexene oxide + 2-cyclohexene-1-ol + 2-cyclohexene-1-one + 2-cyclohexene-1-hydroperoxide

FIGURE 4.40 Example of allylic oxidation of cyclohexene.

atives of chitosan for the aerobic oxidation of cyclohexene (Figure 4.40).[392] Phenylenediamine is reacted with substituted benzaldehydes under reflux to form a chelating cage, which is used for the binding of manganese acetate to form a Salophen Mn(III) complex. Chitosan is dissolved in acetic acid solution at 50°C, and the pH is controlled to 7 with sodium carbonate before addition of the Salophen Mn(III) complex in ethanol solution. The washed and dried solid is used for the aerobic oxidation of cyclohexene, which is converted into cyclohexene oxide, 2-cyclohexene-1-ol, 2-cyclohexene-1-one, and 2-cyclohexene-1-hydroperoxide, in the absence of any solvent or reducing agent. The selective formation of 2-cyclohexene-1-one is improved by increasing both temperature (up to 70°C) and reaction time (up to 12 h).

Kramareva et al. discuss the preparation of Pd catalysts supported on chitosan through different methods (cited above, i.e., coprecipitation and adsorption).[318,319] They test these materials for the oxidation of terminal olefins. In neutral media, the catalyst prepared by the coprecipitation method causes olefin isomerization and almost no oxidation of the substrate; in the case of the catalyst prepared by the adsorption method, the main reaction is oxidation. The reactivity of the catalyst is thus strictly controlled by the interaction mode of Pd with the support. The formation of a chelate complex (involving palladium coordination with amino and hydroxyl groups) in the case of the coprecipitation method is less favorable than the coordination model involving the formation of a $Pd(NH_2R)_2Cl_2$ complex. The chemical reduction of these metal-chitosan assemblies influences their structure: Pd_3 triangular structures and Pd_2 dimers for coprecipitation and adsorption methods, respectively. The coprecipitation method causes more metal aggregation, which in turn reduces the activity of the catalyst.

The oxidation of catecholamines, such as 3-(3,4-dihydroxyphenyl)-L-alanine (L-DOPA) or 3,4-dihydroxy-α-(methylaminomethyl)benzyl alcohol (adrenaline) and hydroquinone, has been carried out using iron, copper, and cobalt catalysts supported on chitosan (Figure 4.41). Chiessi and Pispisa prepared carboxylic derivatives of chitosan for the immobilization of Cu and Fe: the sorbents are mixed in buffered solutions of $CuCl_2$ and $FeCl_3$.[371] Catecholamines (adrenaline and DOPA) are aerobically oxidized to the corresponding *o*-quinones (i.e., adrenochrome and dopaquinone, respectively) in pH 8.5 buffered solutions [*N*-2-hydroxyethylpiperazine-*N*′-2-ethanesulfonic acid (HEPES) buffer]. The oxidation in alkaline solution of adrenaline and DOPA by Fe(III) and Cu(II) complexes immobilized to chitosan proceeds by different pathways, depending on the oxidation power of the central

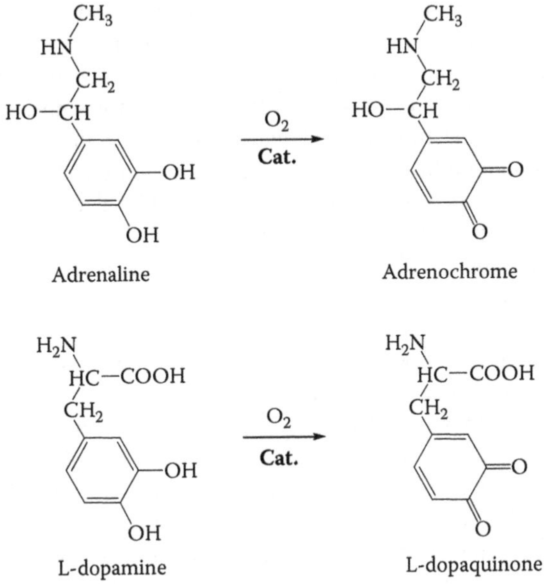

FIGURE 4.41 Example of oxidation of catecholamines.

metal ion. With the Fe(III) immobilized catalyst, catecholamines undergo an electron transfer process with iron(III) ions within a binary adduct, whereas in the case of a Cu(II) catalyst, adrenaline is oxidized by molecular oxygen within a ternary precursor complex, in which Cu(II) ions act as electron mediators.[371] The reduction potential of Cu is significantly decreased by coordination to polymeric ligands (from –80 mV at pH 10 for free copper chloride to –200 mV at pH 8.5 for immobilized copper). In the case of Fe(III), the reduction potential, determined at pH 4.6, is estimated to be around 200 mV (–32 mV for Cu complex under selected experimental conditions). Since the Fe-chitosan complexes have a reduction potential definitely higher than that of Cu(II), it is expectable that they exhibit greater oxidation power in alkaline solution.[371] Paradossi et al. prepare chitosan-based hydrogels using oxidized β-cyclodextrin as a cross-linking agent[372]: β-cyclodextrin is oxidized by potassium periodate to form aldehyde groups that react with amine functions (via a Schiff base reaction). The coupling of chitosan with the oxrid-cyclodextrin molecule allows stabilizing the copper-polysaccharide complex: the structural and catalytic features of Cu(II) bound to chitosan derivative in solution is preserved when copper is immobilized on chitosan-cyclodextrin composite. The oxidation of adrenaline and DOPA was also studied by Finashina et al. and Hu et al. using CoSalen complexes immobilized on Schiff base derivatives of chitosan (Figure 4.41).[312–314] Hu et al. immobilized the tetra-coordinated copper complex of *bis*(salicylideneethylene diamine) on *N*-(4-pyridylmethylidene)-chitosan by contact of CoSalen in aqueous solution.[313] The pyridylmethylidene-chitosan derivative is prepared by reaction of chitosan acetic acid solution with 4-pyridinecarboxaldehyde in methanol under reflux, and final precipitation at pH 6. Cobalt ions Co(II) (from

CoSalen complex) coordinates to nitrogen on pyridyl pending group of modified chitosan. Catecholamines oxidation is performed under oxygen atmosphere, and the performance of CoSalen complex immobilized on chitosan is compared to that of free CoSalen complex. It is reported that the free complex does not bind oxygen strongly, leading to the formation of inactive dimmers; the site isolation effect provided by the immobilization of CoSalen on the chitosan derivative inhibits the formation of dimmers and peroxo-bridged dimmers ($CoSalen_2O_2$, inactive for catalysis), and thereby promotes the formation of active superoxo cobalt complex [$Co(III)O_2^-$]. Finashina et al. also used a 4-pyridinecarboxaldehyde derivative of chitosan for the immobilization of CoSalen through different methods[312]:

(a) CoSalen is deposited on the Schiff base derivative of chitosan by adsorption in aqueous solutions.
(b) A homogeneous CoSalen-chitosan complex is prepared by mixing a CoSalen solution with chitosan dissolved in HCl solution.
(c) The homogeneous CoSalen-chitosan complex is dropped to a 0.5 M NaOH solution to form spherical gel particles.
(d) The homogeneous solution of CoSalen-chitosan complex is mixed with SiO_2 for impregnation of the support with the catalytic complex, prior to neutralization in NaOH solution. A variant of this method consists of the immobilization of CoSalen on preformed chitosan-SiO_2 composite.

The immobilization of CoSalen on a chitosan derivative has a beneficial effect essentially when the complex is immobilized on structured material (i.e., eggshell SiO_2 support) leading to a highly developed surface area. The influence of a high specific surface area is underlined by Kucherov et al. for the oxidation of dihydroxybenzene.[222] Copper chloride is dissolved in chitosan-HCl solution to prepare a homogeneous Cu-chitosan complex. The complex can be precipitated in a NaOH solution to form spherical globules. Alternatively, a copper catalyst is prepared by adsorption of copper on preformed chitosan gel beads. The immobilization of a Cu-chitosan complex on SiO_2 is performed by impregnation of the solid followed by alkaline precipitation and drying (for SiO_2). MCM-41 support (microporous silica) is impregnated with a chitosan-HCl solution before glutaraldehyde cross-linking and further adsorption of copper from aqueous solution. These different catalysts are tested for the aerobic oxidation of hydroquinones. Although homogeneous catalysts are rapidly inactivated by formation of stable copper-hydroquinone complexes, heterogenized Cu-chitosan complexes form stable molecular complexes with intermediate quinhydrone products (produced in the earlier stage of oxidation process). The beneficial effect of an eggshell structure requires using a macroporous material to prevent blocking of micropores due to polymer deposition. Additionally, heterogenized chitosan stabilizes and retains isolated Cu^{2+} ions in highly unsaturated coordinative state, maintaining high catalytic activity. Guibal et al. used Cu-chitosan flakes for the oxidation of hydroquinone to *p*-benzoquinone[393] using oxygen and hydrogen peroxide as oxidizing agents. With hydrogen peroxide at pH 5.8, drastic oxidizing conditions

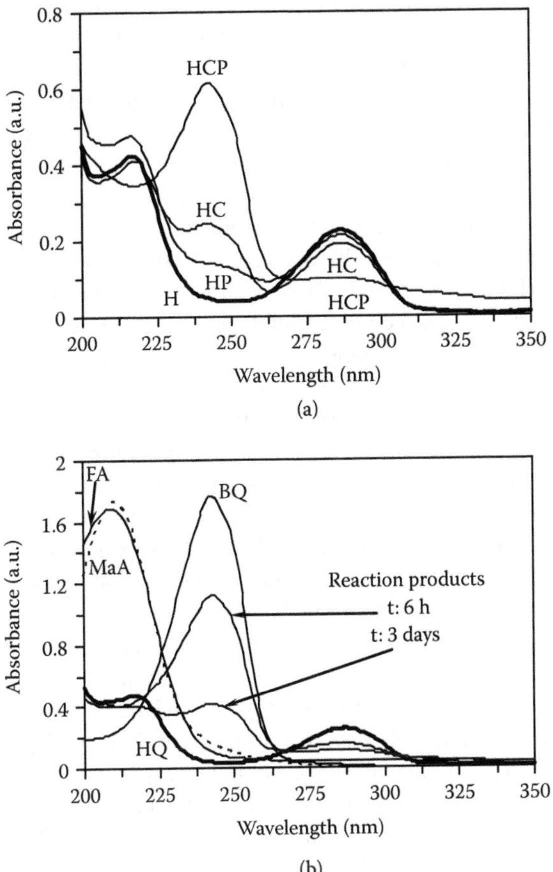

FIGURE 4.42 Effect of oxidation conditions (H: HQ alone; HP: HQ + hydrogen peroxide; HC: HQ + catalyst; HPC: HQ + hydrogen peroxide + catalyst) on the UV spectra of reaction products after 1 h contact (a) and spectra of reaction products (corresponding to HPC conditions) after longer contact time compared to HQ, BQ, and fumaric acid (FA) and maleic acid (MaA) spectra (b). (Reprinted from E. Guibal, T. Vincent, E. Touraud, S. Colombo, and A. Ferguson, *J. Appl. Polym. Sci.*, 100, 3034–3043, 2006. With permission. Copyright 2006, John Wiley & Sons.)

led to the formation of subproducts (Figure 4.42). With a short contact time, together with the use of a low hydrogen peroxide concentration and a small amount of the catalyst, the formation of subproducts could be minimized. The influence of the catalyst:substrate and hydrogen peroxide:substrate ratios is investigated to determine optimum experimental conditions for a high initial oxidation rate and a high production of *p*-benzoquinone. The oxidizing agent has to be carefully selected: dioxygen provides very selective oxidation but low production rates in comparison with hydrogen peroxide. With this drastic oxidizing agent, the kinetics of the oxidation reaction is strongly increased but at the expense of a loss in selectivity (resulting in the synthesis of secondary products, possibly fumaric and maleic acids). The amount of catalyst and the excess of hydrogen peroxide control this decrease in selectivity. These two parameters appear to be dependent. A short contact time of the catalyst with the substrate limits the synthesis of these by-products. The process can therefore be optimized, taking into account (1) the limited hydrogen peroxide concentration suitable for catalyst integrity, and (2) the appropriate amount of the catalyst. An excess of the catalyst contributed to increased synthesis of subproducts. Srivatsan et al. immobilized a nucleobase (9-allyladenine) on divinylbenzene (DVB): the nucleobase is mixed with $CuCl_2$, divinylbenzene, and AIBN (azobisisobutylonitrile) in a chloroform/methanol solution.[391] The metallated nucleobase polymeric matrix is efficiently used for the oxidation of monophenols and *o*-diphenols.

Mirescu and Prüsse prepared gold colloids by reducing tetrachloroauric acid with sodium borohydride in the presence of polymers.[352] These polymers (synthetic or biopolymers) act as stabilizing agents and contribute to preventing the precipitation and the agglomeration of colloidal gold. The 50 mg Au L^{-1} solution of $HAuCl_4$ is mixed with the polymer solution after addition of a small amount of NaOH. Sodium borohydride is finally added to the mixture, which turned almost instantaneously to a red color. The oxidation reaction takes place at 40°C under oxygen atmosphere with a pH controlled to 9 with the automatic addition of NaOH. Some of the polymers were unable to avoid the precipitation of gold colloids during the synthesis procedure: alginate, poly(ethylene oxide), poly(1-vinyl-pyrrolidone-co-vinyl acetate), Brij 35 (polyoxyethylene dedecyl ether detergent). With polyvinyl alcohol, polyethylene imine, poly(*N*-vinyl-2-pyrrolidone) and polymin Type P, the colloids are stable but the catalytic activity is negligible and the best results are obtained with the polymers following the sequence PDADMAC poly(diallyldimethyl ammonium chloride) > chitosan > poly-2,2-trimethyl ammonium methylmethachloride (poly-TMAMMCl). At the pH selected for glucose oxidation (i.e., pH 9), glucose is converted to gluconic acid with selectivity greater than 99%; in the case of chitosan and poly-TMAMMCl, the isomerization of glucose to fructose is detected. Mirescu and Prüsse commented that the isomerization is attributable to the effect of the low alkalinity of the reactive medium; the lower catalytic activity of chitosan- and poly-TMAMMCl-Au catalyst means that more time is available for isomerization of glucose (as a parallel reaction).

Recently, Šuláková et al. described the immobilization of copper on chitosan by the coprecipitation of copper chloride and chitosan acetic acid solution into NaOH.[394] The neutralization of chitosan-solution drops results in the formation of spherical catalyst particles tested for the oxidation of a series of azo textile dyes using hydrogen peroxide (which contributes to reoxidize Cu after dye decomposition). The decolorization of the dyes strongly depends on Cu content, the pH, and the concentration of H_2O_2. The optimum Cu content is reached at 3.6% (in weight); above this value, the efficiency of the process decreases. Šuláková et al. commented that an excess of copper decreases the number of sorption sites available for dye adsorption, and thus the potential of the catalyst for interacting with the substrate.[394] The poor stability of the catalyst in acidic solutions convinced the authors to set the pH to 7, despite lower catalytic activity compared to lower pH values. Despite the high activity of the catalyst with H_2O_2 concentrations greater than 0.1 *M*, the poor stability of the catalyst in the presence of hydrogen peroxide requires working with concentrations as low as 0.05 *M*. The type of substituent on the azo group controls the decomposition of the dyes: the presence of nitro groups (electron withdrawing) in the *para* position facilitates the formation of hydrazone, which is less stable against oxidative species.

Yao et al. described the immobilization of metallophthalocyanine on silk fibers and they tested the activity of supported catalysts for the oxidation of methanethiol and hydrogen sulfur.[395] More specifically, they prepared two binuclear phthalocyanine derivatives for Co(II) and Fe(III) complexation and they carried out their potential for oxidation of mercaptoethanol in aqueous solutions. The cobalt complex exhibits a greater activity than the iron complex; however, the mixture of Co(II) and Fe(III) complexes shows a marked synergistic effect. They immobilized the mixture of metallophthalocyanines on silk fibers (used as produced or modified by grafting tetra-alkyl ammonium): the support and the metal complexes are mixed at 60°C in the presence of a leveling agent. The supported catalysts are suspended in a closed tank filled with sulfur-based gas: the highest degradation rate is obtained with the binary Co(II)-Fe(III) metallophthalocyanine complex immobilized on tetra-alkyl ammonium modified silk fiber. The modification of the fiber with alkyl ammonium allows binding the carboxylic groups of the complex with the quaternary ammonium group of modified fiber; this allows preserving the activity of a central metal ion. Indeed, in the case of nonmodified fibers, the binding of the organometallic compound occurs by coordination of the central metal ion to amino groups.

4.4.3 Polymerization Reactions

Inaki et al. used chitosan and a glycol derivative of chitosan for the immobilization of copper and use the complex for the polymerization of vinyl monomers in carbon tetrachloride.[366,367] The polymerization of acrylonitrile and methyl methacrylate occurs through an initiation mechanism: the reduction of Cu(II) to Cu(I), *in situ*, induces the transfer of a free radical to carbon tetrachloride and the formation of trichloromethyl radical; this radical initiates the polymerization of vinyl monomers. The catalytic performance is controlled by the pH. At pH 11, the high catalytic activity correlates to a redox reaction between the reducing end groups of glucosamine and Cu(II) complex. Although the presence of reducing ends is shown to be favorable, this is not sufficient to explain the high catalytic activity of the chitosan material. At slightly acidic pH (pH 4 to 6.5), the complex is formed between two units of glucosamine and Cu(II) by coordination with one amino group and one adjacent hydroxyl group on each monomer; and the initiation of the polymerization is not observed. Increasing the pH increases the catalytic conversion; at alkaline pH, the complex between Cu and glucosamine units involved one amino group and one adjacent oxygen atom on each glucosamine unit. The primary free radical is assumed to be formed by a reaction of the copper(II) complex with a free amine unit or by a redox reaction accompanying ligand substitution to give an amino free radical. With the polymer, the coordination of copper involves three amine groups at slightly acidic pH (pH 5) and the polymerization reaction is not initiated. The complex changes in structure with increasing pH (between 7 and 9) as a chelate is formed between Cu(II) and two amino groups and two hydroxyl groups; the conversion reaches a maximum in this pH range. Indeed, under these conditions chitosan exists in the form of a neutral polymer, which forms the most compact conformation, causing the closest proximity between the units. The presence of NaCl increases the polymerization of methyl methacrylate.[366] This can be explained by the change in the conformation of the polymer due to salt effect. Changing the molar ratio between amino groups and copper leads to considerable changes in polymerization; maximum catalytic activity is obtained when amino groups are in large excess (i.e., 10:1). The presence of uncoordinated glucosamine units on the polymer chains appear to be necessary for efficient polymerization.

More recently, Guan and Cheng used a chitosan-Co complex for the selective polymerization of vinyl acetate.[374] A chitosan film is soaked in a $CoCl_2$ aqueous solution before being dehydrated in ethanol. The polymerization of vinyl acetate (and other vinyl monomers such as methyl methacrylate, styrene, acrylamide, and methyl acrylate) is performed at pH 7 and room temperature in the presence of sodium sulfite. The kinetics is characterized by a first induction period, whose duration depends on the type of vinyl monomer, followed by a dynamic step. The induction period varies from 27 s for vinyl acetate to 2 min with methyl methacrylate and up to 4 h for styrene, acrylamide, and methyl acrylate. The type of vinyl monomer also controls the yield of the reaction: 53% with vinyl acetate, 20% for methyl methacrylate, and below 5% for other substrates. These results demonstrate the relative selectivity of this catalyst for the polymerization of vinyl acetate. Guan and Cheng observed that the addition of *p*-benzenediol stops the reaction; this result confirms that the reaction occurs through free radical polymerization mechanism since *p*-benzenediol is known for inhibiting this polymerization mechanism.[374]

Zeng et al. used rare-earth catalysts (Y, La, Pr, Nd, and Er) supported on chitosan for the ring-opening polymerization of propylene oxide.[341] Metal oxide is dissolved in HCl and mixed with a chitosan solution (the biopolymer being dissolved in acetic acid). The pH is finally controlled to 6.7 using dilute ammonia solution. The metal-chitosan complex is precipitated by adding an acetone-ethanol mixture; the solid is filtered and dried. The metal-chitosan complex is dispersed in toluene, and air is flushed out by nitrogen gas under heating and reflux; triisobutyl aluminum (Al) is added at 60°C (under reflux) for 2 h. Dried acetylacetone (acac) is finally added; the slurry is agitated at room temperature for 1 h. These chitosan-supported catalysts are used to polymerize propylene oxide at 70°C (synthesis of polypropylene oxide, PPO). The type of rare-earth element immobilized on chitosan weakly affects the polymerization performance (conversion yield: 170 kg PPO mol^{-1} rare earth; high molecular weight: close to 10^6 g mol^{-1}) and yttrium, being cheaper, is often preferred for the preparation of

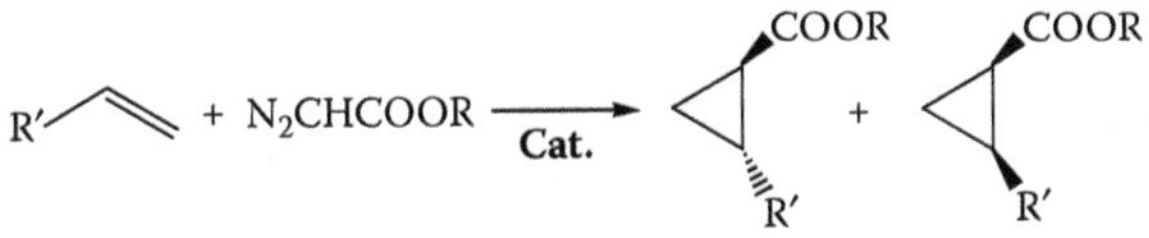

$$R'\text{-}CH{=}CH_2 + N_2CHCOOR \xrightarrow{\textbf{Cat.}} \text{(cyclopropane: COOR, R')} + \text{(cyclopropane: COOR, R')}$$

FIGURE 4.43 Example of cyclopropanation of olefins with diazoacetates.

this catalyst. The best catalytic performance (molecular weight and conversion yield) is obtained with Al/Y and acac/Al, which have ratios around 70 and 0.5, respectively. Increasing the reaction temperature causes the catalytic activity and molecular weight of the resulting polymer to increase. Zeng et al. observed that the chitosan-supported materials have a higher stereoselectivity (based on the percentage of isotactic PPO) than conventional systems.[341] The molecular weight of PPO sharply increases with reaction time and tends to level off above 8 h. The catalytic activity is approximately 16 times higher than that obtained with conventional catalytic systems. The optimum content of Y in the catalyst is in the range 0.5 to 0.6 mmol Y g^{-1} chitosan, above which the embedment of yttrium atoms results in lower efficiency of the catalytic metal.

4.4.4 Cyclopropanation of Olefins

Cyclopropane derivatives are important intermediate reagents in organic chemistry. They are generally produced by carbine transfer from diazo-compounds to alkenes using metal complexes bearing chiral ligands. Xia and coworkers have investigated the cyclopropanation of olefins (styrene, 1-heptene, 1-octene, 1-nonene, 2,5-dimethyl-2,4-hexadiene, α-methylstyrene) with alkyl diazoacetates using copper immobilized on chitosan derivatives obtained by Schiff base reaction with substituted salicylaldehydes.[309,311,336] Metal impregnation is obtained in ethanol medium (under reflux). The generic reaction is shown on Figure 4.43. The yield of the reaction and the optical resolution are controlled by the type of substitutent on salicyladehyde (Schiff base derivative of chitosan), the content of Cu, the temperature, and the solvent. For styrene cyclopropanation, the best experimental conditions (compromise between production and optical yield) correspond to 1,2-dichloroethane as the solvent, a temperature of 60°C, and 5 mol% of copper on the catalyst (copper content hardly affects production yield but strongly controlled optical yield). A slightly better enantioselectivity is obtained when the salicylaldehyde bears two *t*-butyl groups; Xia et al. concluded that the steric effect is more remarkable than the electron-withdrawing effect in improving the enantioselectivity of the reaction.

4.4.5 Hydration Reactions

The synthesis of alcohols can proceed by the reaction of epoxides with Grignard reagent; however, a number of catalysts have been designed for the production of alcohols by alkene hydration. Three biopolymer-supported catalysts have been prepared for the asymmetric hydration of 1-octene,[305] alkenes and allyl

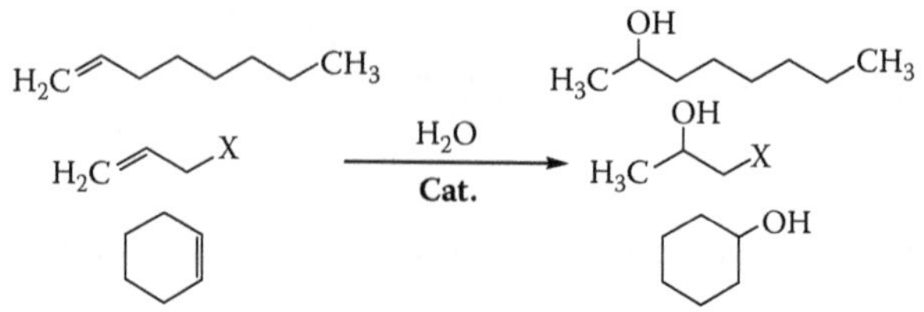

FIGURE 4.44 Examples of hydration reactions.

compounds,[340] or allyl alcohol.[258] Figure 4.44 shows the typical representation of these hydration reactions.

For the hydration of allyl alcohol, Zhang et al. immobilized gelatin on polysulfostyrene for Co binding.[258] When mixed, the two polymers establish ionic and hydrogen bonds to give a stable structure. The composite support is used for Co binding in an ethanol solution; the metal is supposed to bind with sulfonic groups on the synthetic polymer and amine functions of gelatin (Figure 4.17). The optimization of the reaction conditions gives high production yield (close to 98%) for the conversion of allyl alcohol to (*S*)-(–)-1,2-propanediol, with a high optical yield (higher than 98%). The enantioselectivity of the reaction is improved in the presence of acetic acid, at 90°C, under 1 atm N_2 pressure, with a Co content on the catalyst close to 0.4 mmol g^{-1}. The optimum weight ratio for polysulfostyrene-gelatin is obtained with a little excess of synthetic polymer, that is, a 4:3 mass ratio. The hydration of cyclohexene and of a series of allyl derivatives (allyl halogens, allyl amine, acrylonitrile, and lactonitrile) was carried out by Jia et al. using a bimetallic catalyst (made of palladium and iron) immobilized on wool.[340] Metal impregnation is performed in an ethanol solution containing appropriate amounts of palladium and ferric salts, under nitrogen atmosphere and reflux. Metal binding occurs through interactions of the metals with sulfur, sulfonic, and primary and secondary amine groups. The hydration reaction is carried out under reflux and under nitrogen atmosphere with mixing of the catalyst and the substrate in a water-butyl ether-phenol solution. The conversion yield is strongly influenced by Pd:Fe molar ratio; for example, for allyl bromide hydration (to 1-bromo-2-propanol), the optimum molar ratio is close to 0.33. The reaction temperature should be maintained higher than 90°C for optimum conversion, which is reached after 24 h of contact.

Xue et al. investigated the hydration of 1-octene to (*S*)-(+)-2-octanol using cobalt deposited on SiO_2-chitosan composite.[305] Cobalt binds to amine groups of chitosan in ethanol solution under reflux. The optimum conditions for the preparation of the catalysts are a SiO_2-chitosan ratio close to 10%, and a Co content close to 0.5 mmol g^{-1}. The optimization of the process allows obtaining high conversion (i.e., 98%) and high optical resolution (close to 98%) when using the appropriate substrate:Co molar ratio (i.e., 50:1). The recycling of the catalysts has been successfully tested for three to seven cycles, with yields that systematically remain higher than 90%.

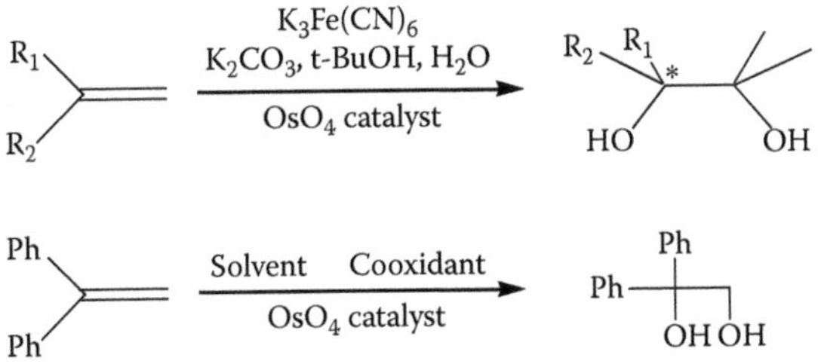

FIGURE 4.45 Examples of hydroxylation reactions.

4.4.6 Hydroxylation Reactions

Osmium-tetroxide complexes have been used for hydroxylation of olefins, and this type of reaction has been tested with chitosan-supported materials in conventional solvents (Figure 4.45).[332] Huang et al. prepared SiO_2-chitosan by the precipitation-drying procedure; the support is impregnated with osmium tetroxide in *tert*-butyl alcohol, under agitation and reflux.[332] They use the chitosan-supported OsO_4 catalyst for vicinal hydroxylation of olefins. The olefin is dissolved in *tert*-butyl alcohol:water (1:1) completed with cooxidants [$K_3Fe(CN)_6$, K_2CO_3] and the mixture is mixed with supported catalyst at room temperature under nitrogen atmosphere. Several olefins (styrene, α-methyl styrene, 1-heptene, 1-octene, and 1-decene) have been converted by dihydroxylation to get (*R*)-(−)-1-phenyl-1,2-ethanediol, (*R*)-(−)-2-phenyl-1,2-propanediol, (*S*)-(−)-1,2-heptanediol, (*S*)-(−)-1,2-octanediol, and (*R*)-(+)-1,2-decanediol, respectively, with production yields ranging between 50 and 90% and optical yields in the range 38 to 81%.

4.4.7 Carbonylation Reactions

Chitosan has been used as a support for carbonylation reactions.[365,396,397] Zhang and Xia studied the effect of hydroesterification of olefins in the presence of CO and alcohol for the synthesis of carboxylic esters.[365] They prepared bimetallic catalysts supported on chitosan for the carbonylation of 6-methoxy-2-vinylnaphtalene to esters of naproxen (a nonsteroidal antiinflammatory drug). Palladium chloride is mixed with 4 *M* HCl:ethanol solution (1:19 v/v). The solution is mixed with a SiO_2-chitosan support (prepared by the precipitation-drying procedure) under reflux. The carbonylation of 6-methoxy-2-vinylnaphtalene is operated under CO atmosphere, at fixed temperature, in the presence of triphenylphosphine (PPh_3), HCl, methanol, dioxane, and biphenyl, and a second metal chloride salt. The reaction leads to the formation of three different products (Figure 4.46): the methyl ester of naproxen, its linear isomer [methyl 3-(6-methoxy-2-naphtyl] propanoate), and the product of the etherification reaction of methanol with α-(6-methoxy-2-naphtyl) ethanol. Compared to more conventional materials ($PdCl_2$/$NiCl_2$, alone or combined with polyvinylpyrrolidone, used as a stabilizer of Pd/Ni nanoparticles), the conversion yields remain comparable, while the selectivity for ester of

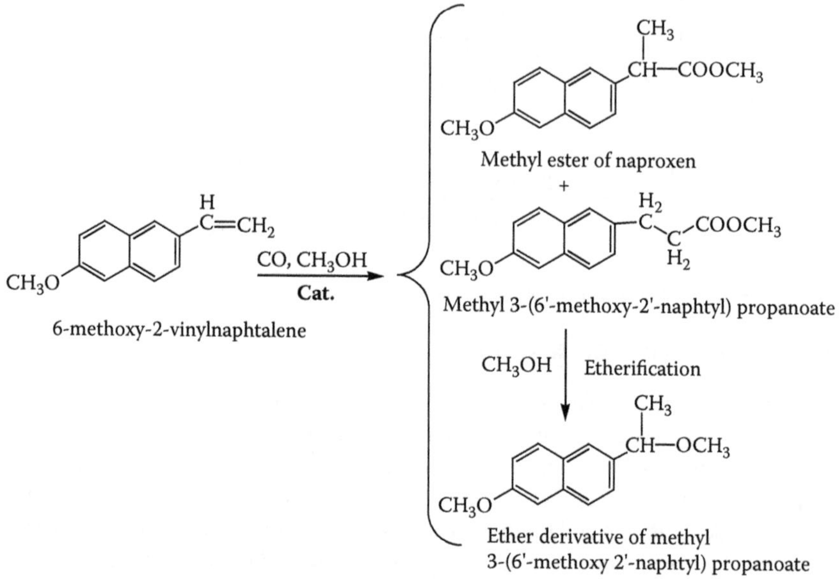

6-methoxy-2-vinylnaphtalene

CO, CH_3OH / Cat.

Methyl ester of naproxen

+

Methyl 3-(6'-methoxy-2'-naphtyl) propanoate

CH_3OH Etherification

Ether derivative of methyl 3-(6'-methoxy 2'-naphtyl) propanoate

FIGURE 4.46 Hydroesterification of vinylnaphtalene derivative.

naproxen is strongly increased for chitosan-supported material. The reaction is controlled by the temperature (optimum at 100°C), the pressure of CO (optimum at 40 bar), the presence of an acid promoter (0.8 *M* HCl), the solvent (methanol is preferred), and the type of second metal (Ni, with molar ratio Ni:Pd close to 2). The addition of triphenylphosphine to the reactive media contributes to palladium stabilization and to preventing the formation of inactive bulk Pd particles (the optimum molar ratio between PPh_3 and Pd is close to 3). The production yield reaches 96% with regioselectivity greater than 95%. The same reaction is tested with styrene (and various derivatives). Zhang and Xia concluded that the catalyst is more efficient and more selective for styrene than for its derivatives.[365] The selectivity and the stability of the catalyst are significantly improved by the addition of *p*-benzoquinone, due to its protective effect against aggregation of Pd particles.

Kolesnichenko et al. carried out the carbonylation of methyl acetate using rhodium supported on polymeric materials (chitosan, chitin, polyglucin, and synthetic polymers bearing pyrrolidinopyridine or pyridylmaleimide groups).[396] Methyl acetate is converted to acetic anhydride. The reaction is operated under 50 bar pressure of CO/H_2 gas (95:5 v/v) at 190°C, using $RhCl_3$ stabilized by polymeric ligands in the presence of iodides (MeI or anhydrous LiI) and acetic acid. The high activity of the catalyst is retained upon the repeated use of the catalyst separated from reaction products. The nitrogen-containing supports (such as chitosan) serve as cocatalysts. The induction period is considerably reduced and the catalytically active species are stabilized. This is especially important since the introduction of chitosan-based ligands makes possible the replacement of expensive LiI for cheaper salts of Li (such as AcOLi or $LiCO_3$). Chitosan is

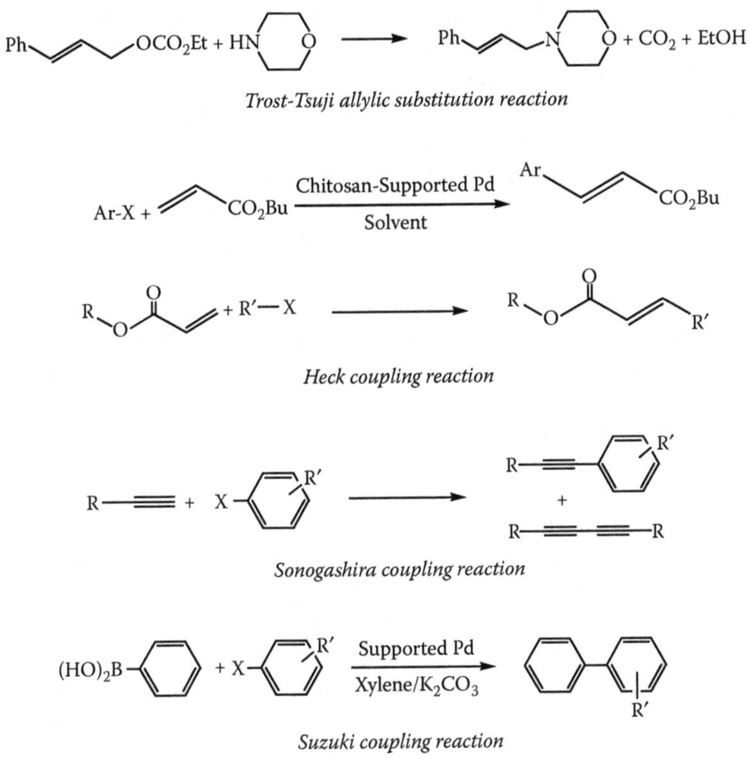

Trost-Tsuji allylic substitution reaction

Heck coupling reaction

Sonogashira coupling reaction

Suzuki coupling reaction

FIGURE 4.47 Examples of cross-coupling reactions.

able to alkylate MeI on the nitrogen atom; therefore, the coordination of the alkylated ligand with the catalytically active $Rh(CO_2)I_2^-$ favors the oxidation of MeI to the Rh atom. Kolesnichenko et al. also investigated the carbonylation of benzyl alcohol to phenylacetic acid in the presence of $RhCl_3$ and HI. The reaction is enhanced by the addition of KI as the promoter and benzyl chloride as co-reactant. The addition of chitosan improves the reactivity and the selectivity of the reaction even at high temperature.[397]

4.4.8 Cross-Coupling Reactions: Heck, Sonogashira, Suzuki, and Trost-Tsuji Reactions

Palladium-catalyzed reactions involving the formation of new C–C bonds are extremely useful for the chemical industry. The use of these reactions for synthesizing better and more complicated molecules suitable as intermediates for the preparation of natural compounds, pharmaceuticals, and molecular organic materials has been widespread.[398] The emblematic reactions for C–C cross-bonding are known as Heck, Sonogashira, Suzuki, and Trost-Tsuji reactions (Figure 4.47). They have received a great attention for the past 5 years in the field of biopolymer-supported materials.[327,328,335,376,397,398]

Quignard et al. used chitosan as a support for the immobilization of a chiral Pd compound (i.e., sulfonated triphenylphosphine derivative, TPPTS) for the allylic substitution of (*E*)-cinnamylethyl carbonate by morpholine (i.e., Trost-Tsuji reaction).[376,400] Palladium acetate is mixed with an excess of TPPTS, and the ready reduction of palladium at 40°C leads to the formation of $Pd(TPPTS)_3$ and partial oxidation of TPPTS.[376] Chitosan is mixed with a Pd complex solution by the incipient wetness method. The catalytic reaction is performed under H_2 atmosphere with direct dissolving of the substrates in benzonitrile. The adjustment of water content of the catalyst solid and the addition of a cationic surfactant (cetyl tri-ammonium bromide, CTAB) both contribute a drastic improvement in the efficiency of the catalytic reaction. The cationic surfactant increases the lipophilic character of the outer surface of the small solid particles (contributing to better dispersion in benzonitrile) and enhances the access of water (and substrates) to the amorphous part of the solid. Buisson and Quignard used the same methodology for other polysaccharide-supported Pd catalysts [i.e., cellulose, polygalacturonic acid (PGA)].[377] They correlate the highest activity of PGA to a higher surface area and lower degree of crystallinity. The strong hydrophilicity of these polysaccharides forms the basis of these promising supported aqueous-phase catalysts. Further studies based on the same methodology have shown that alginate materials dried under supercritical CO_2 conditions have much better catalytic properties than carrageenan and chitosan beads.[227] Ethanol can be used instead of nitrile solvent of the reaction. In the case of alginate-supported materials, the catalytic activity is not affected; this allows using a more environmental friendly solvent.

Liu et al. immobilized Pd on chitosan flakes in ethanol solution under reflux and they tested the catalyst for the Heck reaction of iodobenzene (ArI) cross-coupling with acrylic acid and acrylate (AA).[327,328] The conversion and the regioselectivity are strictly controlled by experimental conditions. Optimum performance (yield close to 94% and 99% for acrylic acid and acrylate, respectively) is obtained in dimethylformamide (DMF) solvent, at the temperature of 80°C, with the addition of 25 mmol of triethylamine, and a 50% molar excess of AA (catalyst dosage: 50 mg in 6 mL DMF).

Recently, Cui et al. compared the catalytic activity of a Pd catalyst immobilized on cross-linked chitosan and cross-linked chitosan-salilylaldehyde derivative for the Heck arylation of aryl iodides with acrylic acid (into *trans*-cinnamic acid) and styrene (into *trans*-stilbene).[401] Chitosan is dissolved in acetic acid solution and mixed at room temperature overnight with epoxy chloropropane (epichlorhydrine) in alkaline solution to prepare cross-linked chitosan. The salilylaldehyde derivative of chitosan is prepared by contact of chitosan with salilylaldehyde in ethanol completed with acetic acid at 80°C. For the cross-linking of this derivative, the modified polymer is mixed in a NaOH solution with cethyltrimethyl ammonium bromide prior to successive additions of epoxy chloropropane and NaOH at 50°C. The supports are loaded with the catalytic metal by contact with $PdCl_2$ solutions in acetone under reflux (60°C for 3 days). The supported catalyst is mixed with

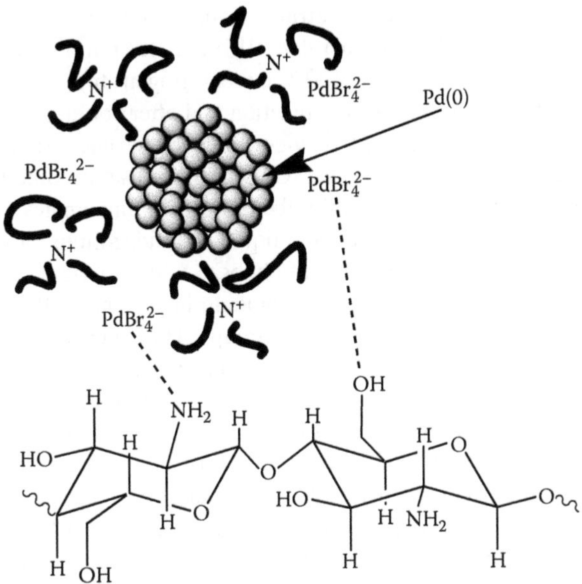

FIGURE 4.48 Modeling of nanosized palladium chitosan immobilized on chitosan in ionic liquids. (Reprinted with permission from V. Calo, A. Nacci, A. Monopoli, A. Fornaro, L. Sabbatini, N. Cioffi, & N. Ditaranto, *Organometallics,* 23, 5154–5158, 2004. With permission. Copyright 2004, American Chemical Society.)

acrylic acid (or styrene), iodobenzene, tributylamine, and dimethylformamide at 90°C for 3 h. The salilylaldehyde derivative of chitosan is more efficient than the simple cross-linked material as a support for Pd and for tested Heck reactions. The modification of chitosan allows significant diminishing of the reaction temperature (to 60°C compared to the standard temperature of 100°C) and the amount of catalyst required to reach a given conversion yield (the catalyst dosage can be four times lower compared to standard cross-linked chitosan, while maintaining high conversion), and substantially increasing the number of reuse cycles.

Calo et al. developed an original method for the Heck cross-coupling reaction (aryl bromides and aryl chloride) using chitosan as a support in ionic liquids (IL) [tributyl ammonium bromide (TBAB)] (Figure 4.48).[399] The nanocomposite was electrosynthesized in a three-electrode cell by means of the sacrificial anode technique. The working and counter electrodes were twin palladium sheets while the reference electrode was a homemade Ag/AgNO3 (0.1 *M* in acetonitrile) one. In a typical experiment, 200 mg of chitosan were suspended in 5 mL of an acetonitrile:tetrahydrofuran (1:3) mixture containing TBAB as a base electrolyte (0.1 *M*) under nitrogen atmosphere. A potential of +1.5 V (vs. reference) was applied to the working electrode during the process, which was coulometrically controlled and stopped at an electrolysis charge value of 200 C. Evaporation under vacuum of the solvent leaves a gray solid that is stored under nitrogen. To

TBAB (3.6 mmol) were added tetrabutylammonium acetate (2.4 mmol), bromobenzene (1.2 mmol), butyl acrylate (0.192 g, 1.8 mmol), and Pd catalyst. The reaction medium was heated at 130°C under nitrogen atmosphere with stirring for the proper time. At the end of the reaction and after cooling to room temperature, the solid mixture was extracted twice with cyclohexane, leaving the catalyst in the ionic liquid that can be recycled. Only 15 min reaction time is required to convert bromobenzene or *p*-nitrochlorobenzene into cinnamates. The reaction does not occur when changing the ionic liquid to imidazolium-based IL.

Hardy et al. prepared a derivative of chitosan with 2-pyridinecarboxaldehyde (in ethanol solution, under reflux) for Pd binding in acetone.[335] Palladium acetate is bound to chitosan through pyridylimine groups. The Pd catalyst supported on a chitosan derivative is tested for Heck and Suzuki reactions (Figure 4.47). In the Suzuki coupling reaction of benzene boronic acid and bromobenzene (in xylene solvent and in the presence of potassium carbonate and *n*-dodecane), optimum conditions for conversion correspond to reflux temperature (143°C) and a substrate:Pd molar ratio lower than 1200. The yield of the reaction is compared for different substrates: best results (conversion yield exceeding 80%) are obtained for substrates bearing electron donating substituents. In the case of nitro-containing compounds, a decomposition of the catalyst is observed. While bromo- and iodo-compounds have comparable activity, chlorobenzene is virtually inert. In the Heck coupling of (a) iodobenzene and *n*-butyl acrylate, and (b) styrene and bromobenzene (in dioxane solvent and in the presence of triethylamine), the yield exceeds 80% and 88% for production of butyl cinnamate and stilbene, respectively.

Gronnow et al. synthesized an amino derivative of expanded starch; the starch is swollen in dry toluene and mixed under argon with 3-aminopropyltriethoxysilane under reflux for 24 h.[398] The amino-modified starch is reacted with 2-acetyl pyridine in ethanol solution under reflux to prepare a Schiff base derivative. This derivative is mixed with palladium acetate under reflux in acetone solvent. The catalyst is used for (a) Suzuki coupling of bromobenzene and benzene boronic acid (xylene/K_2CO_3 solution under argon), (b) Keck coupling of iodobenzene and methyl acrylate (xylene/triethylamine under argon), and (c) Sonogashira coupling of phenylacetylene (alkyne) and iodoarene [in the presence of 1,4-diazabicyclo[2.2.2]octane (DABCO)] at 100°C. In the case of Heck and Suzuki reactions, the coupling is optimized upon increasing the temperature to 140°C (compared to the usual temperature, in the range 80 to 90°C). The yield reaches 100% with no side-reaction products for the Suzuki reaction and 77% for the Heck reaction [with a 100% selectivity for (*E*)-methy cinnamate]. In the case of Sonogashira coupling reactions, the yield and the selectivity are influenced by the types of substituents on the phenyl group (Figure 4.47). It is noteworthy that microwave irradiation during the Sonogashira reaction significantly increases the turnover frequency compared to standard thermal activation.

Cellulose has been used by Reddy et al. for the immobilization of Pd(0) for Heck and Sonogashira coupling reactions (reaction of aryl iodides with alkenes and phenylacetylene, respectively).[402] Microcrystalline cellulose is mixed with $PdCl_2$ in

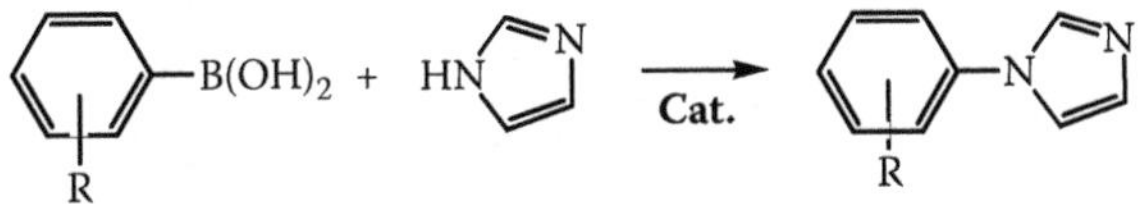

Arylboronic acid Imidazole

R: H, Cl, CH_3, OCH_3, $COCH_3$, CF_3

FIGURE 4.49 Example of coupling reaction between arylboronic acid and imidazole.

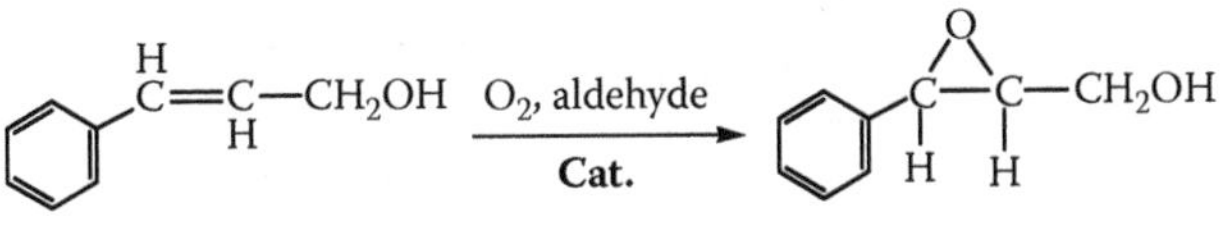

O_2, aldehyde

Cat.

Cyannamyl alcohol (2*R*, 3*R*)(+)-3-phenylglycidol

FIGURE 4.50 Example of epoxidation of cinnamyl alcohol.

methanol for 15 min before addition of hydrazine hydrate to reduce immobilized palladium. The Heck–Sonogashira reactions are carried out by mixing aryl iodide, triethylamine, and olefin-phenylacetylene in acetonitrile or dimethylformamide with the catalyst under reflux. They discuss the cross-coupling reactions for different substrates and they point out the importance of steric effect on the conversion yield (especially when the substituent is in *ortho* position). The *N*-arylation of nitrogen heterocycles (imidazole) with aryl halides and arylboronic acids has been carried out by Reddy et al. using copper immobilized on cellulose (Figure 4.49).[402] Microcrystalline cellulose is mixed with a copper nitrate solution before addition of hydrazine hydrate to reduce copper to Cu(0). The reaction of *N*-arylation of arylboronic acid (or alternatively aryl halides) is performed by contact of the catalyst with imidazole or triethylamine (or alternatively potassium carbonate) in methanol (or alternatively dimethylsulfoxide) under reflux. Reddy and coworkers tested several arylboronic acids and aryl halides; the advancement of *N*-arylation reaction (and the time required for achieving substantial conversion) strongly depends on the nature of the substituents. The catalysts are recovered by filtration at the end of the reaction; after washing with acetone and drying in an oven, they are successfully reused for a series of cycles with minimal loss of activity.

4.4.9 Miscellaneous

Zhang et al. immobilized cobalt chloride on SiO_2-casein (prepared by impregnation of silica with casein dissolved in water at 60°C and drying) in ethanol solution under reflux.[337] They use the SiO_2-casein-$CoCl_2$ catalyst for the epoxydation of cinnamyl alcohol under O_2 atmosphere in 1,2-dichloroethane solvent at 60°C in the presence of aldehyde compound for the synthesis of (2*R*,3*R*)-(+)-3-phenylglycidol (Figure 4.50). The process has been tested with several metals (including Fe, Ni, Mn, V, Mo, Pt, and Co) and best results (in terms of both production and

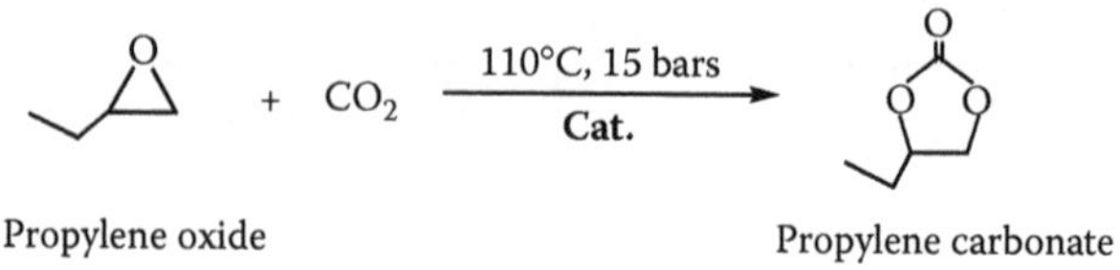

Propylene oxide + CO_2 → (110°C, 15 bars; Cat.) Propylene carbonate

FIGURE 4.51 Example of cycloaddition of carbon dioxide on epoxide.

Lauric acid + glycidol → (Chitosan) α-monolaurin

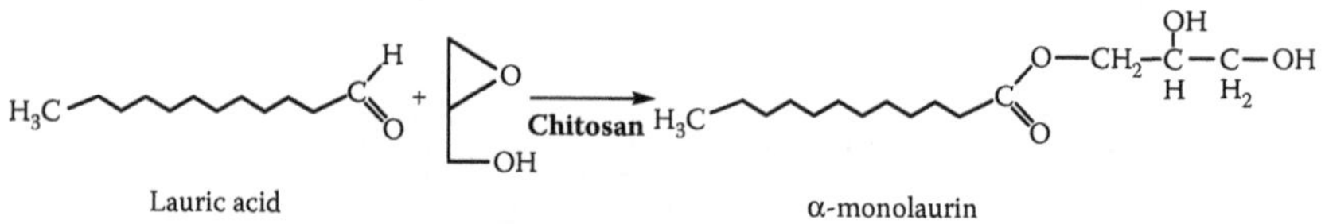

FIGURE 4.52 Example of synthesis of monoglyceride.

optical yields) are obtained with cobalt. Different experimental conditions (temperature, metal content, and aldehyde type) have been tested; best conditions, taking into account both the production and the optical yields, correspond to temperatures in the range 60 to 70°C, Co content close to 0.2 mmol Co g^{-1}, and 2,4-dihydroxybenzaldehyde as the aldehyde co-reactant. The catalyst is reused three times without significant loss of activity and selectivity.

The coupling of carbon dioxide to a series of cyclic carbonate has been investigated by Xiao et al. using a chitosan-supported zinc catalyst (Figure 4.51).[404] The catalyst is prepared by the reflux method mixing chitosan with $ZnCl_2$ in ethanol. The coupling reaction is performed by mixing the catalyst with the epoxide and an ionic liquid (1-butyl-3-methylimidazole halides BMImX, preferentially BMImBr) as the cocatalyst, under fixed pressure of CO_2 at 110°C. The cycloaddition of carbon dioxide on epoxides is very efficient and very selective (Figure 4.51). The optimum molar ratio catalyst:cocatalyst is found close to 1:8. The catalyst immobilized on chitosan is significantly more efficient than Pd immobilized on polyvinyl pyrrolidone, especially when using mild conditions [T: 110°C, P(CO_2): 15 bar] compared to conventional operating conditions. The conversion yield progressively decreases when reusing the catalyst; however, even after six cycles, the yield exceeds 87% and the selectivity remains greater than 99% for CO_2 coupling on propylene oxide.

Valentin et al. showed that chitosan microspheres (prepared by coagulation of chitosan acetic acid solution into NaOH) can be directly used for the synthesis of α-monoglycerides by fatty acid addition to glycidol (Figure 4.52).[226] The reaction (conversion of lauric acid to α-monolaurin) takes place in toluene as the solvent, with the reaction temperature fixed to 70°C. The conversion yield and the selectivity are strongly increased when using the material dried under supercritical CO_2 conditions, compared to oven-dried or freeze-dried supports. The catalytic behavior is close to that obtained with amine-functionalized silica (grafting of primary amine groups onto silica support). However, the stability of the chitosan material compared to SiO_2-NH_2 material in alkaline solutions constitutes a strong advantage compared to conventional material.

4.5 CHITOSAN-SUPPORTED Pd CATALYST FOR HYDROGENATION REACTIONS, FROM FLAKES TO CATALYTIC HOLLOW FIBER

4.5.1 Chitosan Flakes for Hydrogenation Transfer

The hydrogenation transfer reactions based on chitosan-supported catalysts has been investigated by Guibal and coworkers using sodium formate as the hydrogen donor, and chitosan flakes as the support.[282,342–346]

4.5.1.1 Synthesis of Chitosan-Supported Pd Catalyst

4.5.1.1.1 Metal Sorption

Palladium is immobilized on the support through adsorption mechanisms in aqueous solutions involving electrostatic attraction of tetrachloropalladate species on protonated amine groups or chelation on amine groups. The optimum pH for palladium sorption is close to pH 2; the protonation of amine groups (pK_a close to 6.4) allows the electrostatic attraction of metal anions. In more acidic solutions (pH below 2), the competition of counterions decreases Pd sorption capacity. The pH should be controlled using HCl to bring enough chloride ions to displace the equilibrium of palladium toward the formation of chloropalladate species (Figure 4.10). In other acidic solutions, the sorbent has a lower affinity and the addition of chloride ions in the acidic solutions improves the sorption of Pd (Figure 4.11), as a complementary evidence of the necessary conjunction of two favorable parameters: (a) the protonation of amine groups, and (b) the speciation of the metal. Since chitosan is soluble in most mineral acids (and especially HCl), it is necessary to reinforce biopolymer stability using a cross-linking treatment, such as the reaction of amine groups with glutaraldehyde. The formation of imine groups between aldehyde and amine groups (through a Schiff base reaction) induces supplementary bonds between polymer chains contributing to polymer stabilization in acidic solutions. The amine groups are occupied by these linkages, so their availability and the steric hindrance around these reactive groups decrease; however, the impact of polymer modification is less detrimental than in the case of chelation mechanism (Figure 4.53). The molar ratio between amine groups and aldehyde functions on the cross-linking agent is fixed to 1, to achieve good chemical stabilization and to maintain sorption capacity higher than 200 mg Pd g^{-1} (i.e., higher than 2 mmol Pd g^{-1}). Alternatively, chitosan can be mixed with sodium sulfate (3 g L^{-1}, in pH 2 sulfuric acid solution) to bind sulfate anions with protonated amine groups. The choice of the cross-linking process is very important since it may affect the mechanical properties of the material. In the case of chitosan flakes, the formation of supplementary linkages does not significantly affect the mechanical resistance of

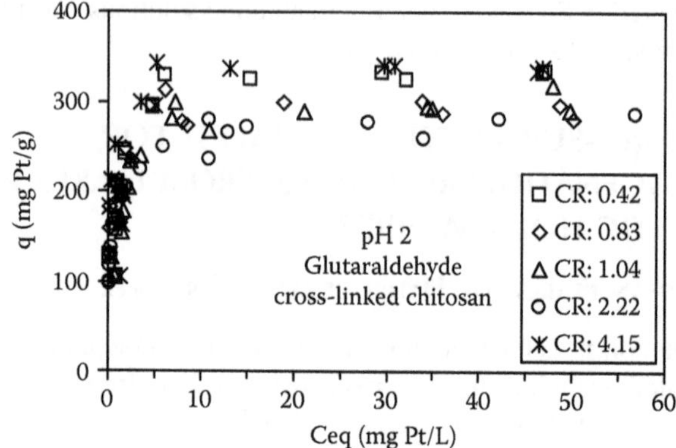

FIGURE 4.53 Influence of the excess of cross-linking agent on Pt sorption using chitosan flakes cross-linked with glutaraldehyde (cross-linking ratio, CR: molar ratio between aldehyde functions in the cross-linking bath and amine functions on the polymer). (Reprinted from E. Guibal, T. Vincent, A. Larkin, and J.M. Tobin, *Ind. Eng. Chem. Res.*, 38, 4011–4022, 1999. With permission. Copyright 1999, American Chemical Society.)

the flakes, while in the case of hollow fibers the glutaraldehyde cross-linking causes fiber brittleness. This means that the cross-linking ratio (aldehyde:amine molar ratio) should be selected the lowest possible, taking into account chemical stability, or alternatively preferring the sulfate treatment for soft cross-linking.

The SEM-EDAX (scanning electron microscopy coupled with x-ray diffraction analysis) analysis of loaded particles shows that the metal is distributed in the whole mass of the sorbent (Figure 4.19). However, the weak porosity of the biopolymer considerably affects the diffusion properties (and, more specifically, the kinetic profiles), as evidenced by the control of sorption kinetics by particle size; the equilibrium sorption capacities are weakly affected by this parameter. This means that palladium can be bound until the center of the particle; however, the diffusion of the reactants to the internal reactive groups may be a critical parameter for the catalytic process.

Metal interactions are quite strong and the metal is quite stable on the polymer in usual media; Pd desorption requires very destructive conditions (high acid concentrations) or very specific media (thiourea, ammonium hydroxide solutions) to reach significant desorption. Metal desorption is less than 10% when using 0.5 *M* HCl solutions. This means that the release of the catalytic metal from the support is very limited in conventional experimental conditions.

4.5.1.1.2 Metal Reduction

Although chitosan has the potential to partially reduce bound metals (depending on the normal redox potential, see above), this reducing effect is not sufficient in the case of palladium. The color of loaded sorbent remains unchanged, contrary to the process described in the literature for metal binding in ethanol solution under reflux.

The reduction of palladium can be performed using different processes, alone or in combination: (a) sodium borohydride, (b) sodium formate, and (c) hydrogen gas (generated *in situ* by reaction of sulfuric acid with zinc powder). Several parameters can be used for the optimization of the process such as the weight loss (especially in processes involving the use of strongly acidic solutions) and the catalytic activity of Pd crystals. For the optimization of a synthesis procedure, Vincent and Guibal investigated the reduction of chromate using sodium formate as the hydrogen donor.[344] This simple reaction has no practical interest except for the ready analysis and testing of catalytic activity. Best results are obtained when using glutaraldehyde cross-linking agent; sulfate-treated support may not be sufficiently stable (especially when reduction is operated in sulfuric acid solutions, for H_2 *in situ* generation). Despite higher sorption capacities, the catalytic materials prepared with sulfate-treated support are less active than glutaraldehyde-treated materials. The most appropriate reduction process combines in this case the treatment with sodium borohydride, followed by reduction with *in situ* generated H_2. However, palladium reduction is not complete as evidenced by XPS analysis: about 40 to 50% of Pd remains in the Pd(II) original form. The catalytic activity of chitosan-supported Pd is significantly higher than that of palladium microparticles prepared by reduction and precipitation.

The analysis of Pd-loaded chitosan flakes (after reduction, embedment in a resin, and cutting in thin slices) by transmission electron microscopy shows the formation of nanoparticles characterized by sizes of 3 to 5 nm. However, some large aggregates are formed with sizes of 150 nm. Although the nanoparticles are homogeneously distributed at the center of chitosan flakes, the peripheral areas are more densely loaded (Figure 4.19).

4.5.1.2 Characterization of Catalytic Performance

These materials have been tested for environmental applications, including the treatment of aqueous solutions containing phenol derivatives, bearing halogen (chloride) or amino groups, at low concentration in the presence of sodium formate as the hydrogen donor (Figure 4.36).

4.5.1.2.1 Chlorophenol Dehalogenation

A chitosan-supported Pd catalyst has been used for the dehalogenation of chlorophenol [2-chlorophenol (2-CP)] in aqueous solutions using sodium formate as the hydrogen donor (hydrogenation transfer) (Figure 4.36). The hydrogenation involves the conversion of chlorophenol to phenol during the first 90 min; however, in a second phase of the reaction, phenol is converted to cyclohexanone or cyclohexanol. Chlorophenol degradation proceeds by a two-step procedure involving, successively, dehalogenation and dearomatization of the substrate.

Simultaneous with chlorophenol dehalogenation, the pH of the solution increases due to formate conversion to CO_2. The half-reaction time, the turnover frequency, and the kinetic coefficient (pseudo first-order equation) depend on the pH of the solution (Figure 4.54). Optimum initial pH is found close to 3 (i.e., close to the pK_a of sodium formate, 3.7); the adsorption of sodium formate appears to be the critical step in the hydrogenation process, since in acidic solutions the substrate remains in its molecular form (pK_a of 2-CP: 8.55).

4.5.1.2.2 Nitrophenol Hydrogenation

The chitosan-supported Pd catalyst described in the preceding section has been tested for nitrophenol hydrogenation using sodium formate as the hydrogenation transfer agent (Figure 4.36). The only change in the synthesis procedure consists of the palladium reduction step: the reduction is performed using only H_2 generated *in situ*. Two steps (at least) have been identified in the hydrogenation of nitrophenol (NP) with sodium formate in open atmosphere (aerobic conditions): (a) the reductive hydrogenation of nitrophenol into aminophenol (AP), followed after 45 to 60 min of reaction by (b) the oxidation of aminophenol into subproducts (for example, muconic acid and oxidized derivatives), which depend on experimental conditions (complete elimination of sodium formate, aerobic conditions, etc.). The catalyst has been carried out for the hydrogenation of aminophenol, and the products at long reaction time are similar to the final products of nitrophenol degradation. It is noteworthy that under selected experimental conditions, phenol is not substantially degraded; the formation of by-products does not proceed by a degradation of aminophenol into phenol, followed by its subsequent degradation. It is probable that the final degradation is simultaneous to the conversion of nitro- to amino-moieties. The reaction is oriented by the position of the substituents on the phenolic ring; indeed 3-NP is converted to the same by-products of 3-AP hydrogenation, while 2-NP and 4-NP are altered into the same products of degradation as 2-AP. Aramendia et al. reported that the efficiency of the hydrogenation process (halobenzene dehalogenation) is controlled by the relative adsorption coefficient between sodium formate and halobenzene at the catalyst surface.[405] The strong sorption of the substrate may control the sorption of the hydrogen donor, and then reduces the reaction rate. The following simplified mechanism has been proposed:

1. Adsorption of both sodium formate and substrate at the surface of the catalyst
2. Decomposition of formate into hydrogen and carbon dioxide
3. Reaction of hydrogen with the substrate at the surface of the catalyst
4. Release of degradation products (and subsequent degradation)

Based on this hypothesis, the effect of pH can be explained. The optimum pH is in the range pH 3 to 4; because the pK_a of sodium formate is close to 3.7, the anionic hydrogen donor can be adsorbed in the pH range 3 to 5.5 (where more than 90% of amine groups are protonated), while the pK_a of nitrophenol is close to 7.1. Nitrophenolates are only predominant in neutral or alkaline solutions, where chitosan is not protonated and thus not available for nitrophenolate adsorption. The kinetic profiles have been modeled in this case using the variable-order kinetic equation (Equation [4.3], see above).[12]

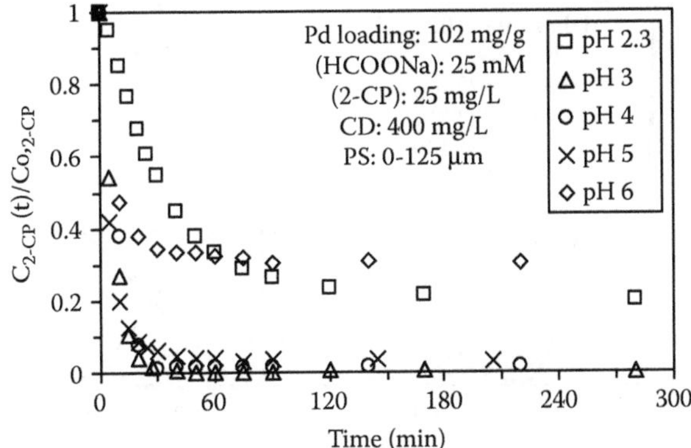

FIGURE 4.54 Influence of pH on the hydrogenation of 2-chlorophenol using Pd-immobilized on chitosan flakes and HCOONa as the hydrogen donor. (Reprinted from T. Vincent, S. Spinelli, and E. Guibal, *Ind. Eng. Chem. Res.*, 42, 5968–5976, 2003. With permission. Copyright 2003, American Chemical Society.)

4.5.1.2.3 Aniline hydrogenation

The catalyst prepared with the single H_2 gas (*in situ* generated) reduction procedure was tested for the hydrogenation of nitroaniline (4-nitroaniline), which is converted into 1,4-phenylenediamine, using sodium formate as the hydrogen donor (hydrogen transfer reagent) (Figure 4.36). Several studies on the hydrogenation of nitro-aromatic compounds (involving iron oxide/hydroxide catalyst or Pd supported on vinylpyridine-based polymers) have shown that phenylenediamine is the only product obtained in the reduction of nitroaniline.[406–408] Xu et al. observed that the rate of hydrogenation of nitro-compounds is strongly controlled by the structure of the substrate and more specifically by the type of substituent close to nitro-moiety: electron-donating groups are more favorable for degradation than electron-withdrawing groups.[408] Amino and methoxy groups may increase the electron density of the nitro group, which in turn favors substrate coordination with palladium. Conversely, electron-withdrawing groups are not favorable to the activation of the nitro group. The reaction may be also controlled by steric effects. Hence, at low conversion the rate order was found to be *p*- > *m*- > *o*-substituted nitrobenzenes, while at higher conversion the rate order was *p*- > *o*- > *m*-substituted nitrobenzenes.[408]

The influence of pH on aniline hydrogenation using chitosan-supported Pd catalyst has been studied by Vincent and coworkers.[345] Under selected experimental conditions (sodium formate as the hydrogen donor), the substrate is not fully converted when the pH is greater than 5. At low acidity, the conversion of sodium formate to carbon dioxide substantially increases the pH of the solution and strongly impacts catalytic efficiency. It is noteworthy that the change of hydrogen donor, using hydrogen gas (at 1 *M* pressure), halves the time required for converting nitroaniline (Figure 4.55); the kinetics is controlled by the interfacial surface between the gas and the liquid phase. Hydrogen transfer to the liquid phase is the limiting step.

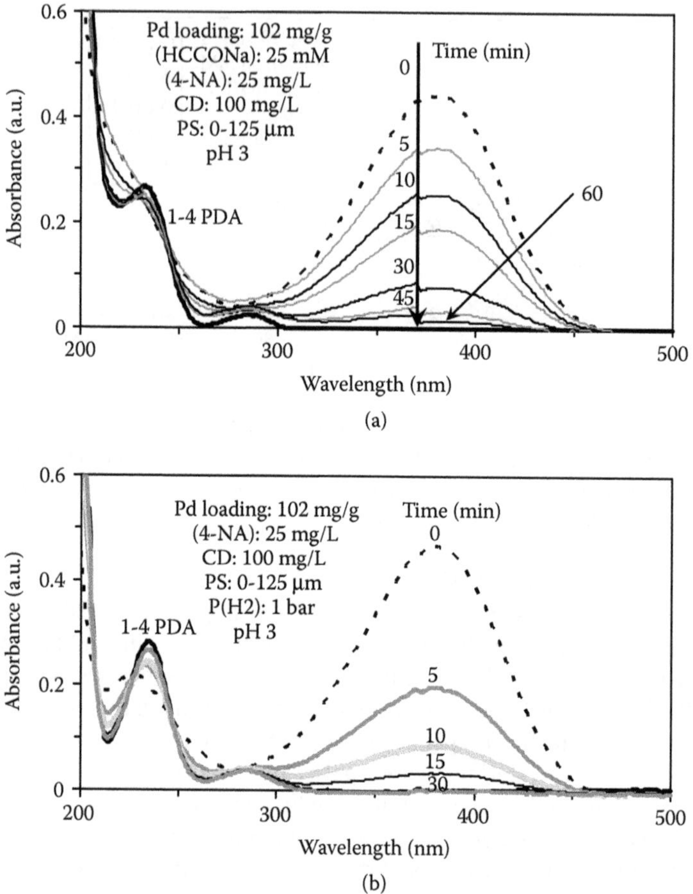

FIGURE 4.55 Hydrogenation of 4-nitroaniline using Pd immobilized on chitosan flakes and HCOONa (a) or hydrogen gas (b) as the hydrogen donor. (Reprinted from T. Vincent, F. Peirano, and E. Guibal, *J. Appl. Polym. Sci.*, 94, 1634–1642, 2004. With permission. Copyright 2004, John Wiley & Sons.)

4.5.1.2.4 Critical Parameters

The hydrogenation efficiency and the catalytic properties are controlled by several factors, in addition to the pH effect (see above): agitation speed, substrate concentration, sodium formate concentration (substrate:sodium formate molar ratio), catalyst dosage and catalyst loading (Pd content), particle size, temperature, and so on.

The major impact of agitation velocity turns on film diffusion. In the case of nitrophenol hydrogenation, Vincent and Guibal showed that the half-time of reaction varies by less than 2 min (around an average value of 14 min) when increasing the agitation velocity from 100 rpm to 750 rpm (Figure 4.24).[370] Similar conclusions are reached considering the time required for achieving 90% of total conversion; this reaction time decreases from 35 min to 25 min at increasing agitation velocity from 100 rpm to 750 rpm. The velocity of agitation hardly influences hydrogenation kinetics, which indicates that the resistance to external diffusion has a limited effect on reaction kinetics.

Substrate concentration and hydrogen donor (sodium formate) concentration influence both the conversion yield and the catalytic activity (turnover frequency). Actually, a limit molar ratio between the substrate and the hydrogen donor concentrations is required for achieving the complete conversion of the substrate under selected experimental conditions. In the case of 2-chlorophenol (2-CP) dehalogenation, the required sodium formate/2-CP molar ratio is close to 50[346]; for the hydrogenation of 3-nitrophenol (3-NP) and 4-nitroaniline (4-NA), the excess molar ratio between the hydrogen donor and substrate concentrations drops to 10 to 15 (Figure 4.56).[343,345] However, increasing the temperature of the reaction allows decreasing this required molar excess, as shown in the case of nitrophenol hydrogenation.[342] Both the catalytic activity (or turnover frequency, mmol of substrate converted g^{-1} Pd min^{-1}) and the initial hydrogenation rate increase nonlinearly with sodium formate concentration.[342,343,345,346] The kinetic profiles for the hydrogenation of 2-CP, 3-NP, and 4-NA are weakly affected by the change in initial concentration of substrate; the decay curves almost overlap in the experimental range (12–25 to 50–200 mg L^{-1}), indicating that this parameter has a limited impact on catalytic activity when the hydrogen donor is in excess. The initial hydrogenation rate slightly decreases with increasing substrate concentration.

The activation energy has been calculated for 2-CP and 3-NP hydrogenation using chitosan-supported Pd catalyst and sodium formate as the hydrogen donor, varying the temperature of the reaction (in the range 20 to 60°C) (Figure 4.57). It varies between 18 and 26 kJ mol^{-1}, depending on catalyst dosage for 3-NP hydrogenation,[342] and between 20 and 25 kJ mol^{-1} for 3-NP degradation.[346] Increasing the temperature strongly increases initial hydrogenation rate and turnover frequency.

Increasing the catalyst dosage increases the volumetric density of potential reactive groups in the solution; as expected, this increases the initial hydrogenation rate. However, the enhancement of the initial hydrogenation rate is not proportional to the amount of Pd in the reactive medium; this may be explained by the weak accessibility of Pd at the center of catalytic particles (Figure 4.58).[342,343,345,346] Actually, a fraction of the metal immobilized is not active and the reaction is localized at the surface of the particle. Consequently, despite the increase in the initial rate of hydrogenation, the turnover frequency decreases with catalyst dosage in the solution. This is confirmed by the weak impact of the Pd loading of catalyst particles on the initial hydrogenation rate. For example, in the case of 3-NP hydrogenation,[343] doubling Pd loading on the particles results in a slight increase (around 30%) of the initial rate, while the turnover frequency significantly decreases: the catalyst has poor porosity, and increasing the quasi-homogeneous loading of the particles makes a part of immobilized Pd poorly accessible for reaction. This is also confirmed by the dramatic effect of catalyst particle size on kinetic parameters.[343,345] Chitosan flakes have been ground and sieved into different

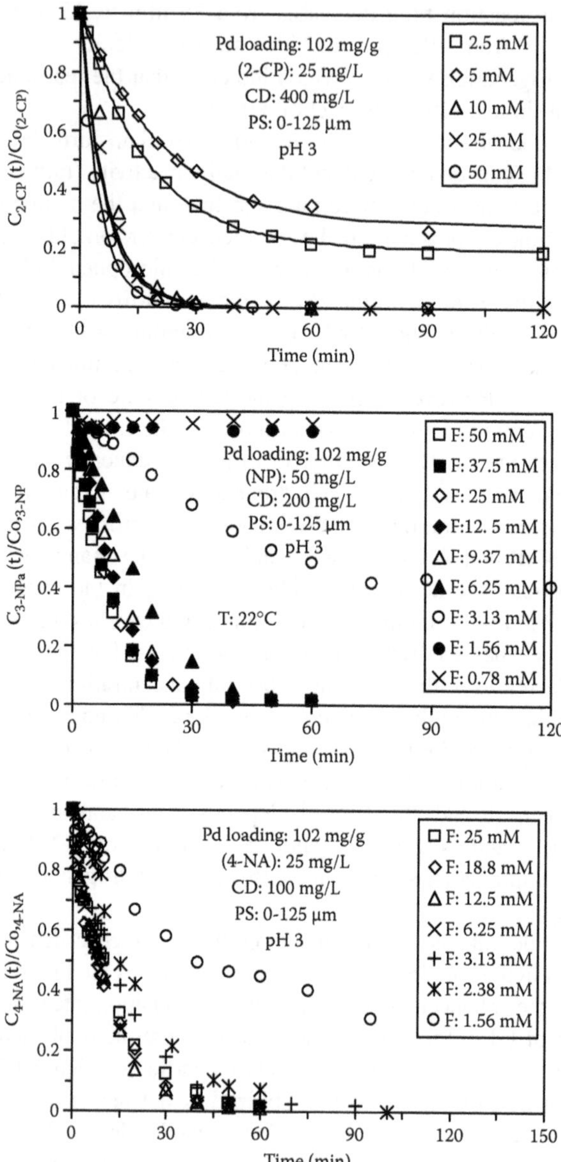

FIGURE 4.56 Influence of hydrogen donor concentration on the hydrogenation of chlorophenol, nitrophenol, and nitroaniline, using Pd immobilized on chitosan flakes and HCOONa as the hydrogen donor. (Reprinted from T. Vincent and E. Guibal, *Langmuir,* 19, 8475–8483, 2003. With permission. Copyright 2003, American Chemical Society; Reprinted from T. Vincent, S. Spinelli, and E. Guibal, *Ind. Eng. Chem. Res.,* 42, 5968–5976, 2003. With permission. Copyright 2003, American Chemical Society; Reprinted T. Vincent, F. Peirano, and E. Guibal, *J. Appl. Polym. Sci.,* 94, 1634–1642, 2004. With permission. Copyright 2004, John Wiley & Sons.)

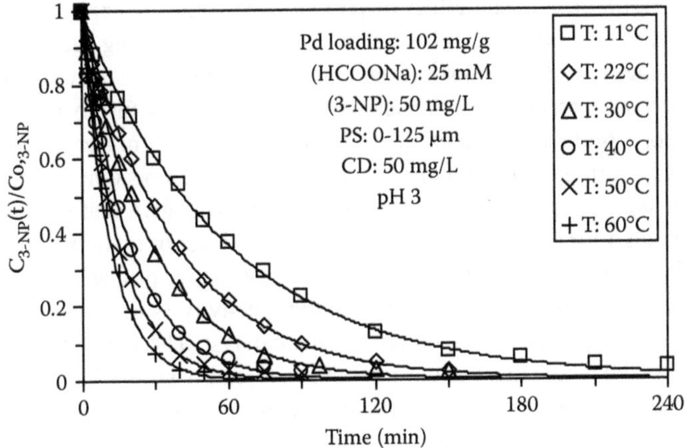

FIGURE 4.57 Influence of temperature on the hydrogenation of nitrophenol using Pd immobilized on chitosan flakes and HCOONa as the hydrogen donor. (Reprinted from E. Guibal and T. Vincent, *J. Environ. Manage.*, 71, 97–106, 2004. With permission. Copyright 2004, Elsevier.)

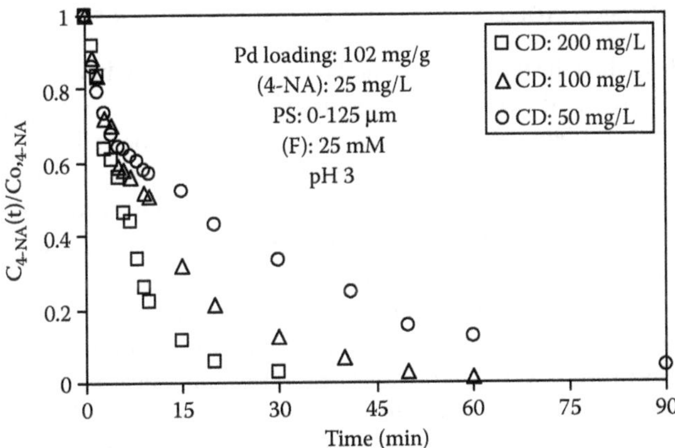

FIGURE 4.58 Influence of catalyst dosage on the hydrogenation of nitroaniline using Pd immobilized on chitosan flakes and HCOONa as the hydrogen donor. (Reprinted from T. Vincent, F. Peirano, and E. Guibal, *J. Appl. Polym. Sci.*, 94, 1634–1642, 2004. With permission. Copyright 2004, John Wiley & Sons.)

fractions before Pd sorption and metal reduction. The different catalysts have been tested for 3-NP and 4-NA hydrogenation. Decreasing the size of catalyst particles strongly increased the initial rate of hydrogenation and the turnover frequency. The limiting effect of particle size is especially important when catalyst diameter exceeds 125 μm; below, the kinetic profiles overlap, indicating that the diffusion layer should not exceed 60 μm for limiting the effect of resistance to intraparticle diffusion (Figure 4.23). The effect of particle size could be explained by several diffusion problems: (a) metal diffusion, (b) diffusion of reducing agent (at the metal reduction stage), and (c) substrate and product diffusion. The weak impact of particle size on sorption performance shows that metal diffusion is not the controlling step. Grinding large particles (500 μm) to 0 to 95 μm size fraction allows substantial restoring of the activity of the catalyst, which becomes comparable to that of the smallest particles 0 to 125 μm. This is indicative that the reduction is efficient enough for converting Pd(II) to Pd(0) at a level of activity comparable to that obtained with small particles. The limiting effect is probably attributable to the weak porosity of the support. The catalytic activity is limited to the external surface (and first layers, below 60 μm) of the catalyst. The metal homogeneously distributed in the whole mass of the sorbent is not readily accessible to the substrate when located at the center of the particle. Therefore, a better use of the metal and an enhanced catalytic activity can be obtained using either chitosan as a coating agent for immobilization of catalytic metals on inert surfaces or thin layers of chitosan (such as obtained when preparing chitosan-based hollow fibers; see below). An alternative may consist of expanding the structure of the support by preparation of aerogels and using a drying procedure that avoids the collapse of the porous structure of the gel (i.e., drying under supercritical CO_2 conditions).[226,227]

4.5.2 Chitosan Hollow Fiber for Catalytic Hydrogenation

Liu et al. described the use of mono- and bimetallic catalytic hollow fiber reactors for the selective hydrogenation of olefins.[315,316,409] They use polysulfone and cellulose acetate fibers coated with synthetic polymers or cellulose derivatives for the binding of Pd alone or associated with cobalt. Metal-polymer-hollow fiber composites are reduced with hydrazine or sodium borohydride and the catalytic systems are used for the hydrogenation of butane, cyclopentadiene, propanediene, and propyne.[315,316,409] These are part of the few papers published on the use of catalytic hollow fibers. Despite the interest in these conditionings for the control of experimental conditions (separation of solution and hydrogen, control of diffusion layers, and so on), these reactors have not been frequently investigated. Hollow fibers made of chitosan have been used for immobilizing palladium for the preparation of supported catalysts.[282,283] They have been tested for the hydrogenation of nitrophenol in aqueous solution[282,283] and for the hydrogenation of nitrotoluene in methanol solutions.[410] The solution to be treated is flowed through the lumen of the catalytic hollow fiber while the hydrogen donor (either sodium formate solution or hydrogen gas) is placed in the outer compartment (Figure 4.59).

4.5.2.1 Manufacturing of Catalytic Hollow Fibers and Characterization

4.5.2.1.1 Metal Impregnation and Reduction

The manufacturing of chitosan hollow fibers has been briefly described above.[274,275,279–281] For the preparation of catalytic hollow fibers, the membranes

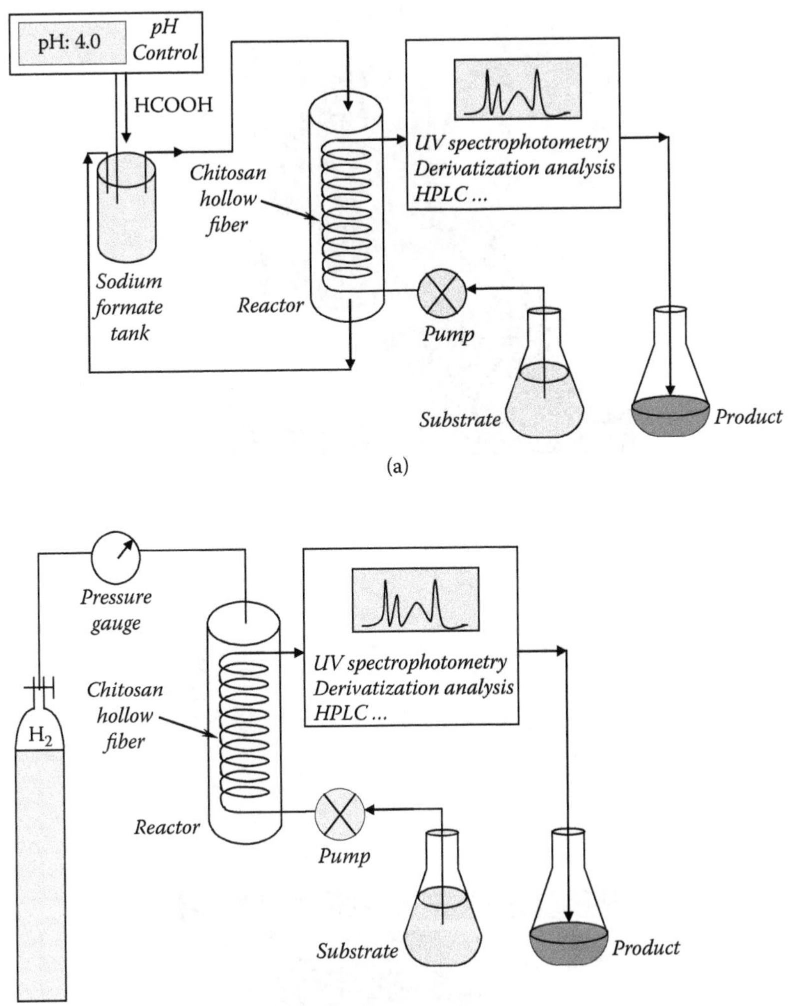

FIGURE 4.59 Experimental setup used for the hydrogenation of nitro compounds using Pd immobilized on chitosan flakes.

are prepared by extrusion of a chitosan-acetic acid solution into an alkaline coagulation bath and further evacuation of the core portion of the fiber that was not coagulated.[280,281] Figure 4.60 shows some typical SEM microphotographs of chitosan hollow fibers. Palladium is immobilized on chitosan by sorption in HCl solutions (at pH 2). Because chitosan dissolves in dilute HCl solutions, the hollow fibers require a stabilizing treatment. They can be cross-linked with either sulfate anions or glutaraldehyde. Glutaraldehyde concentration (and molar ratio with amine groups) should be controlled to prevent an excessive cross-linking, which contributes to making the fibers very brittle. Glutaraldehyde cross-linking needs

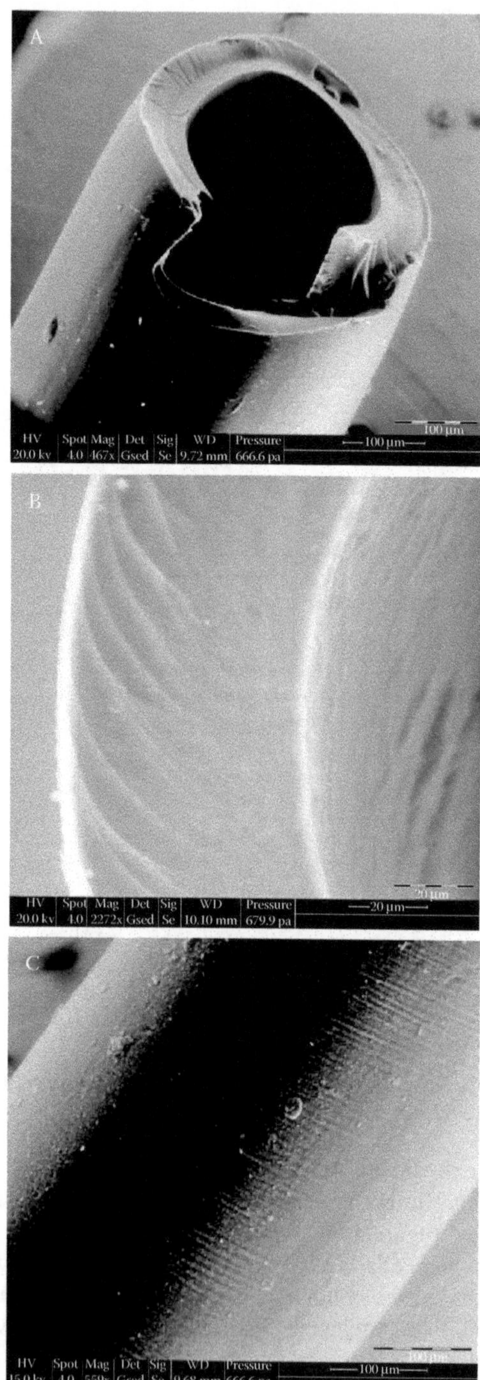

FIGURE 4.60 SEM of chitosan hollow fibers (A): overall view; (B): section; (C): external view.

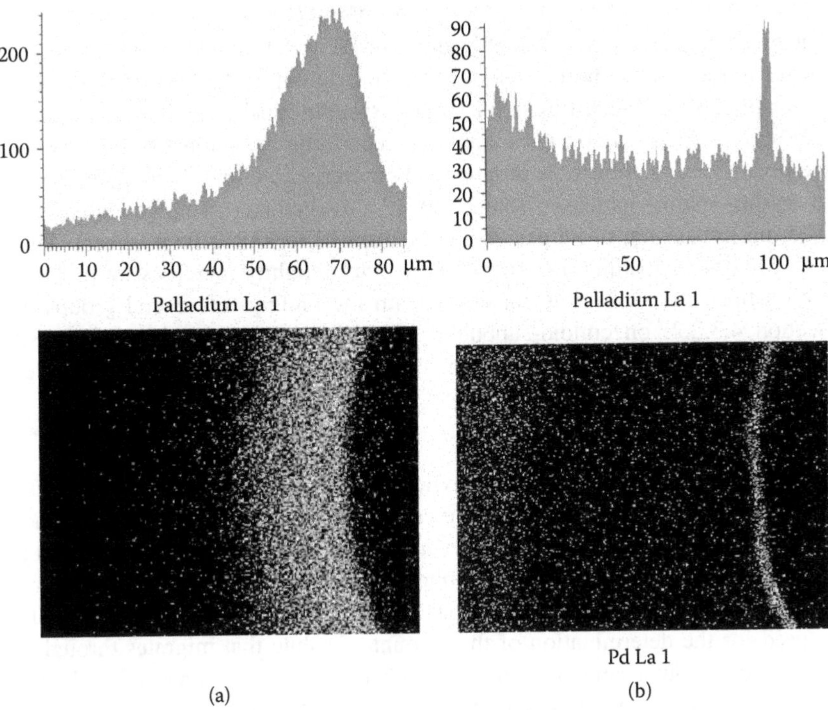

FIGURE 4.61 Dispersive x-ray microanalysis (EDX) and palladium distribution scan lines of palladium catalytic hollow fibers. High palladium content fiber. (a): 2.5 mg Pd/fiber, compared to low palladium content fiber (b): 0.25 mg Pd/fiber.

a balanced selection of experimental conditions for maintaining flexibility and mechanical properties of the fibers and improving chemical resistance.

Practically, an acidic palladium solution is pumped and recirculated through the lumen of the fiber at high superficial velocity (short residence time) to homogeneously disperse the metal along the fibers. A gradient in metal distribution is observed between the outer and inner sides of the fiber (Figure 4.61), depending on the metal loading of the fiber. Immobilized palladium is reduced by *in situ*–generated H_2: a sulfuric acid solution is flowed through a cartridge containing finely divided Zn powder, and the solution containing hydrogen gas is circulated in the fiber. The yellow color of the fiber (due to palladate sorption) rapidly turns to the typical black color of Pd(0). XPS analyses show that palladium is only partially reduced and that a gradient in reduction efficiency may exist between outer and inner sides of the fiber; a significant difference in the percentage of Pd(0) is detected when comparing the XPS spectra of raw and ground fibers. The effect of an electron beam on palladium reduction makes the exact determination of the percentage of metal reduction difficult, but the trends observed on a series of fiber loaded with different amounts of Pd confirm that the reduction process is a key step in the design of these catalytic hollow fibers.

4.5.2.1.2 X-ray Diffraction Analysis: Characterization of Pd Crystallites

The fibers prepared with different loadings of Pd have been characterized by x-ray diffraction analysis. The full width at half maximum of the band at the identification peak allows determining the size of crystallites using the Scherrer equation (see above). Figure 4.20 shows that when increasing the amount of Pd loaded on the fiber, the size of crystals progressively increases. At low Pd loading, the size of crystals sharply increases from 1.7 to 2.9 nm when increasing Pd loading from 0.25 mg to 0.75 mg of Pd per fiber (i.e., from 11 to 33 mg Pd g^{-1} fiber). Above a 0.75 mg/fiber loading, the size of Pd crystals slightly increases up to 3.9 nm with a linear trend. This is consistent with the findings of several groups (see Section 4.3.2.2) on colloid-supported or solid-supported catalysts. Increasing metal concentration on the support may cause the aggregation of nanoparticles with significant impact on the catalytic properties.

4.5.2.1.3 Diffusion Properties

The diffusion properties of the catalytic hollow fibers (at the different stages of the synthesis procedure) have been investigated using vitamin B_{12} as the tracer. The solute is recirculated in the fiber immersed in a closed water tank, and the concentration of vitamin B_{12} is monitored online at the outlet of the fiber by UV-visible spectrophotometric analysis (Figure 4.62). The mass balance equation is used for the determination of the amount of solute that migrates through the fiber into the outer compartment of the fiber. The flux of vitamin B_{12} (F, μmol m^{-2} min^{-1}) is determined in function of the gradient of vitamin B_{12} concentration between the inner (C_{hf}, μmol L^{-1}) and the outer (C_r, vitamin B_{12} tank shell, μmol L^{-1}) compartments (Figure 4.62). Taking into account the thickness of the fiber (l, here 60 μm ± 5 μm), it is possible to approximate the diffusion coefficient in the fiber (D, m^2 min^{-1}), neglecting the diffusion of the solute in the film on both sides of the membrane.

$$F = \frac{D}{l}\left(C_r - C_{hf}\right) \tag{4.4}$$

The Renkin equation is used for the evaluation of pore size (R_p, nm), based on the molecular diffusivity of vitamin B_{12} in water (D_{mol}, 2.274 10^{-8} m^2 min^{-1}) and the hydrodynamic radius of vitamin B_{12} (R_s, 0.85 nm)[288,292–294]:

$$\frac{D}{D_{mol}} = \left(1 - \frac{R_s}{R_p}\right)\left[1 - 2.104\left(\frac{R_s}{R_p}\right) + 2.09\left(\frac{R_s}{R_p}\right)^3 - 0.95\left(\frac{R_s}{R_p}\right)^5\right] \tag{4.5}$$

The Renkin equation should be taken into account as indicative of the order of magnitude of the pore size, since the relation requires R_s/R_p to be greater than 0.5 ($R_p > 1.7$ nm); this is not the case for the smallest pore sizes (Figure 4.63).

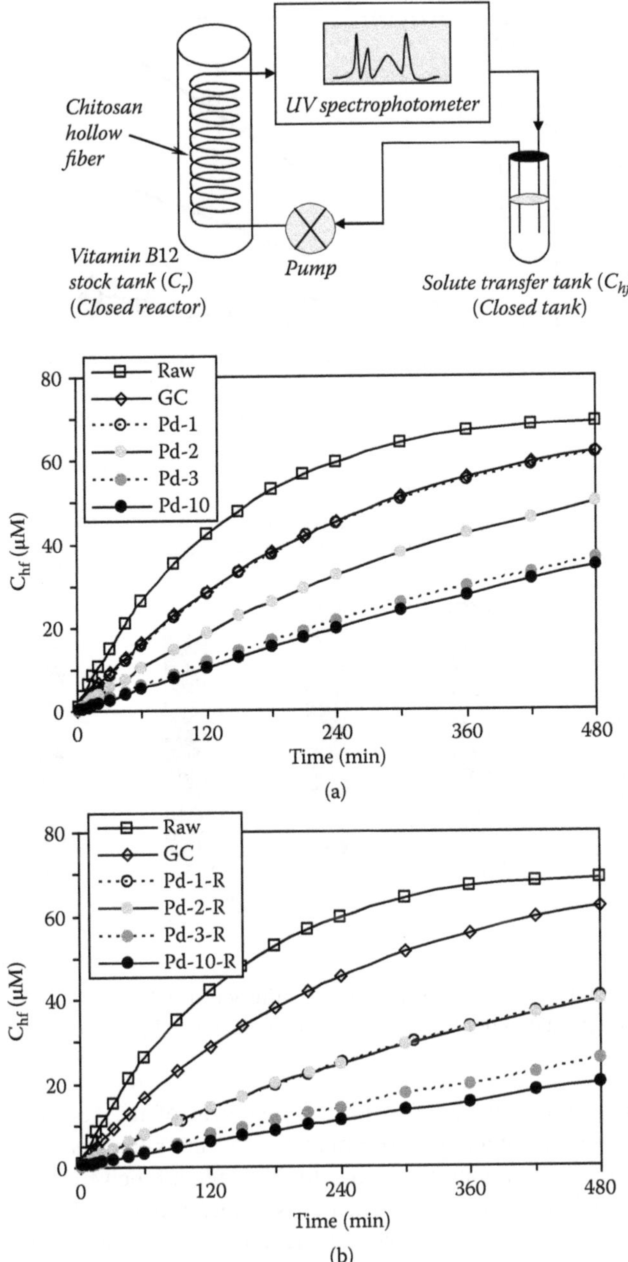

FIGURE 4.62 Diffusion properties of chitosan hollow fibers loaded with Pd at different contents (1–10: 0.25–2.5 mg Pd/fiber): Vitamin B12 diffusion compared for raw, glutaraldehyde cross-linked, Pd loaded (a) and Pd-loaded/reduced fibers (b). Experimental apparatus, top scheme.

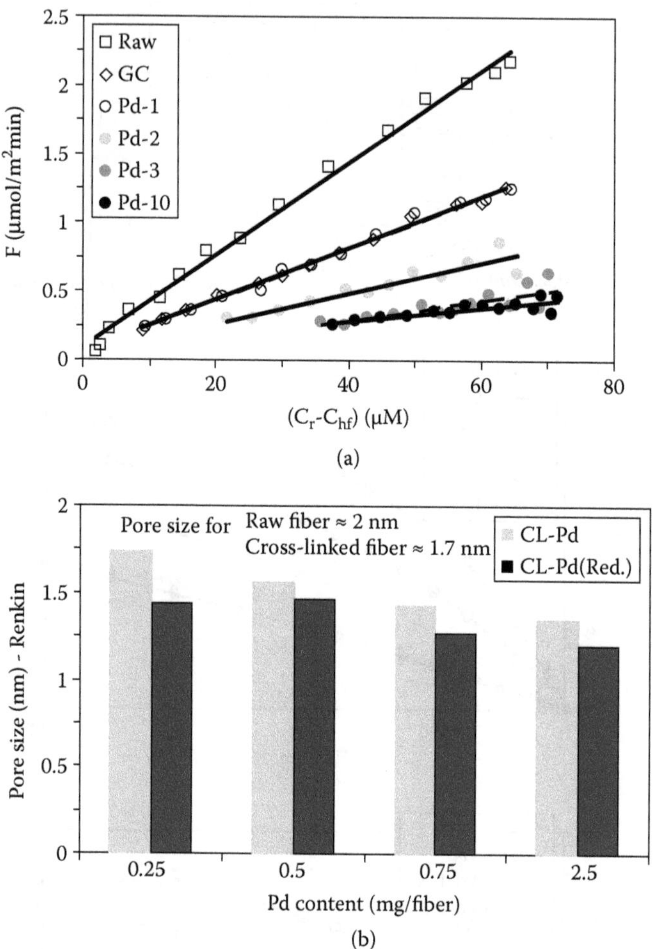

FIGURE 4.63 Influence of Pd loading on the diffusion flux (vs. concentration gradient outer/inner compartments of the fiber) (a) and approximation of pore size using the Renkin equation (b).

These results show that the cross-linking treatment of the fiber decreases the porosity and that Pd binding also diminishes the mass-transfer rate. However, most unfavorable conditions correspond to Pd sorption followed by metal reduction. The reduction in pore size is increased with increasing Pd content, especially above 0.5 mg Pd/fiber (i.e., above 22 mg Pd g^{-1}). This approach should be considered carefully: The weak adsorption of vitamin B_{12} may affect diffusion properties, independently of the actual size of the pores. These results should be taken as indicative of possible pore restrictions. Complementary experiments are currently under investigation. Palladium content in the fibers should be optimized taking into account its impact on diffusion properties and catalytic efficiency.

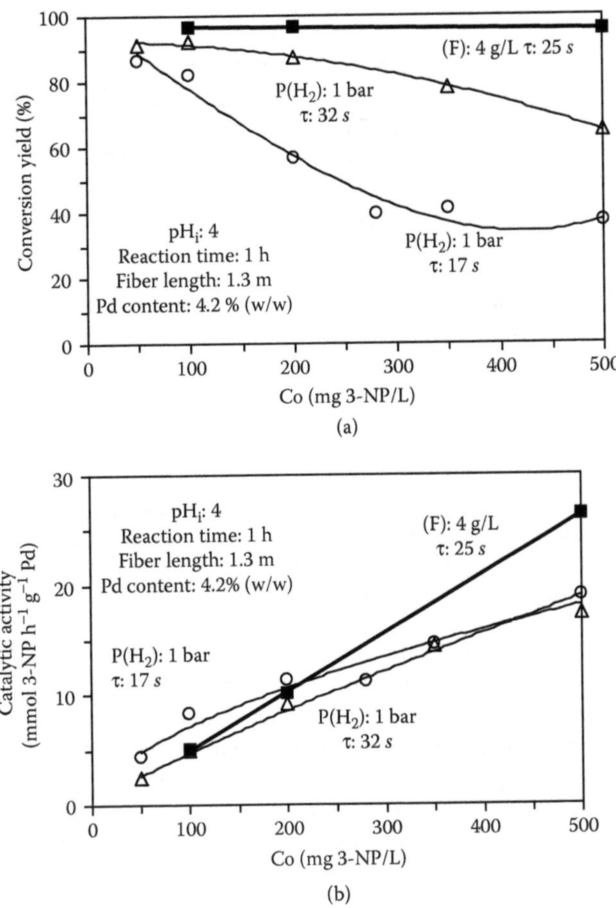

FIGURE 4.64 Influence of nitrophenol concentration on conversion yield (a) and catalytic activity (b) for hydrogenation of nitrophenol using Pd immobilized on chitosan hollow fibers and HCOONa or hydrogen gas as hydrogen donor. (Reprinted from E. Guibal and T. Vincent, *Environ. Sci. Technol.*, 38, 4233–4240, 2004. With permission. Copyright 2004, American Chemical Society.)

4.5.2.2 Nitrophenol Hydrogenation

Palladium catalyst immobilized on chitosan hollow fibers has been tested for the hydrogenation of 3-NP in aqueous solutions using two different hydrogen donors: sodium formate (for hydrogenation transfer) and hydrogen gas (for conventional hydrogenation reaction) (Figure 4.64). Contrary to chitosan-flakes support, a partial oxidation may exist with long contact time (and lack of reducing agent); hollow fibers allow the exclusive production of aminophenol. Short contact time, in a well-controlled reducing system, limits the degradation of intermediary products. To prevent the brittleness of the fibers, the cross-linking of the fibers was operated

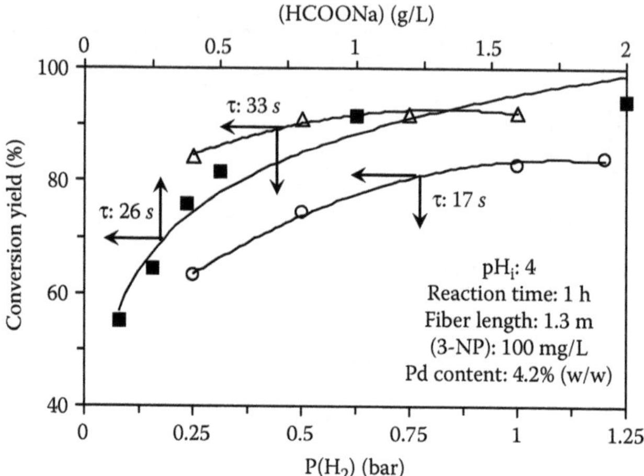

FIGURE 4.65 Influence of HCOONa concentration and hydrogen pressure on conversion yield for nitrophenol hydrogenation using Pd immobilized on chitosan hollow fibers. (Reprinted from E. Guibal and T. Vincent, *Environ. Sci. Technol.*, 38, 4233–4240, 2004. With permission. Copyright 2004, American Chemical Society.)

using sulfate anions or low concentrations of glutaraldehyde; the fibers may be less resistant to acidic media. For this reason, the pH range has been restricted to 4 to 6. The conversion yield and the turnover frequency both decrease when increasing the pH. When increasing the concentration of 3-NP (100 to 500 mg L^{-1}), the conversion yield remains greater than 95%; the molar ratio between sodium formate and the substrate is greater than 25. This large excess may explain the complete conversion of the substrate. The turnover frequency linearly increases with 3-NP concentration to reach 28 mmol 3-NP h^{-1} g^{-1} Pd (Figure 4.64).

In the case of hydrogen gas as the hydrogen donor, the conversion yield depends on the residence time of the solution in the fiber. It systematically decreases with increasing substrate concentration, while the turnover frequency increases almost linearly, reaching 18 mmol 3-NP h^{-1} g^{-1} Pd. The catalytic activity is thus lower for hydrogen gas than for sodium formate. Despite its lower activity, the hydrogen gas system shows the important advantage of limiting the diffusion of the substrate through the membrane as it occurs when using sodium formate; indeed, a small but detectable part of 3-NP (about 3 mg L^{-1}) is found to contaminate the sodium-formate compartment.

The concentration (or pressure) of the hydrogen donor is also a critical parameter for the optimization of the process. The gradient of concentration between the inner and outer compartments enhances the transfer of the hydrogen donor through the membrane. The larger the excess of sodium formate, the greater the conversion yield and the turnover frequency. However, the beneficial effect of increasing hydrogen donor concentration is more significant at low concentration; when sodium formate concentration exceeds 1 g L^{-1}, the improvement of turnover frequency does not exceed 20% (Figure 4.65). The increases of turnover

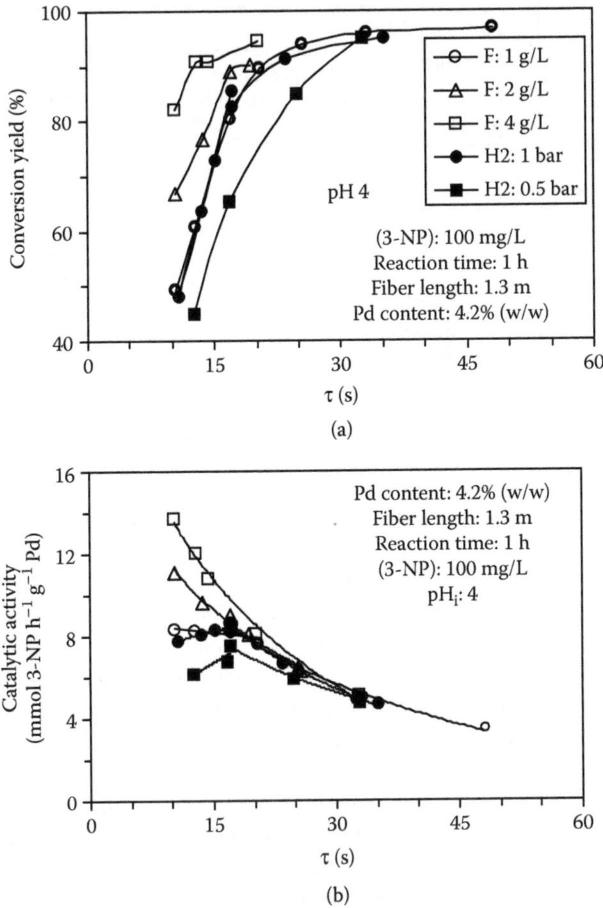

FIGURE 4.66 Influence of residence time on conversion yield (a) and catalytic activity (b) for hydrogenation of nitrophenol using Pd immobilized on chitosan hollow fibers and HCOONa or hydrogen gas as hydrogen donor. (Reprinted from E. Guibal and T. Vincent, *Environ. Sci. Technol.*, 38, 4233–4240, 2004. With permission. Copyright 2004, American Chemical Society.)

frequency and conversion yield with gas pressure are significant only below 0.5 bar and when the residence time is short (less than 20 s). At long residence times, increasing hydrogen pressure has a limited effect; the controlling mechanisms appear to be the diffusion of hydrogen through the membrane or its solubilization in nitrophenol aqueous solution.

The residence time (controlled by flow rate through the fibers) is controlling the overall performance of the catalytic systems, with both sodium-formate and hydrogen donors. Although its impact may be decreased by changing the concentration (or pressure) of hydrogen donor, the appropriate selection of flow rate allows achieving high conversion yield and maximum turnover frequency (Figure 4.66).

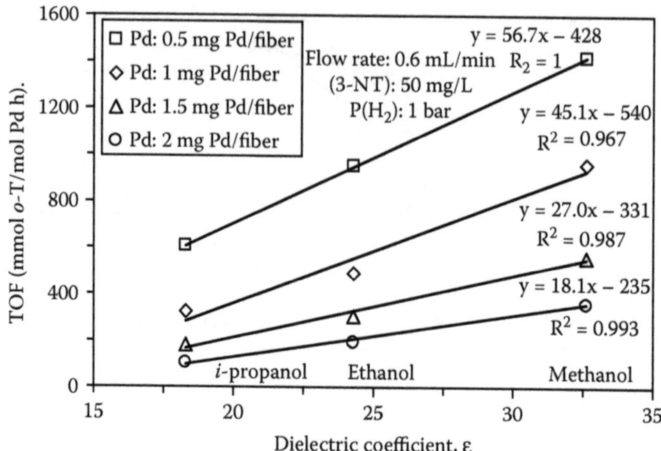

FIGURE 4.67 Influence of solvent on the turnover frequency for the hydrogenation of nitrotoluene using Pd immobilized on chitosan hollow fiber (for different loadings of Pd).

For sodium formate, a 20 s residence time allows reducing the concentration of sodium formate to 1 g L^{-1} while maintaining a conversion yield higher than 90% under selected experimental conditions, and a catalytic activity close to 9 mmol 3-NP h^{-1} g^{-1} Pd. When using hydrogen gas, the required residence time is generally greater (close to 25 to 30 s) when maintaining H_2 pressure at 1 bar. Actually, the curves obtained with a hydrogen pressure of 1 bar and with a sodium-formate concentration of 1 g L^{-1} are perfectly overlapped; the two systems behave similarly regarding the impact of residence time on conversion yield under these experimental conditions. The turnover frequency is generally lower for a hydrogen gas system than for sodium formate; however, the hydrogen gas system avoids the contamination of the second compartment with nitrophenol.

4.5.2.3 Nitrotoluene Hydrogenation

The hydrogenation of 2-nitrotoluene into *o*-toluidine in solvent using hydrogen gas and Pd catalyst supported on chitosan hollow fibers is extensively studied by Peirano et al.[410] The catalytic activity (measured by conversion yield, *o*-toluidine concentration, and turnover frequency) is strongly influenced by parameters such as flow rate (residence time), Pd content in the fiber, substrate concentration, temperature, and solvent. Although increasing hydrogen pressure in the outer compartment increases the catalytic activity, its impact is less pronounced than those of other experimental parameters. Considering the impact of solvent, the turnover frequency (TOF) decreases according the sequence: methanol >> ethanol >> isopropanol (Figure 4.67). The effect of solvent can be due to hydrogen solubility and also to the competition effect of solvent with substrate for the binding on the catalyst or the impact of solvent on the size of pores at the surface of the fiber. Actually, the TOF correlates to the dielectric

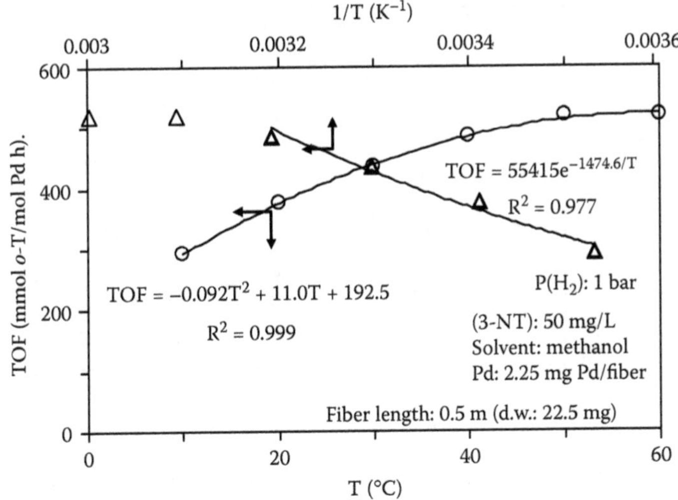

FIGURE 4.68 Influence of temperature on the turnover frequency for the hydrogenation of nitrotoluene using Pd immobilized on chitosan hollow fiber.

constant of these solvents. The effect of temperature in methanol, which increases hydrogen solubility and decreases dielectric constant, shows that the TOF increases at 50°C (Figure 4.68). The dielectric constant of the solvent cannot be positively correlated in this case to the TOF; this suggests that the effect of the solvent may be explained by hydrogen solubility or viscosity of the solution (and its effect on diffusion) rather than by its dielectric properties. The introduction of increasing fractions of water in the solvent induces the continuous decrease of *o*-toluidine production, probably due to a reduction of hydrogen solubility. The dielectric constant of methanol is also diminished by increasing amounts of water in the solvent. The activation energy is close to 12.3 kJ mol^{-1}. This is two times lower than the activation energy found for hydrogenation of chlorophenol and nitrophenol, using different systems (chitosan flakes in aqueous substrate solutions).[342,346]

The turnover frequency is enhanced with decreasing Pd content, and increasing either the concentration of 3-NP or the flow rate (Figure 4.69). Loading the fibers with increasing amounts of Pd contributes to (a) diminishing the pore size of the fibers, and (b) increasing the size of Pd nanoparticles. The accessibility to internal catalytic sites and their reactivity are thus reduced.

While Pd content hardly affects the conversion yield, the yield increases with decreasing 3-NP concentration and the flow rate. The production of *o*-toluidine is also weakly affected by the amount of Pd immobilized on the fibers, while the concentration of *o*-toluidine at equilibrium increases when the initial concentration of the substrate increases or the flow rate decreases (and residence time decreases accordingly: 12.6, 6.3, and 4.2 s for flow rates of 0.3, 0.6, and 0.9 mL min^{-1}, respectively), as shown on Figure 4.70.

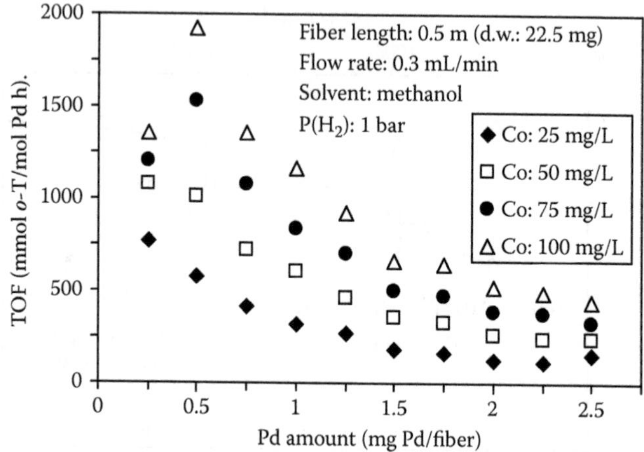

FIGURE 4.69 Influence of nitrotoluene concentration (Co) and Pd content on the turnover frequency for the hydrogenation of nitrotoluene using Pd immobilized on chitosan hollow fiber (Pd loading: 0.25–2.5 mg/fiber, i.e., 1.1–10%, respectively).

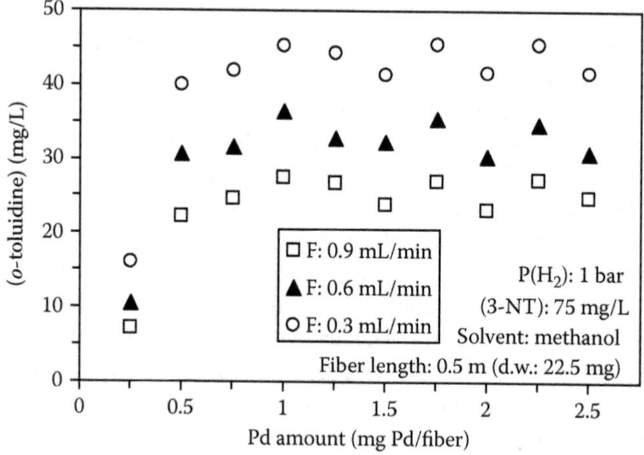

FIGURE 4.70 Influence of flow rate (F) and Pd content on the production of toluidine for the hydrogenation of nitrotoluene using Pd immobilized on chitosan hollow fiber (F: 0.9, 0.6, and 0.3 ml/min corresponding to residence time: 4.2, 6.3, and 12.6, respectively).

A statistical analysis using the methodology of surface response was used for the identification of the main parameters (initial concentration of substrate, flow rate, and Pd content) and simulation of the responses (TOF, *o*-toluidine production, and conversion yield). The optimization of the process is made difficult by the inverse variations of the responses with changes in experimental parameters. Fixing constraints on the levels of the three responses allowed determining the best values for substrate concentration, flow rate, and metal loading. The desirability conditions were fixed to 1500 < TOF [mmol (Pd) $mmol^{-1}$ h^{-1}] < 2500; 65 < conversion yield (%) < 75; 40 < [*o*-toluidine] (mg/L) < 50. The surface analysis tool allowed determining the optimum of the function at Co: 75 mg L^{-1}; F: 0.4 mL min^{-1}; Pd content: 0.74 mg Pd/fiber (i.e., 33 mg Pd g^{-1}).

4.6 CONCLUSION

This review of biopolymer-supported catalysts shows the potential of these materials of biological origin for the design of new catalytic systems. The diversity in functional groups, the stereospecific arrangement of polymer chains, the facility for modifying these materials by grafting new functional groups or changing their conditioning, and the affinity of these materials for catalytic metals are some of the reasons that could explain the increasing interest in using these biopolymers for supporting catalytic metals.

The challenges these materials are facing, which may explain that these supported catalysts have not been widely transferred to industry, are:

(a) *The variability in their composition*, related to the variability in biopolymer sources and in extraction procedures. This causes difficulties in reproducing catalytic performances. Standardization in the production methods and in their characterization is thus required.

(b) *The importance of diffusion limitation.* The weak porosity of these biopolymers makes the development of appropriate conditionings for these supports necessary, depositing, for example, the biopolymers as a coating and reactive layer on materials characterized by high surface area. The preparation of aerogels with a drying step operated under CO_2 supercritical conditions is a promising solution for porosity control. An alternative may consist of designing special biopolymer conditionings, including membranes or hollow fibers, which may bring additional advantages combining large superficial areas and separation properties.

(c) *The chemical and thermal stability.* The denaturizing of these materials in acidic solutions or oxidative conditions may be a problem that can be at least partially solved by using cross-linking treatments. These materials are usually denaturized when the catalysts are heated to temperatures greater than 120 to 150°C. This is probably not a critical point, since the field of applications of these biopolymer-supported catalysts seems to be focused on low-temperature and low-pressure experimental conditions.

Although a majority of published research concerns hydrogenation reactions, the library of reactions catalyzed by metals deposited on biopolymers is continuously growing, with attention paid to conventional industrial reactions. The enantioselectivity brought by these supports may explain these expanding developments.

REFERENCES

1. Zaera, F., Probing catalytic reactions at surfaces, *Prog. Surf. Sci.*, 69, 1, 2001.
2. Augustine, R.L. and Tanielyan, S.K., Enantioselective heterogeneous catalysis. 2. Examination of the formation of the individual (*R*) and (*S*) lactates in the cichonidine modified platinum hydrogenation of pyruvate, *J. Mol. Catal. A: Chem.*, 112, 93, 1996.
3. Baiker, A., Progress in asymmetric heterogeneous catalysis: design of novel chirally modified platinum metal catalysts, *J. Mol. Catal. A: Chem.*, 115, 473, 1997.
4. Molnar, A., Sarkany, A., and Varga, M., Hydrogenation of carbon-carbon multiple bonds: chemo-, regio- and stereo-selectivity, *J. Mol. Catal. A: Chem.*, 173, 185, 2001.
5. Tungler, A. and Fogassy, G., Catalysis with supported palladium metal, selectivity in the hydrogenation of C=C, C=O and C=N bonds, from chemo- to enantioselectivity, *J. Mol. Catal. A: Chem.*, 173, 231, 2001.
6. Tungler, A., Tarnai, T., Hegedûs, L., Fodor, K., and Máthé, T., Palladium-mediated heterogeneous catalytic hydrogenations: selectivity of liquid-phase reactions for the fine chemicals industry, *Platinum Metals Rev.*, 42, 108, 1998.
7. Kragl, U. and Dwars, T., The development of new methods for the recycling of chiral catalysts, *Trends Biotechnol.*, 19, 442, 2001.
8. Toshima, N., Nanostructured metal clusters in polymeric field as a model of artificial enzyme, *Supramol. Sci.*, 5, 395, 1998.
9. Choplin, A. and Quignard, F., From supported homogeneous catalysts to heterogeneous molecular catalysts, *Coord. Chem. Rev.*, 178–180, 1679, 1998.
10. Lashdaf, M., Krause, A.O.I., Lindblad, M., Tiita, M., and Venäläinen, T., Behaviour of palladium and ruthenium catalysts on alumina and silica prepared by gas and liquid phase deposition in cinnamaldehyde hydrogenation, *Appl. Catal. A: Gen.*, 241, 65, 2003.
11. Mahata, N., Raghavan, K.V., and Vishwanathan, V., Influence of alkali promotion on phenol hydrogenation activity of palladium/alumina catalysts, *Appl. Catal. A: Gen.*, 182, 183, 1999.
12. Schüth, C. and Reinhard, M., Hydrodechlorination and hydrogenation of aromatic compound over palladium on alumina in hydrogen-saturated water, *Appl. Catal. B.: Environ.*, 18, 215, 1998.
13. Zhang, Q. and Chuang, K.T., Kinetics of wet oxidation of black liquor over a Pt-Pd-Ce/alumina catalyst, *Appl. Catal. B: Environ.*, 17, 321, 1998.
14. Bae, J.W., Kim, I.G., Lee, J.S., Lee, K.H., and Jang, E.J., Hydrodechlorination of CCl4 over Pt/Al2O3: effects of platinum particle size on product distribution, *Appl. Catal. A: Gen.*, 240, 129, 2003.
15. Yu, K.Q. and Jones, C.W., Elucidating the role of silica surfaces in the ring-opening polymerization of lactide: catalytic behavior of silica-immobilized zinc beta-diiminate complexes, *J. Catal.*, 222, 558, 2004.
16. Shore, S.G., Ding, E., Park, C., and Keane, M.A., Vapor phase hydrogenation of phenol over silica supported Pd and Pd-Yb catalysts, *Catal. Comm.*, 3, 77, 2002.
17. Ma, Z., Sun, W.H., Zhu, N., Li, Z.L., Shao, C.X., and Hu, Y.L., Preparation of silica-supported late transition metal catalyst and ethylene polymerization, *Polym. Int.*, 51, 349, 2002.

18. Macquarrie, D.J., Gotov, B., and Toma, S., Silica-supported palladium-based catalysts for clean synthesis, *Platinum Metals Rev.*, 45, 102, 2001.
19. Rodrigues, S., Silveira, F., dos Santos, J.H.Z., and Ferreira, M.L., An explanation for experimental behavior of hybrid metallocene silica-supported catalyst for ethylene polymerization, *J. Mol. Catal. A.: Chem.*, 216, 19, 2004.
20. Phan, N.T.S., Brown, D.H., and Styring, P., A facile method for catalyst immobilisation on silica: nickel-catalysed Kumada reactions in mini-continuous flow and batch reactors, *Green Chem.*, 6, 526, 2004.
21. Ray, S., Galgali, G., Lele, A., and Sivaram, S., *In situ* polymerization of ethylene with *bis*(imino)pyridine iron(II) catalysts supported on clay: the synthesis and characterization of polyethylene-clay nanocomposites, *J. Polym. Sci. A: Polym. Chem.*, 43, 304, 2005.
22. Yadav, G.D. and Naik, S.S., Clay-supported liquid-liquid-solid phase transfer catalysis: synthesis of benzoic anhydride, *Org. Proc. Res. Dev.*, 4, 141, 2000.
23. Takasu, Y., Kawaguchi, T., Sugimoto, W., and Murakami, Y., Effects of the surface area of carbon support on the characteristics of highly-dispersed Pt-Ru particles as catalysts for methanol oxidation, *Electrochim. Acta*, 48, 3861, 2003.
24. Stevens, D.A., Zhang, S., Chen, Z., and Dahn, J.R., On the determination of platinum particle size in carbon-supported platinum electrocatalysts for fuel cell applications, *Carbon*, 41, 2769, 2003.
25. Khilnani, V.L. and Chandalia, S.B., Selective hydrogenation. II. m-Dinitrobenzene to m-nitroaniline using palladium on carbon as catalyst, *Org. Proc. Res. Dev.*, 5, 263, 2001.
26. Halligudi, S.B. and Khaire, S.S., Kinetics of hydrogenation of 4-chloro-2-nitrophenol catalyzed by Pt/carbon catalyst, *J. Chem. Tech. Biotechnol.*, 77, 25, 2001.
27. Choudhary, V.R., Sane, M.G., and Tambe, S.S., Kinetics of hydrogenation of *o*-nitrophenol to *o*-aminophenol on Pd/carbon catalysts in a stirred three-phase slurry reactor, *Ind. Eng. Chem. Res.*, 37, 3879, 1998.
28. Gallezot, P., Laurain, N., and Isnard, P., Catalytic wet-air oxidation of carboxylic acids on carbon-supported platinum catalysts, *Appl. Catal. B: Environ.*, 9, 11, 1996.
29. Yuan, G. and Keane, M.A., Liquid phase catalytic hydrodechlorination of 2,4 dichlorophenol over carbon supported palladium: an evaluation of transport limitations, *Chem. Eng. Sci.*, 58, 257, 2003.
30. Altava, B., Burguete, M.I., García-Verdugo, E., Luis, S.V., Vicent, M.J., and Mayoral, J.A., Supported chiral catalysts: the role of the polymeric network, *React. Funct. Polym.*, 48, 25, 2001.
31. Antony, R., Tembe, G.L., Ravindranathan, M., and Ram, R.N., Polymer-supported palladium(II) complexes and their catalytic study, *J. Appl. Polym. Sci.*, 90, 370, 2003.
32. Aumo, J., Lilja, J., Maki-Arvela, P., Salmi, T., Sundell, M., Vainio, H., and Murzin, D.Y., Hydrogenation of citral over a polymer fibre catalyst, *Catal. Lett.*, 84, 219, 2002.
33. Duff, D.G., Mallat, T., Schneider, M., and Baiker, A., Catalysts derived from polymer-stabilised colloidal platinum: effects of support and calcination on the catalytic behaviour in hydrogenation, *Appl. Catal. A: Gen.*, 133, 133, 1995.
34. Fisera, R., Kralik, M., Annus, J., Kratky, V., Zecca, M., and Hronec, M., Deactivation of polymer-supported palladium catalysts in the hydrogenation of 4-nitrotoluene, *Collect. Czech. Chem. Commun.*, 62, 1763, 1997.

35. Ford, W.T., Catalysis by colloidal polymers in aqueous media, *React. Funct. Polym.*, 8, 3, 2001.
36. Hirai, H., Yakura, N., Seta, Y., and Hodoshima, S., Characterization of palladium nanoparticles protected with polymer as hydrogenation catalyst, *React. Funct. Polym.*, 37, 121, 1998.
37. Hodge, P., Synthesis of organic compounds using polymer-supported reagents, catalysts, and/or scavengers in benchtop flow systems, *Ind. Eng. Chem. Res.*, 44, 8542, 2005.
38. Jansson, A.M., Grotli, M., Halkes, K.M., and Meldal, M., Palladium on carbon encapsulated in POEPOP1500: a resin-supported catalyst for hydrogenation reactions, *Org. Lett.*, 4, 27, 2002.
39. Kralik, M. and Biffis, A., Catalysis by metal nanoparticles supported on functional organic polymers, *J. Mol. Catal. A: Chem.*, 177, 113, 2001.
40. Kralik, M., Corain, B., and Zecca, M., Catalysis by metal nanoparticles supported on functionalized polymers, *Chem. Papers*, 54, 254, 2000.
41. Kralik, M., Fisera, R., Zecca, M., D'Archivio, A.A., Galantini, L., Jerabek, K., and Corain, B., Modelling of the deactivation of polymer-supported palladium catalysts in the hydrogenation of 4-nitrotoluene, *Collect. Czech. Chem. Commun.*, 63, 1074, 1998.
42. Kralik, M., Kratky, V., De Rosso, M., Tonelli, M., Lora, S., and Corain, B., Metal catalysis inside polymer frameworks: evaluation of catalyst stability and reusability, *Chem. Eur. J.*, 9, 209, 2003.
43. Magdalene, R.M., Leelamani, E.G., and Gowda, N.M.N., Hydrogenation of nitroarenes using polybenzimidazole-le supported rhodium catalyst, *J. Mol. Catal. A: Chem.*, 223, 17, 2004.
44. Michalska, Z.M., Ostaszewski, B., Zientarska, J., and Rynkowski, J.M., Novel polymer-supported platinum catalyst for selective hydrogenation of crotonaldehyde, *J. Mol. Catal. A: Chem.*, 185, 279, 2002.
45. Moghadam, M., Tangestaninejad, S., Habibi, M.H., and Mirkhani, V., A convenient preparation of polymer-supported manganese porphyrin and its use as hydrocarbon monooxygenation catalyst, *J. Mol. Catal. A: Chem.*, 217, 9, 2004.
46. Patel, D.R. and Ram, R.N., Hydrogenation of nitrobenzene using polymer anchored Pd(II) complexes as catalyst, *J. Mol. Catal. A: Chem.*, 130, 57, 1998.
47. Primavera, A., Zecca, M., and Corain, B., Interpenetrating organometallic polymer networks in heterogeneous metal catalysis: a preliminary evaluation, *J. Mol. Catal. A.: Chem.*, 108, 131, 1996.
48. Puglisi, A., Benaglia, M., Cinquini, M., Cozzi, F., and Celentano, G., Enantioselective 1,3-dipolar cycloadditions of unsaturated aldehydes promoted by A poly(ethylene glycol)-supported organic catalyst, *Eur. J. Org. Chem.*, 567, 2004.
49. Seckin, T., Koytepe, S., Demir, S., Ozdemir, I., and Centinkaya, B., Novel type of metal-containing polyimides for the Heck and Suzuki-Miyaura cross-coupling reactions as highly active catalysts, *J. Inorg. Organomet. Polym.*, 13, 223, 2003.
50. Zecca, M., Fisera, R., Palma, G., Lora, S., Hronec, M., and Kralik, M., Activity enhancement by the support in the hydrogenation of C=C bonds over polymer-supported palladium catalysts, *Chem. Eur. J.*, 6, 1980, 2000.
51. Abe, T. and Kaneko, M., Reduction catalysis by metal complexes confined in a polymer matrix, *Prog. Polym. Sci.*, 28, 1441, 2003.

52. Biffis, A., Landes, H., Jerabek, K., and Corain, B., Metal palladium dispersed inside macroporous ion-exchange resins: the issue of the accessibility to gaseous reactants, *J. Mol. Catal. A.: Chem.*, 151, 283, 2000.
53. Biffis, A., Ricoveri, R., Campestrini, S., Kralik, M., Jerabek, K., and Corain, B., Highly chemoselective hydrogenation of 2-ethylanthraquinone to 2-ethylanthrahydroquinone catalyzed by palladium metal dispersed inside highly lipophilic functional resins, *Chem. Eur. J.*, 8, 2962, 2002.
54. Clapham, B., Reger, T.S., and Janda, K.D., Polymer-supported catalysis in synthetic organic chemistry, *Tetrahedron*, 57, 4637, 2001.
55. Corain, B. and Kralik, M., Dispersing metal nanoclusters inside functional synthetic resins: scope and catalytic prospects, *J. Mol. Catal. A: Chem.*, 159, 153, 2000.
56. D'Archivio, A.A., Galantini, L., Tettamanti, E., Panatta, A., and Corain, B., Metal palladium dispersed inside macroporous ion-exchange resins: rotational and translational mobility inside the polymer network, *J. Mol. Catal. A.: Chem.*, 157, 269, 2000.
57. Deprele, S. and Montchamp, J.L., Environmentally benign synthesis of H-phosphinic acids using a water-tolerant, recyclable polymer-supported catalyst, *Org. Lett.*, 6, 3805, 2004.
58. Gelbard, G., Organic synthesis by catalysis with ion-exchange resins, *Ind. Eng. Chem. Res.*, 44, 8468, 2005.
59. Kim, D.Y. and Huh, S.C., Poly(ethylene glycol) supported chiral quaternary ammonium salts as phase-transfer catalysts for catalytic enantioselective synthesis of alpha-amino acids, *Bull. Korean Chem. Soc.*, 25, 347, 2004.
60. Kinoshita, S., Shibahara, F., and Nozaki, K., Comparison of two preparative methods: a polymer-supported catalyst by metal-complexation with a polymeric ligand or by polymerization of a metal complex, *Green Chem.*, 7, 256, 2005.
61. Kralik, M., Hronec, M., Jorik, V., Lora, S., Palma, G., Zecca, M., Biffis, A., and Corain, B., Microporous poly-(*N,N*-dimethyl-acrylamide)-(1-methacryloyl-ethylene-sulphonate)-(*N,N*-methylene-*bis*-acrylamide) resins as hydrophilic supports for metal catalysts, *J. Mol. Catal. A: Chem.*, 101, 143, 1995.
62. Owsik, I. and Kolarz, B.N., The oxidation of hydroquinone to *p*-benzoquinone catalysed by Cu(II) ions immobilized on acrylic resins with aminoguanidyl groups: part 1, *J. Mol. Catal. A: Chem.*, 178, 63, 2002.
63. Phan, N.T.S., Brown, D.H., and Styring, P., A polymer-supported Salen-type palladium complex as a catalyst for the Suzuki-Miyaura cross-coupling reaction, *Tetrahedr. Lett.*, 45, 7915, 2004.
64. Anyanwu, U.K. and Venkataraman, D., Effect of spacers on the activity of soluble polymer supported catalysts for the asymmetric addition of diethylzinc to aldehydes, *Tetrahedr. Lett.*, 44, 6445, 2003.
65. Burguete, M.I., Fraile, J.M., Garcia-Verdugo, E., Luis, S.V., Martinez-Merino, V., and Mayoral, J.A., Polymer-supported *bis*(oxazolines) and related systems: toward new heterogeneous enantioselective catalysts, *Ind. Eng. Chem. Res.*, 44, 8580, 2005.
66. Degni, S., Wilen, C.E., and Leino, R., Asymmetric C–C bond formation with L-prolinol derived chiral catalysts immobilized on polymer fibers, *Tetrahedr. Asymm.*, 15, 231, 2004.
67. Valodkar, V.B., Tembe, G.L., Ram, R.N., and Rama, H.S., Catalytic asymmetric epoxidation of unfunctionalized olefins by supported Cu(II)-amino acid complexes, *Catal. Lett.*, 90, 91, 2003.

68. Yoon, T.J., Kim, J.I., and Lee, J.K., Rh-based olefin hydroformylation catalysts and the change of their catalytic activity depending on the size of immobilizing supporters, *Inorg. Chim. Acta*, 345, 228, 2003.
69. Bergbreiter, D.E., The use of soluble polymers to effect homogeneous catalyst separation and reuse, *Catal. Today*, 42, 389, 1998.
70. Takeuchi, M., Akiyama, R., and Kobayashi, S., Polymer-micelle incarcerated scandium as a polymer-supported catalyst for high-throughput organic synthesis, *J. Am. Chem. Soc.*, 127, 13096, 2005.
71. Berkovich, Y. and Garti, N., Catalytic colloidal Pd dispersions in water-organic solutions of quaternary ammonium salt, *Colloids Surf. A*, 128, 91, 1997.
72. Desforges, A., Deleuze, H., Mondain-Monval, O., and Backov, R., Palladium nanoparticle generation within microcellular polymeric foam and size dependence under synthetic conditions, *Ind. Eng. Chem. Res.*, 44, 8521, 2005.
73. Felix, G., Regioselectively modified polysaccharide derivatives as chiral stationary phases in high-performance liquid chromatography, *J. Chromatogr. A*, 906, 171, 2001.
74. Franco, P., Senso, A., Minguillon, C., and Oliveros, L., 3,5-dimethylphenylcarbamates of amylose, chitosan and cellulose bonded on silica gel: comparison of their chiral recognition abilities as high-performance liquid chromatography chiral stationary phases, *J. Chromatogr. A*, 796, 265, 1998.
75. Franco, P., Senso, A., Oliveros, L., and Minguillon, C., Covalently bonded polysaccharide derivatives as chiral stationary phases in high-performance liquid chromatography, *J. Chromatogr. A*, 906, 155, 2001.
76. Haginaka, J., Protein-based chiral stationary phases for high-performance liquid chromatography enantioseparations, *J. Chromatogr. A*, 906, 252, 2001.
77. Kim, J.H., Kim, J.H., Jegal, J., and Lee, K.-H., Optical resolution of *a*-amino acids through enantioselective polymeric membranes based on polysaccharides, *J. Membr. Sci.*, 213, 273, 2003.
78. Malinowska, I. and Rózyló, J.K., Separation of optical isomers of amino acids on modified chitin and chitosan layers, *Biomed. Chromatogr.*, 11, 272, 1997.
79. Senso, A., Oliveros, L., and Minguillon, C., Chitosan derivatives as chiral selectors bonded on allyl silica gel: preparation, characterisation and study of the resulting high-performance liquid chromatography chiral stationary phases, *J. Chromatogr. A*, 839, 15, 1999.
80. Thoelen, C., De Bruyn, M., Theunissen, E., Kondo, Y., Vankelecom, I.F.J., Grobet, P., Yoshikawa, M., and Jacobs, P.A., Membranes based on poly(g-methyl-L-glutamate): synthesis, characterization and use in chiral separations, *J. Membr. Sci.*, 186, 153, 2001.
81. Yashima, E., Polysaccharide-based chiral stationary phases for high-performance liquid chromatographic enantioseparation, *J. Chromatogr. A*, 906, 105, 2001.
82. Guzman, J., Saucedo, I., Revilla, J., Navarro, R., and Guibal, E., Copper sorption by chitosan in the presence of citrate ions: influence of metal speciation on sorption mechanism and uptake capacities, *Int. J. Biol. Macromol.*, 33, 57, 2003.
83. Roberts, G.A.F., *Chitin Chemistry*, Macmillan, London, 1992, p. 350.
84. Sorlier, P., Denuzière, A., Viton, C., and Domard, A., Relation between the degree of acetylation and the electrostatic properties of chitin and chitosan, *Biomacromolecules*, 2, 765, 2001.
85. Okuyama, K., Noguchi, K., Kanenari, M., Egawa, T., Osawa, K., and Ogawa, K., Structural diversity of chitosan and its complexes, *Carbohydr. Polym.*, 41, 237, 2000.

86. Okuyama, K., Noguchi, K., Miyazawa, T., Yui, T., and Ogawa, K., Molecular and crystal structure of hydrated chitosan, *Macromolecules*, 30, 5849, 1997.
87. Yui, T., Imada, K., Okuyama, K., Obata, Y., Suzuki, K., and Ogawa, K., Molecular and crystal structure of the anhydrous form of chitosan, *Macromolecules*, 27, 7601, 1994.
88. Ogawa, K. and Yui, T., Crystallinity of partially *N*-acetylated chitosans, *Biosci. Biotech. Biochem.*, 57, 1466, 1993.
89. Ogawa, K., Oka, K., and Yui, T., X-ray study of chitosan-transition metal complexes, *Chem. Mater.*, 5, 726, 1993.
90. Ogawa, K., Effect of heating an aqueous suspension of chitosan on the crystallinity and polymorphs, *Agric. Biol. Chem.*, 55, 2375, 1991.
91. Zugenmaier, P., Conformation and packing of various crystalline cellulose fibers, *Prog. Polym. Sci.*, 26, 1341, 2001.
92. Gyurcsik, B. and Nagy, L., Carbohydrates as ligands: coordination equilibria and structure of the metal complexes, *Coord. Chem. Rev.*, 203, 81, 2000.
93. Simkovic, I., Quaternization cross linking of starch with choline chloride/epichlorohydrin, *Carbohydr. Polym.*, 34, 21, 1997.
94. Simkovic, I., Laszlo, J.A., and Thompson, A.R., Preparation of a weakly basic ion exchanger by crosslinking starch with epichlorohydrin in the presence of NH_4OH, *Carbohydr. Polym.*, 30, 25, 1996.
95. Simkovic, I., One-step quaternization/crosslinking of starch with 3-chloro-2-hydroxypropylammonium chloride/epichlorohydrin in the presence of NH_4OH, *Carbohydr. Polym.*, 31, 47, 1996.
96. Lehninger, A.L., *Biochemistry*, 2nd ed., Worth Publishers, Inc., New York, 1975, p. 1104.
97. Sajiki, H., Ikawa, T., Yamada, H., Tsubouchi, K., and Hirota, K., Preparation of silk fibroin-supported Pd(0) catalyst for chemoselective hydrogenation: reduction of palladium(II) acetate by methanol on the protein, *Tetrahedron Lett.*, 44, 171, 2003.
98. Panouillé, M., Durand, D., Nicolai, T., Larquet, E., and Boisset, N., Aggregation and gelation of micellar casein particles, *J. Colloid Interface Sci.*, 287, 85, 2005.
99. Dauphas, S., Mouhous-Riou, N., Metro, B., Mackie, A.R., Wilde, P.J., Anton, M., and Riaublanc, A., The supramolecular organisation of *b*-casein: effect of interfacial properties, *Food Hydrocolloids*, 19, 387, 2005.
100. Horne, D.S., Casein structure, self-assembly and gelation, *Curr. Opi. Colloid Interface Sci.*, 7, 456, 2002.
101. Guo, C., Campbell, B.E., Chen, K., Lenhoff, A.M., and Velev, O.D., Casein precipitation equilibria in the presence of calcium ions and phosphates, *Colloids Surf., B*, 29, 297, 2003.
102. Gotoh, T., Matsushima, K., and Kikuchi, K.I., Preparation of alginate-chitosan hybrid gel beads and adsorption of divalent metal ions, *Chemosphere*, 55, 135, 2004.
103. Jodra, Y. and Mijangos, F., Ion exchange selectivities of calcium alginate gels for heavy metals, *Water Sci. Technol.*, 43, 237, 2001.
104. Nestle, N. and Kimmich, R., Heavy metal uptake of alginate gels studied by NMR microscopy, *Colloids Surf. A*, 115, 141, 1996.
105. Nestle, N. and Kimmich, R., NMR microscopy of heavy metal absorption in calcium alginate beads, *Appl. Biochem. Biotechnol.*, 56, 9, 1996.
106. Gregor, J.E., Fenton, E., Brokenshire, G., Van den Brink, P., and O'Sullivan, B., Interactions of calcium and aluminium ions with alginate, *Water Res.*, 30, 1319, 1996.

107. Jang, L.K., Geesey, G.G., Lopez, S.L., Eastman, S.L., and Wichlacz, P.L., Use of a gel-forming biopolymer directly dispensed into a loop fluidized bed reactor to recover dissolved copper, *Water Res.,* 24, 889, 1990.
108. Jang, L.K., Nguyen, D., and Geesey, G.G., Selectivity of alginate gel for Cu vs. Co, *Water Res.,* 29, 307, 1995.
109. Jang, L.K., Nguyen, D., and Geesey, G.G., Selectivity of alginate gel for Cu over Zn when acidic conditions prevail, *Water Res.,* 33, 2817, 1999.
110. Abu Al-Rub, F.A., El-Naas, M.H., Benyahia, F., and Ashour, I., Biosorption of nickel on blank alginate beads, free and immobilized algal cells, *Process Biochem.,* 39, 1767, 2004.
111. Dhakal, R.P., Ghimire, K.N., Inoue, K., Yano, M., and Makino, K., Acidic polysaccharide gels for selective adsorption of lead(II) ion, *Sep. Purif. Technol.,* 42, 219, 2005.
112. Gotoh, T., Matsushima, K., and Kikuchi, K.I., Adsorption of Cu and Mn on covalently cross-linked alginate gel beads, *Chemosphere,* 55, 57, 2004.
113. Ibanez, J.P. and Umetsu, Y., Potential of protonated alginate beads for heavy metals uptake, *Hydrometallurgy,* 64, 89, 2002.
114. Ibanez, J.P. and Umetsu, Y., Uptake of trivalent chromium from aqueous solutions using protonated dry alginate beads, *Hydrometallurgy,* 72, 327, 2004.
115. Torres, E., Mata, Y.N., Blazquez, A.L., Munoz, J.A., Gonzalez, F., and Ballester, A., Gold and silver uptake and nanoprecipitation on calcium alginate beads, *Langmuir,* 21, 7951, 2005.
116. Veglio, F., Esposito, A., and Reverberi, A.P., Copper adsorption on calcium alginate beads: equilibrium pH-related models, *Hydrometallurgy,* 65, 43, 2002.
117. Chen, J.P., Tendeyong, F., and Yiacoumi, S., Equilibrium and kinetic studies of copper ion uptake by calcium alginate, *Environ. Sci. Technol.,* 31, 1433, 1997.
118. Konishi, Y., Shimaoka, J.-I., and Asai, S., Sorption of rare-earth ions on biopolymer gel beads of alginic acid, *React. Funct. Polym.,* 36, 197, 1998.
119. Jeon, C., Yoo, Y.J., and Hoell, W.H., Environmental effects and desorption characteristics on heavy metal removal using carboxylated alginic acid, *Bioresource Technol.,* 96, 15, 2005.
120. Jeon, C., Park, J.Y., and Yoo, Y.J., Removal of heavy metals in plating wastewater using carboxylated alginic acid, *Korean J. Chem. Eng.,* 18, 955, 2001.
121. Jeon, C., Park, J.Y., and Yoo, Y.J., Characteristics of metal removal using carboxylated alginic acid, *Water Res.,* 36, 1814, 2002.
122. Chen, J.P., Hong, L.A., Wu, S.N., and Wang, L., Elucidation of interactions between metal ions and Ca alginate-based ion-exchange resin by spectroscopic analysis and modeling simulation, *Langmuir,* 18, 9413, 2002.
123. Chen, J.P. and Wang, L., Characterization of a Ca-alginate based ion-exchange resin and its applications in lead, copper, and zinc removal, *Sep. Sci. Technol.,* 36, 3617, 2001.
124. Smidsrød, O. and Haug, A., Dependence upon uronic acid composition of some ion-exchange properties of alginates, *Acta Chem. Scand.,* 22, 1989, 1968.
125. Haug, A. and Smidsrød, O., Selectivity of some anionic polymers for divalent metal ions, *Acta Chem. Scand.,* 24, 843, 1970.
126. Wu, J.M., Wang, Y.Y., and Yan, C.L., Selective adsorption of copper ion by copper ion templated alginate gel, *Chin. J. Anal. Chem.,* 30, 1414, 2002.
127. Simsek-Ege, F.A., Bond, G.M., and Stringer, J., Polyelectrolye complex formation between alginate and chitosan as a function of pH, *J. Appl. Polym. Sci.,* 88, 346, 2003.

128. Zhang, L.N., Cai, J., Zhou, J.P., and Tang, Y.R., Adsorption Cd2+ and Cu2+ on ion-exchange beads from cellulose/alginic acid blend, *Sep. Sci. Technol.*, 39, 1203, 2004.
129. Guibal, E., Metal ion interactions with chitosan: a review, *Sep. Purif. Technol.*, 38, 43, 2004.
130. Varma, A.J., Deshpande, S.V., and Kennedy, J.F., Metal complexation by chitosan and its derivatives: a review, *Carbohydr. Polym.*, 55, 77, 2004.
131. Guibal, E., Chassary, P., and Vincent, T., Metal anion sorption on chitosan and derivative materials: a strategy for polymer modification and optimum use, *React. Funct. Polym.*, 60, 137, 2004.
132. Gibbs, G., Tobin, J.M., and Guibal, E., Influence of chitosan pre-protonation on Reactive Black 5 sorption isotherms and kinetics, *Ind. Eng. Chem. Res.*, 43, 1, 2004.
133. Gibbs, G., Tobin, J.M., and Guibal, E., Sorption of Acid Green 25 on chitosan: influence of experimental parameters on uptake kinetics and sorption isotherms, *J. Appl. Polym. Sci.*, 90, 1073, 2003.
134. Ruiz, M., Sastre, A.M., and Guibal, E., Palladium sorption on glutaraldehyde-crosslinked chitosan, *React. Funct. Polym.*, 45, 155, 2000.
135. Monteiro, O.A.C.J. and Airoldi, C., Some studies of crosslinking chitosan-glutaraldehyde interaction in a homogeneous system, *Int. J. Biol. Macromol.*, 26, 119, 1999.
136. Hsien, T.-Y. and Rorrer, G.L., Effect of acylation and crosslinking on the material properties and cadmium ion adsorption capacity of porous chitosan beads, *Sep. Sci. Technol.*, 30, 2455, 1995.
137. Inoue, K., Baba, Y., Yoshizuka, K., Noguchi, H., and Yoshizaki, M., Selectivity series in the adsorption of metal ions on a resin prepared by crosslinking copper(II)-complexed chitosan, *Chem. Lett.*, 17, 1281, 1988.
138. Cao, Z., Ge, H., and Lai, S., Studies on synthesis and adsorption properties of chitosan cross-linked by glutaraldehyde and Cu(II) as template under microwave irradiation, *Eur. Polym. J.*, 37, 2141, 2001.
139. Mi, F.-L., Kuan, C.-Y., Shyu, S.-S., Lee, S.-T., and Chang, S.-F., The study of gelation kinetics and chain-relaxation properties of glutaraldehyde-cross-linked chitosan gel and their effects on microspheres preparation and drug release, *Carbohydr. Polym.*, 41, 389, 2000.
140. Tojima, T., Katsura, H., Nishiki, M., Tokura, S., and Sakairi, N., Chitosan beads with pendant *a*-cyclodextrin: preparation and inclusion property to nitrophenolates, *Carbohydr. Polym.*, 40, 17, 1999.
141. Sreenivasan, K., Synthesis and preliminary studies on a *b*-cyclodextrin-coupled chitosan as a novel adsorbent matrix, *J. Appl. Polym. Sci.*, 69, 1051, 1998.
142. Paradossi, G., Cavalieri, F., and Crescenzi, V., 1H-NMR relaxation study of a chitosan-cyclodextrin network, *Carbohydr. Res.*, 300, 77, 1997.
143. Juang, R.-S. and Ju, C.-Y., Equilibrium sorption of copper(II)-ethylenediamine-tetraacetic acid chelates onto cross-linked, polyaminated chitosan beads, *Ind. Eng. Chem. Res.*, 36, 5403, 1997.
144. Juang, R.-S. and Ju, C.-Y., Kinetics of sorption of Cu(II)- ethylenediamine-tetraacetic acid chelated anions on cross-linked, polyaminated chitosan beads, *Ind. Eng. Chem. Res.*, 37, 3463, 1998.
145. Juang, R.-S., Tseng, R.-L., Wu, F.-C., and Lee, S.-H., Adsorption behavior of reactive dyes from aqueous solutions on chitosan, *J. Chem. Technol. Biotechnol.*, 70, 391, 1997.

146. Wong, Y.C., Szeto, Y.S., Cheung, W.H., and McKay, G., Adsorption of acid dyes on chitosan: equilibrium isotherm analyses, *Proc. Biochem.,* 39, 695, 2004.
147. Wu, F.-C., Tseng, R.-L., and Juang, R.-S., Kinetic modeling of liquid-phase adsorption of reactive dyes and metal ions on chitosan, *Water Res.,* 35, 613, 2001.
148. Yoshida, H., Okamoto, A., and Kataoka, T., Adsorption of acid dye on cross-linked chitosan fibers: equilibria, *Chem. Eng. Sci.,* 48, 2267, 1993.
149. Maghami, G.G. and Roberts, G.A.F., Studies on the adsorption of anionic dyes on chitosan, *Makromol. Chem.,* 189, 2239, 1988.
150. McCarrick, P., Tobin, J., and Guibal, E., Comparative sorption of dyes on chitosan and activated carbon, *Sep. Sci. Technol.,* 38, 3049, 2003.
151. Inoue, K., Ohto, K., and Baba, Y., Adsorptive separation behavior of chitosan and chemically modified chitosan for metal ions. 4. Adsorption of metal ions on complexane type of chemically modified chitosan, *J. Ion Exch.,* 9, 115, 1998.
152. Inoue, K., Application of chitosan in separation and purification of metals, in *Recent Advances in Marine Biotechnology, Vol. 2, Environmental Marine Biotechnology,* Fingerman, M., Nagabhushanam, R. and Thompson, M.-F., Eds., Oxford & IBH Publishing PVT. Ltd., New Delhi, 1998, p. 63.
153. Inoue, K., Yoshizuka, K., and Baba, Y., Adsorption of metal ions on chitosan and chemically modified chitosan and their application to hydrometallurgy, in *Biotechnology and Bioactive Polymers,* Gebelein, C. and Carraher, C., Eds., Plenum Press, New York, 1994, p. 35.
154. Dambies, L., Vincent, T., Domard, A., and Guibal, E., Preparation of chitosan gel beads by ionotropic molybdate gelation, *Biomacromolecules,* 2, 1198, 2001.
155. Guibal, E., Milot, C., and Roussy, J., Influence of hydrolysis mechanisms on molybdate sorption isotherms using chitosan, *Sep. Sci. Technol.,* 35, 1021, 2000.
156. Guibal, E., Milot, C., Eterradossi, O., Gauffier, C., and Domard, A., Study of molybdate ion sorption on chitosan gel beads by different spectrometric analyses, *Int. J. Biol. Macromol.,* 24, 49, 1999.
157. Milot, C., McBrien, J., Allen, S., and Guibal, E., Influence of physico-chemical and structural characteristics of chitosan flakes on molybdate removal, *J. Appl. Polym. Sci.,* 68, 571, 1998.
158. Jaworska, M., Kula, K., Chassary, P., and Guibal, E., Influence of chitosan characteristics on polymer properties. 2. Platinum sorption properties, *Polym. Int.,* 52, 206, 2003.
159. Guibal, E., Von Offenberg Sweeney, N., Zikan, M.C., Vincent, T., and Tobin, J.M., Competitive sorption of platinum and palladium on chitosan derivatives, *Int. J. Biol. Macromol.,* 28, 401, 2001.
160. Guibal, E., Ruiz, M., Vincent, T., Sastre, A., and Navarro Mendoza, R., Platinum and palladium sorption on chitosan derivatives, *Sep. Sci. Technol.,* 36, 1017, 2001.
161. Inoue, K., Yamaguchi, T., Iwasaki, M., Ohto, K., and Yoshizuka, K., Adsorption of some platinum group metals on some complexane types of chemically modified chitosan, *Sep. Sci. Technol.,* 30, 2477, 1995.
162. Ruiz, M., Sastre, A., Zikan, M.C., and Guibal, E., Palladium sorption on glutaraldehyde-crosslinked chitosan in fixed-bed systems, *J. Appl. Polym. Sci.,* 81, 153, 2001.
163. Baba, Y. and Hirakawa, H., Selective adsorption of palladium(II), platinum(IV), and mercury(II) on a new chitosan derivative possessing pyridyl group, *Chem. Lett.,* 21, 1905, 1992.

164. Guzman, J., Saucedo, I., Revilla, J., Navarro, R., and Guibal, E., Vanadium(V) interactions with chitosan: influence of polymer protonation and metal speciation, *Langmuir,* 18, 1567, 2002.
165. Hoshi, S., Konuma, K., Sugawara, K., Uto, M., and Akatsuka, K., The simple and rapid spectrophotometric determination of trace chromium(VI) after preconcentration as its colored complex on chitin, *Talanta,* 47, 659, 1998.
166. Hoshi, S., Konuma, K., Sugawara, K., Uto, M., and Akatsura, K., The spectrophotometric determination of trace molybdenum(VI) after collection and elution as molybdate ion on protonated chitin, *Talanta,* 44, 1473, 1997.
167. Jaworska, M., Sakurai, K., Gaudon, P., and Guibal, E., Influence of chitosan characteristics on polymer properties. I. Crystallographic properties, *Polym. Int.,* 52, 198, 2003.
168. Guibal, E., Vincent, T., and Navarro Mendoza, R., Synthesis and characterization of a thiourea-derivative of chitosan for platinum recovery, *J. Appl. Polym. Sci.,* 75, 119, 2000.
169. Guibal, E., Von Offenberg Sweeney, N., Vincent, T., and Tobin, J.M., Sulfur derivatives of chitosan for palladium sorption, *React. Funct. Polym.,* 50, 149, 2002.
170. Kawamura, Y., Yoshida, H., Asai, S., and Tanibe, H., Recovery of HgC12 using polyaminated highly porous chitosan beads: effect of salt and acid, *J. Chem. Eng. Jpn.,* 31, 1, 1998.
171. Lasko, C.L. and Hurst, M.P., An investigation into the use of chitosan for the removal of soluble silver from industrial wastewater, *Environ. Sci. Technol.,* 33, 3622, 1999.
172. Baes, C.F., Jr. and Mesmer, R.E., *Hydrolysis of Cations,* Wiley, New York, 1976, p. 489.
173. Brack, H.P., Tirmizi, S.A., and Risen, W.M.J., A spectrophotometric and viscometric study of the metal-ion induced gelation of the biopolymer chitosan, *Polymer,* 38, 2351, 1997.
174. Draget, K.I., Vårum, K.M., Moen, E., Gynnild, H., and Smidsrød, O., Chitosan cross-linked with Mo(VI) polyoxyanions: a new gelling system, *Biomaterials,* 13, 635, 1992.
175. Mi, F.-L., Shyu, S.-S., Lee, S.-T., and Wong, T.-B., Kinetic study of chitosan-tripolyphosphate complex reaction and acid-resistive properties of the chitosan-tripolyphosphate gel beads prepared by in-liquid curing method, *J. Polym. Sci. Part B.: Polym. Phys.,* 37, 1551, 1999.
176. Lee, S.-T., Mi, F.-L., Shen, Y.-J., and Shyu, S.-S., Equilibrium and kinetic studies of copper(II) ion uptake by chitosan-tripolyphosphate chelating resin, *Polymer,* 42, 1879, 2001.
177. Delval, F., Crini, G., Morin, N., Vebrel, J., Bertini, S., and Torri, G., The sorption of several types of dye on crosslinked polysaccharides derivatives, *Dyes Pigments,* 53, 79, 2002.
178. Hosny, W.M., Hadi, A., Elsaied, H., and Basta, A.H., Metal-chelates with some cellulose derivatives. 3. Synthesis and structural chemistry of nickel(II) and copper(II) complexes with carboxymethyl cellulose, *Polym. Int.,* 37, 93, 1995.
179. Kabay, N., Demircioglu, M., Yayli, S., Yuksel, M., Saglam, M., and Levison, P.R., Removal of metal ions from aqueous solution by cellulose ion exchangers, *Sep. Sci. Technol.,* 34, 41, 1999.

180. Simkovic, I., Hricovini, M., and Sasinkova, V., Preparation of ion-exchangers by cross-linking of starch or polygalacturonic acid with 1,3-*bis*(3-chloro-2-hydroxypropyl)imidazolium hydrogen sulphate, *Carbohydr. Polym.*, 47, 131, 2002.
181. Simkovic, I. and Jakab, E., Thermogravimetry/mass spectrometry study of weakly basic starch-based ion exchanger, *Carbohydr. Polym.*, 45, 53, 2001.
182. Matsumoto, K., Hirayama, C., and Motozato, Y., Preparation of bead-shaped starch ion exchangers and their properties, *Nippon Kagaku Kaishi*, 10, 657, 1998.
183. Pearson, R.G., Hard and soft acids and bases, *J. Am. Chem. Soc.*, 85, 3533, 1963.
184. Marcus, Y., *Ion Properties,* Marcel Dekker, Inc., New York, 1997, p. 259.
185. Braccini, I. and Pérez, S., Molecular basis of Ca2+-induced gelation in alginates and pectins: the egg-box model revisited, *Biomacromolecules,* 2, 1089, 2001.
186. DeRamos, C.M., Irwin, A.E., Nauss, J.L., and Stout, B.E., 13C NMR and molecular modeling studies of alginic acid binding with alkaline earth and lanthanide metal ions, *Inorg. Chim. Acta,* 256, 69, 1997.
187. Lamelas, C., Avaltroni, F., Benedetti, M., Wilkinson, K.J., and Slaveykova, V.I., Quantifying Pb and Cd complexation by alginates and the role of metal binding on macromolecular aggregation, *Biomacromolecules,* 6, 2756, 2005.
188. Sreeram, K.J., Shrivastava, H.Y., and Nair, B.U., Studies on the nature of interaction of iron(III) with alginates, *Biochem. Biophys. Acta,* 1670, 121, 2004.
189. Ferreira, M.L. and Gschaider, M.E., Theoretical and experimental study of Pb2+ and Hg2+ adsorption on biopolymers. 1. Theoretical study, *Macromol. Biosci.,* 1, 233, 2001.
190. Dzul Erosa, M.S., Saucedo Medina, T.I., Navarro Mendoza, R., Avila Rodriguez, M., and Guibal, E., Cadmium sorption on chitosan sorbents: kinetic and equilibrium studies, *Hydrometallurgy,* 61, 157, 2001.
191. Hsien, T.-Y. and Rorrer, G.L., Heterogeneous cross-linking of chitosan gel beads: kinetics, modeling, and influence on cadmium ion adsorption capacity, *Ind. Eng. Chem. Res.*, 36, 3631, 1997.
192. Park, J.W., Park, M.-O., and Park, K.K., Mechanism of metal ion binding to chitosan in solution: cooperative inter- and intramolecular chelations, *Bull. Kor. Chem. Soc.*, 5, 108, 1984.
193. Rhazi, M., Desbrières, J., Tolaimate, A., Rinaudo, M., Vottero, P., and Alagui, A., Contribution to the study of the complexation of copper by chitosan and oligomers, *Polymer,* 43, 1267, 2002.
194. Rhazi, M., Desbrières, J., Tolaimate, A., Rinaudo, M., Vottero, P., Alagui, A., and El Meray, M., Influence of the nature of the metal ions on the complexation with chitosan: application to the treatment of liquid waste, *Eur. Polym. J.,* 38, 1523, 2002.
195. Schlick, S., Binding sites of Cu2+ in chitin and chitosan: an electron spin resonance study, *Macromolecules,* 19, 192, 1986.
196. Chiessi, E., Paradossi, G., Venanzi, M., and Pispisa, B., Copper complexes immobilized to chitosan, *J. Inorg. Biochem.,* 46, 109, 1992.
197. Domard, A., pH and C.D. measurements on a fully deacetylated chitosan: application to Cu(II)-polymer interactions, *Int. J. Biol. Macromol.,* 9, 98, 1987.
198. Nieto, J.M., Peniche-Covas, C., and Del Bosque, J., Preparation and characterization of a chitosan-Fe(III) complex, *Carbohydr. Polym.,* 18, 221, 1992.
199. Shahgholi, M., Callahan, J.H., Rappoli, B.J., and Rowley, D.A., Investigation of copper-saccharide complexation reactions using potentiometry and electrospray mass spectrometry, *J. Mass Spectrom.,* 32, 1080, 1997.

200. Choi, K.S. and Ahn, H.S., A study on synthesis of crosslinked chitosan phosphate and adsorption characteristics of metallic ions, *Polym. Korea,* 14, 516, 1990.
201. Heras, A., Rodriguez, N.M., Ramos, V.M., and Agullo, E., N-methylene phosphonic chitosan: a novel soluble derivative, *Carbohydr. Polym.,* 44, 1, 2001.
202. Nishi, N., Ebina, A., Nishimura, S.-I., Tsutsumi, A., Hasegawa, O., and Tokura, S., Highly phosphorylated derivatives of chitin, partially deacetylated chitin and chitosan as new functional polymers: preparation and characterization, *Int. J. Biol. Macromol.,* 8, 311, 1986.
203. Nishi, N., Maekita, Y., Nishimura, S.-I., Hasegawa, O., and Tokura, S., Highly phosphorylated derivatives of chitin, partially deacetylated chitin and chitosan as new functional polymers: metal binding property of the insolubilized materials, *Int. J. Biol. Macromol.,* 9, 109, 1987.
204. Ramos, V.M., Rodriguez, N.M., Rodriguez, M.S., Heras, A., and Agullo, E., Modified chitosan carrying phosphonic and alkyl groups, *Carbohydr. Polym.,* 51, 425, 2003.
205. Demarger, S. and Domard, A., Chitosan behaviours in a dispersion of undecylenic acid: morphological aspects, *Carbohydr. Res.,* 27, 101, 1995.
206. Piron, E. and Domard, A., Formation of a ternary complex between chitosan and ion pairs of strontium carbonate, *Int. J. Biol. Macromol.,* 23, 113, 1998.
207. Kurita, K., Controlled functionalization of the polysaccharide chitin, *Prog. Polym. Sci.,* 26, 1921, 2001.
208. Dambies, L., Guimon, C., Yiacoumi, S., and Guibal, E., Characterization of metal ion interactions with chitosan by x-ray photoelectron spectroscopy, *Colloids Surf. A,* 177, 203, 2000.
209. Guibal, E., Roussy, J., and Le Cloirec, P., Photochemical reaction of uranium with glucosamine, acetylglucosamine and related polymers: chitin and chitosan, *Water S.A.,* 22, 19, 1996.
210. Yonezawa, Y., Kawabata, I., and Sato, T., Photochemical formation of colloidal gold particles in chitosan films, *Ber. Bunsenges. Phys. Chem.,* 100, 39, 1996.
211. Yonezawa, Y., Sato, T., and Kawabata, I., Photo-induced formation of gold metal film from metal salt of chitosan, *Chem. Lett.,* 23, 355, 1994.
212. Wang, R., Liao, X.P., and Shi, B., Adsorption behaviors of Pt(II) and Pd(II) on collagen fiber immobilized bayberry tannin, *Ind. Eng. Chem. Res.,* 44, 4221, 2005.
213. Ogata, T. and Nakano, Y., Mechanisms of gold recovery from aqueous solutions using a novel tannin gel adsorbent synthesized from natural condensed tannin, *Water Res.,* 39, 4281, 2005.
214. Kim, Y.H. and Nakano, Y., Adsorption mechanism of palladium by redox within condensed-tannin gel, *Water Res.,* 39, 1324, 2005.
215. Liao, X.P., Zhang, M., and Shi, B., Collagen-fiber-immobilized tannins and their adsorption of Au(III), *Ind. Eng. Chem. Res.,* 43, 2222, 2004.
216. Liao, X.P., Lu, Z., Du, X., Liu, X., and Shi, B., Collagen fiber immobilized Myrica rubra tannin and its adsorption to UO22+, *Environ. Sci. Technol.,* 38, 324, 2004.
217. Zhan, X., Miyazaki, A., and Nakano, Y., Mechanisms of lead removal from aqueous solutions using a novel tannin gel adsorbent synthesized from natural condensed tannin, *J. Chem. Eng. Jpn.,* 34, 1204, 2001.
218. Nakano, Y., Takeshita, K., and Tsutsumi, T., Adsorption mechanism of hexavalent chromium by redox within condensed-tannin gel, *Water Res.,* 35, 496, 2001.
219. Nakano, Y., Tanaka, M., Nakamura, Y., and Konno, M., Removal and recovery system of hexavalent chromium from waste water by tannin gel particles, *J. Chem. Eng. Jpn.,* 33, 747, 2000.

220. Köckritz, A., Bischoff, S., Morawsky, V., Prusse, U., and Vorlop, K.D., A novel strategy for heterogenisation of homogeneous and colloidal chiral catalysts and their application in enantioselective reactions, *J. Mol. Catal. A: Chem.,* 180, 231, 2002.
221. Kucherov, A.V., Finashina, E., Kramareva, N.V., Rogacheva, V., Zezin, A., Said-Galiyev, E., and Kustov, L.M., Comparative study of Cu(II) catalytic sites immobilized onto different polymeric supports, *Macromol. Symp.,* 204, 175, 2003.
222. Kucherov, A.V., Kramareva, N.V., Finashina, E., Koklin, A.E., and Kustov, L.M., Heterogenized redox catalysts on the basis of the chitosan matrix. 1. Copper complexes, *J. Mol. Catal. A: Chem.,* 198, 377, 2003.
223. Kar, P. and Misra, M., Use of keratin fiber for separation of heavy metals from water, *J. Chem. Technol. Biotechnol.,* 79, 1313, 2004.
224. Piron, E., Accominotti, M., and Domard, A., Interaction between chitosan and uranyl ions: role of physical and physicochemical parameters on the kinetics of sorption, *Langmuir,* 13, 1653, 1997.
225. Piron, E. and Domard, A., Interaction between chitosan and uranyl ions. 1. Role of physicochemical parameters, *Int. J. Biol. Macromol.,* 21, 327, 1997.
226. Valentin, R., Molvinger, K., Quignard, F., and Brunel, D., Supercritical CO2 dried chitosan: an efficient intrinsic heterogeneous catalyst in fine chemistry, *New J. Chem.,* 27, 1690, 2003.
227. Valentin, R., Molvinger, K., Viton, C., Domard, A., and Quignard, F., From hydrocolloids to high specific surface area porous supports for catalysis, *Biomacromolecules,* 6, 2785, 2005.
228. Ouiminga, S.A., Brandebourger, M., Chaufer, B., Deratani, A., and Sebille, B., Preparation of water-soluble chelating aminated starch derivatives and their use for the concentration of metal ions by ultrafiltration, *React. Polym.,* 5, 111, 1987.
229. Juang, R.-S. and Chiou, C.-H., Ultrafiltration rejection of dissolved ions using various weakly basic water-soluble polymers, *J. Membr. Sci.,* 177, 207, 2000.
230. Juang, R.-S. and Shiau, R.-C., Metal removal from aqueous solutions using chitosan-enhanced membrane filtration, *J. Membr. Sci.,* 165, 159, 2000.
231. Rivas, B.L., Pereira, E., and Moreno-Villoslada, I., Water-soluble polymer-metal ion interactions, *Prog. Polym. Sci.,* 28, 173, 2003.
232. Guibal, E., Touraud, E., and Roussy, J., Chitosan interactions with metal ions and dyes: dissolved-state versus solid-state application, *World J. Microbiol. Biotechnol.,* 21, 913, 2005.
233. Kuncoro, E.K., Roussy, J., and Guibal, E., Mercury recovery by polymer-enhanced ultrafiltration: comparison of chitosan and poly(ethylenimine) used as macroligand, *Sep. Sci. Technol.,* 40, 659, 2005.
234. Ishizuki, N., Torigoe, K., Esumi, K., and Meguro, K., Characterization of precious metal particles prepared using chitosan as a protective agent, *Colloids Surf.,* 55, 15, 1991.
235. Belapurkar, A.D., Kapoor, S., Kulshreshtha, S.K., and Mittal, J.P., Radiolytic preparation and catalytic properties of platinum nanoparticles, *Mater. Res. Bull.,* 36, 145, 2001.
236. Shim, I.W., Noh, W.T., Kwon, J., Cho, J.Y., Kim, K.S., and Kang, D.H., Preparation of copper nanoparticles in cellulose acetate polymer and the reaction chemistry of copper complexes in the polymer, *Bull. Kor. Chem. Soc.,* 23, 563, 2002.
237. Huang, H. and Yang, X., Synthesis of chitosan-stabilized gold nanoparticles in the absence/presence of tripolyphosphate, *Biomacromolecules,* 5, 2340, 2004.

238. Huang, H. and Yang, X., Synthesis of polysaccharide-stabilized gold and silver nanoparticles: a green method, *Carbohyd. Res.*, 339, 2627, 2004.
239. Huang, H., Yuan, Q., and Yang, X., Preparation and characterization of metal-chitosan nanocomposites, *Colloids Surf. B*, 39, 31, 2004.
240. Huang, H., Yuan, Q., and Yang, X., Morphology study of gold-chitosan nanocomposites, *J. Colloid Interf. Sci.*, 282, 26, 2005.
241. Liu, H.W., Li, D.Y., Huang, M.Y., and Jiang, Y.Y., Catalytic behaviors of silica-supported natural biopolymer-oxalic acid-Pt complexes in hydrogenation, *Polym. Adv. Technol.*, 11, 258, 2000.
242. Zhang, X., Han, B., Huang, M.Y., and Jiang, Y.Y., Asymmetric hydroformylation of vinyl acetate catalyzed by zeolite-supported gelatin-Co-Ru complex, *Polym. Adv. Technol.*, 12, 693, 2001.
243. Liu, H.W., Xin, F., Wu, L.M., Huang, M.Y., and Jiang, Y.Y., Catalytic behavior of silica-supported methylcellulose-oxalic acid-Pt complex in asymmetric hydrogenation of 2-butanone and itaconic acid, *Polym. Adv. Technol.*, 13, 210, 2002.
244. Liu, Q., Zhang, Z., Bo, H., and Sheldon, R.A., Direct separation of the enantiomers of cetirizine and related compounds by reversed-phase chiral HPLC, *Chromatographia*, 56, 233, 2002.
245. Xue, L., Li, B.Y., Huang, M.Y., and Jiang, Y.Y., Preparation of silica-supported alginic acid-L-glutamic acid-Pt complex and its catalytic behaviors in asymmetric hydrogenation of 4-methyl-2-pentanone, *Polym. Adv. Technol.*, 14, 282, 2003.
246. Wei, W.L., Zhu, H.Y., Zhao, C.L., Huang, M.Y., and Jiang, Y.Y., Asymmetric hydrogenation of furfuryl alcohol catalyzed by a biopolymer-metal complex, silica-supported alginic acid-amino acid-Pt complex, *React. Funct. Polym.*, 59, 33, 2004.
247. Huang, K., Xue, L., Hu, Y.-C., Huang, M.-Y., and Jiang, Y.-Y., Catalytic behaviors of silica-supported starch-polysulfosiloxane-Pt complexes in asymmetric hydrogenation of 4-methyl-2-pentanone, *React. Funct. Polym.*, 50, 199, 2002.
248. Isaeva, V., Ivanov, A., Kozlova, L., and Sharf, V., Preparation of the chitosan based catalysts for several hydrogenation reaction in the liquid phase, in *Scientific Bases for the Preparation of Heterogeneous Catalysts*, 143 in the series "Studies in Surface Sciences and Catalysis," Gaigneaux, E., De Vos, D.E., Grange, P., Jacobs, P.A., Martens, J.A., Ruiz, P., and Poncelet, G., Eds., Elsevier, Amsterdam, 2002, p. 435.
249. Zhou, D.-Q., He, M., Zhang, Y.-H., Huang, M.-Y., and Jiang, Y.-Y., Asymmetric hydrogenation of diketones catalyzed by magnesium oxide-supported chitosan-Rh complex, *Polym. Adv. Technol.*, 14, 287, 2003.
250. Shi, Q.-H., Tian, Y., Dong, X.-Y., Bai, S., and Sun, Y., Chitosan-coated silica beads as immobilized metal affinity support for protein adsorption, *Biochem. Eng. J.*, 16, 317, 2003.
251. Tang, L.M., Huang, M.Y., and Jiang, Y.Y., Synthesis of a silica-supported carboxymethyl cellulose platinum complex and its catalytic behaviors for hydrogenation of ethyl benzoate, *Chin. Chem. Lett.*, 6, 255, 1995.
252. Jin, J.-J., Chen, G.-C., Huang, M.-Y., and Jiang, Y.-Y., Catalytic hydrogenation behaviours of palladium complexes of chitosan-polyacrylic acid and chitosan polymethacrylic acid, *React. Polym.*, 23, 95, 1994.
253. Huang, A., Liu, Y., Chen, L., and Hua, J., Synthesis and property of nanosized palladium catalysts protected by chitosan/silica, *J. Appl. Polym. Sci.*, 85, 989, 2002.

254. Liu, X.D., Tokura, S., Nishi, N., and Sakairi, N., A novel method for immobilization of chitosan onto nonporous glass beads through a 1,3-thiazolidine linker, *Polymer,* 44, 1021, 2003.
255. Liu, H.W., Huang, K., Mi, J.-D., Jia, Y., Huang, M.Y., and Jiang, Y.Y., Hydrogenation of nitrobenzene, cyclohexanone and octene catalyzed by silica-supported starch-L-glutamic acid-Fe complex, *Polym. Adv. Technol.,* 14, 355, 2003.
256. Zhang, X., Geng, Y., Han, B., Ying, M.Y., Huang, M.Y., and Jiang, Y.Y., Asymmetric hydrogenation of ketones catalyzed by zeolite-supported gelatin-Fe complex, *Polym. Adv. Technol.,* 12, 642, 2001.
257. Shen, L., Ye, J.-L., Huang, M.-Y., and Jiang, Y.-Y., Preparation and catalytic behaviors of silica-supported casein-Fe complex in hydrogenation of cyclohexanone, *Polym. Adv. Technol.,* 13, 173, 2002.
258. Zhang, X., Li, Y.-J., Huang, M.-Y., and Jiang, Y.-Y., Asymmetric hydration of allyl alcohol catalyzed by polysulfostyrene-gelatin-Co complex, *Polym. Adv. Technol.,* 13, 305, 2002.
259. Denkbas, E.B., Kiliçay, E., Birlikseven, C., and Oztürk, E., Magnetic chitosan microspheres: preparation and characterization, *React. Funct. Polym.,* 50, 225, 2002.
260. Denkbas, E.B. and Odabasi, M., Chitosan microspheres and sponges: preparation and characterization, *J. Appl. Polym. Sci.,* 76, 1637, 2000.
261. Rorrer, G.L., Hsien, T.-Y., and Way, J.D., Synthesis of porous-magnetic chitosan beads for removal of cadmium ions from waste water, *Ind. Eng. Chem. Res.,* 32, 2170, 1993.
262. Fundueanu, G., Nastruzzi, C., Carpov, A., Desbrieres, J., and Rinaudo, M., Physico-chemical characterization of Ca-alginate microparticles produced with different methods, *Biomaterials,* 20, 1427, 1999.
263. Potter, K. and McFarland, E.W., Ion transport studies in calcium alginate gels by magnetic resonance microscopy, *Solid State Nucl. Magn. Reson.,* 6, 323, 1996.
264. Velings, N.M. and Mestdagh, M.M., Physico-chemical properties of alginate gel beads, *Polym. Gels Networks,* 3, 311, 1995.
265. Khairou, K.S., Kinetics and mechanism of the non-isothermal decomposition. I. Some divalent cross-linked metal-alginate ionotropic gels, *J. Therm. Anal. Calorim.,* 69, 583, 2002.
266. Khairou, K.S., Al-Gethami, W.M., and Hassan, R.M., Kinetics and mechanism of sol-gel transformation between sodium alginate polyelectrolyte and some heavy divalent metal ions with formation of capillary structure polymembranes ionotropic gels, *J. Membr. Sci.,* 209, 445, 2002.
267. Khairou, K.S., Al-Gethami, W.M., and Hassan, R.M., Kinetics and mechanism of sol-gel transformation on polyelectrolytes of divalent metal-alginate gel complexes with formation of capillary ionotropic polymembranes, *Bull. Pol. Acad. Sci., Chem.,* 50, 299, 2002.
268. Dambies, L., Vincent, T., and Guibal, E., Treatment of arsenic-containing solutions using chitosan derivatives: uptake mechanism and sorption performances, *Water Res.,* 36, 3699, 2002.
269. Mi, F.-L., Shyu, S.-S., Wong, T.-B., Jang, S.-F., Lee, S.-T., and Lu, K.-T., Chitosan-polyelectrolyte complexation for the preparation of gel beads and controlled release of anticancer drug. II. Effect of pH-dependent ionic crosslinking or interpolymer complex using tripolyphosphate or polyphosphate as reagent, *J. Appl. Polym. Sci.,* 74, 1093, 1999.

270. Molvinger, K., Quignard, F., Brunel, D., Boissière, M., and Devoiselle, J.M., Porous chitosan-silica hybrid microspheres as a potential catalyst, *Chem. Mater.,* 16, 3367, 2004.
271. Valentin, R., Horga, R., Bonelli, B., Garrone, E., Di Renzo, F., and Quignard, F., FTIR spectroscopy of NH_3 on acidic and ionotropic alginate aerogels, *Biomacromolecules,* 7, 877, 2006.
272. Ruiz, M.A., Sastre, A.M., and Guibal, E., Pd and Pt recovery using chitosan gel beads. I. Influence of drying process on diffusion properties, *Sep. Sci. Technol.,* 37, 2143, 2002.
273. Kurokawa, Y. and Hanaya, K., Functionality of cellulose by impregnation of inorganic substances, *Carbohydr. Polym.,* 27, 313, 1995.
274. Agboh, O.C. and Quin, Y., Chitin and chitosan fibers, *Polym. Adv. Technol.,* 8, 355, 1997.
275. Kaminski, W., Eckstein, W., Modrzejewska, Z., and Sroda, Z., Chitosan flat and hollow-fiber membranes, in *Chitin World,* Karnicki, Z.S., Wojtasz-Pajak, A., Brzeski, M.M., and Bykowski, P.J., Eds., Wirtschaftsverlag NW, Bremerhaven, Germany, 1995, p. 600.
276. Kim, Y.-J., Yoon, K.-J., and Ko, S.-W., Preparation and properties of alginate superabsorbent filament fibers crosslinked with glutaraldehyde, *J. Appl. Polym. Sci.,* 78, 1797, 2000.
277. Amaike, M., Senoo, Y., and Yamamoto, H., Sphere, honeycomb, regularly spaced droplet and fiber structures of polyion complexes of chitosan and gellan, *Macromol. Rapid Commun.,* 19, 287, 1998.
278. Knaul, J.Z., Hudson, S.M., and Creber, K.A.M., Improved mechanical properties of chitosan fibers, *J. Appl. Polym. Sci.,* 72, 1721, 1999.
279. Modrzejewska, Z. and Eckstein, W., Chitosan hollow fiber membranes, *Biopolymers,* 73, 61, 2004.
280. Vincent, T. and Guibal, E., Cr(VI) extraction using Aliquat 336 in hollow fiber module made of chitosan, *Ind. Eng. Chem. Res.,* 40, 1406, 2001.
281. Vincent, T. and Guibal, E., Non-dispersive liquid extraction of Cr(VI) by TBP/Aliquat 336 using chitosan-made hollow fiber, *Solvent Extr. Ion Exch.,* 18, 1241, 2000.
282. Guibal, E. and Vincent, T., Chitosan-supported palladium catalyst. V. Nitrophenol degradation on catalytic chitosan hollow fiber, *Environ. Sci. Technol.,* 38, 4233, 2004.
283. Guibal, E., Vincent, T., and Spinelli, S., Chitosan-supported catalysts for environmental application: catalytic hollow fibers for the degradation of phenolic derivatives, *Sep. Sci. Technol.,* 40, 633, 2005.
284. Wang, X. and Spencer, H.G., Preparation and properties of sodium alginate formed-in-place membranes, *J. Appl. Polym. Sci.,* 61, 827, 1996.
285. Yeom, C.K. and Lee, K.-H., Characterization of sodium alginate membrane crosslinked with glutaraldehyde in pervaporation separation, *J. Appl. Polym. Sci.,* 67, 209, 1998.
286. Lee, K.Y., Rowley, J.A., Eiselt, P., Moy, E.M., Bouhadir, K.H., and Mooney, D.J., Controlling mechanical and swelling properties of alginate hydrogens independently by cross-linker type and cross-linking density, *Macromolecules,* 33, 4291, 2000.
287. Mi, F.-L., Shyu, S.-S., Wu, Y.-B., Lee, S.-T., Shyong, J.-Y., and Huang, R.-N., Fabrication and characterization of a sponge-like asymmetric chitosan membrane as a wound dressing, *Biomaterials,* 22, 165, 2001.

288. Krajewska, B., Diffusion of metal ions through gel chitosan membranes, *React. Funct. Polym.*, 47, 37, 2001.
289. Wang, X.-P. and Shen, Z.-Q., Studies on the effects of copper salts on the separation performance of chitosan membranes, *Polym. Int.*, 49, 1426, 2000.
290. Tual, C., Espuche, E., Escoubes, M., and Domard, A., Transport properties of chitosan membranes: influence of crosslinking, *J. Polym. Sci. Pol. Phys.*, 38, 1521, 2000.
291. Modrzejewska, Z. and Kaminski, W., Separation of Cr(VI) on chitosan membranes, *Ind. Eng. Chem. Res.*, 38, 4946, 1999.
292. Krajewska, B. and Olech, A., Pore structure of gel chitosan membranes. II. Modelling of pore size distribution from solute diffusion measurements: Gaussian distribution-mathematical limitations, *Polym. Gels Networks,* 4, 45, 1996.
293. Krajewska, B. and Olech, A., Pore structure of gel chitosan membranes. I. Solute diffusion measurements, *Polym. Gels Networks,* 4, 33, 1996.
294. Krajewska, B., Pore structure of gel chitosan membranes. III. Pressure-driven mass transport measurements, *Polym. Gels Networks,* 4, 55, 1996.
295. Urbanczyk, G.W. and Lipp-Symonowicz, B., The influence of processing terms of chitosan membranes made of differently deacetylated chitin on the crystalline structure of membranes, *J. Appl. Polym. Sci.*, 51, 2191, 1994.
296. Arvanitoyannis, I.S., Nakayama, A., and Aiba, S.-I., Chitosan and gelatin based edible films: state diagrams, mechanical and permeation properties, *Carbohydr. Polym.*, 37, 371, 1998.
297. Gushikem, Y. and Campos, E.A., Cobalt(II) hematoporphyrin IX immobilized in a cellulose acetate niobium(V) oxide composite membrane: preparation and oxygen reduction study, *J. Braz. Chem. Soc.*, 9, 273, 1998.
298. Kubota, N., Permeability properties of chitosan-transition metal complex membranes, *J. Appl. Polym. Sci.*, 64, 819, 1997.
299. Zheng, X. and Ruckenstein, E., Control of pore sizes in macroporous chitosan, and chitin membranes, *Ind. Eng. Chem. Res.*, 35, 4169, 1996.
300. Corain, B. and Kralik, M., Generating palladium nanoclusters inside functional cross-linked polymer frameworks, *J. Mol. Catal. A: Chem.*, 173, 99, 2001.
301. Corain, B., Zecca, M., and Jerabek, K., Catalysis and polymer networks: the role of morphology and molecular accessibility, *J. Mol. Catal. A: Chem.*, 177, 3, 2001.
302. Louvier-Hernandez, J.F., Luna-Barcenas, G., Thakur, R., and Gupta, R.B., Formation of chitin nanofibers by supercritical antisolvent, *J. Biomed. Nanotechnol.*, 1, 109, 2005.
303. Valodkar, V.B., Tembe, G.L., Ravindranathan, M., Ram, R.N., and Rama, H.S., A study of synthesis, characterization and catalytic hydrogenation by polymer anchored Pd(II)-amino acid complexes, *J. Mol. Catal. A: Chem.*, 202, 47, 2003.
304. El-Hamshary, H. and Al-Sigeny, S., Catalytic activities of amino acid modified, starch-grafted acrylamide for the decomposition of hydrogen peroxide, *J. Appl. Polym. Sci.*, 93, 630, 2004.
305. Xue, L., Zhou, D.J., Tang, L., Ji, X.F., Huang, M.Y., and Jiang, Y.Y., The asymmetric hydration of 1-octene to (S)-(+)-2-octanol with a biopolymer-metal complex, silica-supported chitosan-cobalt complex, *React. Funct. Polym.*, 58, 117, 2004.
306. Sashiwa, H. and Aiba, S.-I., Chemically modified chitin and chitosan as biomaterials, *Prog. Polym. Sci.*, 29, 887, 2004.
307. Macquarrie, D.J. and Hardy, J.J.E., Applications of functionalized chitosan in catalysis, *Ind. Eng. Chem. Res.*, 44, 8499, 2005.

308. Chang, Y., Wang, Y., Zha, F., and Wang, R., Preparation and catalytic properties of chitosan bound Schiff base copper complexes, *Polym. Advan. Technol.*, 15, 284, 2004.
309. Sun, W., Xia, C.G., and Wang, A.Q., Cyclopropanation of styrene catalyzed by chitosan Schiff-base copper (II) heterogeneous catalyst, *Acta Chim. Sinica*, 60, 162, 2002.
310. Tong, J., Li, Z., and Xia, C., Highly efficient catalysts of chitosan-Schiff base Co(II) and Pd(II) complexes for aerobic oxidation of cyclohexane in the absence of reductants and solvents, *J. Mol. Catal. A: Chem.*, 231, 197, 2005.
311. Wang, H., Sun, W., and Xia, C., An easily recoverable and efficient catalyst for heterogeneous cyclopropanation of olefins, *J. Mol. Catal. A: Chem.*, 206, 199, 2003.
312. Finashina, E., Kramareva, N.V., and Kustov, L.M., Oxidation of catecholamines on chitosan-immobilized Co(II)-Salen complexes, *Macromol. Symp.*, 204, 205, 2003.
313. Hu, D., Cui, Y., Dong, X., and Fang, Y., Studies of CoSalen immobilized onto *N*-(4-pyridylmethylidene)-chitosan, *React. Funct. Polym.*, 48, 201, 2001.
314. Hu, D.D., Shi, Q.Z., Tang, Z.X., Fang, Y., and Kennedy, J.F., CoSalen immobilized to chitosan and its electrochemical behavior, *Carbohydr. Polym.*, 45, 385, 2001.
315. Liu, C.Q., Xu, Y., Liao, S.J., and Yu, D.R., Mono- and bimetallic catalytic hollow-fiber reactors for the selective hydrogenation of butadiene in 1-butene, *Appl. Catal. A: Gen.*, 172, 23, 1998.
316. Liu, C., Xu, Y., Liao, S., and Yu, D., Selective hydrogenation of cyclopentadiene in mono- and bimetallic catalytic hollow-fiber reactors, *J. Mol. Catal. A.: Chem.*, 157, 253, 2000.
317. Kramareva, N.V., Finashina, E., Kucherov, A.V., and Kustov, L.M., Copper complexes stabilized by chitosans: peculiarities of the structure, redox, and catalytic properties, *Kinet. Catal.*, 44, 793, 2003.
318. Kramareva, N.V., Koklin, A.E., Finashina, E., Telegina, N.S., Stakheev, A.Y., and Kustov, L.M., Study of palladium complexes with chitosan and its derivatives as potential catalysts for terminal olefin oxidation, *Kinet. Catal.*, 45, 784, 2004.
319. Kramareva, N.V., Stakheev, A.Y., Tkachenko, O., Klementiev, K.V., Grünert, W., Finashina, E.D., and Kustov, L.M., Heterogenized palladium chitosan complexes as potential catalysts in oxidation reactions: study of the structure, *J. Mol. Catal. A: Chem.*, 209, 97, 2004.
320. Inoue, K., Baba, Y., and Yoshizuka, K., Adsorption of metal ions on chitosan and cross-linked copper(II)-complexed chitosan, *Bull. Chem. Soc. Jpn.*, 66, 2915, 1993.
321. Baba, Y., Masaaki, K., and Kawano, Y., Synthesis of a chitosan derivative recognizing planar metal ion and its selective adsorption equilibria of copper(II) over iron(III), *React. Funct. Polym.*, 36, 167, 1998.
322. Guan, H.M. and Cheng, X.S., Study of cobalt(II)-chitosan coordination polymer and its catalytic activity and selectivity for vinyl monomer polymerization, *Polym. Adv. Technol.*, 15, 89, 2004.
323. Wang, X.-X., Huang, M.-Y., and Jiang, Y.-Y., Hydrogenation catalytic behaviors of palladium complexes of chitin and chitosan, *Macromol. Chem., Macromol. Symp.*, 59, 113, 1992.
324. Tang, L.-M., Huang, M.-Y., and Jiang, Y.-Y., Selective hydrogenation of phenol to cyclohexanone catalyzed by a silica-supported chitosan-palladium complex, *Macromol. Rapid Commun.*, 15, 527, 1994.

325. Tang, Y., Chen, B., and Mo, S., Separation and preconcentration of ultratrace lead in biological organisms and its determination by graphite furnace atomic absorption spectrometry, *Talanta,* 43, 761, 1996.
326. Yin, M.Y., Yuan, G.L., Huang, M.Y., and Jiang, Y.Y., Catalytic behavior of a wool-Pd complex in asymmetric hydrogenation of diacetone alcohol and 3-methyl-2-butanone, *J. Mol. Catal. A: Chem.,* 147, 89, 1999.
327. Liu, P., Wang, L., Li, L.M., and Wang, X.Y., Study of the Heck arylation reaction catalyzed by chitosan-immobilized palladium(0) complex, *Chin. J. Org. Chem.,* 24, 59, 2004.
328. Liu, P., Wang, L., and Wang, X.Y., Chitosan-immobilized palladium complex: a green and highly active heterogeneous catalyst for Heck reaction, *Chin. Chem. Lett.,* 15, 475, 2004.
329. Sun, Y., Guo, Y., Lu, Q., Meng, X., Xiaohua, W., Guo, Y., Wang, Y., Liu, X., and Zhang, Z., Highly selective asymmetry transfer hydrogenation of prochiral acetophenone catalyzed by palladium-chitosan on silica, *Catal. Lett.,* 100, 213, 2005.
330. Yuan, G.L., Yin, M.Y., Jiang, T.T., Huang, M.Y., and Jiang, Y.Y., Catalytic behaviors of silica-supported chitin-platinum complex for asymmetric hydrogenation of α-phenylethanol, *J. Mol. Catal. A.: Chem.,* 159, 45, 2000.
331. Wei, W.-L., Hao, S.-J., Zhou, J., Huang, M.-Y., and Jiang, Y.-Y., Catalytic behavior of silica-supported chitosan-platinum-iron complex for asymmetric hydrogenation of ketones, *Polym. Adv. Technol.,* 15, 287, 2004.
332. Huang, K., Liu, H.-W., Dou, X., Huang, M.-Y., and Jiang, Y.-Y., Silica-supported chitosan-osmium tetroxide complex catalyzed vicinal hydroxylation of olefins using hexacyanoferrate(III) ion as a cooxidant, *Polym. Adv. Technol.,* 14, 364, 2003.
333. Han, H.-S., Jiang, S.-N., Huang, M.-Y., and Jiang, Y.-Y., Catalytic hydrogenation of aromatic nitro compounds by non-noble metal complexes of chitosan, *Polym. Adv. Technol.,* 7, 704, 1996.
334. Han, X.-X., Zhou, R.-X., Lai, G.-H., Yue, B.-H., and Zheng, X.M., Hydrogenation catalysts based on platinum complexes with polymer, *React. Kinet. Catal. Lett.,* 81, 41, 2004.
335. Hardy, J.J.E., Hubert, S., Macquarrie, D.J., and Wilson, A.J., Chitosan-based heterogeneous catalysts for Suzuki and Heck reactions, *Green Chem.,* 6, 53, 2004.
336. Sun, W., Xia, C.-G., and Wang, H.-W., Efficient heterogeneous catalysts for the cyclopropanation of olefins, *New J. Chem.,* 26, 755, 2002.
337. Zhang, X., Han, B., Hua, Y.-N., Huang, M.-Y., and Jiang, Y.-Y., Asymmetric epoxidation of cinnamyl alcohol catalyzed by silica-supported casein-Co complex, *Polym. Adv. Technol.,* 13, 216, 2002.
338. Yuan, G.-L., Yin, M.-Y., Huang, M.-Y., and Jiang, Y.-Y., Catalytic behavior of wool-Pt complex in asymmetric hydrogenation of ketones, *Polym. Adv. Technol.,* 10, 442, 1999.
339. He, M., Zhou, D.-Q., Ge, H.-L., Huang, M.-Y., and Jiang, Y.-Y., Catalytic behavior of wool-Rh complex in asymmetric hydrogenation of 2-methyl furan, *Polym. Adv. Technol.,* 14, 273, 2003.
340. Jia, B., Yang, X., Huang, M.Y., and Jiang, Y.Y., Hydration of alkenes catalyzed by wool-palladium-iron complex, *React. Funct. Polym.,* 57, 163, 2003.
341. Zeng, X., Zhang, Y., and Shen, Z., Ring opening polymerization of propylene oxide by chitosan-supported rare earth catalytic system and its kinetics, *J. Polym. Sci. A.: Polym. Chem.,* 35, 2177, 1997.

342. Guibal, E. and Vincent, T., Chitosan-supported palladium catalyst. IV. Influence of temperature on nitrophenol degradation, *J. Environ. Manage.*, 71, 97, 2004.
343. Vincent, T. and Guibal, E., Chitosan-supported palladium catalyst. III. Influence of experimental parameters on nitrophenol degradation, *Langmuir*, 19, 8475, 2003.
344. Vincent, T. and Guibal, E., Chitosan-supported palladium catalyst. 1. Synthesis procedure, *Ind. Eng. Chem. Res.*, 41, 5158, 2002.
345. Vincent, T., Peirano, F., and Guibal, E., Chitosan supported palladium catalyst. VI. Nitroaniline degradation, *J. Appl. Polym. Sci.*, 94, 1634, 2004.
346. Vincent, T., Spinelli, S., and Guibal, E., Chitosan-supported palladium catalyst. II. Chlorophenol dehalogenation, *Ind. Eng. Chem. Res.*, 42, 5968, 2003.
347. Huang, H. and Yang, X., Chitosan mediated assembly of gold nanoparticles multilayer, *Colloids Surf. A*, 226, 77, 2003.
348. Esumi, K., Takei, N., and Yoshimura, T., Antioxidant-potentiality of gold-chitosan nanocomposites, *Colloids Surf. B*, 32, 117, 2003.
349. Kundu, S., Mandal, M., Ghosh, S.K., and Pal, T., Photochemical deposition of SERS active silver nanoparticles on silica gel and their application as catalysts for the reduction of aromatic nitro compounds, *J. Colloid Interf. Sci.*, 272, 133, 2004.
350. Pal, A., Esumi, K., and Pal, T., Preparation of nanosized gold particles in a biopolymer using UV photoactivation, *J. Colloid Interf. Sci.*, 288, 396, 2005.
351. Adlim, M., Abu Bakar, M., Liew, K.Y., and Ismail, J., Synthesis of chitosan-stabilized platinum and palladium nanoparticles and their hydrogenation activity, *J. Mol. Catal. A: Chem.*, 212, 141, 2004.
352. Mirescu, A. and Prüsse, U., Selective oxidation on gold colloids, *Catal. Commun.*, 7, 11, 2004.
353. Ascensio, J.A., Mejia, Y., Liu, H.B., Angeles, C., and Canizal, G., Bioreduction synthesis of Eu-Au nanoparticles, *Langmuir*, 19, 5882, 2003.
354. Satterfield, C.N. and Sherwood, T.K., *The Role of Diffusion in Catalysts*, Addison-Wesley Publishing Company, Inc., Reading, MA, 1963, p. 118.
355. Neri, G., Musolino, M.G., Milone, C., Pietropaolo, D., and Galvagno, S., Particle size effect in the catalytic hydrogenation of 2,4-dinitrotoluene over Pd/C catalysts, *Appl. Catal. A: Gen.*, 208, 307, 2001.
356. Douidah, A., Marécot, P., Szabo, S., and Barbier, J., Evaluation of the metal-support interactions; case of platinum-supported catalysts: effect of the support nature and the metallic dispersion, *Appl. Catal. A: Gen.*, 225, 21, 2002.
357. Vogel, B., Schneider, C., and Klemm, E., The synthesis of cresol from toluene and N2O on HAlZSM-5: minimizing the product diffusion limitation by the use of small crystals, *Catal. Lett.*, 79, 107, 2002.
358. Chen, C.-W., Tano, D., and Akashi, M., Colloidal platinum nanoparticles stabilized by vinyl polymers with amide side chains: dispersion stability and catalytic activity in aqueous electrolyte solutions, *J. Colloid Interf. Sci.*, 225, 349, 2000.
359. Harada, M. and Einaga, H., Photochemical deposition of platinum on TiO2 by using poly(vinyl alcohol) as an electron donor and a protecting agent., *Catal. Today*, 5, 63, 2004.
360. Schimpf, S., Lucas, M., Mohr, C., Rodemerck, U., Brückner, A., Radnik, J., Hofmeister, H., and Claus, P., Supported gold nanoparticles: in-depth catalyst characterization and application in hydrogenation and oxidation reactions, *Catal. Today*, 72, 63, 2002.

361. Karhu, H., Kalantar, A., Väyrynen, I.J., Salmi, T., and Murzin, D.Y., XPS analysis of chlorine residues in supported Pt and Pd catalysts with low metal loading, *Appl. Catal. A: Gen.*, 247, 283, 2003.
362. Tang, L.-M., Huang, M.-Y., and Jiang, Y.-Y., Hydrogenation of phenol and cresols catalyzed by chitosan supported palladium complex at mild conditions, *Chin. J. Polym. Sci.*, 14, 57, 1996.
363. Yin, M.-Y., Yuan, G.-L., Wu, Y.-Q., Huang, M.-Y., and Jiang, Y.-Y., Asymmetric hydrogenation of ketones catalyzed by a silica-supported chitosan-Palladium complex, *J. Mol. Catal. A: Chem.*, 147, 93, 1999.
364. Chang, Y., Wang, Y., and Su, Z., Preparation of chitosan-bound nitrobenzaldehyde metal complexes and studies on its catalytic oxidative activity and reactive mechanism, *J. Appl. Polym. Sci.*, 83, 2188, 2002.
365. Zhang, J. and Xia, C.G., Natural biopolymer-supported bimetallic catalyst system for the carbonylation to esters of naproxen, *J. Mol. Catal. A.: Chem.*, 206, 59, 2003.
366. Inaki, Y., Otsuru, M., and Takemoto, K., Vinyl polymerization by metal complexes. 31. Initiation by chitosan-copper(II) complex, *J. Macromol. Sci., Chem, A*, 12, 953, 1978.
367. Inaki, Y., Otsuru, M., and Takemoto, K., Vinyl polymerization by metal complexes. 32. Formation of glycolchitosan-copper(II) complex and the initiation of vinyl polymerization by using the complex, *J. Macromol. Sci., Chem, A*, 14, 823, 1980.
368. Zhou, D.Q., Zhou, D.J., Cui, X.H., Wang, F.M., Huang, M.Y., and Jiang, Y.Y., Asymmetric hydrogenation of ketones catalyzed by silica-supported chitosan-iron-nickel complex, *Polym. Adv. Technol.*, 15, 350, 2004.
369. Jiang, Y.Y., Studer, M., and Blaser, H.-U., The enantioselective hydrogenation of 1-phenyl ethanol and of ketones using Pd and Pt supported on natural polymers, *J. Mol. Catal. A.: Chem.*, 177, 307, 2002.
370. Guibal, E., Vincent, T., and Spinelli, S., Environmental application of chitosan-supported caralysts: Catalytic hollow fibers for the degradation of phenolic derivatives, *Sep. Sci. Technol.*, 40, 633, 2005.
371. Chiessi, E. and Pispisa, B., Polymer-supported catalysis: oxidation of catecholamines by Fe(III) and Cu(II) complexes immobilized to chitosan, *J. Mol. Catal.*, 1994.
372. Paradossi, G., Chiessi, E., Cavalieri, F., Moscone, D., and Crescenzi, V., Networks based on chitosan and oxidized cyclodextrin. II. Structural and catalytic features of a copper(II)-loaded network, *Polym. Gels Networks*, 5, 525, 1997.
373. Yin, J.B., Chen, X.S., Zhang, L., Liu, A., Yang, Y.N., Yin, M., and Lou, L., Liquid-phase hydrogenation of furfural to tetrahydrofurfuryl alcohol catalyzed by polymer resin supported palladium catalysts, *Chem. J. Chin. Univ.*, 23, 1363, 2002.
374. Guan, H.-M. and Cheng, X.-S., Study of cobalt(II)-chitosan coordination polymer and its catalytic activity and selectivity for vinyl monomer polymerization, *Polym. Adv. Technol.*, 15, 89, 2004.
375. Köckritz, A., Bartoszek, M., Döbler, C., Beller, M., Mägerlein, W., and Militzer, H.-C., Development of protocols for the separation of Os catalysts from organic products in the catalytic dihydroxylation of olefins, *J. Mol. Catal. A: Chem.*, 218, 55, 2004.
376. Quignard, F., Choplin, A., and Domard, A., Chitosan: a natural polymeric support of catalysis for the synthesis of fine chemicals, *Langmuir*, 16, 9106, 2000.
377. Buisson, P. and Quignard, F., Polysaccharides: natural polymeric supports for aqueous phase catalysts in allylic substitution reactions, *Austr. J. Chem.*, 55, 73, 2002.

378. Guibal, E., Heterogeneous catalysis on chitosan-based materials: A review, *Prog. Polym. Sci.*, 30, 71, 2005.
379. An, Y., Yuan, D., Huang, M.-Y., and Jiang, Y.-Y., Selective hydrogenation of chloronitrobenzenes catalyzed by palladium complexes of silica-supported chitosan, *Macromol. Symp.*, 80, 257, 1994.
380. Zhu, H., Mizugaki, T., Ebitani, K., and Kaneda, K., Dimethylaminoethylated hydroxypropyl-chitosan: preparation and application as polymeric ligand to form Rh6 cluster complexes for the reduction of benzaldehyde and nitrobenzene, *J. Appl. Polym. Sci.*, 80, 447, 2001.
381. Zhu, H., Mizugaki, T., and Kaneda, K., Synthesis of dimethylaminoethyl chitin and applications as a polymeric ligand to form Rh cluster complexes for the reduction of benzaldehyde and nitrobenzene, *Macromol. Chem. Phys.*, 201, 1431, 2000.
382. Sajiki, H., Ikawa, T., and Hirota, K., Markedly chemoselective hydrogenation with retention of benzyl ester and *N*-Cbz functions using a heterogeneous Pd-fibroin catalyst, *Tetrahedr. Lett.*, 44, 8437, 2003.
383. Tang, L.M., Meng, Z.H., Huang, M.Y., and Jiang, Y.Y., Selective hydrogenation of anisol to cyclohexyl methyl-ether catalyzed by a silica-supported carboxymethyl cellulose platinum complex under mild reaction conditions, *React. Polym.*, 24, 251, 1995.
384. Kurita, K., Hayakawa, M., Nishiyama, Y., and Harata, M., Polymeric asymmetric reducing agents: preparation and reducing performance of chitosan/dihydronicotinamide conjugates having L- and D-phenylalanine spacer arms, *Carbohyd. Polym.*, 47, 7, 2002.
385. Nishiyama, Y., Yoshida, T., Mori, T., Ishii, S., and Kurita, K., Asymmetric reduction with chitosan/dihydronicotinamide conjugates: influence of L-alanine spacer arms on reducing performance, *React. Funct. Polym.*, 37, 83, 1998.
386. Kapoor, S., Belapurkar, A.D., Mittal, J.P., and Mukherjee, T., Catalytic oxidation of carbon monoxide over radiotically prepared Pt nanoparticles supported on glass, *Mater. Res. Bull.*, 40, 1654, 2005.
387. Qin, L.S., Yin, D.H., Liu, J.F., and Li, C.Y., Mesoporous alumina-supported nano-Au catalysts for carbon monoxide oxidation at low temperature, *Chin. J. Catal.*, 26, 714, 2005.
388. Guo, C.-C., Huang, G., Zhang, X.-B., and Guo, D.-C., Catalysis of chitosan-supported iron tetraphenylporphyrin for aerobic oxidation of cyclohexane in absence of reductants and solvents, *Appl. Catal. A: Gen.*, 247, 261, 2003.
389. Huang, G., Li, X.J., and Guo, C.C., Aerobic oxidation of cyclohexane catalyzed by chitosan-supported cobalt tetraphenylporphyrin, *Chin. J. Catal.*, 26, 765, 2005.
390. Guo, C.C., Huang, G., and Guo, D.C., Preparation of nitrogenous polysaccharide-supported ironporphyrins and their catalysis for the aerobic oxidation of cyclohexane, *Kinet. Catal.*, 47, 93, 2006.
391. Srivatsan, S.G., Nigam, P., Rao, M.S., and Verma, S., Phenol oxidation by copper-metallated 9-allyladenine-DVP polymer: reaction catalysis and polymer recycling, *Appl. Catal. A: Gen.*, 209, 327, 2001.
392. Tong, J., Zhang, Y., Li, Z., and Xia, C., Highly effective catalysts of natural polymer supported Salophen Mn(III) complexes for aerobic oxidation of cyclohexene, *J. Mol. Catal. A.: Chem.*, 249, 47, 2006.
393. Guibal, E., Vincent, T., Touraud, E., Colombo, S., and Ferguson, A., Oxidation of hydroquinone to p-benzoquinone catalyzed by Cu(II) supported on chitosan flakes, *J. Appl. Polym. Sci.*, 100, 3034, 2006.

394. Šuláková, R., Hrdina, R., and Soares, G.M.B., Oxidation of azo textile soluble dyes with hydrogen peroxide in the presence of Cu(II)-chitosan heterogeneous catalysts, *Dyes Pigments,* in press, 2006.
395. Yao, Y., Chen, W., Lu, S., and Zhao, B., Binuclear metallophthalocyanine supported on treated silk fibres as a novel air-purifying material, *Dyes Pigments,* 2005.
396. Kolesnichenko, N.V., Batov, A.E., Markova, N.A., and Slivinsky, E.V., Carbonylation of methyl acetate in the presence of polymeric rhodium-containing catalysts, *Russ. Chem. Bull.,* 51, 259, 2002.
397. Kolesnichenko, N.V., Markova, N.A., Batov, A.E., Voronina, Z.A., Panina, O.A., Frantsuzov, V.K., and Slivinsky, E.V., Carbonylation of benzyl alcohol in the presence of rhodium catalysts, *Petrol Chem,* 43, 150, 2003.
398. Gronnow, M.J., Luque, R., Macquarrie, D.J., and Clark, J.H., A novel highly active biomaterial supported palladium catalyst, *Green Chem.,* 7, 552, 2005.
399. Calo, V., Nacci, A., Monopoli, A., Fornaro, A., Sabbatini, L., Cioffi, N., and Ditaranto, N., Heck reaction catalyzed by nanosized palladium on chitosan in ionic liquids, *Organometallics,* 23, 5154, 2004.
400. Quignard, F. and Choplin, A., Cellulose: a new bio-support for aqueous phase catalysts, *Chem. Commun.,* 21, 2001.
401. Cui, Y., Zhang, L., and Li, Y., Synthesis of chitosan derivatives supported palladium complexes and their catalytic behavior in the Heck reaction, *Polym. Adv. Technol.,* 16, 633, 2005.
402. Reddy, K.R., Kumar, N.S., Sreedar, B., and Kantam, M.L., Cellulose supported palladium(0) catalyst for Heck and Sonogashira coupling reactions, *J. Mol. Catal. A: Chem.,* 252, 12, 2006.
403. Reddy, K.R., Kumar, N.S., Sreedar, B., and Kantam, M.L., *N*-arylation of nitrogen heterocycles with aryl halides and arylboronic acid catalyzed by cellulose supported copper(0), *J. Mol. Catal. A: Chem.,* 252, 131, 2006.
404. Xiao, L.-F., Li, F.-W., and Xia, C.-G., An easily recoverable and efficient natural biopolymer-supported zinc chloride catalyst system for the chemical fixation of carbon dioxide to cyclic carbonate, *Appl. Catal. A: Gen.,* 279, 125, 2005.
405. Aramendia, M.A., Borau, V., Garcia, I.M., Jiménez, C., Marinas, A., Marinas, J.M., and Urbano, F.J., Hydrogenolysis of aryl halides by hydrogen gas and hydrogen transfer over palladium-supported catalysts, *C.R. Acad. Sci., Paris, Série IIc, Chem.,* 3, 465, 2000.
406. Lauwiner, M., Rys, P., and Wissman, J., Reduction of aromatic nitro compounds with hydrazine hydrate in the presence of an iron oxide hydroxide catalyst. I. The reduction of monosubstituted nitrobenzenes with hdyrazine hydrate in the presence of ferrihydrite, *Appl. Catal. A: Gen.,* 172, 141, 1998.
407. Xi, X., Liu, Y., Shi, J., and Cao, S., Palladium complex of poly(4-vinylpyridine-co-acrylic acid) for homogeneous hydrogenation of aromatic nitro compounds, *J. Mol. Catal. A: Chem.,* 192, 1, 2003.
408. Xu, S., Xi, X., Shi, J., and Cao, S., A homogeneous catalyst made of poly(4-vinylpyridine-co-*N*-vinylpyrrolidone)-Pd(0) complex for hydrogenation of aromatic nitro compounds, *J. Mol. Catal. A: Chem.,* 160, 287, 2000.
409. Liu, C., Xu, Y., Liao, S., Yu, D., Zhao, Y., and Fan, Y., Selective hydrogenation of propanediene and propyne in propene with catalytic polymeric hollow-fiber reactor, *J. Membr. Sci.,* 137, 139, 1997.
410. Peirano, F., Vincent, T., and Guibal, E., Hydrogenation of nitrotoluene using palladium supported on chitosan hollow fibers: II Influence of experimental parameters, in preparation.

5 Ion Exchange Selectivity as a Surrogate Indicator of Relative Permeability of Homovalent Ions in Reverse Osmosis Processes

Parna Mukherjee and Arup K. SenGupta

CONTENTS

5.1 Introduction 294
5.1.1 General Premise 295
5.2 Electrolyte Permeation Mechanism through Reverse Osmosis and Nanofiltration Membranes 296
5.2.1 Theory 298
5.2.1.1 Relationship of Salt Flux to Feed Concentrations 298
5.2.1.2 Ion-Pair Mechanism 298
5.2.1.3 Coupled Transport Mechanism 300
5.2.2 Results 302
5.2.3 Discussion 305
5.3 Ion Exchange Selectivity as a Surrogate Indicator of Relative Permeability of Homovalent Ions in Reverse Osmosis Processes 306
5.3.1 Relationship between Permeation, Interdiffusion Coefficient, and Ion Exchange Selectivity 306
5.3.2 Materials and Methods 310
5.3.3 Results 311
5.3.3.1 Monovalent Ions 311
5.3.3.2 Effects of pH, Solvent Dielectric Constant, and Conjugate Ion 315
5.3.3.3 Nanofiltration Membrane 318
5.3.4 Discussion 319

5.4 Hierarchy of Permeation of Ions among Multivalent Species 321
5.5 Special Case: Permeation Behavior of Trace Species through Membranes 323
5.5.1 Theoretical Approach 324
5.5.1.1 Calculation of k_s from Salt Permeation Experiments 324
5.5.1.2 Interdiffusion Coefficient of a 1-1 Single Electrolyte Permeating through the Membrane 324
5.5.1.3 Interdiffusion Coefficient of a Multicomponent 1-1 Electrolyte with Trace Species 324
5.5.1.4 Salt Permeability Coefficient, k_s 325
5.5.2 Experimental Methods 326
5.5.2.1 Salt Permeation Experiments 326
5.5.2.2 Membranes Used 326
5.5.2.3 Experimental Protocol 327
5.5.2.4 Analytical Methods 327
5.5.3 Results 327
5.5.4 Discussion 332
5.6 Concluding Remarks 334
Acknowledgments 335
References 335

5.1 INTRODUCTION

Reverse osmosis (RO) is a pressure-driven membrane process applied primarily to separate dissolved solids from water. It is, however, well documented that RO membranes reject different electrolytes or ions to different degrees; that is, permeation of different ions through RO membranes is different, all other conditions remaining identical. Also, the rejection of a specific ion is influenced by the accompanying electrolytes in feed water.[1–5] With an increased application of RO and nanofiltration (NF) processes in the area of water treatment and wastewater reuse, there is now a greater need in predicting the relative degree of rejection (or permeation) of various ions, including the environmentally regulated species. The existing bodies of laboratory and field data clearly establish that divalent cations and anions are more strongly rejected by RO and NF membranes compared to their monovalent counterparts.[4,5] Also, both cellulose acetate and polyamide membranes often contain weak-acid functional groups[6,7]; hence, the charge density on the membrane surface is influenced by pH. Donnan exclusion effect thus controls solute permeability as a function of pH. Progress has been made in developing models that integrate the Nernst-Planck equation with the Donnan effect to predict the relative permeations of ions of dissimilar valence, that is, heterovalent ions.[4–10] The Nernst-Planck model, with or without Donnan effect, is, however, unable to differentiate the permeation between ions of identical valence, that is, homovalent ions. Marinas and Selleck[11] attempted to predict the electrolyte permeation based on the formation constants of various ion pairs, especially divalent ions. Hodgson[2] was the first to undertake an elaborate study to show differences in the relative permeation of various ions through cellulose acetate membrane. Hodgson's experimental work also substantiated how the permeation of a specific cation or anion was influenced by its conjugate ions and other accompanying electrolytes. Although the experimental data resulting from this study are very useful, the scientific approach was quite empirical and lacks generalization. Strikingly absent in all these studies[8–12] is the absence of any scientific protocol to predict the differential permeation of homovalent ions such as monovalent anions Cl^-, Br^-, F^-, NO_3^-, and ClO_4^-; monovalent cations Li^+, Na^+, K^+, and NH_4^+; divalent anions SO_4^{2-}, HPO_4^{2-}, and SeO_4^{2-}; and similar others.

Specific diffusivity of an ion in the aqueous phase is inversely correlated to its hydrated ionic radius. *A priori* knowledge of hydrated ionic radii data for various ions can be seemingly useful in predicting permeation rates of strongly ionized electrolytes. Hydrated ionic radii data have also been computed from free energy of solvation of parent ions.[13] Nevertheless, a careful search and comparison of the hydrated ionic radii data from different sources[10,13,14] quickly indicate a lack of consistency, especially among polyatomic ions, such as nitrate, nitrite, sulfate, ammonium, phosphate, and so on. Furthermore, in environmental separation, we often encounter relatively uncommon species, namely, perchlorate, chloroacetate, radium, or selenate, for which hydrated ionic radii data are not readily available. Equivalent conductance data available in the literature were used by researchers[8,10] for computing ionic diffusion coefficients that were then used to predict the relative permeation rates of monovalent anions for RO membranes. The approach, however, failed even to qualitatively predict the relative permeation rates of chloride and nitrate.[8]

5.1.1 General Premise

The general objective of the present study is to present ion exchange selectivity as a surrogate parameter to predict the relative permeability of different ions of identical valence in RO and NF processes. One key advantage pertaining to this approach is the ease and reliability with which ion exchange selectivity data can be rapidly determined even for exotic ionic species. According to information available in the open literature, ion exchange selectivity data have never been used as surrogate process parameters for predicting the hierarchy of permeation of ions in semipermeable membrane processes. The hierarchy of permeation of ions through membranes afterward follows this discussion. This chapter shows how the knowledge of ion exchange selectivity acts as a predictive tool in a real-life field situation, where there are myriad competing ions in the feed of the membrane process.

The initial part of the chapter is devoted to establishing the mechanism of transport of electrolytes across the membrane. For this purpose, two possible modes of transmembrane electrolyte transport have been considered and one has been validated using the experimental data obtained from different sources.

Finally, there is a discussion regarding how the permeation behavior of an ionic species changes when it appears in the feed as a trace species as opposed to being a bulk from a concentration viewpoint.

5.2 ELECTROLYTE PERMEATION MECHANISM THROUGH REVERSE OSMOSIS AND NANOFILTRATION MEMBRANES

For reverse osmosis (RO) membranes, the solution-diffusion mechanism has been widely used to characterize the transport of solvent and permeating solutes through the active asymmetric membrane.[15–19] In accordance with the basic premise of the solution-diffusion model, the entire process of electrolyte transport across the membrane can be broken down into the following three successive steps: first, convective transport of electrolytes from the bulk aqueous phase to the membrane–water interface; second, partitioning at the active membrane layer; and third, transport through the membrane to the product water side. Note that the dielectric constant monotonically decreases from the bulk solution to the concentration polarization zone near the membrane–water interface and then into the asymmetric active layer of the membrane. Consequently, the hydrated ionic radii of individual cations and anions shrink, thus increasing the possibility of ion pair formation. From a mechanistic viewpoint, there remain two distinct possibilities for electrolyte transport through the RO membrane:

(i) Neutral ion pairs formed at the concentration-polarization zone get partitioned into the membrane phase and then diffuse as ion pairs to the low-pressure side.
(ii) Individual ions (not ion pairs) get partitioned into the membrane phase and then migrate through the membrane active layer in a manner such that the negative charge of an anion is balanced by the positive charge of an accompanying cation; that is, it is strictly a case of coupled transport.

Considering sodium chloride to be the predominant electrolyte in the feed, Figure 5.1a illustrates the effect of concentration polarization effect near the membrane–water interface. Figure 5.1b and Figure 5.1c depict the transmembrane salt transport pathways in accordance with the two aforementioned mechanisms. In an earlier work, Reusch and Cussler[20] showed that the passage of concentrated potassium chloride across a polyether–chloroform liquid membrane follows the ion-pair mechanism. In that investigation, the transmembrane electrolyte transport was essentially a case of facilitated diffusion where polyether served as the carrier of KCl^0 ion pair. No transmembrane pressure differential was present for the liquid membrane investigated; that is, while the transport of potassium chloride was governed solely by the concentration gradient, water flux was essentially absent. In comparison, RO and nanofiltration (NF) processes always operate under trans-membrane pressure gradient accompanied by the flow of solvent (i.e., water)

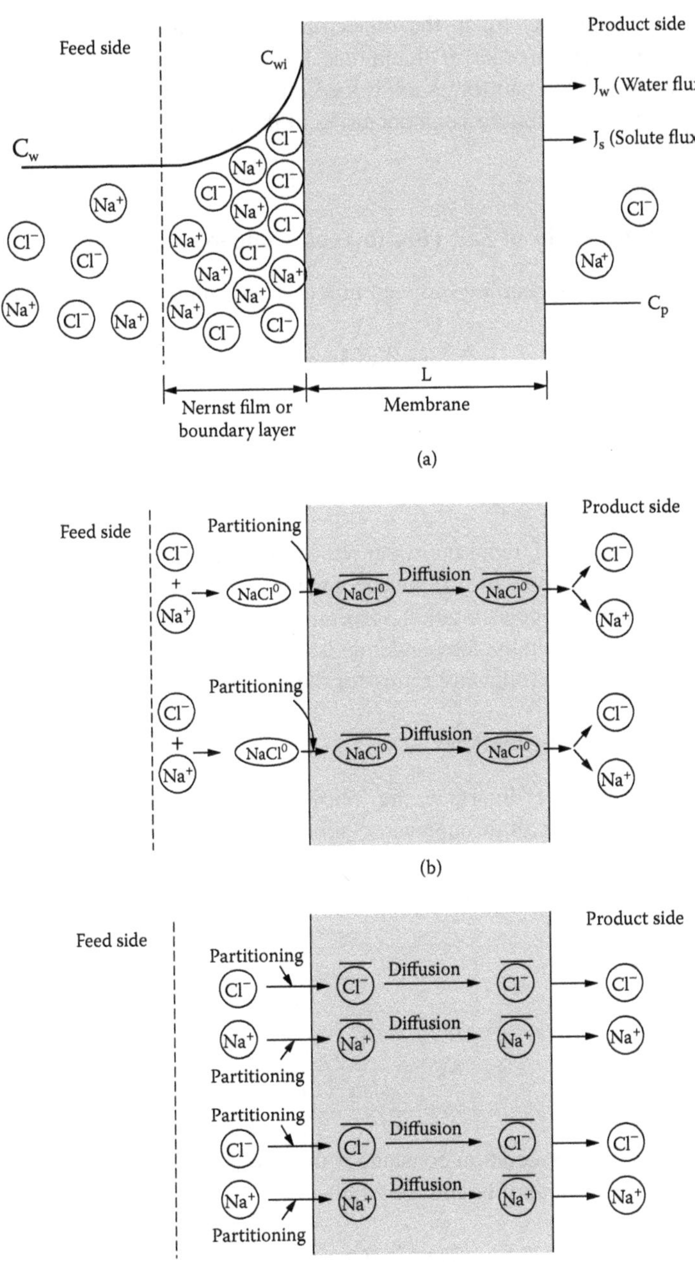

FIGURE 5.1 (a) Concentration polarization near the membrane water interface; (b) diffusion of neutral ion pairs through the membrane; and (c) coupled transport of individual ions through the membrane.

through the membrane. Since the dielectric constant of solvated RO and NF membranes is over an order of magnitude lower than water,[11,21] permeation of electrolyte through ion pair formation is theoretically feasible. Nevertheless, information in the open literature is inconclusive in this regard.

5.2.1 Theory

5.2.1.1 Relationship of Salt Flux to Feed Concentrations

In the absence of any coupling, solvent flux (J_w) and solute flux (J_s) are given by:

$$J_w = W_p\,(\Delta p - \Delta\pi) \tag{5.1}$$

and

$$J_s = k_s\,(\Delta C) = J_w C_p \tag{5.2}$$

where W_p, k_s, $\Delta\pi$, and C_p represent water permeability coefficient, salt permeability constant, and osmotic pressure difference of the two solutions across the membrane and product water–salt concentration, respectively. Considering sodium chloride to be the only electrolyte in the feed under an isothermal operating condition, the salt flux equation takes the following forms for the two proposed mechanisms.

5.2.1.2 Ion-Pair Mechanism

At the membrane–water interface, the following equilibrium leading to the formation of a $NaCl^0$ ion pair through association of Na^+ and Cl^- is operative, that is,

$$Na^+ + Cl^- \qquad NaCl^0 \tag{5.3}$$

Considering ideality,

$$K_a = \frac{[NaCl^0]}{[Na^+]\,[Cl^-]} \tag{5.4}$$

where K_a is the ion association constant at the operating temperature.

From an electroneutrality consideration,

$$[Na^+] = [Cl^-] = C_{Na} \tag{5.5}$$

From Equations (5.4) and (5.5),

$$[NaCl^0] = K_a\,[Na^+]\,[Cl^-] = K_a C_{Na}^2 \tag{5.6}$$

considering a simple one-dimensional description of concentration polarization based on film theory, the ratio of the concentration of an ion (say, Na^+) at the membrane–water interface and that in the feed for low solvent recovery is given by References 12 and 22,

$$\ln\frac{C_{\text{Na}}^{m}-C_{\text{Na}}^{p}}{C_{\text{Na}}^{f}-C_{\text{Na}}^{p}}=\left(\frac{J_{w}}{k_{b}}\right) \tag{5.7}$$

where C_{Na}^{m}, C_{Na}^{f}, and C_{Na}^{p} represent Na^+ concentrations at the membrane–water interface, in the feed and the product sides, respectively, while k_b and J_w denote the liquid-phase mass-transfer coefficient and product water flux. Under the experimental conditions of the investigation as presented later,

$$C_{\text{Na}}^{m} >> C_{\text{Na}}^{p} \text{ and } C_{\text{Na}}^{f} >> C_{\text{Na}}^{p}$$

Thus from Equation (5.7),

$$\frac{C_{\text{Na}}^{m}}{C_{\text{Na}}^{f}}=\exp\left(\frac{J_{w}}{k_{b}}\right) \tag{5.8}$$

For a given hydrodynamic condition and constant product water flux, the parameters k_b and J_w are essentially constant. Under such operating conditions,

$$\frac{C_{\text{Na}}^{m}}{C_{\text{Na}}^{f}}=\text{constant }\left(\text{say, } K^{1}\right) \tag{5.9}$$

The diffusion of ion-pair $NaCl^\circ$ through the membrane of thickness L is given by the following flux equation:

$$J_{\text{NaCl}}=-D^{0}\frac{d\overline{\text{NaCl}^{\circ}}}{dZ}=-D^{\circ}K^{11}\frac{d\text{NaCl}^{\circ}}{dZ} \tag{5.10}$$

where K^{11} is the partitioning coefficient for the $NaCl^\circ$ ion pair and D^0 is the diffusion coefficient of the ion pair through the membrane.

The boundary conditions are:

$$\text{at } Z=0,\ C_{\text{NaCl}^{\circ}}=C_{\text{NaCl}}^{m} \tag{5.11}$$

$$\text{at } Z=L,\ C_{\text{NaCl}^{\circ}}\cong 0 \tag{5.12}$$

Using the equalities in Equation (5.6) and Equation (5.9),

$$C^{m}_{\mathrm{NaCl}} = K_a K^{1} K^{11} (C^{f}_{\mathrm{Na}})^2 \tag{5.13}$$

Integration of Equation (5.10) within the limits and consideration of the equality from Equation (5.13) yields

$$J_{\mathrm{NaCl}} = \frac{D^0 K_a K^1}{L} K^{11} (C^{f}_{\mathrm{Na}})^2 \tag{5.14}$$

$$\text{or } J_{\mathrm{NaCl}} = \text{constant } (C^{f}_{\mathrm{Na}})^2 \tag{5.15}$$

Again from Equation (5.2),

$$J_{\mathrm{NaCl}} = J_w C_p$$

where J_w is the water flux and C_p is the sodium chloride concentration in the permeate side. Thus,

$$J_w C_p = \text{constant } (C^{f}_{\mathrm{Na}})^2 \tag{5.16}$$

By varying feed concentration and transmembrane pressure for a single electrolyte, J_w can be held nearly constant. Then

$$C_p = \text{constant } (C^{f}_{\mathrm{Na}})^2 \tag{5.17}$$

5.2.1.3 Coupled Transport Mechanism

In accordance with this mechanism, the steady-state diffusive flux of NaCl through the membrane is always equal to that of Na^+ or Cl^- and given by

$$J_{\mathrm{NaCl}} = J_{\mathrm{Na}^+} = J_{\mathrm{Cl}^-} = -\overline{D} \frac{d\overline{C}_{\mathrm{Na}}}{dZ} = -\overline{D} K^{111} \frac{dC_{\mathrm{Na}}}{dZ} \tag{5.18}$$

where K^{111} is the partition coefficient for monovalent ions, and $\overline{D}$ is the interdiffusion coefficient of Na^+ and Cl^- and given by

$$\overline{D} = \frac{D_{\mathrm{Na}} D_{\mathrm{Cl}} (C_{\mathrm{Na}} + C_{\mathrm{Cl}})}{D_{\mathrm{Na}} C_{\mathrm{Na}} + D_{\mathrm{Cl}} C_{\mathrm{Cl}}} \tag{5.19}$$

From electroneutrality,

$$C_{\text{Na}} = C_{\text{Cl}} \tag{5.20}$$

Thus,

$$\overline{D} = \frac{2}{\dfrac{1}{D_{\text{Na}}} + \dfrac{1}{D_{\text{Cl}}}} \tag{5.21}$$

The boundary conditions for Equation (5.18) are as follows:

$$\text{at } Z = 0,\ C_{\text{Na}} = C_{\text{Na}}^{m} = K^{1} C_{\text{Na}}^{f} \tag{5.22}$$

$$\text{at } Z = L,\ C_{\text{Na}} \approx 0 \tag{5.23}$$

upon integration,

$$J_{\text{NaCl}} = \frac{\overline{D}}{L} K^{111} C_{\text{Na}}^{m} = \frac{\overline{D}\,K^{1}}{L} K^{111} C_{\text{Na}}^{f} = \text{constant } C_{\text{Na}}^{f} \tag{5.24}$$

Again, under experimental conditions when J_w is kept constant by varying the transmembrane pressures for different feed concentrations, Equation (5.24) becomes

$$C_p = \text{constant } \mathrm{C}_{\text{Na}}^{f} \tag{5.25}$$

Plotted on a logarithm scale, Equation (5.17) and Equation (5.25) from the two proposed mechanisms take the following forms at constant J_w values:

Ion-pair:

$$\log C_p = \log \text{constant} + 2 \log C_{\text{Na}}^{f} \tag{5.26}$$

Coupled transport:

$$\log C_p = \log \text{constant} + \log C_{\text{Na}}^{f} \tag{5.27}$$

The distinction between Equation (5.26) and Equation (5.27) can be readily noted. At constant water flux, the salt concentration in the permeate is linearly dependent on the feed concentration according to the coupled transport mechanism, while the permeate salt concentration varies with the square of the feed concentration according to ion pair mechanism. Thus, C_p vs. C_s^f plot on a log-log scale will yield a slope of 2 according to ion pair mechanism, while the slope will be equal to 1 for the coupled transport mechanism.

5.2.2 Results

Experimental results presented and used in the study were obtained from five independent research investigations, including our own.[11,23–28] We have used the data without any bias to analyze the transmembrane electrolyte transport mechanism. Detailed experimental protocols of these investigations are already available in the open literature[11,23–28] and are not duplicated here.

The mechanism of electrolyte transport through the membrane should be independent of the valences of the constituent cations and anions. That is why experimental data were collected for five different single-component electrolytes with ions of different valences, for example, NaCl (1-1), Na_2SO_4 (1-2), $MgSO_4$ (2-2), $MgCl_2$ (2-1), and $CdSO_4$ (2-2).

Lipp et al.[23] carried out extensive flat-leaf RO test runs using single-component NaCl and Na_2SO_4 electrolytes. A commercially available RO membrane (Filmtec FT 30, Dow Chemical Co.) was used in the study; further details about the membrane are available elsewhere.[24] Figure 5.2 shows log C_p vs. log C_f plots for separate RO runs with NaCl and Na_2SO_4 as the feed; experimental J_w values were fairly close to each other, as shown. Since sulfate is a divalent ion, permeation of Na_2SO_4 is always lower than NaCl under identical conditions. With an increase in feed concentration (C_f) for a constant water flux (J_w), salt concentrations in the permeate increase for both NaCl and Na_2SO_4 in a manner that log C_p vs. log C_f plots always have a slope equal to unity. Superimposed on the figure is a best-fit dashed line for slope equal to 2 for NaCl permeation data. Note that the experimental results are in complete disagreement with the log C_p vs. log C_f plot having a slope equal to 2.

Figure 5.3 shows log C_p vs. log C_f plot for another independent RO test run using NaCl as the feed electrolyte as replotted from Marinas and Selleck.[11] For each test run, the feed concentration was varied and the transmembrane pressures were adjusted to maintain a nearly constant J_w value; experimental protocol and other details have been provided by Marinas and Selleck.[11] Note that log C_p vs. log C_f plot had a slope nearly equal to unity.

Figure 5.4 shows a plot of log C_p vs. log C_f during a NF run by Garba et al.,[25] using $CdSO_4$ as the feed electrolyte. The experimental data were obtained from a test run using a spirally wound membrane (SWM) unit. Note that the slope of C_p vs. C_f plot in log-log scale is essentially equal to unity.

In a similar vein, Figure 5.5 shows log C_p vs. log C_f plot for a flat-leaf NF test run using $MgSO_4$ as the feed electrolyte. Flat-leaf membrane unit was fabricated at Lehigh University and the details of the experimental protocol are available elsewhere.[27] This particular "loose" NF membrane yields low electrolyte rejection but produces high solvent flux at relatively low transmembrane pressures. Samples

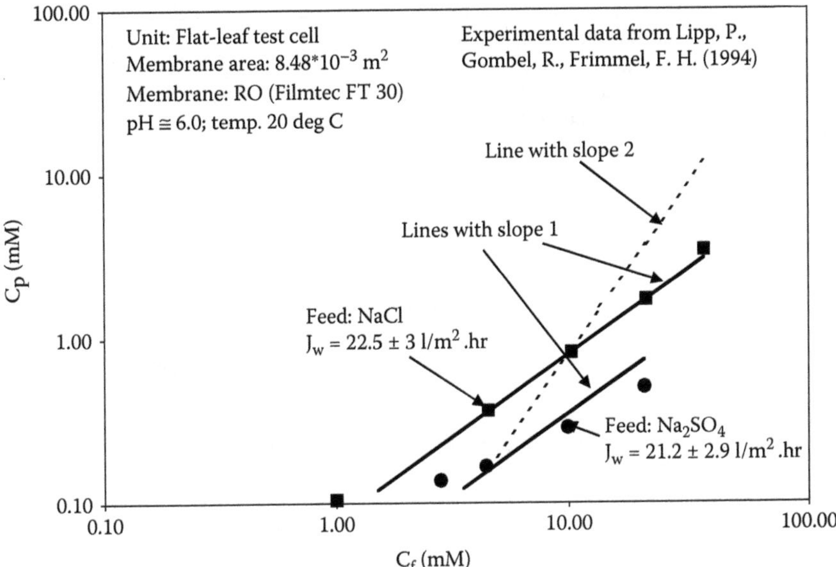

FIGURE 5.2 Plot of log C_p vs. log C_f for two separate reverse osmosis runs using NaCl (1-1 electrolyte) and Na_2SO_4 (1-2 electrolyte) in the feed. Water fluxes were kept nearly constant by altering feed pressures. Solid lines show the best fit lines with slope equal to unity; the dashed line shows the best fit line for NaCl run with a slope equal to 2.

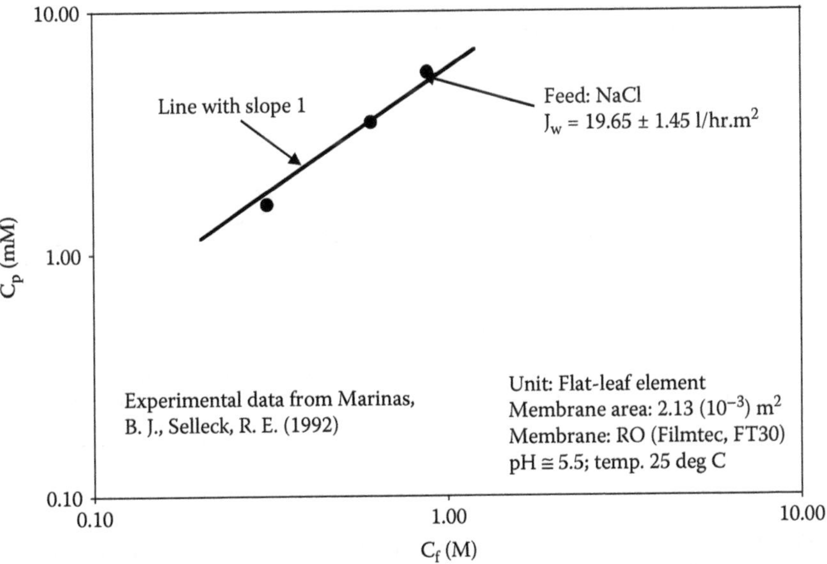

FIGURE 5.3 Plot of log C_p vs. log C_f for a reverse osmosis test run at near constant water flux with NaCl (1-1 electrolyte) in the feed.

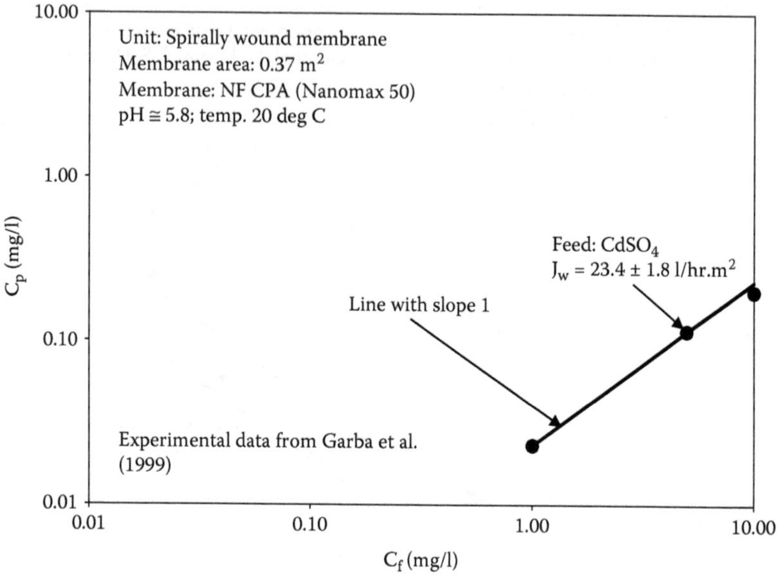

FIGURE 5.4 Plot of log C_p vs. log C_f for a nanofiltration run using $CdSO_4$ (2-2 electrolyte) in the feed.

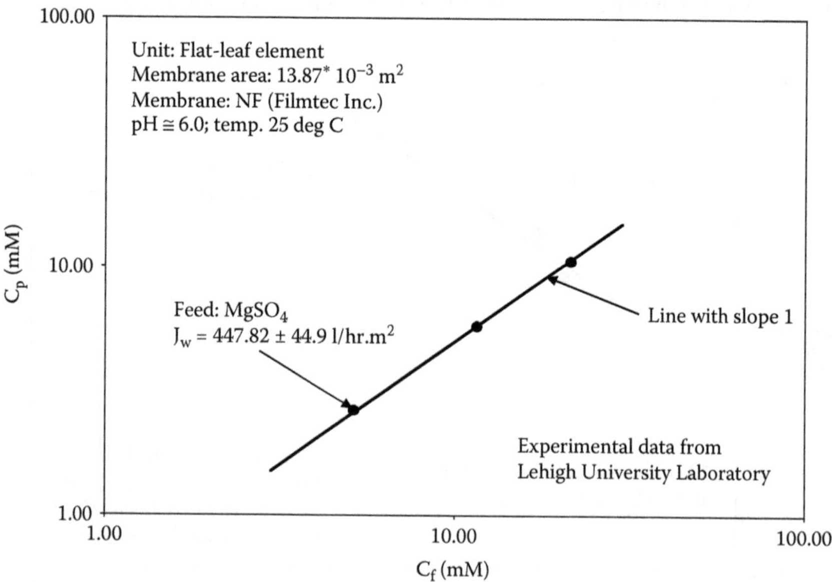

FIGURE 5.5 Plot of log C_p vs. log C_f for a nanofiltration run at near constant water flux with $MgSO_4$ (2-2 electrolyte) in the feed.

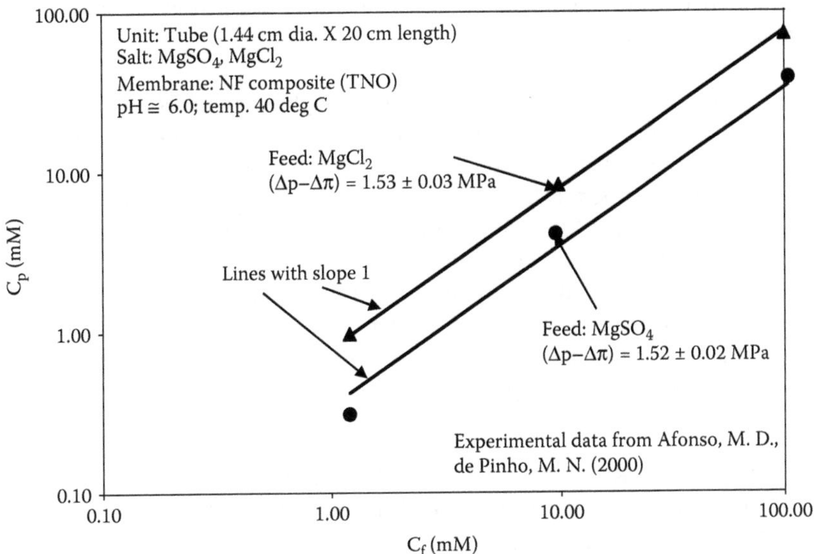

FIGURE 5.6 Plot of log C_p vs. log C_f for separate nanofiltration runs using $MgSO_4$ (2-2 electrolyte) and $MgCl_2$ (2-1 electrolyte) in the feed.

were collected only after J_w values remained constant for 30 min. Again, the slope was close to unity for the log C_p vs. log C_f plot in accordance with Equation (5.26).

Figure 5.6 shows log C_p vs. log C_f plots from a series of flat-leaf test runs carried out using different feed concentrations (C_f) of magnesium chloride and magnesium sulfate from Afonso and de Pinho.[28] For each test run, the feed was changed in a manner that the overall ($\Delta P - \Delta$) remained constant for every feed concentration. Consequently, the solvent flux (J_w) was nearly the same for each run. As expected, the salt passage of $MgCl_2$ (2-1 electrolyte) was always greater than $MgSO_4$ (2-2 electrolyte), all other conditions remaining identical. However, the slopes of log C_p vs. log C_f plots for both $MgCl_2$ and $MgSO_4$ were close to unity, in agreement with the prediction of coupled transport mechanism.

5.2.3 Discussion

Experimental results collected from five independent sources[11,23–28] showed that log C_p vs. log C_f plots at constant solvent or water fluxes always have slopes close to unity for 1-1, 2-1, 1-2, and 2-2 electrolytes. This general observation is in agreement with the proposed mechanism of coupled transport of ions per Equation (5.26). For commercial RO and NF membranes treating water and wastewater, it can be concluded with a high degree of certainty that an ion pair mechanism is not responsible for electrolyte transport across RO and NF membranes.

Some noteworthy ramifications of the findings presented in the study can be summarized as follows:

- Since electrolytes do not permeate through membranes as ion pairs, greater ion pair formation constants of electrolytes do not lead to greater permeability through RO and NF membranes.
- The relative permeability of different electrolytes is governed by their respective interdiffusion coefficients, which in turn are dependent on the hydrated ionic radii of the constituent ions.
- The findings of the study are valid for the commercial RO and NF membranes investigated and similar others. For specialty membranes that are strictly nonpolar and solvent permeation is absent, electrolyte transport through ion pair mechanism is plausible.

5.3 Ion Exchange Selectivity as a Surrogate Indicator of Relative Permeability of Homovalent Ions in Reverse Osmosis Processes

5.3.1 Relationship between Permeation, Interdiffusion Coefficient, and Ion Exchange Selectivity

For RO membranes, a solution-diffusion mechanism has been widely used to characterize the transport of solvent and permeating solute through the active asymmetric membrane layer.[11,12,22,23,29,30] In the absence of any coupling, volumetric solvent flux (J_w) and solute flux (J_s) are given by Equation (5.1) and Equation (5.2), respectively.

For flat-leaf RO test runs used in the present study, the product or permeate flow rate (Q_p) is much lower than the feed rate (Q_f), that is, $Q_p << Q_f$. Figure 5.7a represents a concentration profile near the membrane–water interface. Considering a simple one-dimensional description of concentration polarization based on film theory [Equation (5.7)], the solute permeation has been deduced[11,12,31,32] and is given by Equation (5.28),

$$\frac{C_p}{C_f} = \frac{k_S \exp(J_w / k_b)}{J_w} \tag{5.28}$$

where C_f and k_b are the feed–water salt concentration and the solute back-transport mass-transfer coefficient, respectively. The term $\exp(J_w / k_b)$ is essentially the concentration polarization factor for the solutes governed by the hydrodynamic conditions near the membrane–water interface. Since $C_p << C_f$ or C_w (wastewater salt concentration), all electrolytes essentially undergo concentration polarization to the same extent. Differential permeation of different solutes through RO membranes, as observed universally, is thus governed solely by the difference in the magnitude of k_S in Equation (5.2).

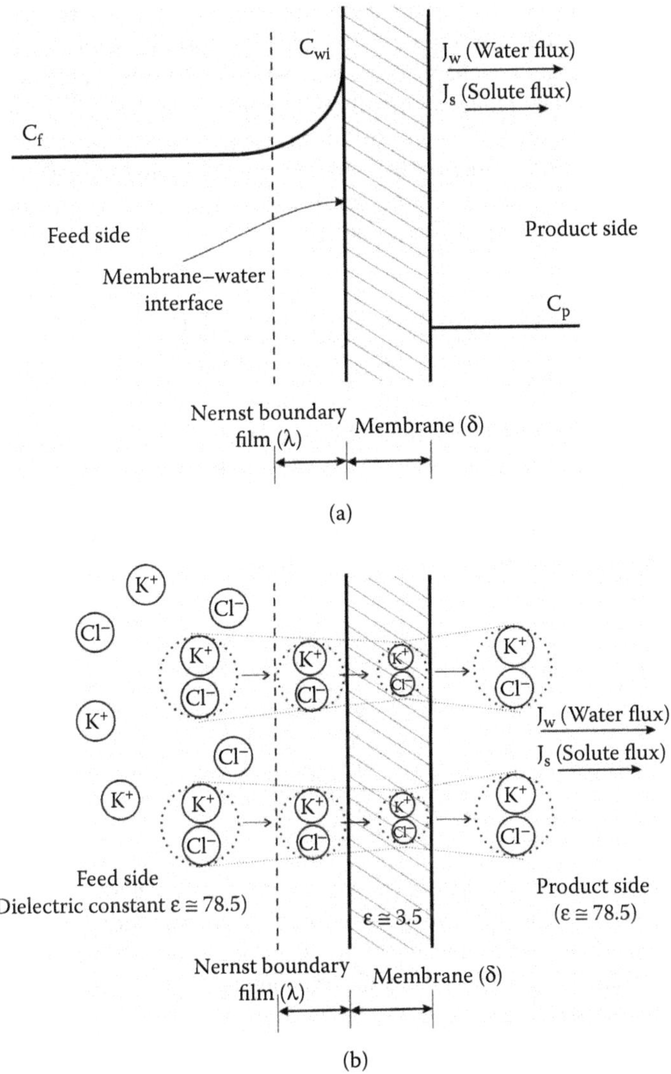

FIGURE 5.7 (a) Concentration profile near membrane–water interface depicting the effect of polarization, and (b) an illustration depicting the gradual shrinkage of K^+ and Cl^- during the coupled transport through RO membrane.

In accordance with the basic premise of the solution-diffusion model, the electrolytes (cations and anions) first get dissolved at the membrane–water interface and then migrate through the active layer toward the product water in a manner that the negative charge of an anion is balanced by the positive charge of the accompanying cation. Note that the dielectric constant monotonically decreases from the bulk solution to the concentration polarization zone and then into the asymmetric active layer of the membrane. Consequently, the effective hydrated ionic radius of the electrically coupled ions (say K^+ and Cl^-) shrinks, as schematically shown in Figure 5.7b. Since ion-solvent interaction is predominantly dependent on the charge of an ion, the ratio of the hydrated ionic radii of two homovalent ions remains virtually unchanged from the bulk solution to the membrane phase. Interdiffusion coefficients of permeating electrolytes are the primary determinants of their relative permeation rates and proportional to k_S in Equation (5.28). For an electrolyte consisting of a cation 1 and an anion 2, the interdiffusion coefficient, D_{12}, is given as

$$D_{12} = \frac{D_1 D_2 (Z_1^2 C_1 + Z_2^2 C_2)}{D_1 Z_1^2 C_1 + D_2 Z_2^2 C_2} \tag{5.29}$$

where D_i is the diffusion coefficient of the ith ion; Z_i is the charge of the ion, and C_i is its molar concentration. Since the relative permeation of monovalent cations and anions is the primary focus of the present study, that is, $Z_1 = Z_2 = 1$, let us consider two electrolytes with the same cation 1 but with anions i and j. The interdiffusion coefficients of these two electrolytes are

$$D_{1i} = \frac{D_1 D_i (C_1 + C_i)}{D_1 C_1 + D_i C_i} \tag{5.30}$$

$$D_{1j} = \frac{D_1 D_j (C_1 + C_j)}{D_1 C_1 + D_j C_j} \tag{5.31}$$

Without any loss of generality and for mathematical simplicity, we may consider a multicomponent feed-water solution where both i and j are trace anions; that is, the high concentration of cation 1 is balanced by other anions present in solutions. Thus, $C_1 >> C_i$ and $C_1 >> C_j$. Equation (5.30) and Equation (5.31) then degenerate into the following:

$$D_{1i} \simeq D_i \tag{5.32}$$

$$D_{1j} \simeq D_j \tag{5.33}$$

Again, for an ion i in a solvent, Stokes-Einstein equation relates its diffusivity to its solvated ionic radius, R_i, as follows[33]:

$$D_i = \frac{k_B T}{6\pi\mu R_i} \tag{5.34}$$

where k_B is Boltzmann's constant and μ is solvent viscosity. Under identical conditions, that is, at the membrane–water interface and within the solvent permeated membrane, the ratio of the diffusivities of ions i and j is inversely proportional to their hydrated ionic radii, that is,

$$\frac{D_i}{D_j} = \frac{R_j}{R_i} \tag{5.35}$$

It is well established that the separation factor (α_{ij}) value for ions of identical valence (e.g., i and j) toward an ion exchanger due to electrostatic interaction varies inversely with their hydrated ionic radii, that is, the greater the value of α_{ij}, the lower the ratio R_i:R_j.[34,35] Large and diverse bodies of experimental data in the open literature attest to the foregoing relationship without any major exception.[34–36] For the subject study, the following semiempirical equation adequately describes the relationship between the ion exchange separation factor α_{ij}, and the ratio of the hydrated ionic radii, R_j:R_i, for all values other than zero:

$$\frac{R_j}{R_i} = \text{constant } (\log \alpha_{ij}) \tag{5.36}$$

Note that the separation factor, α_{ij}, is essentially the thermodynamic equilibrium constant under ideal conditions for the following ion exchange reaction involving anions i and j of identical valence:

$$\overline{Z^+j^-} + i^-(aq) \leftrightarrows \overline{Z^+i^-} + j^-(aq) \tag{5.37}$$

where the overbar denotes the exchanger phase and Z^+ is an anion exchange resin with fixed positive charges. Thus,

$$\alpha_{ij} = \frac{q_i C_j}{q_j C_i} \tag{5.38}$$

where q represents the concentration in the exchanger phase while C is the concentration in the solution. Under identical operating conditions, relative permeation of anions i and j for the case described can be presented as follows using Equation (5.28) to Equation (5.36):

$$\frac{(C_p/C_f)_i}{(C_p/C_f)_j} = \frac{k_{s(i)}}{k_{s(j)}} = \frac{D_{1i}}{D_{1j}} \simeq \frac{D_i}{D_j} \simeq \frac{R_j}{R_i} = \text{constant } (\log \alpha_{ij}) \tag{5.39}$$

Therefore, for ions $i, j, k, m,$ and so forth of identical valence with a selectivity sequence of $\alpha_{ij} > \alpha_{jk} > \alpha_{km}$, the relative sequence of permeability can be predicted as

$$(C_p/C_f)_i > (C_p/C_f)_j > (C_p/C_f)_k > (C_p/C_f)_m \tag{5.40}$$

A priori knowledge of separation factor values can thus characteristically predict the relative permeation sequence of ions of identical valence in RO and NF processes. When i and j are not trace anions, the foregoing mathematical deductions will understandably be complex, but the predicted sequence of relative permeation as shown in Equation (5.39) and Equation (5.40) will remain unchanged; that is, the magnitude of the interdiffusion coefficient will be in the following order: $D_{1i} > D_{1j} > D_{1k} > D_{1m}$. For feed waters with different or multiple cations, the absolute permeating fluxes of anions $i, j, k, m,$ and so on will vary, but the hierarchy of relative permeability will still follow the ion exchange selectivity sequence; that is, the higher the selectivity, the greater the permeability.

The underlying assumptions leading to the deduction of Equation (5.40) are:

- A strongly ionized polymeric exchanger can be viewed as a condensed electrolyte that is characteristically similar to the microenvironment at the membrane–water interface and solvent permeated membrane.
- Ion exchange selectivity is based on Coulombic interaction only.
- Membrane surface charge and the resulting Donnan effect will influence two ions of identical valence equally.
- Any complex formation due to metal–ligand interaction and hydrophobic interactions between ions and membrane surface are absent.
- The polyatomic ions upon hydration tend to be spherical.

Ion exchange selectivity or separation factor values can be determined rapidly and reliably even for unusual species such as perchlorate, selenate, haloacetate, radium, nitrite, and so forth by using ion chromatography or batch isotherm technique.[35,36] The specific objectives of the present study are to: first, experimentally validate that ion exchange selectivity can be an effective tool in predicting the relative permeation of ions in RO processes; second, confirm that the hierarchy or sequence of relative permeation of ions is independent of the type of membrane, pH, feed composition, and dielectric constant of the solvent medium; and third, present convincing evidences that such predictions can also be extended to divalent ions.

5.3.2 Materials and Methods

A flat-leaf (FL) test cell with a membrane area 9.5 cm × 14.6 cm was fabricated and used to perform RO and NF experiments. The unit was operated from 100 psi to 400 psi transmembrane pressure using three different types of commercial membranes. Table 5.1 summarizes salient information about these membranes. The flow sheet of the experimental setup and a cross-section of membrane cell are shown in Figure 5.8a and Figure 5.8b, respectively. The flat-leaf unit, the

TABLE 5.1
Properties of BW30, DS-CA, and NF90 Membrane

Membrane	Manufacturer	Type	Flux (Cm/S)
BW30	Filmtec Inc., Minneapolis, MN	CPA (RO)	0.001[a]
DS-CA	Osmonics Desal Inc., Vista, CA	CA (RO)	0.0007[a]
NF90	Filmtec Inc.	CPA (NF)	0.006[b]

CPA: composite polyamide; CA: cellulose acetate.
[a] 5000 mg/1 of KNO_3, 200 psi transmembrane pressure;
[b] 1000 mg/1 of $MgSO_4$, 100 psi transmembrane pressure.

high-pressure pump, and the tubings were all made of SS-316 to avoid corrosion. New membranes were adequately conditioned by running the test cell for a prolonged period of time (hours). Samples of feed, product, and reject streams were taken only under steady-state conditions. Conductivity and pH of each sample were measured immediately after collection. Since the volume of the reservoir (35.0 liters) was sufficiently large compared to the rest of the system volume (less than 0.5 liter), temperature of all the test runs remained between 22 and 26°C. We closely examined the stability and durability of the membranes; no significant change in flux or rejection was observed after several hours of testing.

The anions were analyzed using an ion chromatograph (Dionex Inc. DX 120). Monovalent cations were analyzed in the ion chromatograph (DX 120) and their concentrations were also often cross-checked using the atomic absorption (AA) spectrometer (Perkin Elmer AA 100). Divalent cations were always analyzed with the AA.

5.3.3 Results

5.3.3.1 Monovalent Ions

Figure 5.9 shows an ion chromatograph of several monovalent anions using HCO_3^-/CO_3^2 eluent (Dionex Model 120). Since elution at a later time represents higher sorption affinity, the selectivity sequence of monovalent anions is $NO_3^- > Br^- > NO_2^- > Cl^- > Ac^- > F^-$. Separation factor, α_{i/Cl^-}, of an ion i with respect to a reference species, say chloride, is given by

$$\alpha_{i/Cl^-} = \frac{t_i - t_o}{t_{Cl^-} - -t_o} \tag{5.41}$$

where t_o corresponds to the elution time of water. Figure 5.10 shows a plot of relative elution time $(t_i - t_o)$ vs. separation factor with respect to chloride; a linear relationship is noted.

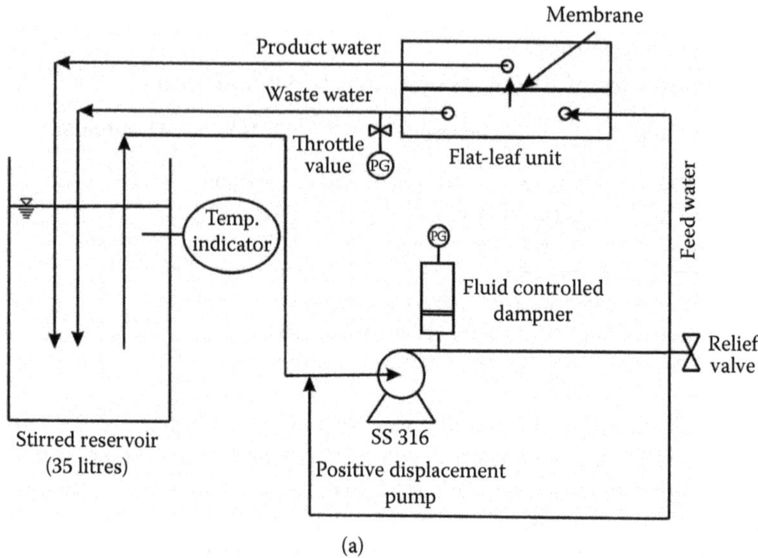

(a)

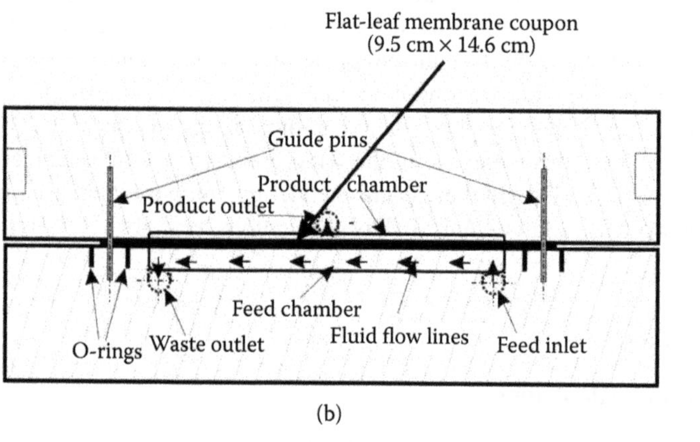

(b)

FIGURE 5.8 (a) Flow sheet of the experimental setup, and (b) cross-sectional view of the flat-leaf unit.

A RO run was carried out with a feed solution containing three monovalent anions, namely, fluoride, bromide, and nitrate. Sodium was the conjugate cation and the exact feed composition is provided in Figure 5.11. Under the hydrodynamic conditions within the flat-leaf test cell, it may be assumed that the solution phase was uniformly mixed, that is, that a continuous stirred-tank condition prevailed. Figure 5.11 shows the percentage relative permeation of fluoride, bromide, and nitrate for three transmembrane pressures, and they are significantly different from one another. However, note that in accordance with the prediction of Equation (5.40), the sequence of relative permeability follows the order of ion exchange selectivity at all transmembrane pressures; that is, (C_p/C_f) is the greatest

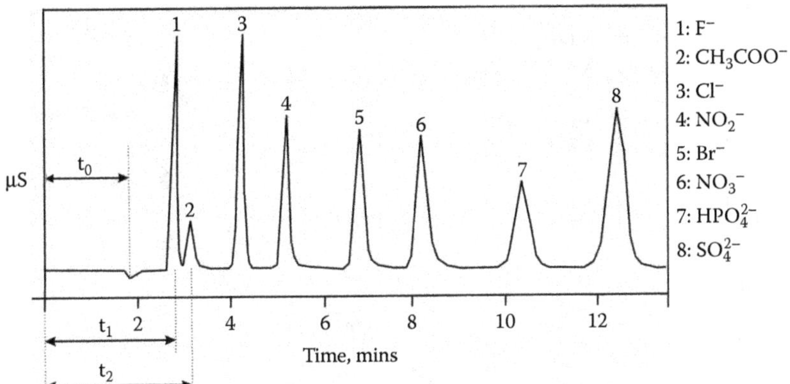

FIGURE 5.9 Ion chromatograph of monovalent anions; t_0: elution time of water; t_i: elution time of ion i.

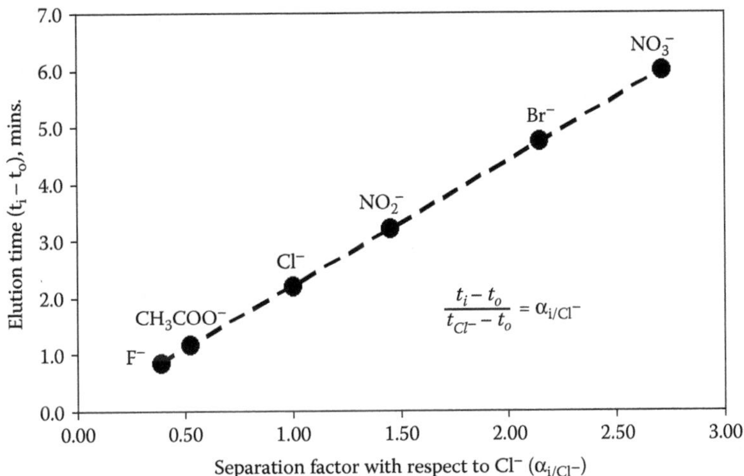

FIGURE 5.10 Relative elution time vs. separation factor with respect to chloride.

for nitrate, followed by bromide and then fluoride. The water permeability coefficient (W_p) at three different transmembrane pressures was nearly the same and equal to 2.3 to $2.4E^{-7}$ $\frac{\text{cm}}{\text{sec KPa}}$.

Figure 5.12 shows percentage permeation rates of chloroacetate, chloride, and nitrate from a similar RO flat-leaf test run. At the inset, an ion chromatograph of these three anions is also provided. Chloroacetate or the parent chloroacetic acid ($CH_2ClCOOH$, $pK_a = 2.7$) is the most commonly encountered haloacetic acid formed following prechlorination of surface water containing natural organic matter. Since chloroacetate is a potential carcinogen, its removal by subsequent treatment processes is of significant interest. Note that the percent permeation of

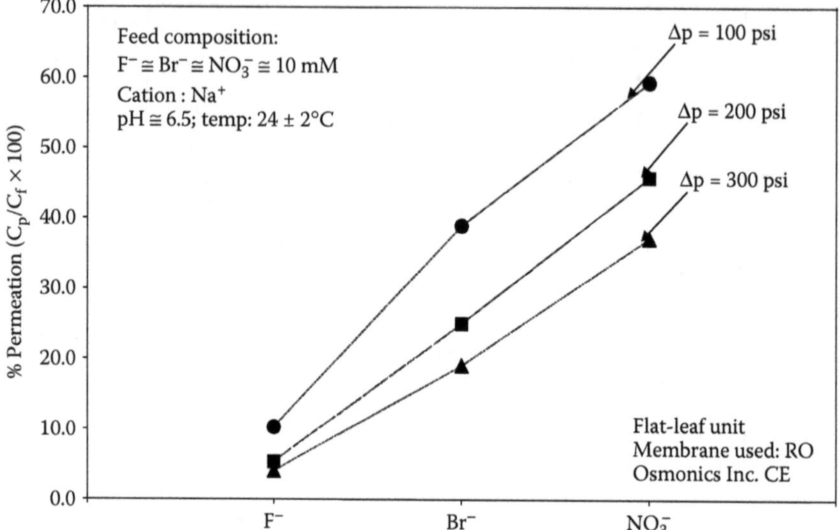

FIGURE 5.11 Comparison of percentage permeation of fluoride, bromide, and nitrate at three different transmembrane pressures.

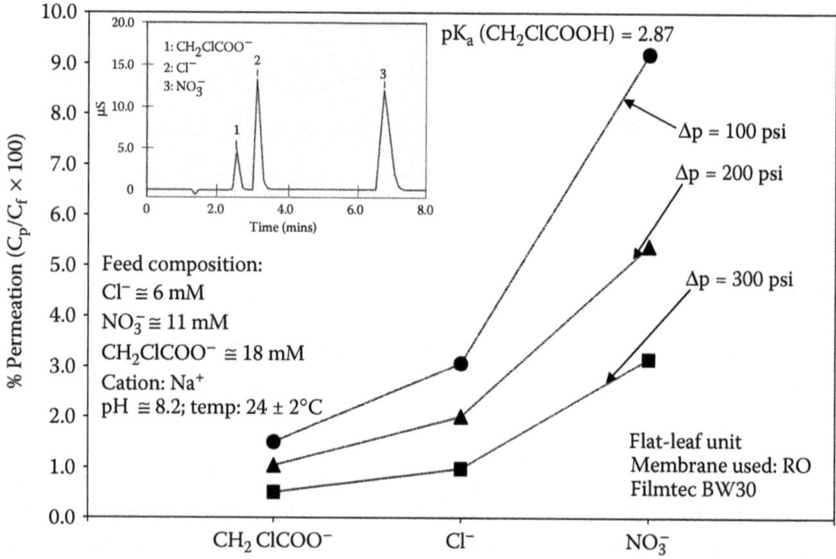

FIGURE 5.12 Comparison of percentage permeation of chloroacetate, chloride, and nitrate at three different transmembrane pressures; inset shows ion chromatograph of anions.

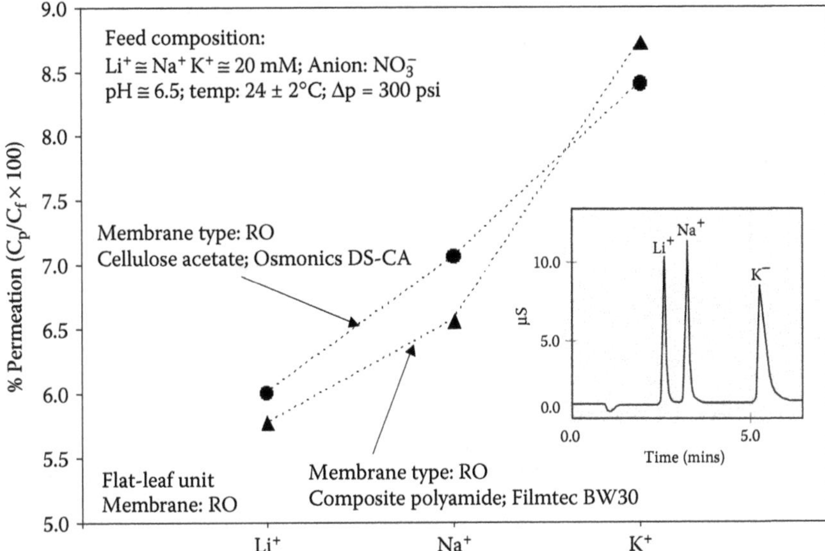

FIGURE 5.13 Comparison of percentage permeation of lithium, sodium, and potassium through two different RO membranes under otherwise identical conditions; inset shows ion chromatograph of cations.

chloroacetate is lower compared with both chloride and nitrate; this observation is in agreement with the relative order of their ion exchange selectivity, that is, $CH_2C1COO^- < C1^- < NO_3^-$.

Figure 5.13 shows the percentage permeations of Li^+, Na^+, and K^+ during flat-leaf runs using two RO membranes, namely, polyamide BW 30 (Filmtec Inc.) and cellulose acetate CE (Osmonics Inc.). For both runs, nitrate was the conjugate anion. Although the percent permeations for each ion are different for two membranes, the descending order of permeability (that is, $K^+ > Na^+ > Li^+$) is identical for both the membranes and follows the same order with respect to ion exchange selectivity. The chemical makeup of the RO membranes (that is, cellulose acetate or polyamide) influenced the overall solvent flux and salt permeability but the relative permeability of two ions is predictable from their ion exchange separation factor.

5.3.3.2 Effects of pH, Solvent Dielectric Constant, and Conjugate Ion

It is well recorded that weak-acid surface charges are inevitably present in most of the commercially available RO membranes.[11,23,37,38] With an increase in pH, the membrane surfaces get increasingly negatively charged. As a result, the anions get rejected by the Donnan co-ion exclusion effect, which in turn causes rejection of cations for maintenance of electroneutrality and leads to lower overall permeability at an elevated pH. Flat-leaf RO test runs were carried out

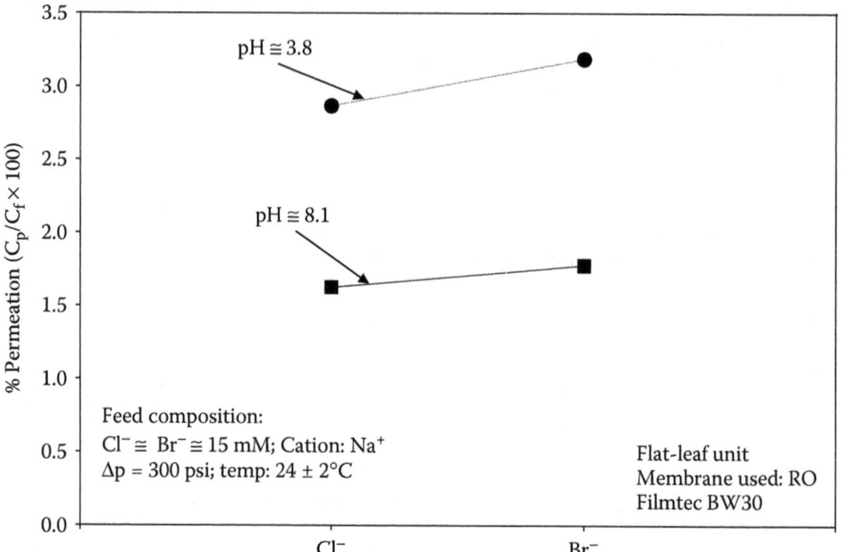

FIGURE 5.14 Relative percentage permeation of chloride and bromide at two different pH levels for a RO membrane with weak acid functional groups.

at two different pHs, namely, 3.5 and 8.5 under otherwise identical conditions; the RO membrane used in the study was Filmtec BW 30. Previous studies have confirmed that BW 30 has carboxylate groups that become negatively charged at around neutral pH.[11] Figure 5.14 shows the percentage relative permeation of Cl^- and Br^- at two different pH levels. It may be readily noted that the permeations of all the ions were significantly reduced at alkaline pH due to the Donnan exclusion effect. However, the order of relative permeability between Cl^- and Br^- remained unchanged with changes in pH, in accordance with the predictions of Equation (5.40).

Figure 5.15a shows the relative permeation rates Li^+, Na^+, and K^+ with solvents at two different dielectric constants, namely, water (ε = 80) and an ethanol–water mixture (ε = 62) for a transmembrane pressure of 200 psi. With a lowering of dielectric constant, the hydrated ionic radii of ions shrink, resulting in increased interdiffusion coefficient values. Consequently, the permeation increases for every ion under identical operating conditions as validated by the results of Figure 5.15a. However, the sequence of relative permeation, that is, $K^+ > Na^+ > Li^+$, remains unchanged with a change in the dielectric constant of the solvent medium. In order to assess the quantitative effect of solvent dielectric constant on the permeation of individual monovalent cation, the ratio of the relative permeation of Li^+, Na^+, and K^+ at two different dielectric constants were plotted for several transmembrane pressures in Figure 5.15b. Almost identical effects of dielectric constants were observed for all three cations.

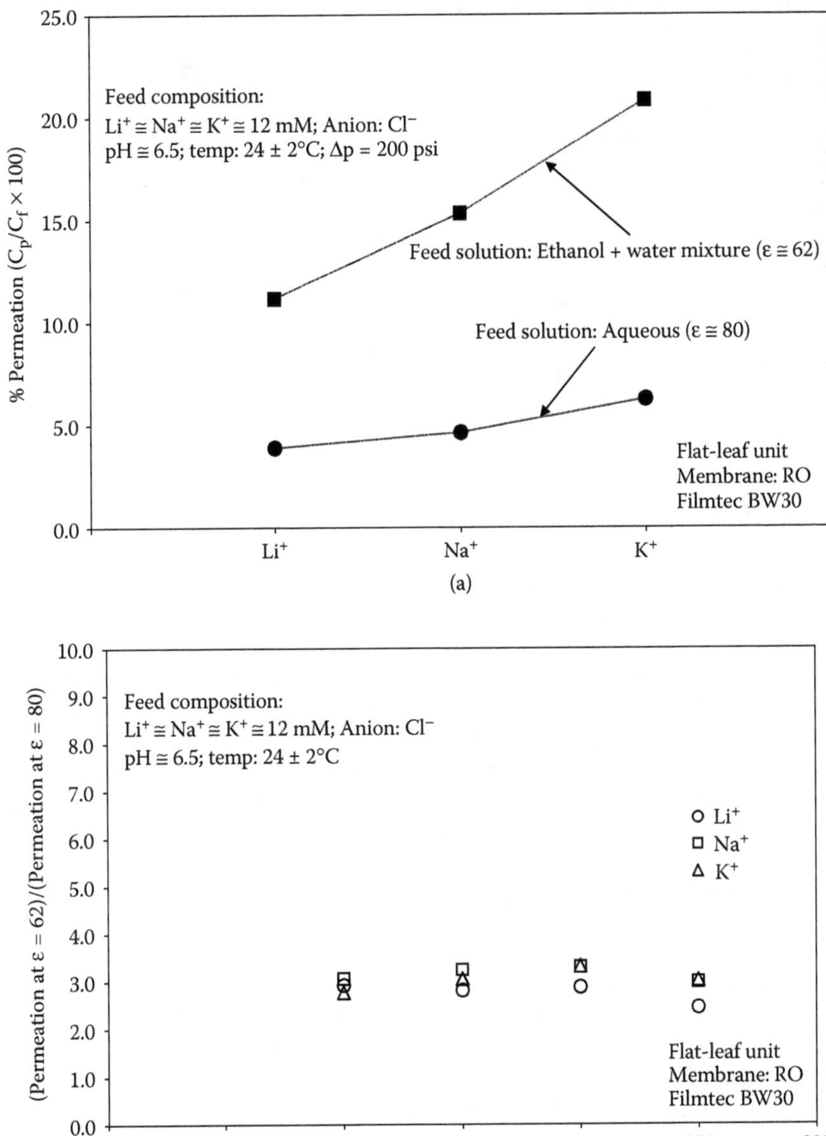

FIGURE 5.15 (a) Comparison of percentage permeation of lithium, sodium, and potassium in two solvents with different dielectric constants, and (b) ratio of the relative permeation of lithium, sodium, and potassium at two different dielectric constants vs. transmembrane pressure.

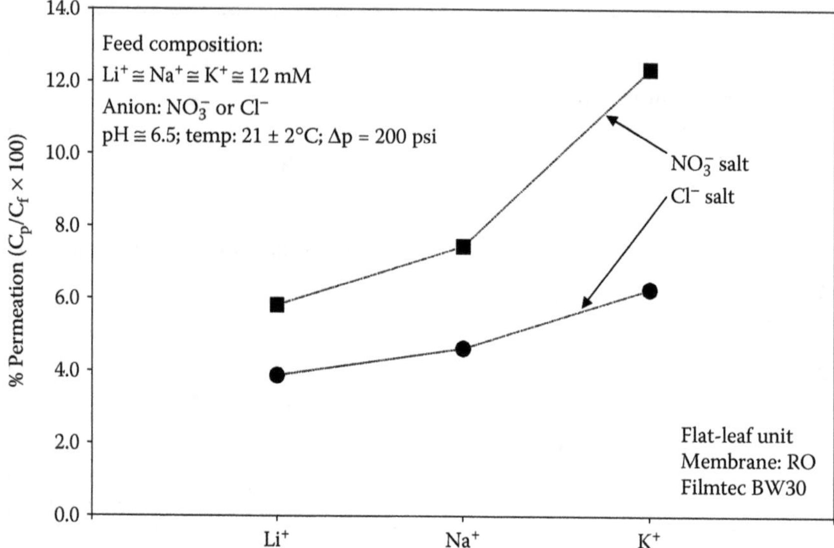

FIGURE 5.16 Comparison of percentage permeation of lithium, sodium, and potassium with two different conjugate ions, namely chloride and nitrate.

Figure 5.16 shows permeations of Li^+, Na^+, and K^+ for two runs using two different conjugate anions, namely, Cl^- and NO_3^-. From the ion exchange selectivity data presented earlier, nitrate is more permeable compared to chloride. Experimental results in Figure 5.16 illustrate that, first, the changeover from Cl^- to NO_3^- increases the permeability of all the cations; and second, the relative permeability sequence for Li^+, Na^+, and K^+ remains unchanged independent of the conjugate anion.

5.3.3.3 Nanofiltration Membrane

Nanofiltration (NF) membranes have recently found major acceptance in the industrial community for the low transmembrane pressure requirement in efficient rejection of divalent ions. Similar to RO, solvent and solute permeation through NF membranes can be explained using solution diffusion or pore-transport models.[9,12] Figure 5.17 shows the relative permeation of divalent sulfate, selenate, and phosphate through a nanofiltration membrane. The conjugate cation was sodium. Note that selenate or Se(VI), which is regulated in drinking water, has higher permeability than phosphate and sulfate. Again, the sequence of relative permeability is in agreement with the ion exchange selectivity as shown in the ion chromatograph, that is, $SeO_4^{2-} > SO_4^{2-} > HPO_4^{2-}$, at the inset of Figure 5.17. It is worth noting that Marinas and Selleck[11] studied selenate rejection in the presence of sulfate using reverse osmosis; selenate permeation was found to be consistently greater than sulfate. Hydrated ionic radii or equivalent conductance data in the open literature[13,14] are unable to predict greater selenate permeation compared to sulfate in semipermeable membrane processes.

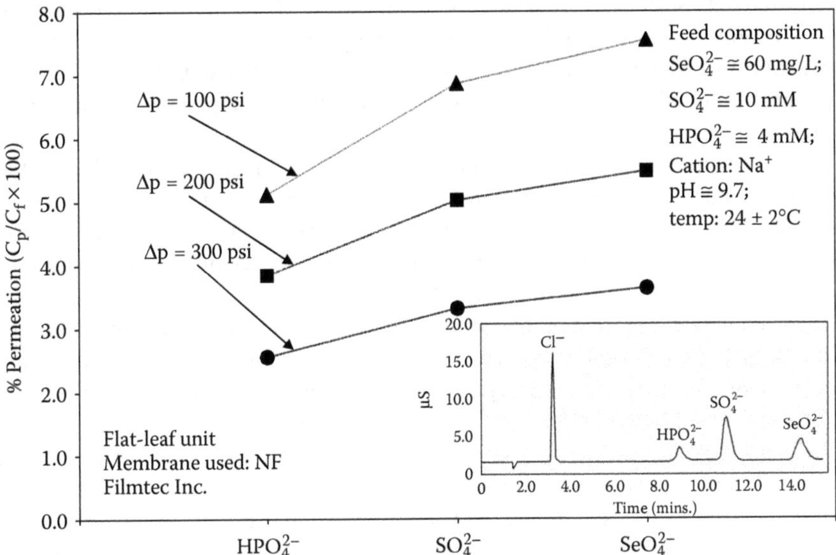

FIGURE 5.17 Comparison of percentage permeation of phosphate, sulfate, and selenate for a NF membrane at three different transmembrane pressures; inset shows the ion chromatograph of anions.

5.3.4 Discussion

Relative permeability of ions of identical valence in RO and NF processes, although of significant interest from an environmental separation viewpoint, cannot be predicted using existing models based on the Nernst-Planck equation with or without incorporation of the Donnan potential. Experimental results presented in this study show a characteristic correlation between ion exchange selectivity and relative permeation; that is, a higher ion exchange selectivity always leads to a greater permeability. A large body of experimental data in the open literature,[2,5,8,10,11,23,39] when carefully reviewed, also validates the scientific premise of the present study. Experimental results presented in Figure 5.14 through Figure 5.16 indicate that the solvent dielectric constant, pH, and the type of co-ion influence the absolute permeabilities of ions. However, the sequence of relative permeation of various ions still follows their ion exchange selectivity order. Theoretically, higher ion exchange selectivity for a specific ion represents a smaller hydrated ionic radius, which in turn leads to a greater interdiffusion coefficient or salt permeability coefficient.

The solvent dielectric constant (ε) around an ion in RO and NF processes changes as it passes from the bulk to the membrane–water interface and finally through the asymmetric membrane layer, as depicted in Figure 5.7b. As a result, the solvated ionic radius decreases and hence, the effective interdiffusion coefficient increases at different segments of the membrane process. Figure 5.15b illustrates that the relative permeation of homovalent ions, as predicted from the ion exchange selectivity data, will be independent of the dielectric constants of the solvent-mediated membrane. Theoretically, the use of the Stokes-Einstein equation provides an underlying reason about the relationship between relative permeabilities of ions and their ion exchange selectivities as presented in Equation (5.34) through Equation (5.39).

In principle, equivalent conductance or ionic mobility data available in the open literature for various ions can be used to compute the interdiffusion coefficients of permeating electrolytes. Ratanatamskul et al.[8] attempted to predict the differential permeation between nitrate and chloride for nanofiltration membranes using ionic mobility data. However, contrary to experimental results, the model predicted greater chloride permeability than nitrate. It is recognized that the availability of hydrated ionic radii data may aid in computing relative interdiffusion coefficients for different electrolytes. However, obtaining reliable hydrated ionic radii data from the open literature is questionable especially for polyatomic ions. Recently, Lee and Lueptow[10] extensively investigated rejection or permeation of nitrogen compounds for RO membranes. The authors reported hydrated ionic radii of both nitrite (NO_2^-) and nitrate (NO_3^-) to be equal to 0.11 nm. The interdiffusion coefficients of their sodium salts in water were reported as 15.7×10^{-10} m^2/s. Based on their equal hydrated ionic radii data and interdiffusion coefficient values, permeation of nitrite and nitrate is expected to be nearly equal. However, according to the RO run data presented, nitrate permeation was significantly greater than nitrite. An ion exchange selectivity approach, as presented in this study, can, however, correctly predict the higher relative permeability of nitrate with respect to nitrite, as evidenced from the ion chromatogram in Figure 5.9. It is postulated that the model predictions presented will attain greater precision if hydrated ionic radii data are refined using ion exchange selectivity. Contrary to experimental observations, the ionic diffusivities of nitrate and perchlorate reported in the open literature[14,33] are lower than chloride. These discrepancies raise concerns about the accuracy of the hydrated ionic radii data and ionic diffusivities of especially polyatomic ions reported in the open literature.

Central to this study is the underlying scientific premise that ion exchange selectivity can successfully predict the relative permeability of cations and anions in RO and NF processes. The use of ion exchange selectivity as an indicator of relative permeability of different ions through semipermeable membranes is meaningful only if the same can be determined rapidly and reliably through easy-to-perform experiments. Use of an ion chromatograph with a conductivity detector is an elegant technique in this regard, and ion chromatograms have been used in the present study to compute relative selectivity. Ion exchange selectivity of ionic species, including environmentally significant ones, may also be conveniently determined by carrying out conventional batch isotherm experiments with strong polymeric cation and anion exchangers. To validate the same, ion exchange selectivities of various monovalent anions were determined by a standard batch isotherm test[35] and the relative selectivity (α_{ij}) between ions i and j were determined using Equation (5.38) following mass balance calculations. Figure 5.18

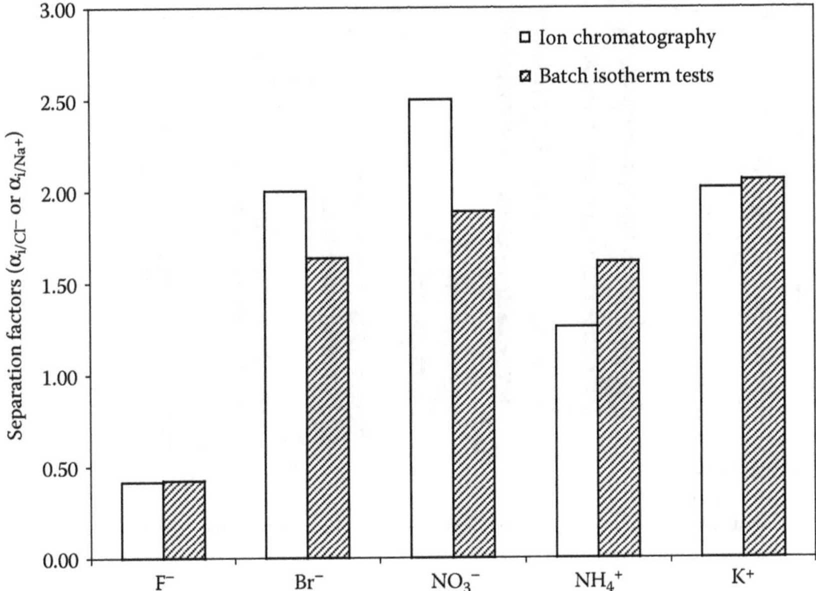

FIGURE 5.18 Separation factors of various anions and cations with respect to chloride and sodium respectively, using ion chromatography and batch isotherm tests.

shows the separation factor values of various anions and cations with respect to Cl^- and Na^+ respectively, determined independently using both ion chromatography and batch isotherm experiments. Note that a good agreement is observed between the two. Ion exchange selectivity of any exotic cation or anion can be readily determined using a batch isotherm technique in the absence of an ion chromatography instrument.

5.4 HIERARCHY OF PERMEATION OF IONS AMONG MULTIVALENT SPECIES

It is well known currently that the lower valent ions permeate more through the membrane than the higher valent ions. According to the permeation model for homovalent ions proposed in previous section, the higher the ion exchange selectivity (or the lower the hydrated ionic radius) for an ion, the higher the permeation for that ion through the membrane. Ion exchange selectivity can be estimated by noting the time of elution in an ion chromatogram; the higher the time of elution, the higher the ion exchange selectivity among homovalent ions.

An experiment was performed in the environmental laboratory of Lehigh University with four different cations, Mg^{2+}, Ba^{2+}, Na^+, and K^+, and two anions, Cl^- and NO_3^-, in the feed solution and their permeation was observed through a NF (Filmtec Inc.) membrane. The materials and methods are the same as that

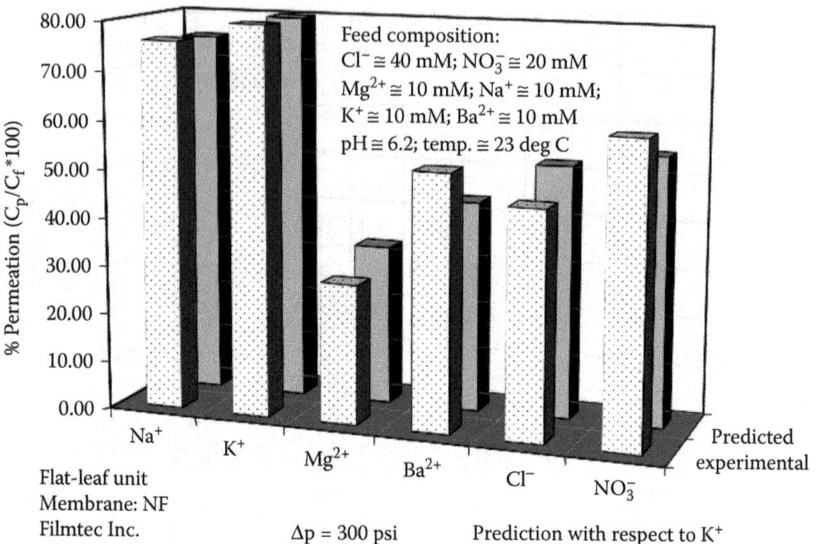

FIGURE 5.19 Experimental and predicted results for percent permeation of Na, K^+, Mg^{2+}, Ba^{2+}, Cl, and NO_3 through NF membrane.

given in Section 5.3.2. The results of the experiment are shown in Figure 5.19. It can be observed from Figure 5.19 that for all transmembrane pressures for which the experiment was carried out, NO_3^- permeated higher than Cl^- between the anions, and the monovalent cations permeate higher than the divalent cations. K^+ permeates more than Na^+ between the two monovalent cations, and between the two divalent cations Ba^{2+} permeates higher than Mg^{2+}.

$$\text{Anions:} \quad NO_3^- > Cl^-$$

$$\text{Cations:} \quad K^+ > Na^+ > Ba^{2+} > Mg^{2+}$$

The order of permeation of all ions is consistent with the predictions from the model. The monovalent cations were expected to permeate more than their divalent counterparts. Observing the cation chromatogram (Figure 5.20), it can be noted that Ba^{2+} has a higher time of elution (and therefore higher selectivity) than Mg^{2+}, which means Ba^{2+} has a smaller hydrated ionic radii than Mg^{2+} and therefore permeates higher through the membrane. K^+ has a higher elution time than Na^+ and therefore permeates higher than Na^+. The anion chromatograph (Figure 5.9) shows that NO_3^- has a longer elution time than Cl^-; that is, NO_3^- has a smaller hydrated ionic radius than Cl^- and, therefore, NO_3^- is more permeable than Cl^-. When myriad ions are present for permeation through a RO or NF

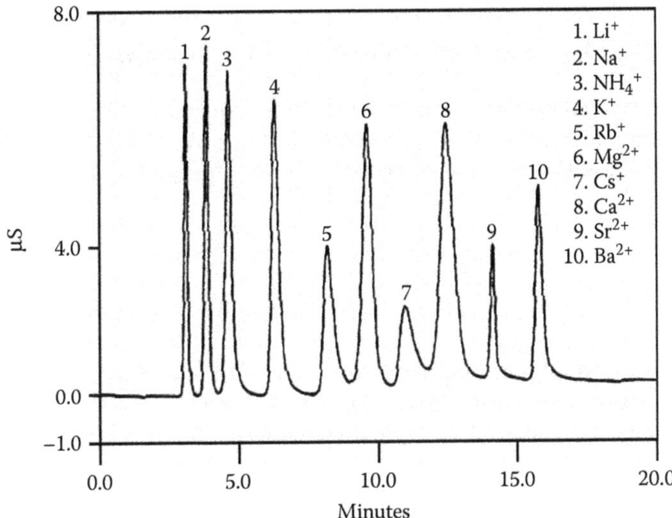

FIGURE 5.20 Cation chromatograph of common cations.

membrane under an applied transmembrane pressure, monovalent anions and cations will permeate more than higher valent ions. Among homovalent ions, the cation with the smallest hydrated ionic radii (longest elution time in the ion chromatograph) will pair up with the anion with the smallest hydrated ionic radius and permeate through the membrane while maintaining electroneutrality.

5.5 SPECIAL CASE: PERMEATION BEHAVIOR OF TRACE SPECIES THROUGH MEMBRANES

In recent years, it has been observed that a number of contaminants in water pose a significant health risk at trace concentrations, many of them ionic in nature, for example, perchlorate (ClO_4^-), nitrate (NO_3^-), arsenic, fluoride (F^-), haloacetic acids, and so forth. Reverse osmosis (RO) and nanofiltration (NF) have emerged as major treatment techniques in water industries. An increase in permeation of some of these regulated species at trace concentrations will have serious consequences. The goal of this part of the study is to investigate and compare the permeation behaviors of four different ionic contaminants, namely, perchlorate (ClO_4^-), nitrate (NO_3^-), fluoride (F^-), and monochloroacetate (CH_2ClCOO^-), through RO and NF membranes when they are present as trace as well as bulk species in the feed water. The anion chromatogram (Figure 5.9) shows that perchlorate (ClO_4^-) and nitrate (NO_3^-) have relatively high elution times (higher diffusivity) while fluoride (F^-) and monochloroacetate (CH_2ClCOO^-) have shorter elution times (lower diffusivity) among monovalent ions. These four different monovalent ions were chosen for the trace species study due to their environmental significance.

5.5.1 Theoretical Approach

5.5.1.1 Calculation of k_s from Salt Permeation Experiments

Combining the solvent flux and the solute flux equations [Equations (5.1) and (5.2)] of the solution-diffusion model along with the film theory model [Equation (5.7)], the salt permeation coefficient (k_s) can be expressed by Equation (5.42).

$$k_s = \frac{C_p J_w}{C_f} \times \frac{1}{\exp\left(J_w / k_b\right)} \tag{5.42}$$

where k_b is the liquid-phase diffusion coefficient. The term $\exp(J_w/k_b)$ is a measure of concentration polarization factor. All our experiments were performed with a flat-leaf unit setup, where $C_f \approx C_w$ and the concentration polarization factor can be assumed to be negligible ($\approx$ 1).

5.5.1.2 Interdiffusion Coefficient of a 1-1 Single Electrolyte Permeating through the Membrane

When a strong electrolyte (dissociated anions and cations) diffuses through a semipermeable membrane, an anion must pair up with a cation electrovalently to maintain electroneutrality to transport from the high-pressure side to the low-pressure side. Thus the interdiffusion coefficient of the anion-cation pair dictates the permeation of the electrolyte through the membrane. The interdiffusion coefficient of an anion and cation pair is given by Equation (5.29).

For a single component of such an anion–cation [(A)–(C)] pair, where $C_A = C_C$, 1-1 ($Z_A = Z_C = 1$) electrolyte, Equation (5.29) reduces to Equation (5.43).

$$D_{AC} = \frac{2}{\dfrac{1}{D_A} + \dfrac{1}{D_C}} \tag{5.43}$$

Equation (5.43) gives the interdiffusion coefficient (D_{AC}) of a single component 1-1 electrolyte, which is the harmonic mean of the individual diffusion coefficients of the anion (D_A) and the cation (D_C). This means that the value of the interdiffusion coefficient is dominated by the diffusion coefficient of the slower ion.

5.5.1.3 Interdiffusion Coefficient of a Multicomponent 1-1 Electrolyte with Trace Species

For a multicomponent solution — for example, when a trace anion species diffuses along with a bulk cation species — that is, $C_A/C_C << 0.01$ for a 1-1 anion-cation system, Equation (5.29) reduces to Equation (5.44).

$$D_{AC} = \frac{D_A D_C \left(\frac{C_A}{C_C} + 1 \right)}{D_A \left(\frac{C_A}{C_C} \right) + D_C} \approx \frac{D_A D_C}{D_C} = D_A \tag{5.44}$$

It is observed from Equation (5.44) that the permeation of a trace species through the membrane depends on its individual diffusion coefficient rather than the interdiffusion coefficient with the associated cation.

5.5.1.4 Salt Permeability Coefficient, k_s

The salt permeability coefficient k_s, which dictates the permeability of an electrolyte through a membrane, is a constant for the membrane type and the electrolyte permeating through the membrane, and described in Equation (5.45).

$$k_s = \frac{D}{\delta} \tag{5.45}$$

where D is the interdiffusion coefficient of the anion and cation of the permeating electrolyte and δ is the thickness of the membrane. Equation (5.45) shows that for a particular type of membrane where the membrane thickness (δ) is constant, the salt permeability coefficient (k_s) can give a measure for the interdiffusion coefficient of the permeating electrolyte. Therefore, according to the aforementioned theoretical discussion from Equation (5.43), for an electrolyte when it is permeating as a bulk species, k_s could be expressed by Equation (5.46).

$$k_s = \frac{D_{AC}}{\delta} = \frac{1}{\delta} \times \left(\frac{2}{\frac{1}{D_A} + \frac{1}{D_C}} \right) \tag{5.46}$$

However, when one of the components (either the anion or the cation) of the same electrolyte is trace compared to the other, then according to Equation (5.44), k_s is expressed by Equation (5.47), when the experimental conditions are unchanged.

$$k_s = \frac{D_{AC}}{\delta} = \frac{D_{trace}}{\delta} \tag{5.47}$$

The ions studied in this paper are monovalent anions: (ClO_4^-), (NO_3^-), (F^-), and (CH_2ClCOO^-). The associated cation was either Na^+ or K^+. It is assumed that the hydrated ionic radii of K^+ and Cl^- are almost identical[40,41] and hence the diffusion coefficient of K^+ is that equal to Cl^- ($D_{K+} \approx D_{Cl-}$). Out of the four anions chosen for this study, F^- and CH_2ClCOO^- have a much lower diffusion coefficient than that of Cl^-, and the other two anions (NO_3^- and ClO_4^-) have a much higher diffusion coefficient than that of Cl^-. Figure 5.9 shows the anion chromatograph of some common anions.

According to the theory presented above, when F^- and CH_2ClCOO^- are found as trace species in the water supply, they would permeate to a lesser extent than they would as a bulk species. But for the ions, NO_3^- and ClO_4^-, which have high individual diffusion coefficient, trace conditions would make them permeate more than they would have if present in bulk.

5.5.2 Experimental Methods

5.5.2.1 Salt Permeation Experiments

Salt permeation experiments were performed with the SEPA CF II cell (GE Osmonics, Minnetonka, Minnesota) in the environmental engineering laboratory at California Polytechnic State University. This unit uses flat-leaf membranes with an effective area of 14.6 cm × 9.5 cm. The unit was set up in a closed-loop membrane operation process. A similar flow sheet of the closed-loop experimental setup is shown in Mukherjee and SenGupta[26] (Figure 5.8a). Membrane coupons were soaked in deionized (DI) water for 24 h prior to experiment and examined before and after experiment to identify any tearing or fouling.

Feed, product, and waste samples were collected at transmembrane pressures of 100, 150, 200, and 250 psi. Samples were collected after the system achieved steady-state conditions (10 min after the last change in parameter). Sample volumes of approximately 25 mL were stored in LDPE (low-density polyethylene) containers with any excess permeate returned to feed tank. Flow rate (ml/s), pH, and conductivity were measured for each sample.

Transmembrane pressure ranging from 100 to 250 psi was controlled by a liquid-filled pressure gage. pH was measured using the MP 200 pH meter (Mettler Toledo, Inc., Columbus, Ohio) and conductivity was measured by the Sension 5 electrical conductivity meter (Hach Company, Loveland, Colorado). Dayco Imperial Nyloseal high-pressure tubing (Dayco Products, LLC, Tulsa, Oklahoma) was used for the feed and the waste streams, and Parker Paraflex tubing (Parker Hannifin Corporation, Cleveland, Ohio) was used to collect the product water. The closed loop setup returned product and waste streams back to the 15-gallon high-density polyethylene (HDPE) tank to provide a constant feed concentration. American Chemical Society (ACS) grade chemicals were used in all experiments.

5.5.2.2 Membranes Used

Membranes investigated were flat-sheet cellulose acetate NF (CK) and cellulose acetate RO (CD) membranes (both from GE Osmonics). The properties of the membranes used in this study are given in Table 5.2.

TABLE 5.2
Properties of the CD and CK Membranes

Membrane Type	CD		CK	
Manufacturer	GE Osmonics		GE Osmonics	
Classified as	RO		NF	
Material	cellulose acetate		cellulose acetate	
Test Pressure	100 psi	200 psi	100 psi	200 psi
Water Flux (GFD)	5.86	11.95	16.58	31.99
Percent Salt Passage @ 1000 ppm NaCl	8.25	6.15	32.13	23.25

5.5.2.3 Experimental Protocol

Two experiment sets were performed for each contaminant of concern. A single-component experiment was done with the salt (either Na^+ or K^+) of the contaminant; this represented the "bulk" condition where the concentration of the anion is equal to the concentration of the cation.

The second set of experiments was multicomponent experiments that constituted several anions accompanied by one single cation. The feed for the second set of experiments contained the same concentration of the contaminant as that of the bulk and the salt (same as that of the contaminant) of some background ions, such as HCO_3^-, Cl^-, Br^-, NO_3^-, HPO_4^{2-}, SO_4^{2-}. This created a "trace" condition for the contaminant, where the concentration of the contaminant (an anion, for this study) became much less than the concentration of the accompanying cation. The total amount of background ions equaled 0.1 *M* to create a ratio of around 1:100 (contaminant anion:accompanying cation). A blend of ions, instead of a single ion, was used to mimic groundwater conditions.

5.5.2.4 Analytical Methods

The primary method of determining anion concentrations in feed, permeate and waste streams was ion chromatography. The DX 120 ion chromatograph (Dionex, Sunnyvale, CA) was used with the IonPac AS9-HC (4 mm) column for analysis of Cl^-, Br^-, NO_3^-, HPO_4^{2-}, SO_4^{2-}, and AS-16 (4 mm) column for the analysis of ClO_4^-. A degassed eluent solution of 35 m*M* sodium hydroxide (NaOH) and a flow rate of 1.3 mL/min were used with the AS-16 column and 9.0 m*M* Na_2CO_3 eluent with a flow rate of 1.0 mL/min were used with the AS9 analytical column.

5.5.3 Results

Figure 5.21 shows the salt permeability coefficient (k_s) of chloroacetate ion, CH_2ClCOO^-, through RO and NF membranes vs. transmembrane pressure. Each graph in Figure 5.21 has two sets of data points: one where chloroacetate was the single component in the feed (that is, chloroacetate permeating as bulk species) and the other set where chloroacetate is the trace species compared to the accompanying

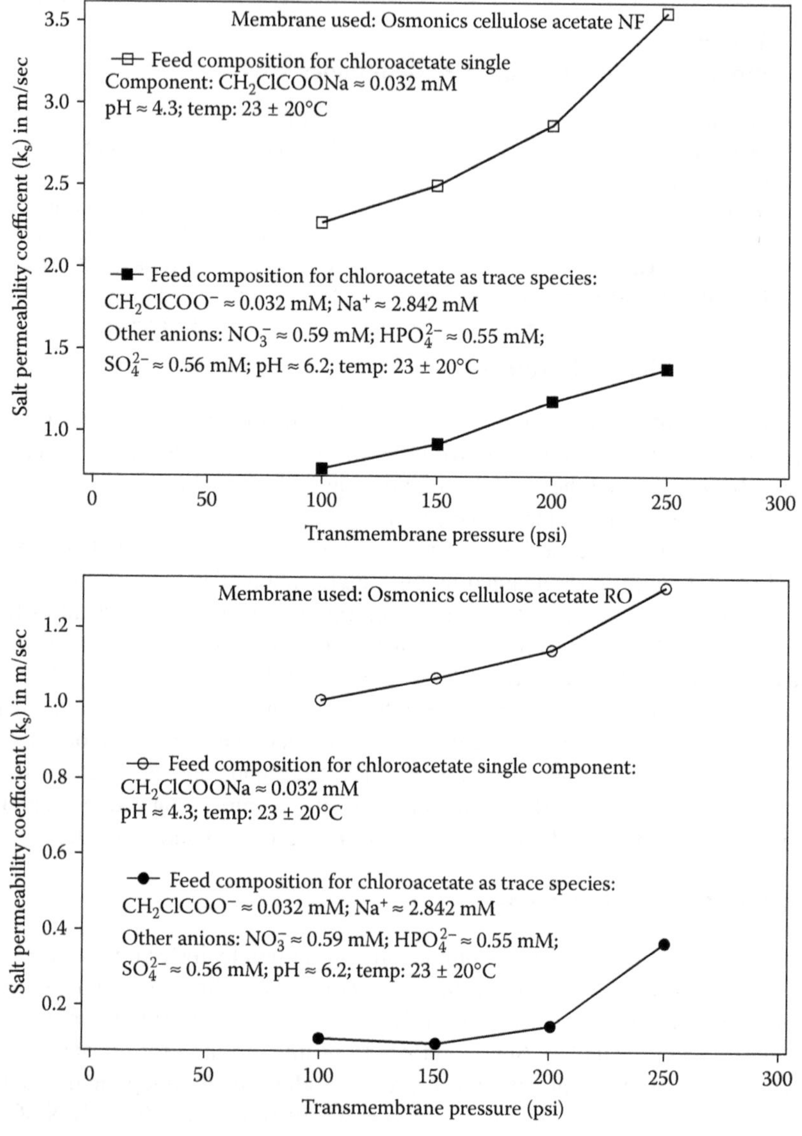

FIGURE 5.21 Salt permeability coefficient (k_S) vs. transmembrane pressure for trace and bulk chloroacetate (CH_2ClCOO) permeation through NF and RO membranes.

cation (here Na^+) (CH_2ClCOO^- = 0.032 m*M*; Na^+ = 2.842 m*M*). The salt permeability coefficient (k_s) is calculated from electrolyte permeation experiments through RO and NF membranes according to Equation (5.42). Since a flat-leaf cell was used for all experiments, where $C_p << C_f$ and $C_f \approx C_w$, the concentration polarization factor is considered to be 1 [$\exp(J_w/k_b) = 1$]. Experimental results from Figure 5.21 show that for all transmembrane pressures the k_s of CH_2ClCOO^- is lower as a trace species compared to that as bulk species, for both RO and NF membranes. According to the Equation (5.47), during trace conditions, the individual diffusion coefficient of the trace species becomes the determinant factor during permeation through membranes. Figure 5.9 shows that the elution time of CH_2ClCOO^- is short, which means that CH_2ClCOO^- has a low individual diffusion coefficient ($D_{CH_2ClCOO^-}$). The individual diffusion coefficient $D_{CH_2ClCOO^-}$ is lower than the interdiffusion coefficient ($D_{CH_2ClCOONa}$) with the associated cation, here Na^+.

$$D_{CH_2ClCOO^-} < D_{CH_2ClCOONa} \tag{5.48}$$

Since CH_2ClCOO^- has a low individual diffusion coefficient, its permeation through membranes becomes even more sluggish when it becomes trace. This condition remains valid for both types of membranes, RO and NF, and this result is in agreement with the theoretical prediction presented earlier. CH_2ClCOO^- is the most commonly formed haloacetic acid during prechlorination of surface waters and is a potential carcinogen. The permeation behavior of CH_2ClCOO^- is rather promising (in that it permeates lower when CH_2ClCOO^- is present as trace species) if membrane processes are used downstream in water treatment processes.

Figure 5.22 shows the salt permeability coefficient (k_s) for F^- vs. transmembrane pressure for RO and NF membranes. Similar to the previous example, each chart in Figure 5.22 contains data points from two sets of experiments, one in which F^- permeates through the membrane as trace species and another in which F^- is the bulk species (single component) in the feed of the process. From Figure 5.22, it is observed that F^- has a lower k_s for all the transmembrane pressures at which the experiment was conducted when F^- is present as a trace species (compared to bulk species) in the experiment feed. Figure 5.9 shows that F^- has a short elution time, which means it has a low diffusion coefficient in water, that is, the individual diffusion coefficient of F^- (D_{F^-}) is lower than the interdiffusion coefficient with the associated cation, here $K^+(D_{KF})$.

$$D_{F^-} < D_{KF} \tag{5.49}$$

Therefore, according to the theoretical prediction presented before, in trace conditions, F^- has a lower permeability than that in bulk conditions. This situation is also a favorable condition for fluoride removal, as F^- is mostly found in water at trace concentrations and hence any membrane treatment is likely to produce product water with a lower F^- concentration.

The next two ions in discussion are NO_3^- and ClO_4^-. As can be observed from Figure 5.9, NO_3^- has a much higher elution time than the reference ion Cl^-, and ClO_4^- has even higher elution time than NO_3^-, the highest among the monovalent ions. This means that both NO_3^- and ClO_4^- have high individual diffusion coefficients. Both NO_3^- and ClO_4^- have similar properties: both are

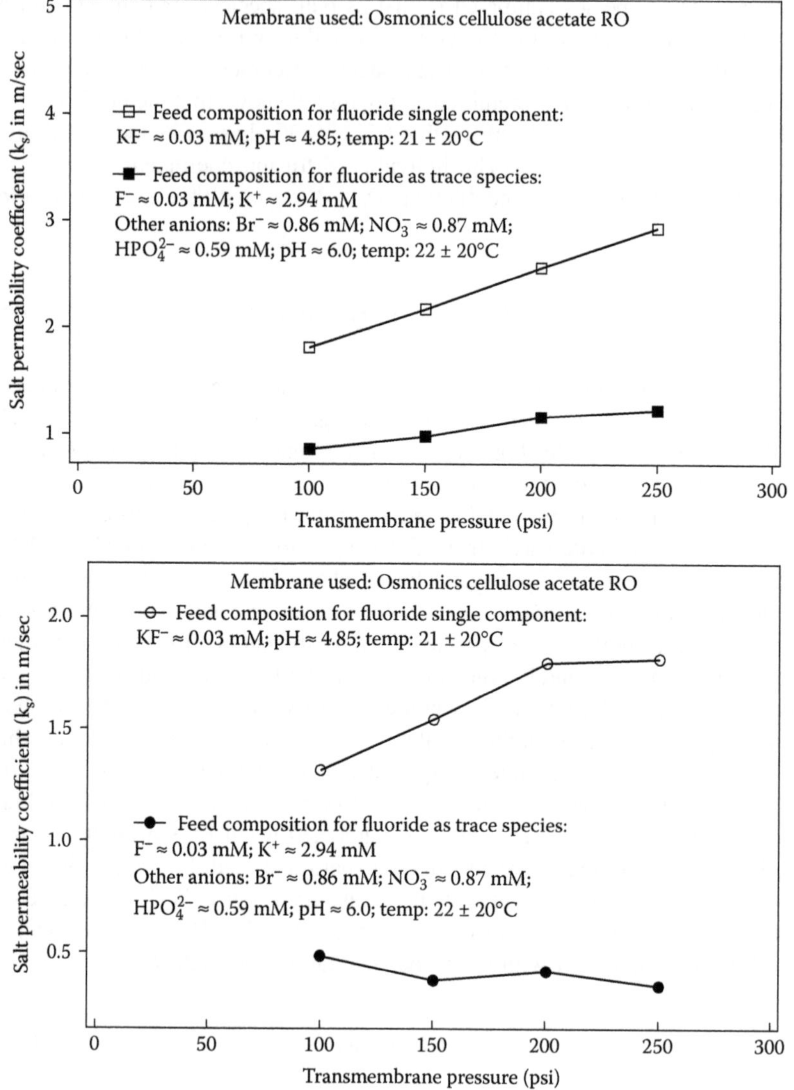

FIGURE 5.22 Salt permeability coefficient (k_S) vs. transmembrane pressure for trace and bulk fluoride (F) permeation through NF and RO membranes.

small hydrophobic ions that are recalcitrant to common treatment methods. Both ions are regulated strictly; NO_3^- has a maximum contaminant level (MCL) of 10 mg/1 and currently there are no MCLs for ClO_4^-. However, EPA's draft drinking water equivalent range varies from 4 to 18 μg/1 for perchlorate.

Figure 5.23 shows the salt permeability coefficient, k_s, of ClO_4^- vs. transmembrane pressure for RO and NF membranes. Each graph in Figure 5.23 contains two sets of data points: one when ClO_4^- is the bulk species in the feed,

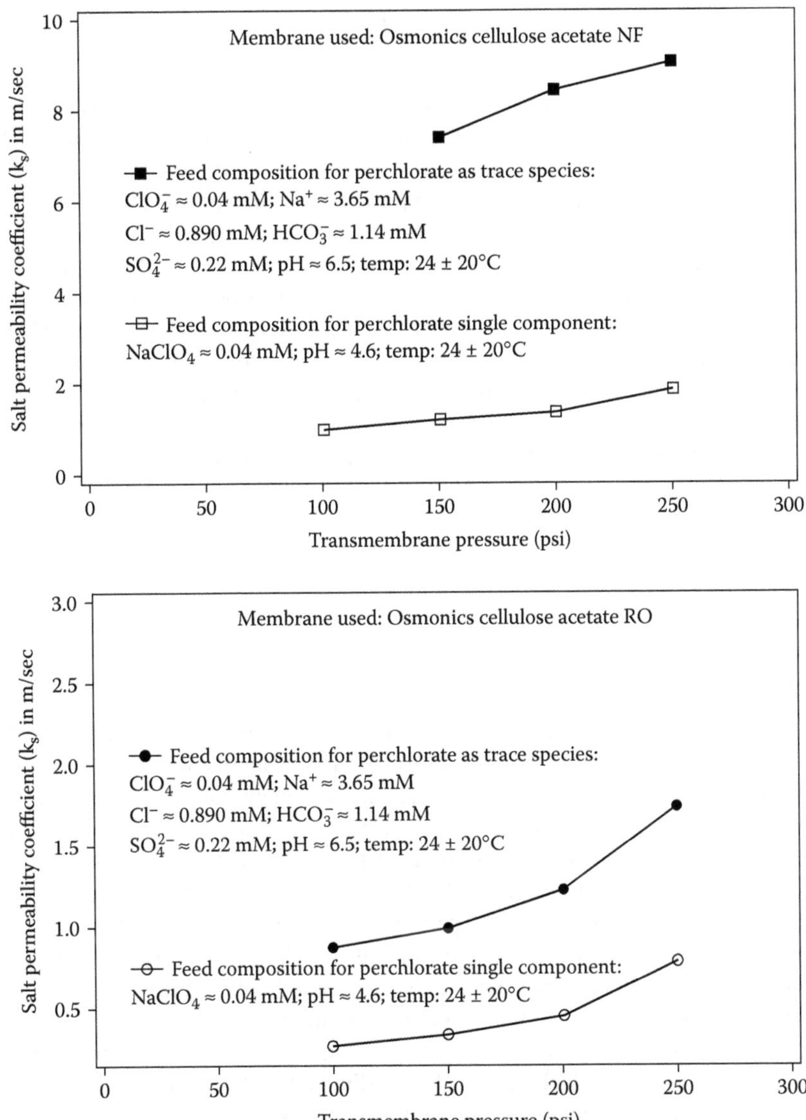

FIGURE 5.23 Salt permeability coefficient (k_S) vs. transmembrane pressure for trace and bulk perchlorate (ClO_4) permeation through NF and RO membranes.

and another where ClO_4^- is the trace species. In both cases of permeation through RO and NF membranes, it can be observed that k_s for ClO_4^- is higher when it permeates as a trace species. The result is in accordance to the theoretical prediction. ClO_4^- has a very high individual diffusion coefficient and the individual diffusion coefficient of ClO_4^- is much higher than the interdiffusion coefficient with the associated cation, Na^+.

$$D_{ClO_4^-} >> D_{NaClO_4} \tag{5.50}$$

$D_{ClO_4^-}$ becomes the determinant factor when ClO_4^- becomes a trace species, as opposed to D_{NaClO_4} when ClO_4^- is the bulk species in the feed. Therefore, ClO_4^- is observed to permeate more as a trace species. This result has serious implications when membrane processes are applied for the treatment of ClO_4^-, as it is always found as a trace species in surface and ground water.

Figure 5.24 shows the salt permeability coefficient (k_s) of NO_3^- vs. transmembrane pressure for RO and NF membranes. From both graphs for RO and NF membrane, it can be observed that the k_s for NO_3^- is higher when NO_3^- is present in trace concentration in the feed, as opposed to being present as a bulk species. This result also is in agreement to the theoretical prediction. NO_3^- has a high individual diffusivity, and in trace conditions $D_{NO_3^-}$ becomes the determinant factor during permeation through membrane as opposed to D_{NaNO_3} in bulk.

$$D_{NO_3^-} > D_{NaNO_3} \tag{5.51}$$

Therefore, NO_3^- permeates higher as a trace species.

5.5.4 Discussion

Experimental results presented in this study show that the k_s and hence the permeation of trace species is significantly different from that for the bulk species. During the permeation of electrolytes through the membrane, the interdiffusion coefficient of the permeating anion-cation pair is the determinant factor. When one of the components of the electrolyte (either anion or cation) becomes small or trace with respect to the other, the diffusion coefficient of the trace species becomes the deciding factor for the permeating electrolyte. Therefore, it is important whether the trace species in question has a high or low individual diffusion coefficient value. When a trace species has a high individual diffusion coefficient value (such as ClO_4^- or NO_3^-), it permeates more through the membrane as opposed to when it is present in bulk in solution. In contrast, ions such as CH_2ClCOO^- and F^-, which have a low individual diffusion coefficient value, permeate to a lesser extent when present as trace species.

Figure 5.25 shows data replotted from perchlorate permeation studies through a NF membrane from Yoon et al.[42] The chart in Figure 5.25 shows percent permeation of perchlorate through a NF membrane as recorded over time. The graph contains data from two separate experiments: the feed solution of one was 100 μg/1 of ClO_4^- in pure water, and the other where the feed was Colorado River water (CRW). CRW contained 3.8 μg/1 of ClO_4^- and it was spiked up to 100 μg/1 to make up the experimental feed. The details of ionic composition of CRW and experimental conditions can be found in Yoon et al.[42] These two experimental

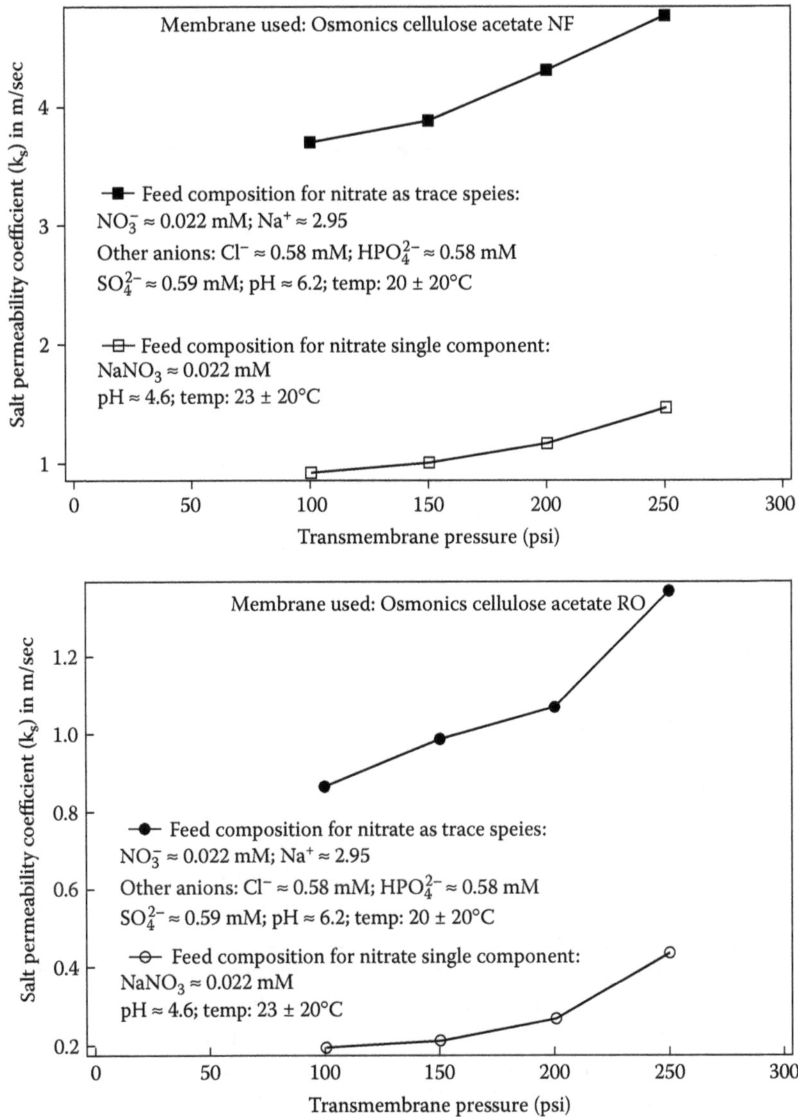

FIGURE 5.24 Salt permeability coefficient (k_S) vs. transmembrane pressure for trace and bulk nitrate (NO_3) permeation through NF and RO membranes.

conditions are similar to that of ours: ClO_4^- in pure water representing the bulk condition and CRW represents ClO_4^- in trace conditions. Observations made from results in Figure 5.25 clearly show that ClO_4^- permeates at much higher rate as trace species throughout the entire duration of the experiment. This result is in agreement with the observations made in Figure 5.23.

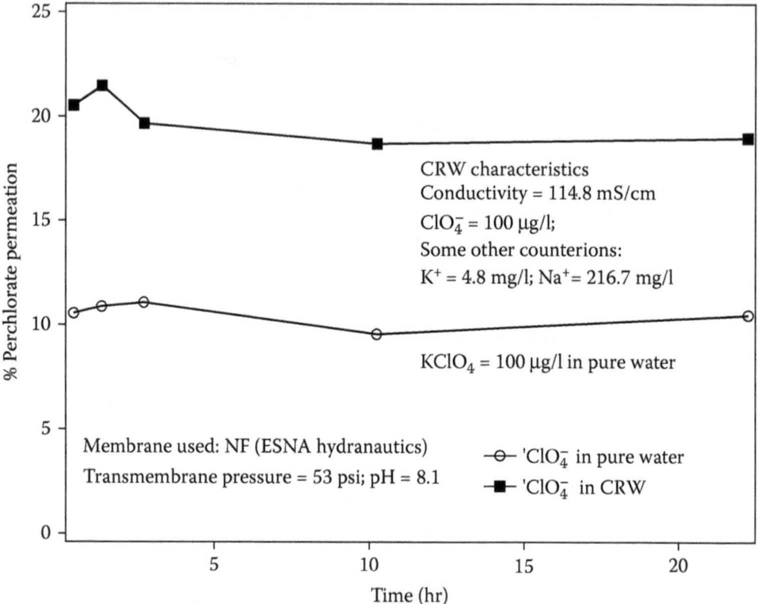

FIGURE 5.25 Percent permeation of perchlorate through NF membrane.[42]

5.6 CONCLUDING REMARKS

The following conclusions can be summarized from the above discussion.

(a) Dissociated electrolytes permeate through the membrane as individual ions (not ion pairs), and they get partitioned into the membrane phase and then migrate through the membrane active layer in a manner such that the negative charge of an anion is balanced by the positive charge of an accompanying cation; that is, it is strictly a case of coupled transport.

(b) Ion exchange selectivity data are related to the hydrated ionic radii, which in turn are related to diffusion coefficients of individual ions: the higher the ion exchange selectivity, the smaller the hydrated radii of ions and the higher the diffusion coefficient of an ion.

(c) Ion exchange selectivity data can be used to estimate the interdiffusion coefficients of individual electrolytes.

(d) Existing solution diffusion model in conjunction with ion exchange selectivity data can be used to predict relative permeation rates of different electrolytes through RO and NF membranes.

(e) The permeation of an ion through the membrane changes when it is present as a trace species. Species with a higher individual diffusion coefficient become more permeable when they are present as a trace species and vice versa. For example, a species such as ClO_4^- becomes more permeable when it is present as a trace species in water.

ACKNOWLEDGMENTS

The Lehigh University Milestone Fellowship awarded to Parna Mukherjee is gratefully acknowledged. Grant No. N00014-04-1-0436 from the Department of the Navy, Office of Naval Research, was indispensable for the trace species study conducted at California Polytechnic State University. The experiments for trace species were conducted by Amanda Schimdt in the environmental engineering laboratory at California Polytechnic State University as part of her master's thesis.

REFERENCES

1. Mukherjee, P. and SenGupta, A.K., Some observations about electrolyte permeation mechanism through reverse osmosis and nanofiltration membranes, *J. Membr. Sci.*, 278, 301, 2006.
2. Hodgson, T.D., Selective properties of cellulose acetate membranes towards ions in aqueous solution, *Desalination*, 8, 99, 1970.
3. Soltanieh, M. and Sahebdelfer, S., Interaction effects in multicomponent separation by reverse osmosis, *J. Membr. Sci.*, 183, 15, 2001.
4. Erickson, D.L., Glater, J., and McCutchan, J.W., Selective properties of high flux cellulose acetate membranes toward ions found in natural waters, *Ind. Eng. Chem. Prod. Res. Develop.*, 5(3), 205, 1966.
5. Lonsdale, H.K., Cross, B.P., Graber, F.M., and Milstead, C.E., Performance of cellulose acetate membranes to selected solutes, *J. Macromolecular Sci.: Phys.*, B5(1), 167, 1971.
6. Hall, M.S., Starov, V.M., and Lloyd, D.R., Reverse osmosis of multicomponent electrolyte solutions, part I and II, *J. Membr. Sci.*, 128, 23, 1997.
7. Childress, A. and Elimelech, M., Relating nanofiltration membrane performance to membrane charge (electrokinetic) characteristics, *Env. Sci. Tech.*, 34, 3710, 2000.
8. Ratantamskul, C., Urase, T., and Yamamoto, K., Description of behaviour in rejection of pollutants in ultra low pressure nanofiltration, *Water Sci. Tech.*, 38(4–5), 453, 1998.
9. Bhattacharya, S., Chen, J.C., and Elimelech, M., Coupled model of concentration polarization and pore transport in crossflow nanofiltration, *AIChE J.*, 47(12), 2001.
10. Lee, S. and Lueptow, R.M., Membrane rejection of nitrogen compounds, *Env. Sci. Tech.*, 35, 3008, 2001.
11. Marinas, B.J. and Selleck, R.E., Reverse osmosis treatment of multicomponent electrolyte solutions, *J. Membr. Sci.*, 72, 211, 1992.
12. Chellam, S. and Taylor, J.S., Simplified analysis of contaminant rejection during ground and surface water nanofiltration under the information collection rule, *Water Res.*, 35(10), 2460, 2001.
13. Marcus, Y., Thermodynamics of solvation of ions, *J. Chem. Soc., Faraday Trans.*, 83, 2985, 1987.
14. Lide, D.R., *CRC Handbook of Chemistry and Physics,* CRC Press, Boca Raton, FL, 2000.
15. Wijmans, J.G. and Baker, R.W., The solution-diffusion model: a review, *J. Membr. Sci.*, 107, 1, 1995.

16. Rangarajan, R., Majid, M.A., Matsuura, T., and Sourirajan, S., Predictability of membrane performance for mixed-solute reverse osmosis systems. 4. System: cellulose acetate-nine seawater ions-water, *Ind. Eng. Chem. Pro. Des. Dev.*, 24, 977, 1985.
17. Williams, M.E., Hestekin, J.A., Smothers, C.N., and Bhattacharyya, D., Separation of organic pollutants by reverse osmosis and nanofiltration membranes: mathematical models and experimental verification, *Ind. Eng. Chem. Res.*, 38, 10, 1999.
18. Sourirajan, S., *Reverse Osmosis*, Academic Press, New York, 1970.
19. Koros, W.J., Membranes: learning a lesson from Nature, *Chem. Eng. Prog.*, 91, 68, 1995.
20. Reusch, C.F. and Cussler, E.L., Selective membrane transport, *AIChE J.*, 19, 4, 1973.
21. Blais, P., Polyamide membranes, in *Reverse Osmosis and Synthetic Membranes: Theory, Technology, Engineering,* Sourirajan, S., Ed., National Research Council Publication No. 15627, Ottawa, Canada, 1977, chapter 7, pp. 167–210.
22. Taniguchi, M. and Kimura, S., Estimation of transport parameters of RO membranes for seawater desalination, *AIChE J.*, 46, 1967, 2000.
23. Lipp, P., Gombel, R., and Frimmel, F.H., Parameters influencing the rejection properties of FT30 membranes, *J. Membr. Sci.*, 95, 185, 1994.
24. Lipp, P., Zum Einflu der Rohwasserzusammensetzung auf das Trennverhalten von Membranen bei der Umkehrosmose wäriger Lösungen, Ph.D. dissertation, 1993, Universität Fridericiana Karlsruhe, Germany.
25. Garba, Y., Taha, S., Gondrexon, N., and Dorange, G., Ion transport modelling through nanofiltration membranes, *J. Membr. Sci.*, 160, 187, 1999.
26. Mukherjee, P. and SenGupta, A.K., Ion exchange selectivity as a surrogate indicator of relative permeability of homovalent ions in reverse osmosis process, *Environ. Sci. Technol.*, 37, 1432, 2003.
27. Mukherjee, P., Elative permeation of ions in reverse osmosis processes using ion exchange selectivity data: theoretical approach and experimental validation, Ph.D. dissertation, 2003, Lehigh University, Bethlehem, PA.
28. Afonso, M.D. and de Pinho, M.N., Transport of MgSO4, MgC12 and Na2SO4 across an amphoteric nanofiltration membrane, *J. Membr. Sci.*, 179, 137, 2000.
29. Williams, M.E., Hestekin, J.A., Smothers, C.N., and Bhattacharyya, D., Separation of organic pollutants by reverse osmosis and nanofiltration membranes: mathematical models and experimental verification, *Ind. Eng. Chem. Res.*, 38(10), 3683–3695, 1999.
30. Bhattacharya, D., Mangum, W.C., and Williams, M.E., Reverse osmosis, in *Encyclopedia of Environmental Analysis and Remediation*, John Wiley, New York, 1998.
31. Taylor, J.S. and Wiesner, M., Membranes, in *Water Quality and Treatment,* 5th ed., McGraw-Hill, Inc., New York, 1999.
32. Mallevialle, J., Odendaal, P.E., and Wiesner, M.R., *Water Treatment Membrane Processes,* McGraw-Hill, New York, 1996, sec. 4.1–4.28.
33. Cussler, E.L., *Diffusion: Mass Transfer in Fluid Systems,* Cambridge University Press, Cambridge, 1984.
34. Kitchener, J.A., *Ion-Exchange Resins*, Metuchen & Co. Ltd., London, 1958.
35. Helfferich, F., *Ion Exchange,* Dover Publications, Inc., New York, 1995.
36. Ramana, A. and SenGupta, A.K., A new class of selective sorbents for arsenic and selenium oxy-anions, *J. Env. Eng.*, 118(5), 755, 1992.

37. Xu, Y. and Lebrun, R.E., Investigation of the solute separation by charged nanofiltration membrane: effect of pH, ionic strength and solute type, *J. Membr. Sci.*, 158, 93, 1999.
38. Petersen, R.J., Cadotte, J.E., and Buettner, J.M., *Report on Water Research Technology,* NTIS Report No. PB83-191775, Washington, DC, 1982.
39. Vrijenhoek, E.M. and Waypa, J.J., Arsenic removal from drinking water by a "loose" nanofiltration membrane, *Desalination,* 130, 265, 2000.
40. Harned, H.S. and Owen, B.B., *The Physical Chemistry of Electrolyte Solutions,* Reinhold Publishing Corporation, New York, 1950.
41. Harris, D.C., *Quantitative Chemical Analysis,* 5th ed., W.H. Freeman, New York, 1999.
42. Yoon, Y., Yoon, J., Amy, G., and Liang, S., *Dominant Potential Mechanism for Perchlorate Rejection by Negatively Charged Nanofiltration and Ultrafiltration Membranes,* AWWA Membrane Conference Proceedings, San Antonio 2001.

6 Chitosan: A Versatile Biopolymer for Separation, Purification, and Concentration of Metal Ions

Katsutoshi Inoue and Yoshinari Baba

CONTENTS

6.1 Introduction ... 340
6.2 Adsorption Behavior of Chitosan for Some Metal Ions ... 340
6.3 Solvent Extraction of Metal Ions with Lipophilic Chitosan ... 344
6.4 Adsorption of Metal Ions on Chemically Modified Chitosan ... 347
6.4.1 Complexane Types of Chemically Modified Chitosan ... 347
6.4.1.1 Adsorption of Base Metals on Complexane Types of Chemically Modified Chitosan ... 347
6.4.1.2 Adsorption of Rare Earths on Complexane Types of Chemically Modified Chitosan ... 349
6.4.2 Oxine Type of Chemically Modified Chitosan ... 359
6.4.2.1 Adsorption of Gallium(III) and Indium(III) on the Ga-Oxine Type of Chitosan ... 360
6.4.2.2 Adsorption of Rhodium(III) on the Fe-Oxine Type of Chitosan ... 362
6.4.3 Chemically Modified Chitosan Containing Pyridine Functional Groups ... 365
6.4.4 Chemically Modified Chitosan with Sulfur-Containing Functional Groups ... 368
6.5 Solvent Extraction of Metal Ions with Chemically Modified Lipophilic Chitosan ... 370
6.6 Conclusion ... 372
References ... 372

6.1 INTRODUCTION

In recent years, it has become very necessary to establish a sustainable society instead of our present culture with its huge amount of production, consumption, and waste, and that also consumes a large amount of limited fossil resources, such as petroleum and coal, and pollutes the global environment. In the sustainable society of the future, we should rely not on limited fossil resources but on unlimited renewable resources such as biomass. Also in such a sustainable society, advanced separation technologies will become much more important for recovering valuable materials from various wastes through recycling and reuse, and for removing toxic or harmful materials from the environment. In such cases and for such purposes, the materials for separation, such as adsorbents, ion exchange materials, membranes, and so on, should be produced from unlimited renewable resources. As will be described in detail later, some natural materials exhibit significant separation characteristics that are superior to artificial materials employed in the present society. Therefore, the unknown and unused excellent characteristics of such natural products need to be discovered in order to be able to use them instead of the artificial materials produced from limited fossil resources.

Chitosan is a basic polysaccharide containing many primary amino groups. It has been well known for its ability to adsorb various metal ions through coordination with the primary amino groups, and much work has been conducted in this respect.[1] However, the detailed adsorption mechanism has not yet been fully elucidated.

Although chitosan itself exhibits interesting adsorption behavior for metal ions, it can be much improved or enhanced by immobilizing a variety of functional groups that have special affinities for some specified metal ions by chemically modifying its primary amino groups to produce high reactivity sites. In addition, it is also possible to make chitosan soluble in some organic diluents (such as kerosene and toluene) by means of chemical modification of its hydroxyl groups with hydrophobic long-chain alkyl radicals, thus facilitating its use as a solvent extraction reagent.

From the above-mentioned viewpoint, we have conducted some fundamental work on the adsorption on chitosan and on some chemically modified chitosans, as well as on solvent extraction with chemically modified lipophilic chitosan. Some of our typical work on chitosan is reviewed in this chapter.

6.2 ADSORPTION BEHAVIOR OF CHITOSAN FOR SOME METAL IONS

In nature, there are two types of chitosan: α-type and β-type chitosans. As shown in Scheme 6.1, the former is a highly crystalline polysaccharide contained in shells of crustaceans such as crabs, shrimps, insects, and so on that are extensively employed for various purposes; the latter, contained in the cartilage of squids, is amorphous (Scheme 6.1).

The former is commercially produced from the shells of crabs and shrimps as follows. These shells consist of chitin (the precursor of chitosan), proteins, and calcium (Scheme 6.2).

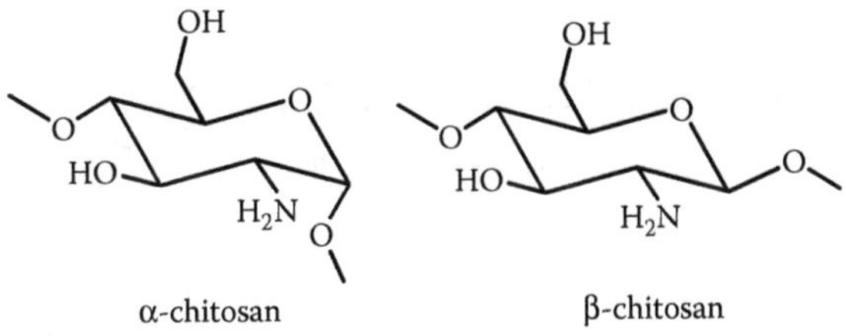

α-chitosan β-chitosan

SCHEME 6.1 Chemical structure of α- and β-chitosan.

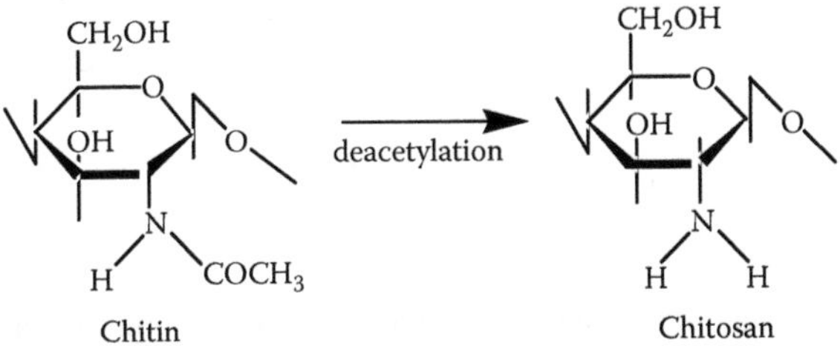

Chitin —deacetylation→ Chitosan

SCHEME 6.2 Chemical structure of chitin and chitosan.

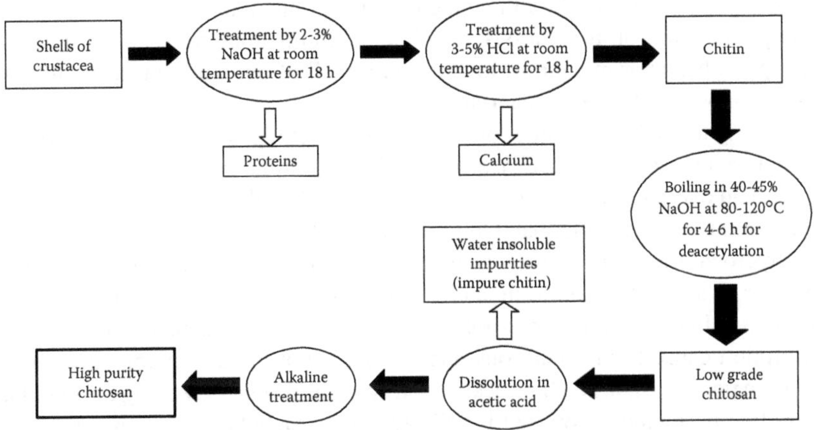

SCHEME 6.3 Flow sheet of commercial production of chitosan.

Scheme 6.3 shows the flowsheet of commercial production of chitosan. After the proteins are removed by treating with dilute sodium hydroxide solution, calcium is removed as calcium chloride by treating with 3 to 5% hydrochloric acid to produce chitin, which is further boiled in concentrated sodium hydroxide solution for hydrolysis of the acetoamide groups of chitin to produce low-grade chitosan. Low-grade chitosan is a mixture consisting of about 50% chitin and 50% chitosan, which is marketed mainly as a coagulating agent for suspensions of fine solid particles. After dissolving the low-grade chitosan in acetic acid solution followed by removal of the insoluble impurity chitin, purified chitosan,

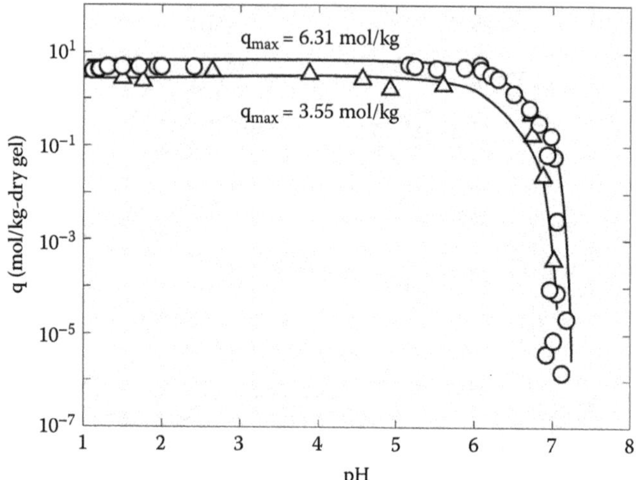

FIGURE 6.1 Amount of adsorbed hydrogen ion on CLC and original chitosan from 1 *M* (= mol/dm^3) ammonium nitrate solution at varying pH.[5]

the purity of which is greater than 95%, is produced. As will be described in the following section in detail, although chitosan exhibits interesting and significant adsorption behavior for various metal ions, chitin does not exhibit practical adsorption behavior in contrast to chitosan.[2] In all of our work, purified chitosan was employed as the feed material.

Owing to the large number of primary amino groups in chitosan, it is soluble in some acidic solutions; it is nearly completely soluble in organic acids such as acetic acid over the whole concentration region, although it is insoluble in sulfuric acid solution. On the other hand, it is soluble in other mineral acids such as nitric and hydrochloric acids in the concentration region between 0.05 and 1 mol/dm^3. In order to avoid dissolution in acidic solutions, it is necessary to cross-link the chitosan. Unfortunately, cross-linking results in a significant decrease in the adsorption capacity of metal ions due to bonding at the adsorption sites on the chitosan polymer matrix. Ohga et al.[3] overcame this problem by using the template cross-linking method proposed by Nishide and Tsuchida for the preparation of poly(4-vinypyridine) resin.[4] By this method, chitosan is cross-linked after complexation with metal ions, which can then be removed after cross-linking by washing with an acidic solution. We prepared the water-insoluble adsorption gel of chitosan, which is abbreviated as CLC hereafter, according to this template cross-linking method using copper(II) ion as the template, and examined its adsorption behavior to some metal ions.

Figure 6.1 illustrates a plot of the amount of adsorbed hydrogen ion on CLC and original chitosan from 1 *M* (= mol/dm^3) ammonium nitrate solution at varying pH.[5] As seen from this figure, hydrogen ion uptake is constant over the whole pH range below about 6, suggesting that both materials are saturated with hydrogen ions or all of their primary amino groups are completely protonated at low pH.

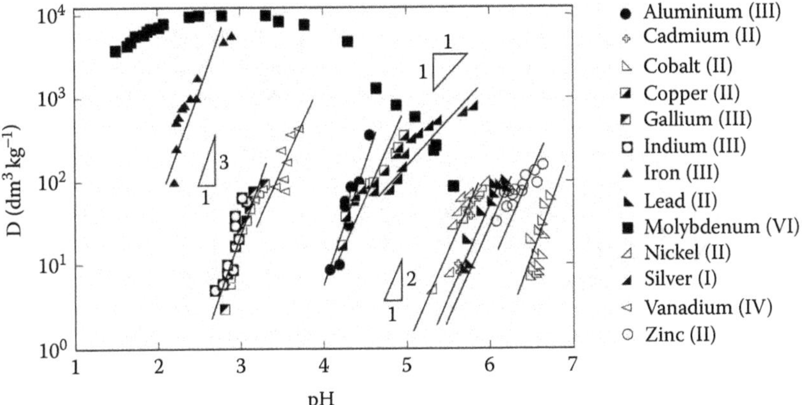

FIGURE 6.2 Distribution ratio of various metal ions (*D*) from 1 *M* aqueous ammonium nitrate solution against equilibrium pH for their adsorption on CLC.

From the constant values, the ion exchange capacities of CLC and original chitosan were evaluated as 3.55 and 6.31 mol/kg dry gel, respectively, which are higher than commercially available weakly basic anion exchange resins. However, it is evident that the capacity is decreased more by cross-linking through the primary amino groups, the adsorption sites, had been protected as mentioned above.

Figure 6.2 illustrates a plot of the distribution ratio of various metal ions (*D*) from 1 *M* aqueous ammonium nitrate solution against equilibrium pH for their adsorption on CLC. As seen from this figure, the selectivity order for the metal ions tested is as follows: Mo(VI) >> Fe(III) >> In(III) > Ga(III) ~ V(IV) >> Al(III) ~ Cu(II) > Ag(I) >> Ni(II) ~ Pb(II) ~ Cd(II) > Zn(II) > Co(II). The distribution ratio of all metal ions tested except for molybdenum increases with increasing pH, which suggests that the adsorption of these cationic metal ions takes place according to a cation exchange mechanism. The decrease in the distribution ratio of molybdenum at pH values higher than around 2.5 is attributable to the formation of its anionic species. The majority of the plots lie on straight lines with slopes equal to the valence of the adsorbed metal ions. That is, the plots for trivalent metal ions such as ferric iron, aluminum, gallium, and indium lie on straight lines with a slope of 3; for monovalent silver ion, it lies on a straight line with a slope of unity; while those for divalent metal ions such as cupric, nickel, cadmium, lead, and zinc are on those with a slope of 2. These tendencies are quite different from those observed in the ion exchange behavior exhibited by commercial synthetic cation exchange resins, including chelating resins, but they are analogous to those observed in the solvent extraction of cationic metal ions with acidic extraction reagents such as acidic organophosphorus compounds and various chelating reagents. This result suggests a cation exchange mechanism, wherein *n*-valent cationic metal ions are adsorbed on chitosan with the release of *n* hydrogen ions. Because all of the amino groups are protonated at pH values less than 6, as mentioned earlier, the released hydrogen ions are considered to be those protonated

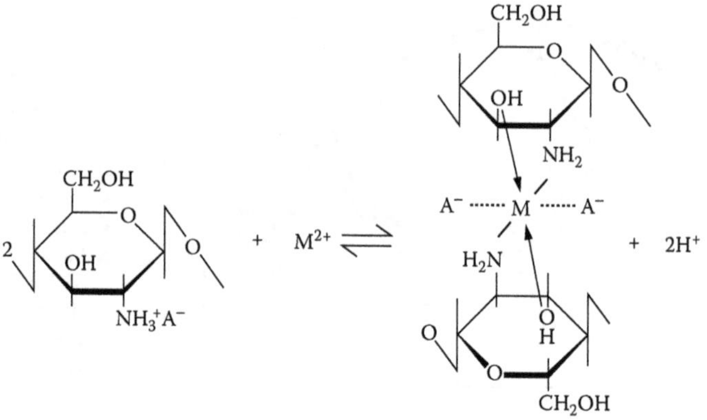

SCHEME 6.4 Adsorption reaction mechanism for divalent metal ions on chitosan.

on the primary amino groups of the chitosan. On the basis of the above discussion, divalent metal ions, for example, are inferred to be adsorbed according to the reaction expressed by Scheme 6.4.

In this adsorption reaction, it is inferred that metal ions are coordinated both by the nitrogen atoms of the primary amino groups and by the oxygen atoms of the hydroxyl groups at the third position to form stable five-membered chelates of the form 1:2 metal:glucosamine. The role of the oxygen atoms will be discussed in detail in a subsequent section. The formation of stable 1:*n* metal:glucosamine chelates is considered to be attributable to the flexible structure of the chitosan polymer chain, which enables a suitable configuration for complexation with metal ions, contrary to the poor flexibility of commercially available chelating resins, which results in an inability to form stable chelates. This is the driving force that provides the high selectivity of chitosans.

From the adsorption isotherm tests, the maximum adsorption capacities of copper(II) were evaluated as 2.31 and 1.37 mol/kg dry gel for the original chitosan and CLC, respectively, which are greater than or similar to that for commercially available chelating resins.

6.3 SOLVENT EXTRACTION OF METAL IONS WITH LIPOPHILIC CHITOSAN

As discussed earlier, it is possible to make chitosan lipophilic by chemical modification of its hydroxyl groups, although chitosan itself is hydrophilic. We prepared O,O′-decanoyl chitosan as a typical lipophilic chitosan[6,7,8] according to the synthetic route proposed by Nishimura et al.[9] As shown in Scheme 6.5, in this method, alcoholic hydroxyl groups are esterified using a long alkyl chain acid chloride. In order to avoid attack by the reagent (acid chloride), the primary amino groups are protected in advance by interacting with phthalic anhydride, which is removed after esterification.

Thus prepared O,O′-decanoyl chitosan is soluble in some organic diluents such as chloroform and toluene but insoluble in hexane and kerosene. Figure 6.3

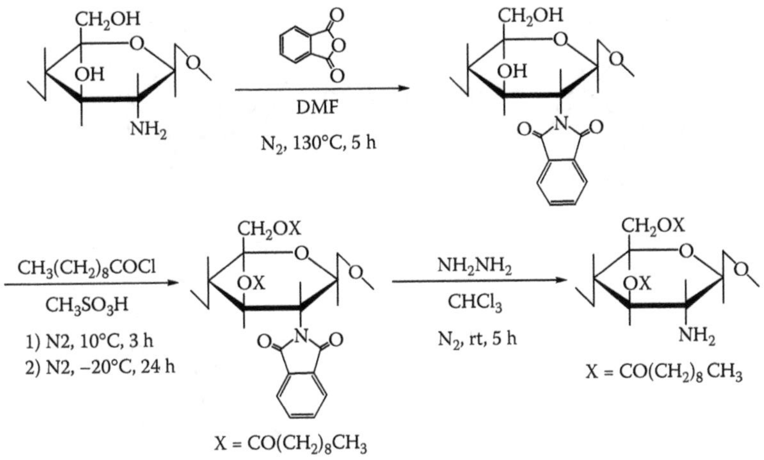

CH2OH
OH
NH2
DMF
N2, 130°C, 5 h
CH3(CH2)8COCl
CH3SO3H
1) N2, 10°C, 3 h
2) N2, −20°C, 24 h
CH2OX
OX
X = CO(CH2)8CH3
NH2NH2
CHCl3
N2, rt, 5 h
NH2
X = CO(CH2)8 CH3

SCHEME 6.5 Synthetic route of O,O′-decanoyl chitosan.

Cu
Ni
Co
Zn
Fe
Pb
% Extraction
pHeq

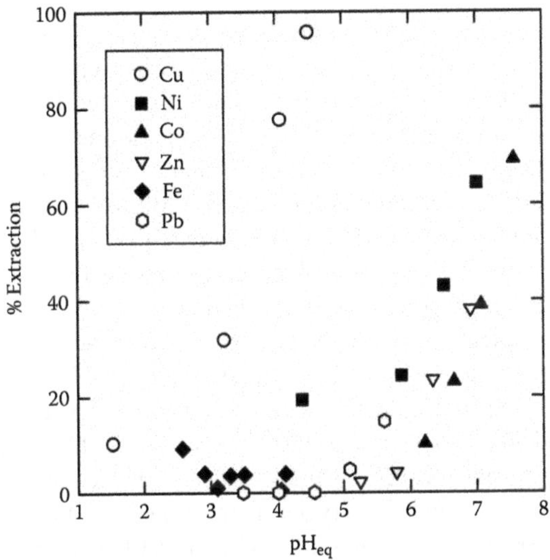

FIGURE 6.3 Percent extraction of various metal ions with O,O′-decanoyl chitosan in chloroform against equilibrium pH from HEPES (2-[4-(2-hydroxylethyl)-1-piperazinyl]ethanesulfonic acid) buffer solution, the pH of which was adjusted by adding small amount of nitric acid.

shows the plot of percent extraction of various metal ions with O,O′-decanoyl chitosan in chloroform against equilibrium pH from HEPES (2-[4-(2-hydroxylethyl)-1-piperazinyl]ethanesulfonic acid) buffer solution, the pH of which was adjusted by adding a small amount of nitric acid. It is noticeable that, contrary to the case of adsorption on CLC, the pH range for copper(II) extraction shifts

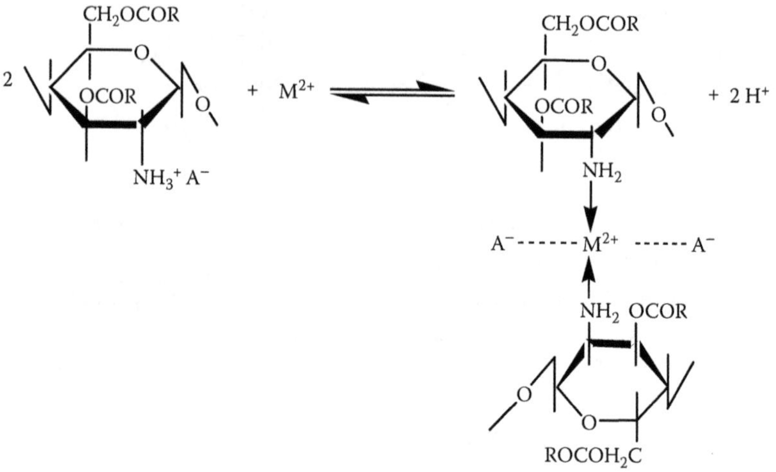

SCHEME 6.6 Solvent extraction reaction mechanism for divalent metal ions by O,O′-decanoyl chitosan.

to lower pH values, while that for molybdenum(VI) and gallium(III) is nearly the same and that for indium(III) and aluminum(III) shifts to higher pH values. The most noticeable feature is that iron(III) is poorly extracted by this reagent. For the purpose of further confirmation, a solvent extraction test was carried out by contacting an aqueous solution containing 1 mg/L copper(II) and 100 mg/L iron(III) with 6 g/L O,O′-decanoyl chitosan in chloroform over the pH range 1.94 to 2.41. It was found that copper(II) was quantitatively extracted while practically no iron(III) was extracted. These results seem to be closely related to the extraction and adsorption mechanism for chitosan derivatives.

In order to theoretically evaluate the optimal structures and chemical properties that give rise to the extraordinary high selectivity for copper(II) over iron (III) with O,O′-decanoyl chitosan, molecular modeling computations were carried out by using the molecular modeling program, MOPAC93 (Fujitsu Co., Ltd.).[8] From the result of the computations, it was elucidated that the distance between the nitrogen atom of the primary amino group and the oxygen atom at the third position is greater in O,O′-decanoyl chitosan than in chitosan itself, and that the electron density on the oxygen atom on chitosan is negatively higher than that on O,O′-decanoyl chitosan, while that on the nitrogen atom is similar for both materials. This result may provide the reason why iron(III), a hard Lewis acid with a high affinity for the oxygen atom, is selectively adsorbed on chitosan or CLC over copper(II), while it is poorly solvent extracted with O,O′-decanoyl chitosan. It can be inferred, therefore, that iron(III) is extracted according to the adsorption reaction mechanism for metal ions as described in the previous section, in which oxygen atoms take part in the formation of metal chelates. On the other hand, the extraction reaction of divalent metal ions including copper(II) ion by O,O′-decanoyl chitosan in which no oxygen atom takes part may be expressed as follows (Scheme 6.6).

6.4 ADSORPTION OF METAL IONS ON CHEMICALLY MODIFIED CHITOSAN

Since chitosan has many highly reactive primary amino groups and hydroxyl groups at the sixth position, it is easy to prepare a variety of chemically modified chitosans. By immobilizing functional groups with a high affinity for some specified metal ions, selectivity for the specified metals and adsorption capacity can be much enhanced. From this viewpoint, much work has been conducted to date. Here, it should be noted that the extent of immobilization of the functional groups on the chitosan polymer matrices is dependent not only on the synthetic route employed but also on various experimental conditions such as temperature, reaction time, and so on, because the preparation reactions are heterogeneous between solid and liquid, and therefore are sensitive to very slight changes in experimental conditions.

6.4.1 Complexane Types of Chemically Modified Chitosan

The authors prepared four complexane types of chemically modified chitosan from the original uncross-linked chitosan as shown in Scheme 6.7 so as to examine their adsorption behaviors for base metals, rare earths, and platinum group metals.[10–16] Among these, the synthetic routes for ethylenediamine-N,N,N′,N′-tetraacetic acid (EDTA) and diethylenetriamine-N,N,N′,N″,N″-pentaacetic acid (DTPA) types of chemically modified chitosan, which have exhibited very interesting adsorption behaviors for base metals and rare earths as will be described later, are very simple and easy, as depicted in Scheme 6.8.

It was found that these complexane types of chemically modified chitosan are insoluble in various acid solutions, including acetic acid solution, in contrast to the original chitosan. This may be attributed to the cross-linking between polymer chains by hydrogen bonds between the carboxyl groups.

6.4.1.1 Adsorption of Base Metals on Complexane Types of Chemically Modified Chitosan

Figure 6.4 shows the plots of the distribution ratio of copper(II) against equilibrium pH for adsorption from 1 M ammonium nitrate solution on EDTA and DTPA types of chitosan and original chitosan for comparison. The pH at which adsorption takes place shifts to lower pH values due to the decrease in pK_a by immobilizing a large number of carboxylic groups in the EDTA and DTPA types of chitosan; that is, the adsorption increases in the order: original chitosan < glycine type << IDA (iminodiacetic acid) type << DTPA type < EDTA type. The difference between the DTPA and EDTA types may be attributable to the difference in the extent of immobilization of the functional groups between them. From the adsorption isotherms, the maximum adsorption capacities of copper(II) were evaluated as 1.7, 2.4, 2.0, and 1.5 mol/kg dry gel for glycine types, IDA types, DTPA types, and EDTA types of chitosan, respectively, compared with 2.3 for the original chitosan as discussed earlier.

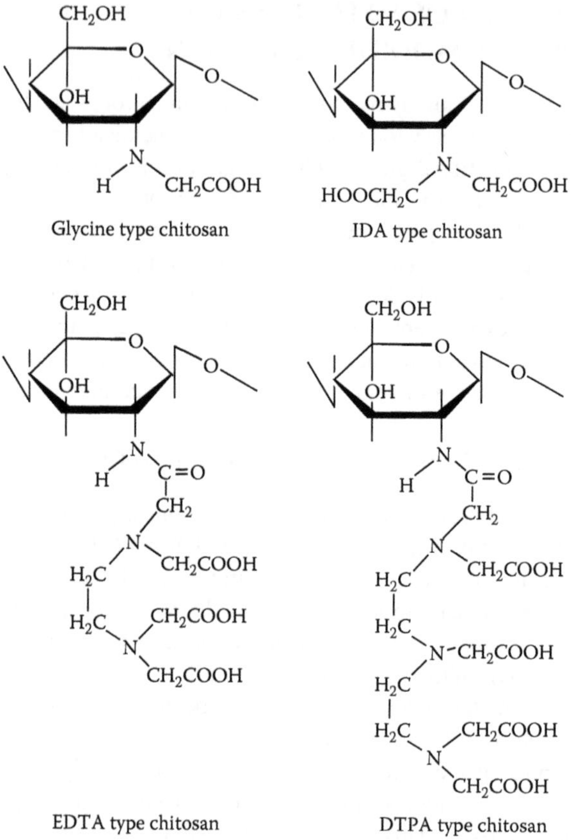

Glycine type chitosan

IDA type chitosan

EDTA type chitosan

DTPA type chitosan

SCHEME 6.7 Chemical structures of complexane type chitosans.

Figure 6.5 and Figure 6.6 depict the plot of the distribution ratio of various base metals against equilibrium pH for their adsorption on EDTA and DTPA types of chitosans, respectively. The sequence of the selectivity series for the metal ions tested is the same in both chemically modified chitosans as follows: Cu(II) > Mo(VI) > Ni(II) > V(IV) >> Zn(II) > Co(II) > Al(III). Although aluminum is very selectively adsorbed over divalent metal ions on the original chitosan, it is poorly adsorbed by these chemically modified chitosans (Scheme 6.9).

Similar adsorption tests were carried out using adsorption gels prepared from polyallylamine, which has many primary amino groups like chitosan and EDTA or DTPA. However, contrary to the EDTA and DTPA types of chitosans, similar tendencies exhibiting good separation characteristics or high selectivity were not observed for these gels, as shown as an example for the EDTA type of chemically modified polyallylamine in Figure 6.7. It is inferred that the excellent separation exhibited by EDTA and DTPA types of chitosans is not attributable to the immobilized chelating functional groups of EDTA or DTPA but to the "synergistic effect" of these functional groups in harmony with the chitosan polymer matrices.

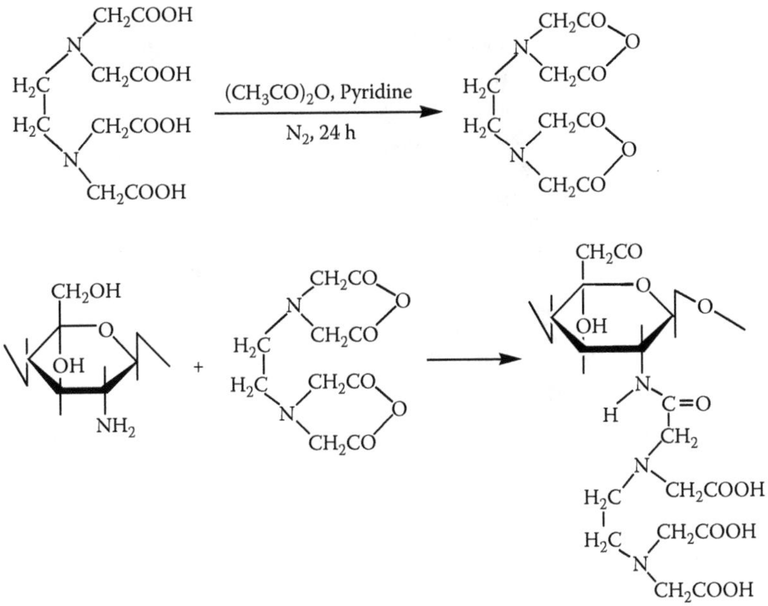

SCHEME 6.8 Synthetic route for the EDTA-type chitosan.

Based on the batch-wise experimental results, breakthrough followed by elution tests were carried out for some pairs of base metals using a column packed with EDTA or DTPA types of chitosan as well as with original chitosan for comparison. Figure 6.8a and Figure 6.8b show the breakthrough and elution profiles of nickel(II) and cobalt(II) from the column packed with EDTA-type chitosan, respectively, under the condition stated in the figure legends.

Since, as shown in Figure 6.5a, nickel is much more selectively adsorbed over cobalt, adsorbed cobalt is expelled from the column by nickel as seen from the breakthrough profile. From the elution profile, it is apparent that cobalt is not practically detected and nickel is eluted in high concentration (as high as about 55 times that of the feed solution). Figure 6.9 shows, for comparison, the breakthrough profiles of nickel(II) and cobalt(II) from the column packed with original chitosan, under the conditions stated in the figure legend. It is apparent that the separation in the breakthrough profile is much inferior to that for the EDTA-type chitosan. From this comparison, it is quite clear that separation between nickel and cobalt is much improved by using the EDTA-type chitosan.

6.4.1.2 Adsorption of Rare Earths on Complexane Types of Chemically Modified Chitosan

Figure 6.10 and Figure 6.11 show the plots of the distribution ratios of some trivalent rare earths against the equilibrium pH for the adsorption on EDTA and DTPA types of chitosans from sulfuric acid solution.[16] As seen from these figures,

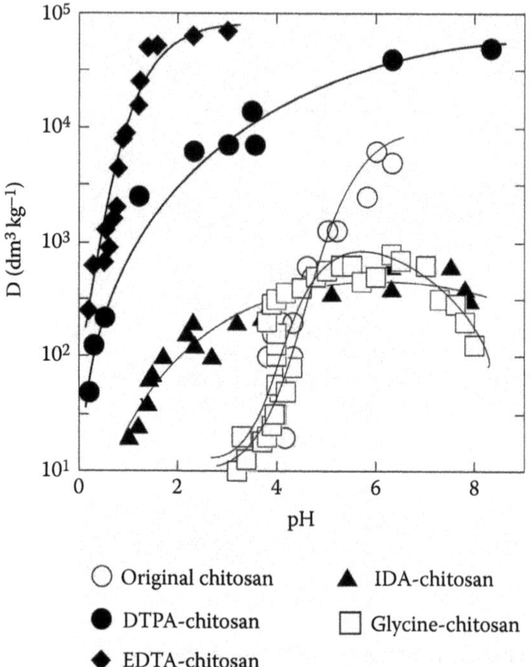

FIGURE 6.4 Distribution ratio of copper(II) against equilibrium pH for adsorption from 1 *M* ammonium nitrate solution on EDTA and DTPA types of chitosan and original chitosan for comparison.

the distribution ratios of rare earths tested are high in both figures, suggesting effective adsorption taking place on both chemically modified chitosans. On the other hand, adsorption of these metals on the original chitosan and on the IDA-type chitosan was found to be quite poor. In both figures, the plots for each rare earth lie on straight lines with a slope of 3 corresponding to each rare earth, suggesting that three hydrogen ions are released for each adsorbed trivalent rare-earth ion, indicating a cation exchange mechanism. The pH at which adsorption takes place shifts toward lower pH as the atomic number of the rare earths increases, which is the same as is observed in solvent extraction with acidic organophosphorus compounds such as 2-ethylhexyl 2-ethylhexylphosphonic acid employed in the commercial-scale purification of rare earths. Apparent equilibrium constants were evaluated for each rare earth from the intercepts of the straight lines with the slope of 3 with the ordinate at pH = 0 in these figures. Figure 6.12 shows the relationships between the evaluated apparent equilibrium constants of each rare earth and their atomic number for EDTA and DTPA types of chitosans. In these figures, the difference in the apparent equilibrium constants between adjacent rare earths represents the difficulty or ease of mutual separation between the adjacent rare earths. For example, mutual separation among light rare earths (La, Ce, Pr, Nd, and Sm) is expected to be easy with DTPA-type chitosan, while

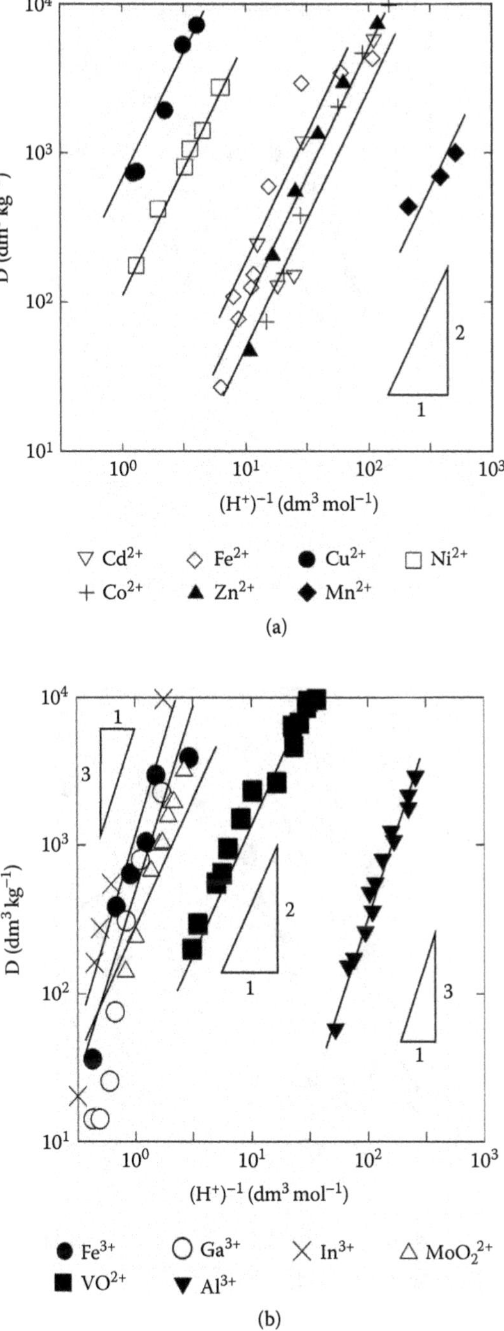

FIGURE 6.5 Distribution ratio of various base metals against equilibrium pH for their adsorption on EDTA-type chitosan.

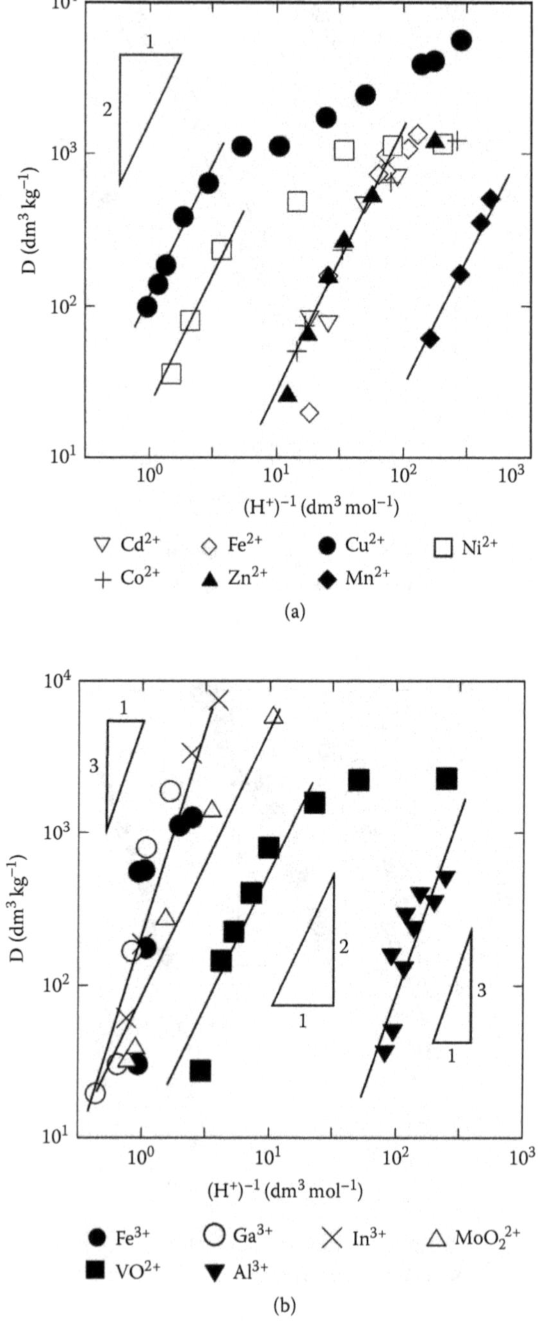

FIGURE 6.6 Distribution ratio of various base metals against equilibrium pH for their adsorption on DTPA-type chitosan.

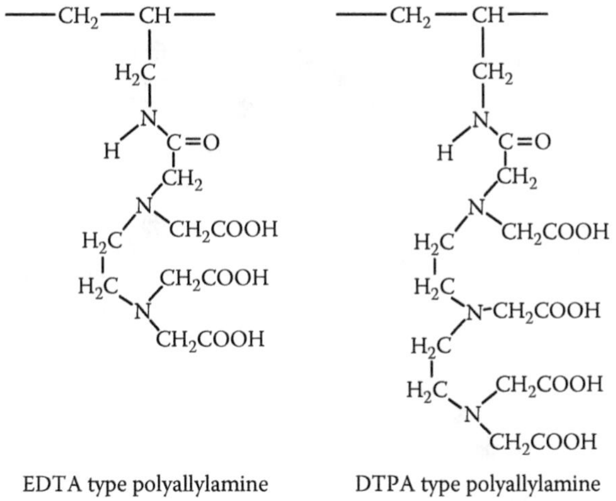

EDTA type polyallylamine DTPA type polyallylamine

SCHEME 6.9 Chemical structures of EDTA and DTPA types of chemically modified polyallylamine.

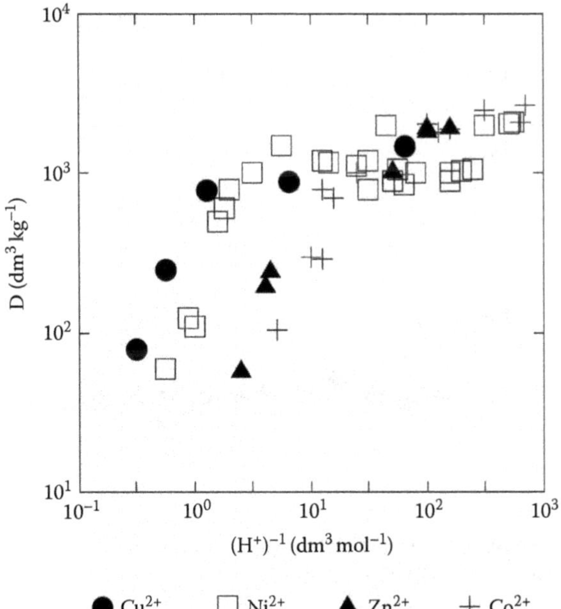

FIGURE 6.7 Distribution ratio of various base metals against equilibrium pH for their adsorption on EDTA type of chemically modified polyallylamine.

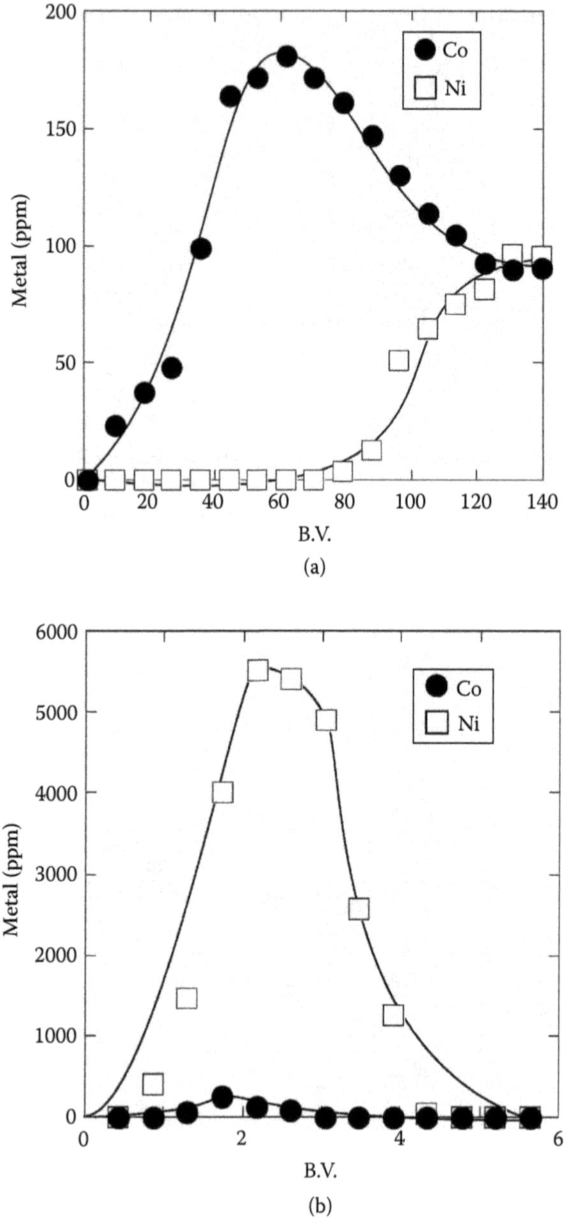

FIGURE 6.8 Breakthrough (a) and elution (b) profiles of nickel(II) and cobalt(II) from the column packed with EDTA-type chitosan.

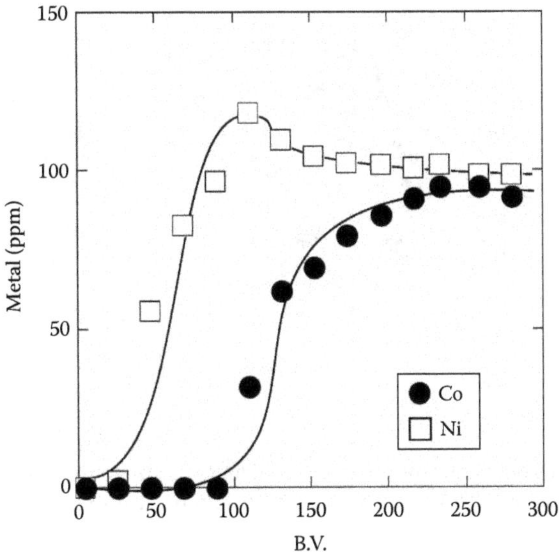

FIGURE 6.9 Breakthrough profiles of nickel(II) and cobalt(II) from the column packed with original chitosan.

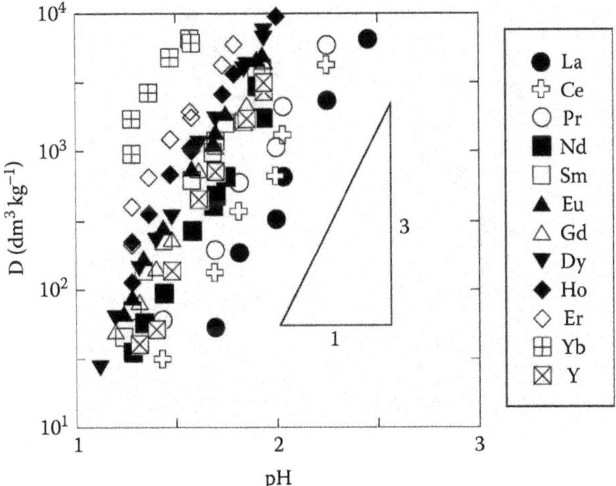

FIGURE 6.10 Distribution ratios of some trivalent rare earths against the equilibrium pH for the adsorption on EDTA-type chitosans from sulfuric acid solution.

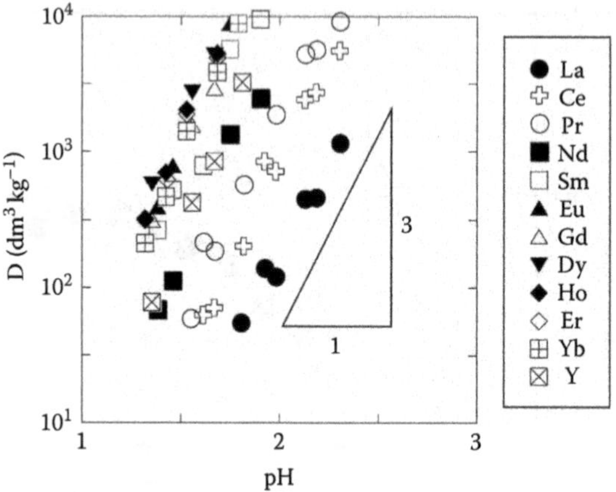

FIGURE 6.11 Distribution ratios of some trivalent rare earths against the equilibrium pH for the adsorption on DTPA-type chitosans from sulfuric acid solution.

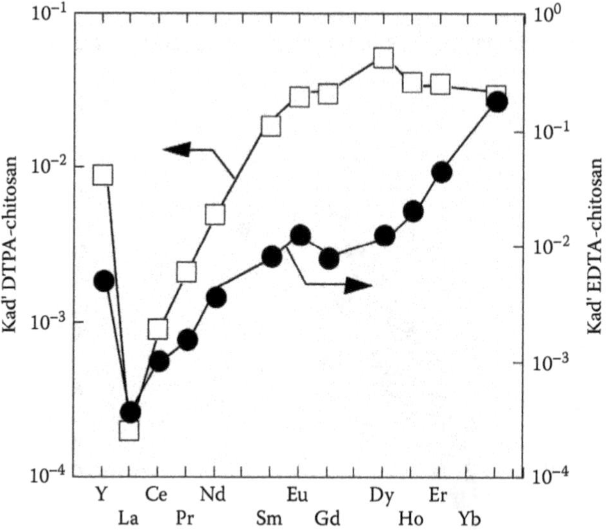

FIGURE 6.12 Relationships between the evaluated apparent equilibrium constants of each rare earth and their atomic number for EDTA- and DTPA-type chitosans.

that among middle and heavy rare earths will be difficult. Figure 6.13 shows the relationship between literature values of the stability constants of these rare earths with EDTA and DTPA and their atomic numbers.[17]

From the comparison of these two figures, it is very interesting that these apparent equilibrium constants are quite similar to the stability constants, suggesting that the chelating characteristics of EDTA and DTPA are still maintained after immobilization of these ligands on chitosan.

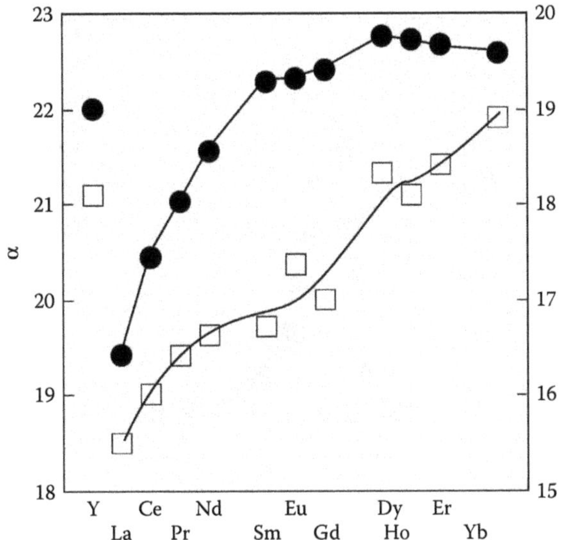

FIGURE 6.13 Relationship between literature values of the stability constants of these rare earths with EDTA and DTPA and their atomic numbers.

On the basis of the results of the batch-wise adsorption tests of rare earths, mutual separation between yttrium and samarium as well as that among lanthanum, cerium, praseodymium, and neodymium was carried out using a column packed with EDTA and DTPA types of chitosans, respectively, using dilute sulfuric acid as an eluting agent.

Figure 6.14 shows the chromatogram for the mutual separation of samarium and yttrium by using a column packed with the EDTA-type chitosan, together with the change in pH in the eluent. As seen from this figure, even these two metal ions, the mutual separation of which appears very difficult from the result of the batch experiments as discussed earlier, can be clearly and easily separated under the conditions described in the figure legend by using this column.

Figure 6.15 shows the chromatogram for the separation of the four selected light rare earths mentioned above by using a column packed with the DTPA-type chitosan together with the change in pH in the effluent. As seen from this figure, the chromatogram peaks corresponding to these four selected light rare earths are satisfactorily separated from each other. Among them, it is noteworthy that even the neodymium and praseodymium pair, the mutual separation of which is the most difficult among rare earths, can be successfully separated from each other by using this column and dilute sulfuric acid as eluting agent.

In recent years, there have been increasing demands for highly purified rare earths for the production of many kinds of novel advanced materials in high-tech industries. Ion exchange is one of the technologies for separating and purifying various metals, including rare earths. According to the conventional ion exchange process for the separation and purification of rare earths, after all of the rare earths

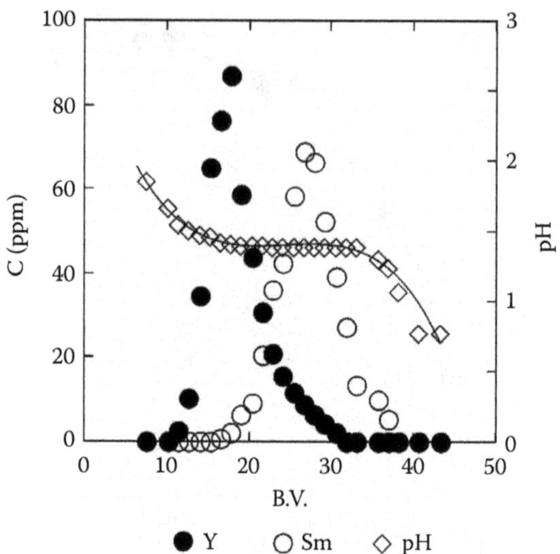

FIGURE 6.14 Chromatogram for the mutual separation of samarium and yttrium by using a column packed with the EDTA-type chitosan together with the change in pH in the eluent.

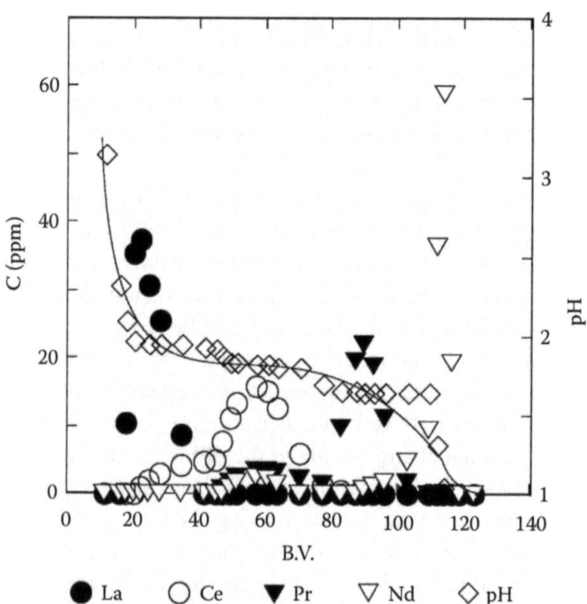

FIGURE 6.15 Chromatogram for the separation of the four selected light rare earths mentioned above by using a column packed with the DTPA-type chitosan together with the change in pH in the effluent.

are collectively loaded on the cation exchange resin packed in an adsorption column, an aqueous solution of chelating reagents such as EDTA or DTPA, which form metal chelates with each of the rare earths but with different stabilities, is passed through the column. On passing through the column, the chelating reagents desorb the rare earths loaded on the resin according to the order of the stability of these metal chelates; that is, that with highest stability with chelating reagents as such is eluted first and that with weakest stability is eluted last. This separation process is dependent not on the recognition or separation abilities of the ion exchange resin but on those of the water-soluble chelating reagents. According to this process, the majority of these expensive chelating reagents are wasted without reuse, resulting in an increase in the production costs of purified rare earths. If some chelating resins with high recognition and separation ability for rare earths were available, each of the rare earths would be mutually separated from each other at high purity and at low cost using cheap mineral acids such as hydrochloric, nitric, or sulfuric acids as the eluting reagents. From this viewpoint, some novel chelating resins having the ability for mutual separation of the rare earths have been developed.[18,19] However, since the synthetic routes for these resins are long and complicated, the production costs will be very expensive and unacceptable to industry.

On the other hand, the production of EDTA and DTPA types of chemically modified chitosans is very simple, easy, and cheap, and in addition to that, they exhibit excellent mutual separation characteristics equivalent to EDTA and DTPA. Consequently, it can be concluded that they are the most advanced and most suitable materials for the purification of rare earths.

6.4.2 Oxine Type of Chemically Modified Chitosan

Oxine (8-quinolinol) is a typical chelating reagent extensively employed in analytical chemistry. Kelex 100 and LIX 26, its derivatives with a hydrophobic long-chain alkyl radical at the seventh position, are unique commercial solvent extraction reagents. For example, Kelex 100 was reported to extract gallium from strong alkaline solution,[20] and rhodium, the most expensive and most refractory metal, from hydrochloric acid solution.[21]

The authors prepared a chemically modified chitosan incorporating oxine functional groups (oxine-type chitosan) to examine its adsorption behavior for metal ions.[22,23] Since oxine has both an acidic phenolic hydroxyl group and a basic quinoline nitrogen atom, the oxine-type chitosan is water soluble at both high and low pH; consequently, it must be cross-linked to be employed as an adsorption gel.

Similar to the case of CLC as discussed in the preceding section, cross-linking was performed using glycerolpolyglycidylether as the cross-linking reagent after complexing gallium(III) or iron(III) with the oxine-type chitosan, which are abbreviated as Ga- or Fe-oxine type chitosan hereafter. Scheme 6.10 depicts the synthetic route of the Ga-oxine type chitosan, as an example.

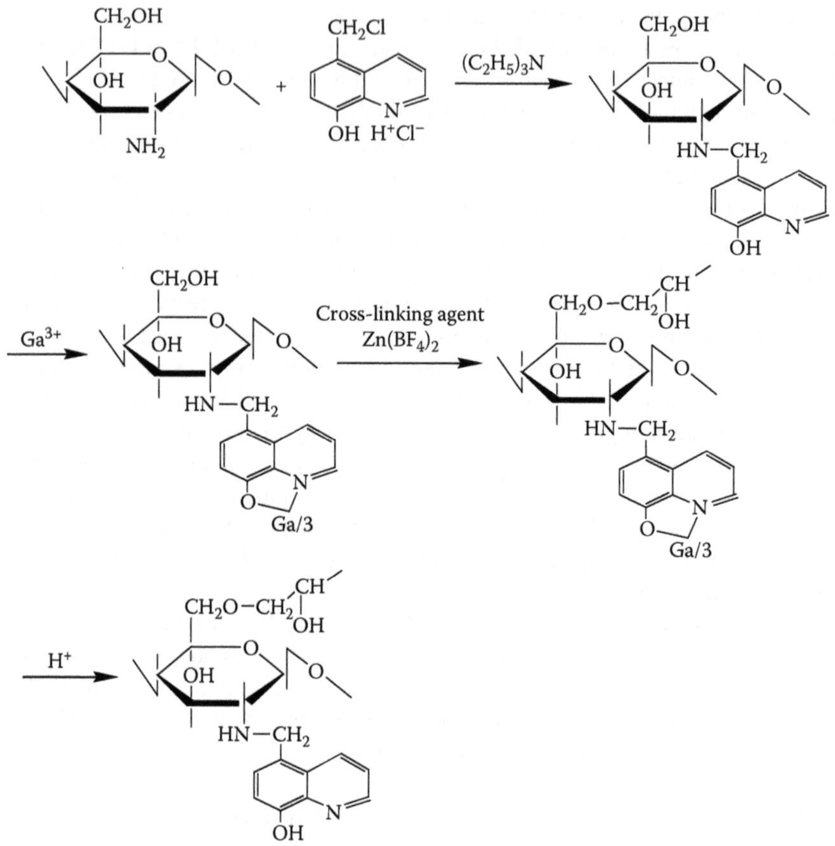

SCHEME 6.10 Synthetic route for the Ga-oxine type of chitosan.

6.4.2.1 Adsorption of Gallium(III) and Indium(III) on the Ga-Oxine Type of Chitosan

Figure 6.16 shows the plot of the distribution ratio of some metal ions against equilibrium pH for the adsorption from sulfuric acid solution on the Ga-oxine type chitosan. Compared to the adsorption on CLC as shown in Figure 6.2, the pH at which adsorption takes place is greatly shifted to lower pH values for all metal ions due to the decrease in pK_a by incorporating oxine functional groups similar to the cases of complexane-type chitosans. Although indium(III) is more selectively adsorbed over gallium(III) on CLC, the latter is slightly more selectively adsorbed over the former on the Ga-oxine type of chitosan, which may be attributable to the fact that the gallium(III) used for the complexation prior to the cross-linking played a "template" role.

Figure 6.17 and Figure 6.18 show the breakthrough and elution profiles, respectively, for the separation of a small amount of gallium and indium (10 mg/L each) from a large excess of zinc (900 mg/L) using a column packed with the

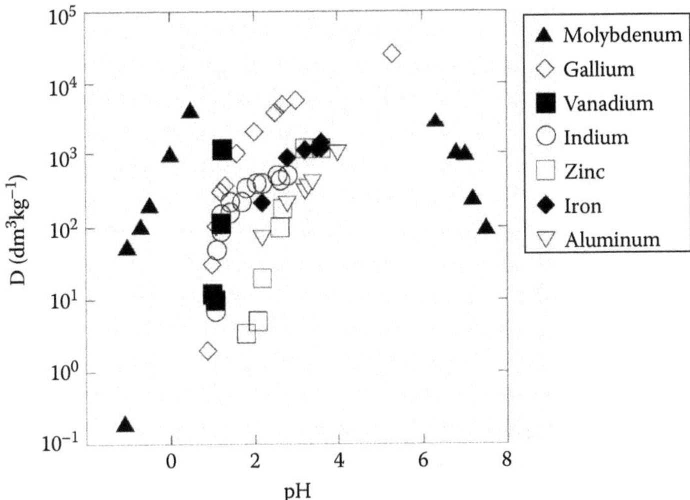

FIGURE 6.16 Distribution ratio of some metal ions against equilibrium pH for the adsorption from sulfuric acid solution on the Ga-oxine type of chitosan.

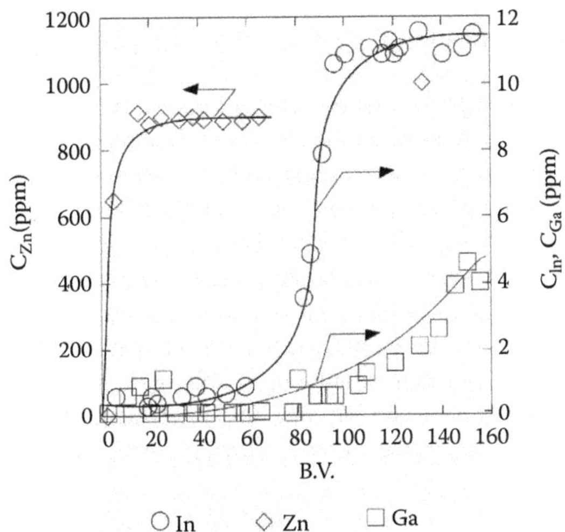

FIGURE 6.17 Breakthrough profile for the separation of a small amount of gallium and indium (10 mg/L each) from a large excess of zinc (900 mg/L) using a column packed with the Ga-oxine type of chitosan.

Ga-oxine type of chitosan. From these figures, it is evident that gallium can be effectively separated not only from a large amount of zinc but also from indium, and eluted free from zinc contamination at a concentration as high as 160 times that of the feed solution, suggesting that the Ga-oxine type of chitosan is suitable for the recovery of gallium and indium from zinc leach liquors.

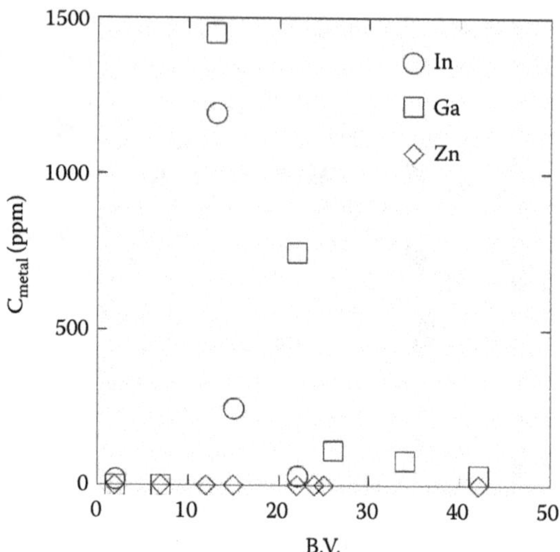

FIGURE 6.18 Elution profile for the separation of a small amount of gallium and indium (10 mg/L each) from a large excess of zinc (900 mg/L) using a column packed with the Ga-oxine type of chitosan.

6.4.2.2 Adsorption of Rhodium(III) on the Fe-Oxine Type of Chitosan

Rhodium is the most expensive metal, as mentioned earlier, and is indispensable for automobile catalytic converters in particular. Due to very limited sources on Earth, it is necessary to develop efficient recovery technology for rhodium from spent catalysts as well as from mined ores. Most of the leach liquors of platinum-group metals are chloride media, and in this media rhodium is coordinated with chloride ions to form various kinetically inert aquachloro complexes.[21] The separation and purification of rhodium is one of the most difficult steps in the processing of the platinum-group metals. It is known that the addition of stannous chloride to an aqueous solution containing rhodium(III) results in the formation of binuclear complexes between Rh and $SnCl_3^-$, reducing trivalent rhodium to the monovalent state[24] according to the following reactions:

$$RhCl_6^{3-} + 6\ SnCl_3^- \rightarrow [Rh(SnCl_3)_5]^{4-} + SnCl_6^{2-} + 3\ Cl^-$$

$$RhCl_5(H_2O)^{2-} + 6\ SnCl_3^- \rightarrow [Rh(SnCl_3)_5]^{4-} + SnCl_6^{2-} + 2\ Cl^- + H_2O$$

Since these complexes are more labile than the aquachloro complexes of Rh(III), these complexes have a high affinity for 8-quinolinol-type ligands. The authors investigated the adsorptive separation of rhodium(III) using the Fe-oxine type of chitosan gel from chloride media containing Sn(II).[23]

Figure 6.19 and Figure 6.20 show the plots of the distribution ratio of rhodium(III), copper(II), and platinum(IV) on the Fe-oxine type of chitosan

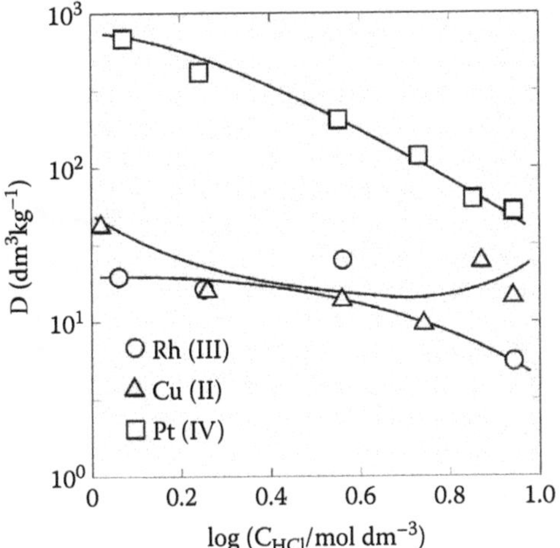

FIGURE 6.19 Distribution ratio of rhodium(III), copper(II), and platinum(IV) on the Fe-oxine type of chitosan against hydrochloric acid concentration in the absence of tin(II).

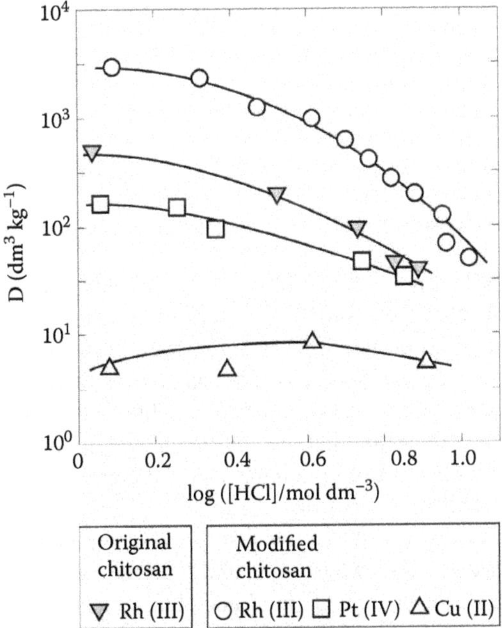

FIGURE 6.20 Distribution ratio of rhodium(III), copper(II), and platinum(IV) on the Fe-oxine type of chitosan against hydrochloric acid concentration in the presence of tin(II).

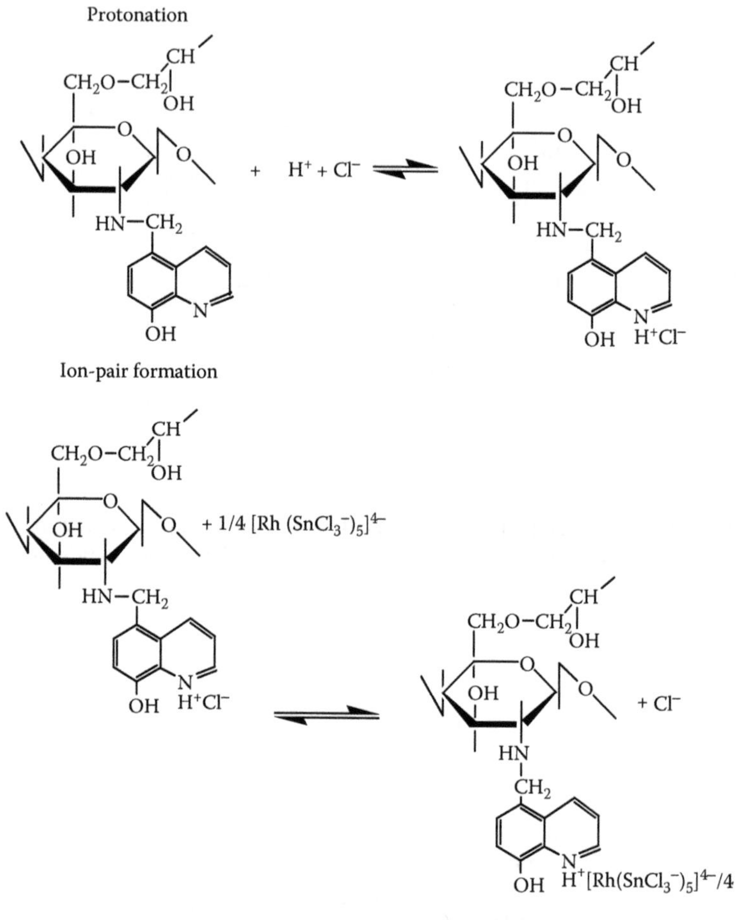

Protonation

$+ \ H^+ + Cl^-$

Ion-pair formation

$+ \ 1/4 \ [Rh \ (SnCl_3^-)_5]^{4-}$

$+ \ Cl^-$

SCHEME 6.11 Adsorption mechanism of rhodium on the Fe-oxine type of chitosan.

against hydrochloric acid concentration in the absence and presence of tin(II), respectively. In the absence of tin(II), platinum(IV) was most selectively adsorbed while only poor adsorption was observed for rhodium(III) and copper(II). On the other hand, in the presence of tin(II), the adsorption of rhodium(III) was significantly increased while platinum(IV) and copper(II) exhibited lower adsorptions than the tin(II)-free system, suggesting that, by adding a large amount of tin(II), selective separation of rhodium(III) from platinum(IV) and copper(II) can be achieved. A similar tendency had been also observed for rhodium solvent extraction with Kelex 100.[25] It was also found that the adsorption of rhodium(III) on the original chitosan is lower as a whole when compared with the Fe-oxine type of chitosan. Better adsorption on the chemically modified chitosan is attributable to the strong affinity of the binuclear complex of Rh with $SnCl_3^-$ for the oxine functional group in the chitosan matrix, on the basis of which the adsorption mechanism described by Scheme 6.11 was inferred.

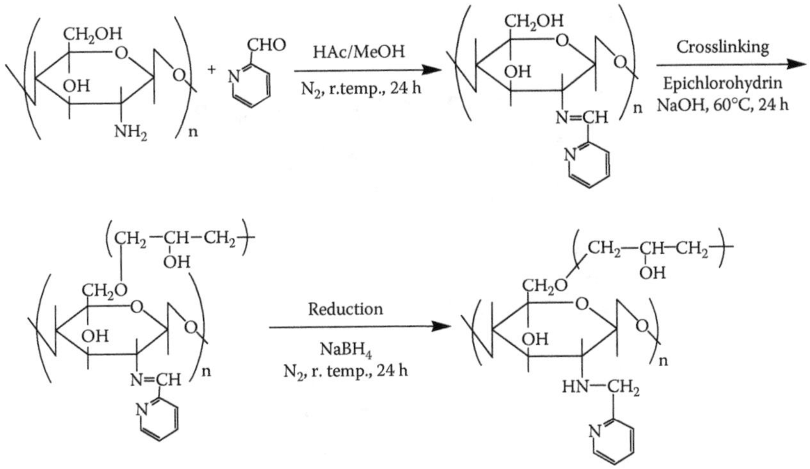

SCHEME 6.12 Synthetic routes of PMC.

The maximum adsorption capacity of rhodium was evaluated as 0.92 mol/kg dry gel at 1.2 M HCl concentration at a molar ratio of [Sn]:[Rh] = 6.

Stripping of loaded rhodium is a difficult problem because of the formation of a very stable complex. Since rhodium is adsorbed in the monovalent state, it was considered that reoxidation to the trivalent state may facilitate stripping. Some oxidation agents including $KMnO_4$, H_2O_2, and HNO_3 were tested. Maximum stripping was observed to be 72.5% for a single contact with 0.5 *M* HCl + 8 *M* HNO_3.

6.4.3 Chemically Modified Chitosan Containing Pyridine Functional Groups

Cross-linked chemically modified chitosan containing pyridine functional groups, abbreviated as PMC hereafter, was prepared by interacting pyridinealdehyde with chitosan according to the reaction described by Scheme 6.12. Here, the cross-linking was carried out by a new method via Schiff's base formation, where the cross-linking reagent, epichlorohydrin, interacts only with hydroxyl groups at the sixth position in the pyranose ring, thus avoiding attack on the Schiff base formed by the interaction of primary amino groups with pyridine aldehyde.[26–28]

Figure 6.21 shows the plots of the distribution ratio of some metal ions against equilibrium pH for their adsorption on PMC from 1 *M* aqueous ammonium nitrate solution. As seen from this figure, adsorption of copper(II), nickel(II), cobalt(II), zinc(II), and cadmium(II) takes place at pH = 0 to 4, which is three to four units lower than that on CLC as described in the previous section, while iron(III) was adsorbed at about one pH unit higher than on CLC.

The high selectivity for palladium(II), copper(II), and nickel(II) over iron(III), cobalt(II), and zinc(II) exhibited by PMC is attributable to the donating atoms that take part in the formation of metal complexes, as observed in solvent extraction with the lipophilic chitosan described in the previous section. That is, the

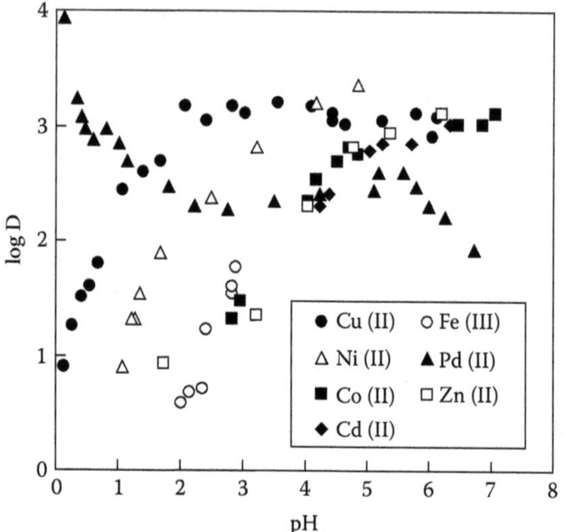

FIGURE 6.21 Distribution ratio of some metal ions against equilibrium pH for their adsorption on PMC from 1 *M* aqueous ammonium nitrate solution.

borderline Lewis base nitrogen atom of the chitosan amino group and that of the immobilized pyridine functional group work as donating atoms in this case to form five-membered chelates more selectively with those metal ions belonging to soft or borderline Lewis acids than with those that are hard Lewis acids.

In order to elucidate the adsorption mechanism of copper(II) in detail, the effects of the chemical species participating in the adsorption reaction, that is, hydrogen and nitrate ions, on the distribution ratio were examined.

Figure 6.22 shows the pH dependence of the adsorption of copper(II) from 1 *M* aqueous ammonium and sodium nitrate solutions. In the adsorption from the former solution, the distribution ratio decreases with increasing pH at high pH, while it monotonously increases in the adsorption from the latter solution, suggesting that higher ammine complexes such as $Cu(NH_3)_4^{2+}$ are not adsorbed on PMC. On the other hand, it was found to exhibit second-order dependency with respect to nitrate concentration at constant pH, suggesting that two nitrate anions participate in the adsorption of copper(II) on PMC. Accordingly, it was inferred from these results that copper(II) is adsorbed on PMC as a stable five-membered chelate, as depicted in the Scheme 6.13.

The adsorption behavior of PMC for precious metals was also examined from hydrochloric acid. From the results as shown in Figure 6.23, it is evident that PMC has a high selectivity for gold(III), palladium(II), platinum(IV), and mercury(II) over copper(II), nickel(II), cobalt(II), zinc(II), cadmium(II), and iron(III). Among these base metals, copper(II), nickel(II), and cadmium(II) were adsorbed to some extent at low hydrochloric acid concentration while iron(III), cobalt(II), and zinc(II) were not practically adsorbed over the whole hydrochloric acid concentration region. In addition to the high selectivity for some precious metals, its high

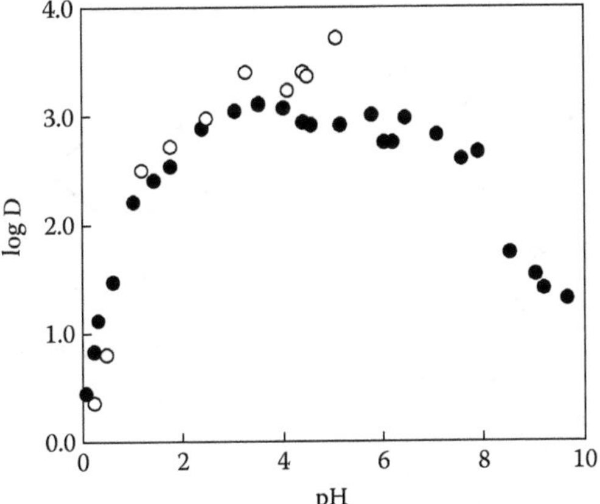

FIGURE 6.22 pH dependence of the adsorption of copper(II) from 1 *M* aqueous ammonium (●) and sodium (○) nitrate solutions.

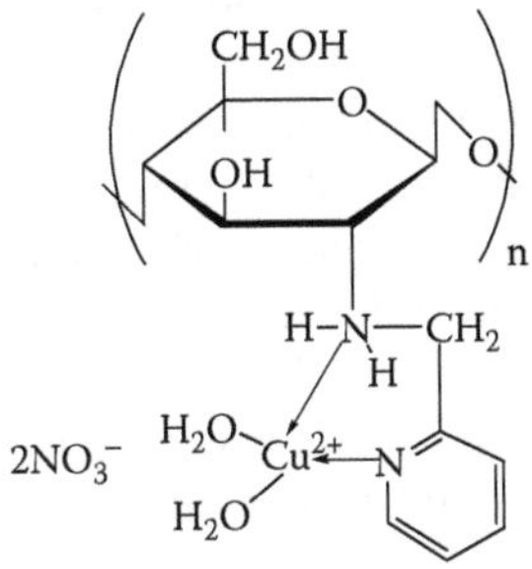

SCHEME 6.13 Inferred chemical structure of the copper(II)-PMC chelate.

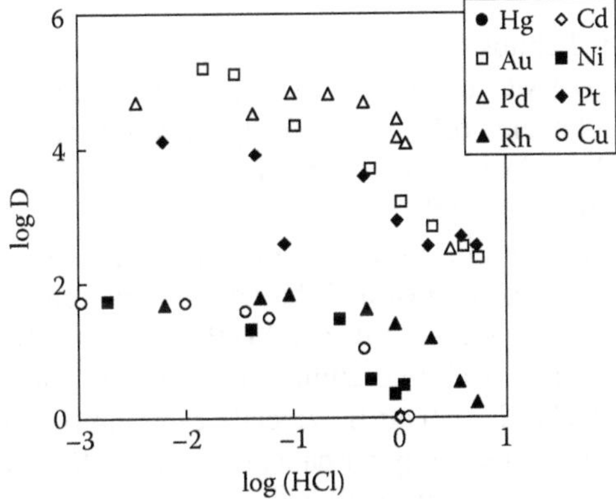

FIGURE 6.23 Adsorption of some metal ions on PMC from hydrochloric acid.

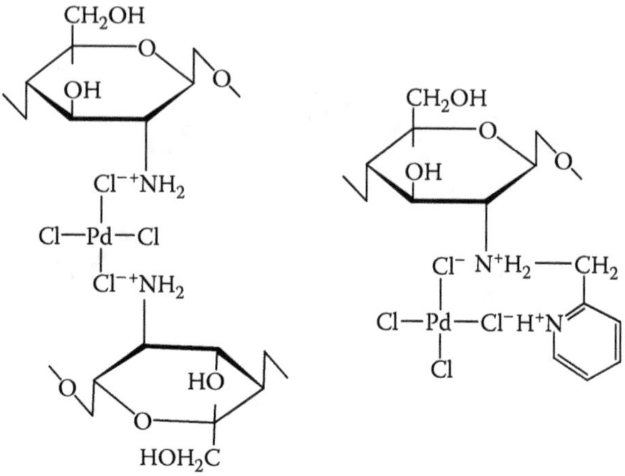

SCHEME 6.14 Inferred chemical structures of the adsorbed species of palladium(II) on CLC (left) and PMC (right).

loading capacity for palladium is noteworthy; that is, the maximum adsorption capacity on PMC at a HCl concentration of 0.1 *M* was as high as 5.8 mol/kg dry gel, while those on CLC and on a commercially available polyethelenepolyamine type of chelating resin, DIAION CR-20, were 2.1 and 1.8 mol/kg, respectively.

From the concentration dependencies of chloride and hydrogen ions on the adsorption of palladium(II) on PMC, it was inferred that adsorption took place according to an anion exchange mechanism wherein the tetrachloro complex forms an ion pair complex with the protonated form of PMC. Scheme 6.14 shows the inferred chemical structure of the ion pair palladium(II) complex with PMC together with that with CLC.

6.4.4 Chemically Modified Chitosan with Sulfur-Containing Functional Groups

Chemically modified chitosan containing functional groups of thiophene or thioether, abbreviated as TMC and MTPC, were prepared according to the synthetic routes described in Scheme 6.15 to selectively adsorb those precious metals that are soft Lewis acids by making use of the high affinity of the sulfur atom, which is a typical soft Lewis base.[29,30]

Figure 6.24 and Figure 6.25 depict the relationship between the distribution ratio of some precious metal ions and the equilibrium concentration of hydrochloric acid in the adsorption on TMC and MTPC, respectively. It was found that they showed little or practically no affinity towards the majority of base metals such as copper(II), nickel(II), cobalt(II), zinc(II), cadmium(II), and iron(III), while they exhibited very high selectivity for palladium(II), gold(III), and platinum(IV). The selectivity to precious metals exhibited by TMC and MTPC was higher than that found for PMC.

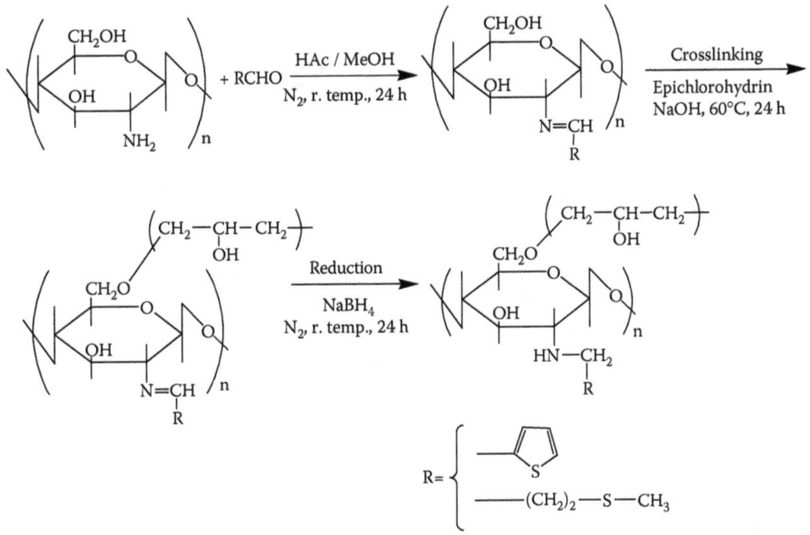

SCHEME 6.15 Synthetic routes for TMC and MTPC.

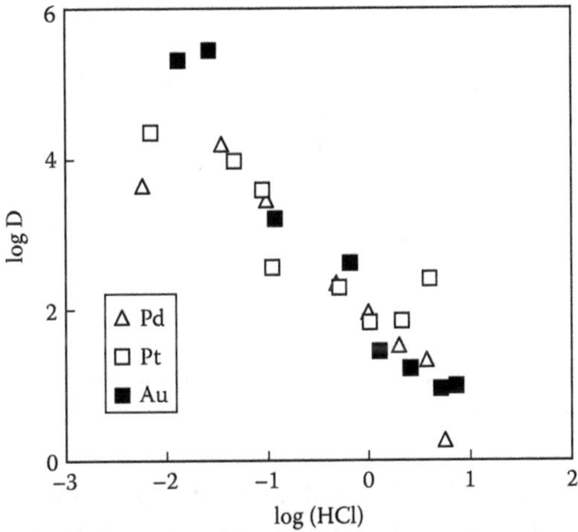

FIGURE 6.24 Relationship between the distribution ratio of some precious metal ions and the equilibrium concentration of hydrochloric acid in the adsorption on TMC.

From the concentration dependencies of hydrogen and chloride ions on the adsorption of palladium(II) on TPC and MTPC, it was inferred that palladium(II) is adsorbed according to an anion exchange mechanism accompanied by the formation of a stable five-membered ion pair ring, as shown in Scheme 6.16.

The maximum adsorption capacity for palladium(II) was remarkably high also in these cases; that is, it was as high as 4.8 and 5.7 mol/kg for TMC and MTPC, respectively.

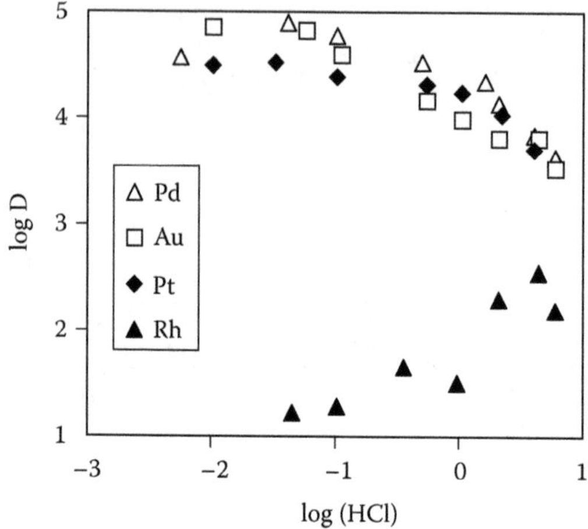

FIGURE 6.25 Relationship between the distribution ratio of some precious metal ions and the equilibrium concentration of hydrochloric acid in the adsorption on MTPC.

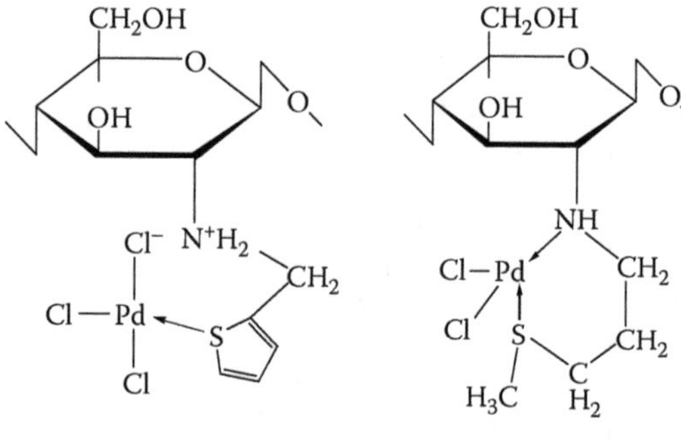

SCHEME 6.16 Inferred chemical structures of the palladium(II) complex adsorbed on TMC (left) and MTPC (right).

6.5 SOLVENT EXTRACTION OF METAL IONS WITH CHEMICALLY MODIFIED LIPOPHILIC CHITOSAN

Similar to the chemically modified chitosan, it is possible to chemically modify lipophilic chitosan by incorporating some functional groups with high affinity for specified metal ions onto its primary amino groups to synthesize novel solvent extraction reagents based on chitosan.[7,31]

The authors prepared lipophilic chitosan chemically modified with functional groups of dithiocarbamate, which is abbreviated as DTC-type lipophilic chitosan, by interacting carbon disulfide with the lipophilic chitosan in ethanol containing ammonia according to the following synthetic route (Scheme 6.17).

The extent of the immobilization of the functional groups of dithiocarbamate onto the glucosamine unit was determined to be as high as 96%, although it was

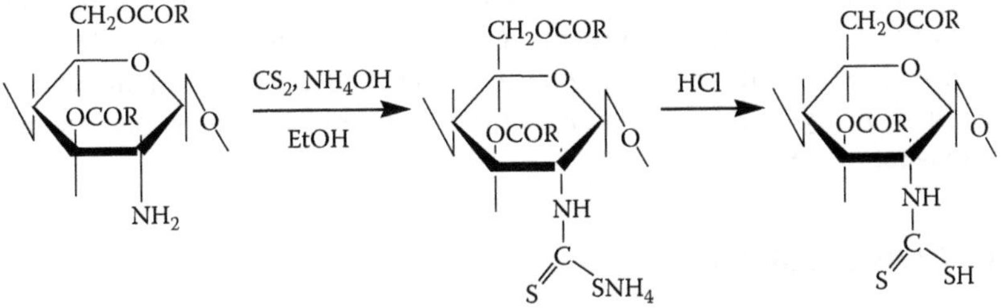

SCHEME 6.17 Synthetic route for the DTC-type lipophilic chitosan.

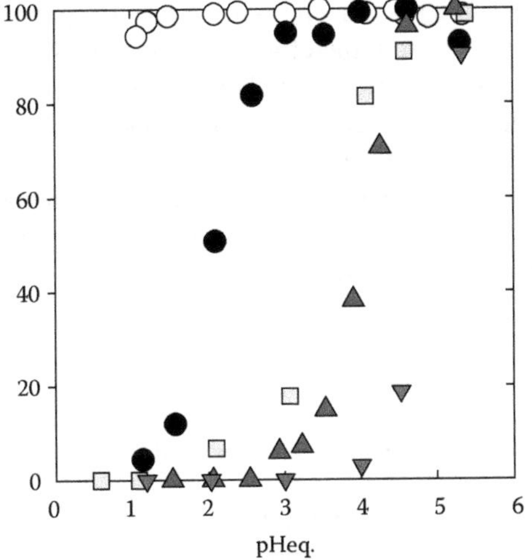

○: Cu (II), ●: Ni (II), □: Fe (III), ▲: Co (II), ▼:Zn (II).

FIGURE 6.26 Percent extraction of some base metals against equilibrium pH for extraction with 2.3 kg/m^3 of the DTC-type lipophilic chitosan in kerosene from an aqueous mixture of 0.1 *M* nitric acid and HEPES buffer solution.

only 25% for immobilization onto the original chitosan, the solid feed material.[32] It was found that the DTC-type lipophilic chitosan is soluble not only in chloroform and toluene but also in hexane and kerosene.

Figure 6.26 shows the plot of percent extraction of some base metals against equilibrium pH for extraction with 2.3 kg/m^3 of the DTC-type lipophilic chitosan in kerosene from an aqueous mixture of 0.1 *M* nitric acid and a HEPES (*N*-2-hydroxyethylpiperazine-*N*′-2-ethanesulfonic acid), buffer solution. From this figure, it is seen that copper(II) and nickel(II) are highly selectively extracted over iron(III), cobalt(II), and zinc(II). In particular, the high selectivity for copper(II) and nickel(II) over iron(III) is noteworthy, which may be attributable to the typical soft Lewis base dithiocarbamate functional groups, which excludes ferric ions, a typical hard Lewis acid. Eighty-two percent stripping of the loaded copper(II) was achieved by a single contact with 1.2 *M* sulfuric acid solution.

Further, a Japanese patent[33] claims that trivalent americium can be highly selectively separated from trivalent lanthanides with this reagent. It can be expected that trivalent actinides, long-lived radioactive nuclides, may be effectively and highly selectively removed from high-level nuclear wastes by using this reagent in the future.

6.6 CONCLUSION

As described in this chapter, chitosan, a natural polysaccharide, can be used in various fields of separation of metals as an adsorption gel and as a solvent extraction reagent. Contrary to artificial synthetic resins and solvent extraction reagents produced from petroleum, chitosan-based adsorption gels and extraction reagents are biodegradable and, therefore, environmentally benign. Although, at present, chitosan is produced from the shells of crabs, shrimps, and prawns or the cartilage of squids, novel technology for producing chitosan-like polymeric materials named "bacteria chitosan" using microorganisms at low cost has been developed in recent years.[34] It can be expected that novel, environmentally benign separation technologies using chitosan or chitosan-like polymeric materials will be commercialized in the near future.

REFERENCES

1. For example, Muzzarelli, R.A.A., in *Natural Chelating Polymers*, Pergamon Press, Oxford, 1973, 144–226.
2. Ghimire, K.N., Inoue, K., Miyajima, T., Yoshizuka, K., and Shoji, T., Adsorption of some metal ions and mineral acids on chitin, *Chitin & Chitosan Res.*, 7, 61–68, 2001.
3. Ohga, K., Kurauchi, Y., and Yanase, H., Adsorption of Cu^{2+} or Hg^{2+} ion on resins prepared by cross-linking metal-complexed chitosans, *Bull. Chem. Soc. Jpn.*, 66, 444–446, 1987.
4. Nishide, H. and Tsuchida, E., Selective adsorption of metal ions on poly(4-vinylpyridine) resins in which the ligand chain is immobilized by cross-linking, *Makromol. Chem.*, 177, 2295–2310, 1976.
5. Inoue, K., Baba, Y., and Yoshizuka, K., Adsorption of metal ions on chitosan and cross-linked copper(II)-complexed chitosan, *Bull. Chem. Soc. Jpn.*, 66, 2915–2921, 1993.
6. Inoue, K., Yoshizuka, K., Ohto, K., and Seki, S., Solvent extraction behavior of O,O′-decanoylchitosan for metal ions, *Kagakukogaku Ronbunshu* (in Japanese), 26, 248–250, 2000.
7. Inoue, K., Yoshizuka, K., and Ohto, K., Development environmentally benign new solvent extraction reagents from chitosan, a natural polysaccharide, in *Proceedings of the International Solvent Extraction Conference, ISEC 2002*, South African Institute of Mining and Metallurgy, Johanesburg, 453–457, 2002.
8. Dhakal, R.P., Inoue, K., Yoshizuka, K., Ohto, K., Yamada, M., and Seki, S., Solvent extraction of some metal ions with lipophilic chitin and chitosan, *Solv. Extr. Ion Exch.*, 23, 529–543, 2005.

9. Nishimura, S., Kohga, O., Kurita, K., Vittavatvong, C., and Kuwahara, H., Synthesis of novel chitosan derivatives soluble in organic solvents by regioselective chemical modifications, *Chem. Lett.*, 19, 243–246, 1990.
10. Inoue, K., Ohto, K., Yoshizuka, K., Shinbaru, R., Baba, Y., and Kina, K., Adsorption behavior of metal ions on some carboxymethylated chitosans, *Bunseki Kagaku* (in Japanese), 42, 725–731, 1993.
11. Inoue, K., Ohto, K., Yoshizuka, K., Shinbaru, R., and Kina, K., Adsorption behaviors of some metal ions on chitosan modified with EDTA-type ligand, *Bunseki Kagaku* (in Japanese), 44, 283–287, 1995.
12. Inoue, K., Ohto, K., Yoshizuka, K., Yamaguchi, T., and Tanaka, T., Adsorption of lead(II) ion on complexane types of chemically modified chitosan, *Bull. Chem. Soc. Jpn.*, 70, 2443–2447, 1997.
13. Inoue, K., Yoshizuka, K., and Ohto, K., Adsorptive separation of some metal ions by complexing agent types of chemically modified chitosan, *Anal. Chim. Acta*, 388, 209–218, 1999.
14. Nagib, S., Inoue, K., Yamaguchi, T., and Tamaru, T., Recovery of Ni from large excess of Al generated from spent hydrodesulfurization catalyst using picolylamine type chelating resin and complexane types of chemically modified chitosan, *Hydrometallurgy*, 51, 73–85, 1999.
15. Inoue, K., Yamaguchi, T., Iwasaki, M., Ohto, K., and Yoshizuka, K., Adsorption of some platinum group metals on some complexane types of chemically modified chitosan, *Sep. Sci. Technol.*, 30, 2477–2489, 1995.
16. Inoue, K., Chromatographic separation of rare earths with complexane types of chemically modified chitosan, in *Advances in Chitin Science*, Vol. 4, Peter, M.G., Domard, A., and Muzzarelli, R.A.A., Eds., Universitaet Potsdam, Potsdam, 2000, 460–465.
17. Powell, J.E., in *The Rare Earths,* Chap. 5, Spedding, F.H. and Daane, A.H., Eds., John Wiley & Sons Inc., New York, 1961; 55–81.
18. Takeda, K., Akiyama, M., Kawakami, F., and Sasaki, M., Recovery of highly-purified rare earth elements using newly-synthesized chelating resins, *Bull. Chem. Soc. Jpn.*, 59, 2225–2232, 1986.
19. Kanesato, M., Yokoyama, T., and Suzuki, T.M., Chromatographic separation of rare earth pairs by a chelating resin having bis(carboxymethyl)amino groups, *Bull. Chem. Soc. Jpn.*, 62, 3451–3456, 1989.
20. Leveque, A. and Helgorsky, J., The recovery of gallium from Bayer process aluminate solutions by liquid–liquid extraction, in *Proceedings of International Solvent Extraction Conference, ISEC 77*, Vol. 2, Lucas, H., Ritcey, G.M., and Smith, H.W., Eds., The Canadian Institute of Mining and Metallurgy, Montreal, 1979, 439–442.
21. Benguerel, E., Demopoulos, G.P., and Harris, G.B., Speciation and separation of rhodium(III) from chloride solutions: a critical review, *Hydrometallurgy*, 40, 135–152, 1996.
22. Inoue, K., Hirakawa, H., Ishikawa, Y., Yamaguchi, T., Nagata, J., Ohto, K., and Yoshizuka, K., Adsorption of metal ions on gallium(III)-templated oxine type of chemically modified chitosan, *Sep. Sci. Technol.*, 31, 2273–2285, 1996.
23. Alam, M.S., Inoue, K., Yoshizuka, K., and Ishibashi, H., Adsorptive separation of rhodium(III) using Fe(III)-templated oxine type of chemically modified chitosan, *Sep. Sci. Technol.*, 33, 655–666, 1998.

24. Zolotov, Yu.A., Petrukhin, O.M., Shevchenko, V.N., Dunina, V.V., and Rukhadze, E.G., Solvent extraction of noble metals with derivatives of thiourea, *Anal. Chim. Acta*, 100, 613–618, 1978.
25. Alam, M.S. and Inoue, K., Extraction of rhodium from other platinum group metals with Kelex 100 from chloride media containing tin, *Hydrometallurgy*, 46, 373–382, 1997.
26. Baba, Y. and Hirakawa, H., Selective adsorption of palladium(II), platinum(IV), and mercury(II) on a new chitosan derivative possessing pyridyl group, *Chem. Lett.*, 21, 1905–1908, 1992.
27. Baba, Y., Koichi, K., and Kawano, Y., Selective adsorption of copper(II) over iron(III) on chitosan derivative introducing pyridyl group, *Chem. Lett.*, 23, 2389–2392, 1994.
28. Baba, Y., Koichi, K., and Kawano, Y., Synthesis of a chitosan derivative recognizing planar metal ion and its selective adsorption equilibria of copper(II) over iron(III), *React. Polym.*, 36, 167–172, 1998.
29. Baba, Y., Hirakawa, H., and Kawano, Y., Selective adsorption of precious metals on sulfur-containing chitosan derivatives, *Chem. Lett.*, 23, 117–120, 1994.
30. Baba, Y., Kawano, Y., and Hirakawa, H., Highly selective adsorption resins. I. Preparation of chitosan derivatives containing 2-pyridylmethyl, 2-thienylmethyl, and 3-(methylthio)prolyl groups and their selective adsorption of precious metals, *Bull. Chem. Soc. Jpn.*, 69, 1255–1261, 1996.
31. Inoue, K., Yoshizuka, K., Ohto, K., and Nakagawa, H., Solvent extraction of some metal ions with lipophilic chitosan chemically modified with functional groups of dithiocarbamate, *Chem. Lett.*, 20, 698–699, 2001.
32. Asakawa, T., Inoue, K., and Tanaka, T., Adsorption of silver on dithiocarbamate type of chemically modified chitosan, *Kagakukogaku Ronbunshu* (in Japanese), 26, 321–326, 2000.
33. Inoue, K., Tachimori, S., and Naganawa, H., Separation of Americium from Lanthanides by Means of Solvent Extraction, Japanese Patent (Kokai), No. 2003–185792, 2003
34. Fujita, M., Ike, M., Tachibana, S., Kitada, G., Kim, S.M., and Inoue, Z., Characterization of a bioflocculant produced by *Citrobacter* sp. TKF04 from acetic and propionic acids, *J. Biosci. Bioeng.*, 89, 40–46, 2000.

7 Short-Bed Ion Exchange

Craig J. Brown

CONTENTS

7.1 Historical Perspective on Ion Exchange Column Design 376
7.2 Conventional Ion Exchange Designs 377
7.2.1 Resin Particle Size 377
7.2.2 Resin Bed Height 377
7.2.3 Freeboard 377
7.2.4 Flow Rates 378
7.2.5 Operating Capacity 378
7.2.6 Cycle Times 378
7.2.7 Column Redundancy 378
7.3 Recoflo® Short-Bed Ion Exchange 378
7.4 Recoflo Features 380
7.4.1 Exchange Zone 380
7.4.2 No Column Freeboard 380
7.4.3 Countercurrent Regeneration 381
7.4.4 Low Resin Loading 382
7.4.5 Fine-Mesh Resin 383
7.4.6 Flow Distribution 385
7.5 Short-Bed Column Design 388
7.6 Prefiltration 389
7.7 Application of Short-Bed Ion Exchange 390
7.7.1 Chromic Acid Recovery 391
7.7.2 Chromic Acid Purification 391
7.7.3 Nickel Salt Recovery 391
7.7.4 Selective Copper Recovery from Printed Circuit Board Wastewaters 393
7.7.5 Acid Retardation 393
7.7.6 Sulfuric Acid Aluminum Anodizing and Steel Pickling 393
7.7.7 Aluminum Etchants 394
7.7.8 Stainless Steel Pickle Liquors 395
7.7.9 Metal Refinery Electrolyte Bleeds 397
7.7.10 Pulp and Paper Industry 397
7.7.11 Amine Purification 398

7.7.12 Water Demineralization 398
7.8 Conclusion 402
References 402

7.1 HISTORICAL PERSPECTIVE ON ION EXCHANGE COLUMN DESIGN

The basic hydromechanical design of ion exchange columns goes back over 70 years and is largely based upon the pressure sand filter, which was well established when synthetic ion exchange resins were first commercialized. The major market for ion exchange plants at that time, the power industry, was already well equipped to design and fabricate the tall steel pressure vessels required, since they were very similar to the designs used in making boilers. There was little incentive to consider other possibilities.

The original designs utilized coflow (i.e., cocurrent) regeneration of the columns since downflow service and regeneration required virtually no modification to the basic equipment. Although the principles and advantages of countercurrent regeneration, notably reduced chemical consumption, were recognized very early, there was little impetus to exploit them at first. Regenerant chemicals were inexpensive and virtually no consideration was given to the treatment of regeneration wastewater. Countercurrently regenerated ion exchange plants have been available since the 1960s and have seen increased application in Europe since that time. Adoption in North America lagged notably, however, probably due to a greater emphasis on minimizing capital costs even with the penalty of higher operating costs.

It is also noteworthy that, to a large extent, development in the field of ion exchange was driven by the ion exchange resin manufacturers until relatively recently. They are indeed to be commended for development of a wide variety of ion exchange resins with outstanding operating characteristics and stability. At the same time, it must be recognized that these commodity chemical companies were interested in selling as much of this material as possible. Any technology that would significantly reduce the quantity of ion exchange resins purchased was not necessarily in their best interest.

The past decade or two have seen the introduction of a new generation of ion exchange designs in North America and Europe. Promoted by several of the major resin manufacturers, these so-called packed-bed designs include Upcore® (Dow Chemical), Puropack® (Purolite Company), and Amberpack® (Rohm & Haas Co.). Although these systems have minor differences in design, they generally tend to share a number of common features. These include:

- Counterflow (i.e., countercurrent) regeneration
- Packed resin beds (i.e., little or no freeboard)
- Somewhat finer or more uniform particle resins
- Somewhat shorter resin columns
- Shorter operating cycles

The major advantages claimed for these systems are lower regenerant chemical consumption, higher demineralized water purity, and smaller equipment size. These systems have indeed sparked renewed interest in ion exchange demineralization. The improved performance has helped reverse, or at least slow, the industry-wide trend towards the use of reverse osmosis in lieu of ion exchange.

A concept coined *process intensification*[1] defines what one might consider a new paradigm in process design. Process intensification refers to technologies that replace large, expensive, energy-intensive equipment or processes with ones that are smaller, less costly, and more efficient. Although the packed-bed designs have advanced the state of the art somewhat by addressing these issues, they should be considered an evolution of the technology, with only incremental improvements. Process intensification as it pertains to ion exchange has been limited by traditional fixed-bed ion exchange design principles, which have been ingrained into the thinking of most water-treatment professionals.

If one can sidestep conventional thinking and extrapolate the same features outlined above into a new paradigm, it is possible to see how more dramatic improvements can be made to ion exchange equipment and process design.

In the mid 1960s, researchers at the University of Toronto outlined the basic principles of a new ion exchange technology called *reciprocating flow ion exchange*, which utilized very short ion exchange columns. This short-bed technology was a radical departure from industry thinking and, indeed, is not fully appreciated by many in the field today.

7.2 CONVENTIONAL ION EXCHANGE DESIGNS

In order to appreciate the differences between conventional and short-bed ion exchange technology, it is necessary to establish a basis of comparison. Although there are indeed small differences between the designs of one manufacturer to the next, most will agree that usual designs in water demineralization (the major application) are based upon the following parameters.

7.2.1 Resin Particle Size

Most resins used industrially today are 0.3 to 1.2 mm (20 to 50 mesh) in diameter with an effective size of 0.4 to 0.8 mm.[2] Recently, monodisperse resins with a uniform particle size within this range have seen increased use.

7.2.2 Resin Bed Height

The minimum recommended resin bed height for most applications is 60 to 90 cm, although 100- to 200-cm beds are more commonly seen in practice.

7.2.3 Freeboard

Conventional designs normally allow at least 50% freeboard space over the settled resin bed to accommodate resin swelling and cleaning of the resin bed by backwash or fluidization. This makes design of countercurrently regenerated systems considerably more complex, as fluidization during upflow must be avoided during service or regeneration. When the freeboard space is included, the total column height is typically 200 to 400 cm. Newer packed-bed designs significantly reduce or eliminate this freeboard space entirely.

7.2.4 Flow Rates

In water demineralization, normal liquid velocities range from 30 to 50 cm/min (7 to 12 gpm/ft^2) for cation resin, down to 20 to 30 cm/min (5 to 7 gpm/ft^2) for strong base anion. For a 100-cm bed height, these translate into maximum space velocities of about 10 to 30 BV/h. Minimum velocities of 8 cm/min (2 gpm/ft^2) are recommended to avoid channeling.

7.2.5 Operating Capacity

Cycles times are normally maximized by applying a sufficient regenerant chemical dosage to utilize at least 75% of the total available resin capacity.

7.2.6 Cycle Times

The normal design philosophy is to maximize the length of the service run so that a column will remain in service for at least one 8-h shift before regeneration. Depending upon the concentration of ions in the feed stream, this is done by varying the bed height, flow rate, and regeneration dosage. Regeneration can take 1 to 3 h for each column.

7.2.7 Column Redundancy

Since it typically takes several hours to regenerate and rinse a conventional ion exchange column, it is common practice to install an extra column that is placed in service while the other is in regeneration. Although this adds additional cost, it is necessary to provide continuous service.

7.3 RECOFLO® SHORT-BED ION EXCHANGE

When comparing a conventional ion exchange demineralizer design such as the one shown in Figure 7.1 with the short-bed unit shown in Figure 7.2, it is readily apparent that this technology is indeed a dramatic departure from conventional thinking. The height of the columns is only a few centimeters compared to several meters. It is not possible to construct a working short-bed ion exchange system by merely reducing the wall height of a conventional system, however. Use of a short bed of ion exchange resin introduces a number of new issues that have an impact upon the performance of the process.

The principles of short-bed ion exchange date back four decades to the University of Toronto with a process originally called *reciprocating flow ion exchange*, which

FIGURE 7.1 Conventional (i.e., tall-bed) ion exchange unit. Ion exchange columns are typically 200 to 400 cm high. (Reprinted with permission.)

FIGURE 7.2 Short-bed ion exchange demineralizer.

has evolved to a process known today as Recoflo (Eco-Tec Inc.). Recoflo remains the only short-bed ion exchange process used on an industrial scale. Although acceptance of the technology has not been universal, well over 1500 Recoflo systems have been installed in over 50 different countries. Recoflo can no longer be considered novel and is indeed now well proven and its advantages well documented.[3] Although the process has been extensively used for chemical purification and waste recovery, the gradual but steady acceptance of Recoflo in water treatment[4,5] is at least partly a reflection of the more conservative nature of that industry.

A detailed explanation of the principles of short-bed ion exchange was presented by Hunter[6] in 1963 and initial patents were issued in 1968.[7,8] After more than 40 years, we can now look at how this technology has been applied with some historical perspective and more easily rationalize why it works and what the limitations are. The following is a somewhat qualitative discussion of the basic features of the Recoflo short-bed process. The reader is referred to the aforementioned writers for a more rigorous treatment.

7.4 RECOFLO FEATURES

7.4.1 Exchange Zone

Although, in principle, the resin bed height can be reduced to as little as about 2 to 3 cm, in practice, the height of the resin beds in industrial Recoflo systems varies between 7.5 and 60 cm, depending on the application. Despite this dramatic reduction in bed height, the performance of these systems often exceeds conventional ion exchange technology. How is this possible?

At any time in a working ion exchange bed, ion exchange is actually taking place in only a portion of the bed called the mass-transfer or exchange zone. The resin ahead of this zone is already exhausted and the resin downstream of this zone is still fresh. This is illustrated in Figure 7.3A. As the cycle proceeds, the exchange zone moves through the bed in the direction of fluid flow. When the leading edge of the exchange zone reaches the exit end of the bed, ions begin to leak into the treated liquid effluent and the ion exchange bed must be regenerated. If regeneration is done countercurrently, the zone is then pushed back up the bed.

In a conventional ion exchange column, the exchange zone typically represents a relatively small portion of the height of the resin bed. The basic premise of short-bed ion exchange is that the length (i.e., height) of the resin bed needs to be only just slightly greater than the length of the exchange zone, as shown in Figure 7.3B. Of course, the duration of the loading cycle will be much shorter, but this is compensated by the fact that the regeneration cycle will also be proportionally shorter.

7.4.2 No Column Freeboard

In a conventional fixed-bed design, 50 to 100% freeboard is present to allow the resin to change volume and to allow the bed to be expanded during backwash. With a short-bed system, it is essential that the exchange zone move up and down the bed during loading and regeneration cycles in a consistent manner. In order for this to occur, the resin bed must be immobilized. This is done by eliminating the freeboard.

Eliminating the freeboard in a conventional ion exchange column is problematic. Resins reversibly change volume as they change ionic form and depending

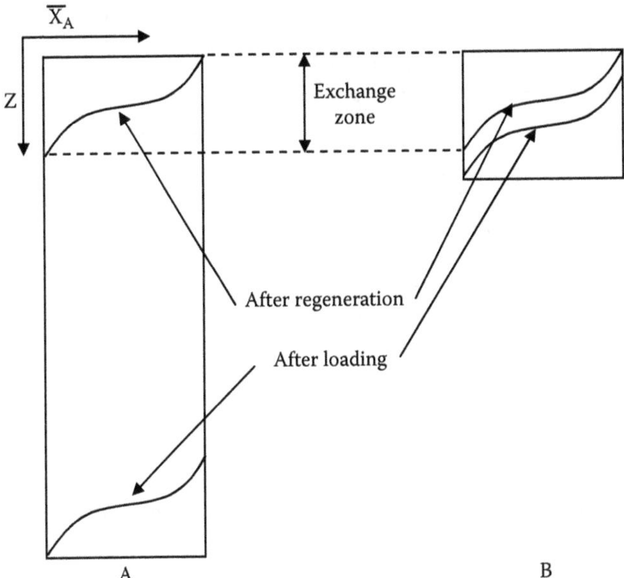

FIGURE 7.3A At any time in a working ion exchange bed, ion exchange is actually taking place in only a portion of the bed called the mass transfer or exchange zone. The resin ahead of this zone is already exhausted and the resin downstream of this zone is still fresh.

FIGURE 7.3B The basic premise of short-bed ion exchange is that the length (i.e., height) of the resin bed needs to be only just slightly greater than the length of the exchange zone.

on the concentration of the surrounding solution. This volume change can easily exceed 10%. For a 100-cm bed, swelling could then be greater than 10 cm. If the freeboard were totally eliminated in such a column, damage to the column's internals or the vessel itself could occur.

7.4.3 Countercurrent Regeneration

The advantages of countercurrent regeneration are well known and accepted. To maintain the cleanest resin at the bottom of the resin beds after regeneration, the regenerant is passed through the bed in the opposite direction of the feed flow. Since the spent regenerant leaves the column having last contacted saturated resin, regenerant utilization is improved.

The difficulty normally encountered with designing a system to accommodate countercurrent regeneration is how to avoid expanding the resin into the freeboard space during upward flow. This would disturb the exchange zone in the bed and obviate any advantage. A number of rather elaborate schemes have been devised over the years to deal with this issue, with varying success. Of course, since no freeboard is present in the Recoflo system, it is not a problem. Flow can be passed through the bed in either direction with equal ease and effectiveness, and the benefits of countercurrent operation can be fully realized.

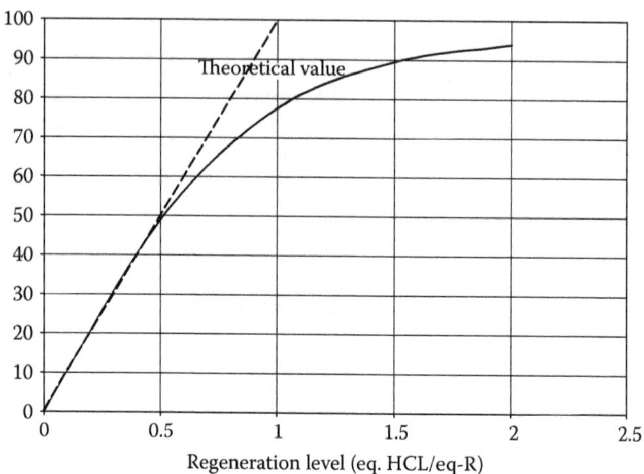

FIGURE 7.4 Increasing the operating capacity of a resin requires more than a proportional increase in regenerant consumption.

7.4.4 Low Resin Loading

Normal practice is to maximize resin working capacity and load the resin to the maximum extent possible each cycle, to maximize the time the unit is in service. Recoflo takes quite the opposite approach. With Recoflo, only the most accessible exchange sites in the resin are used. Recoflo demineralizers, for example, use less than 20% of the total exchange capacity of the resin. By using only the most accessible exchange sites near the surface of the resin beads, the exchange kinetics are improved and regenerant usage is reduced. As shown in Figure 7.4, increasing the operating capacity of a resin requires more than a proportional increase in regenerant consumption. Conversely, by accepting a lower operating capacity, the regenerant consumption is reduced by an even greater amount. As a result, it is possible to approach theoretical (i.e., stoichiometric) regenerant consumption at low resin loadings.

Using low working capacities does reduce the duration of the service run; however, the quantity of the regenerant chemicals is reduced by an even greater amount, so that the percentage of time that the system is in service can be even greater with a Recoflo system.

The low resin loading appreciably decreases resin volume change each cycle and the resulting resin attrition that normally occurs. For example, a typical strong acid cation exchanger will shrink about 10% when completely converted from the hydrogen form to the salt form. If the resin loading is only 15%, the reversible swelling and shrinking will then only be 1.5% each cycle.

The exchange sites that are most accessible are generally the ones located nearest the surface of the resin bead. Because of the three-dimensional bead geometry, approximately 15% of the resin's total capacity is located within 5% of the distance from the surface. On the other hand, utilization of 87.5% of the

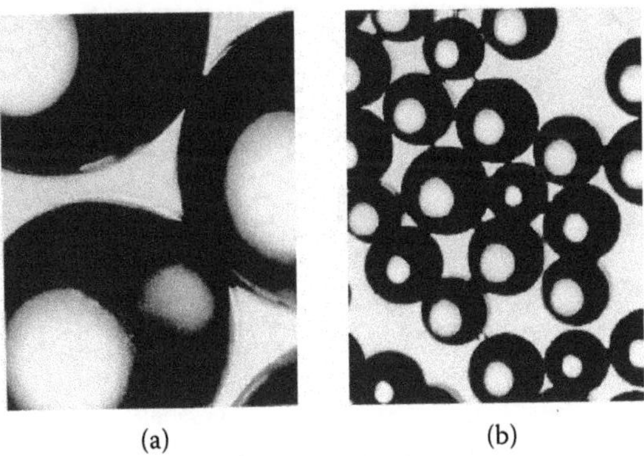

FIGURE 7.5 The size of Recoflo resin particles is approximately one-fifth the size of the resin used in conventional ion exchange systems.

total capacity would require penetration to at least 50% of the distance to the bead center — 10 times as far! This has an appreciable positive impact on particle diffusion rates. By using low exchanger loading, one may choose to accept a shorter exchange zone and shorter bed height or alternatively higher flow rates. This same principle has recently been exploited in a technology called SST (shallow shell technology) wherein only the outer shell of a conventional-size (i.e., coarse) resin bead is functionalized.[9]

7.4.5 Fine-Mesh Resin

The size of Recoflo resin particles is approximately one fifth the size of the resin used in conventional ion exchange systems (see Figure 7.5). This obviously reduces the distance that ions must travel through the resin particles. As discussed above, the reduced diffusion path length further enhances the kinetics of the ion exchange process and allows use of dramatically higher flow rates. Factoring in the reduced particle size as well as the low loadings, the length of the particle diffusional path through the resin particle is reduced by a factor of up to 400.

The improved ion exchange rates in fine resins have been confirmed experimentally. At low concentrations, the rate has been found to be inversely proportional to the resin particle radius, while at high ion concentrations, where particle diffusion is expected to be the rate-limiting step, the exchange rate was found to be inversely proportional to the square of the resin particle radius.[10]

This improvement in diffusion is especially significant for large organic molecules that, because of their extremely low diffusion rates, have a tendency to foul anion resins in many water demineralization applications.

In practice, fluid velocities employed in Recoflo systems are three to five times higher than in conventional systems. For example, in water demineralization, loading velocities are 120 cm/min (30 gpm/ft^2) on a 15 cm (6") bed. This translates into 480 bed volumes per hour (BV/h). For water softening, loading velocities of 200 cm/min (50 gpm/ft^2) on a 7.5 cm (3") bed, or 1600 BV/h, are found.

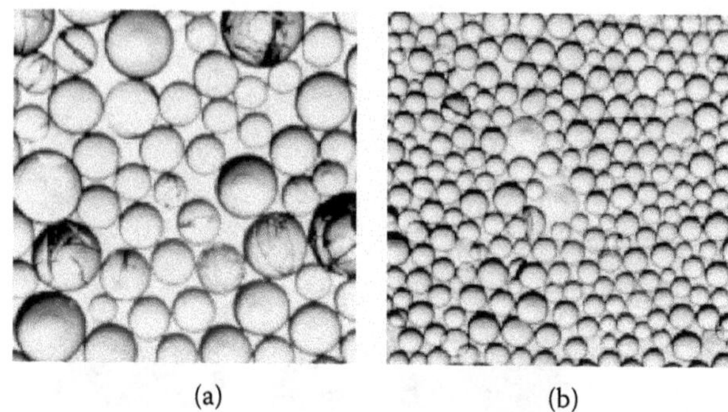

FIGURE 7.6 The smaller particle size resin exhibited greater stability under the test conditions.

In addition to the reduction in bed height, a reduction in column diameter equal to the inverse of the square root of the increase in fluid velocity is achieved. This makes a significant reduction in the floor space required to accommodate the equipment as well as the headroom. Because of the reduction in equipment size, it is feasible to completely factory preassemble and test even large systems before they are shipped to the site. This results in a significant reduction in final cost, since with conventional ion exchange systems, erection of columns, bedding of the resins, and final assembly of piping as well as testing is usually done on site.

Fine resins are known to be much stronger and less susceptible to attrition due to osmotic shock than conventional, coarse resins.[11] Tests have been reported by Dow Chemical showing the effect of particle size on osmotic stability. Conventional 20- to 50-mesh resin as well as 50- to 100-mesh material was cycled with 4 BV of 10% NaOH, 2 BV of water, and 2 BV of 10% HCl. As shown in Figure 7.6, the smaller particle size resin exhibited greater stability under the test conditions. A procedure was developed by Hochmuller to determine and quantitatively measure the resistance of ion exchange resins to osmotic and hydromechanical stress. These tests also showed that beads having a small diameter are more stable than those having a large diameter.[12] These results are not too surprising. Civil engineers have always known that cracking of concrete sidewalks due to thermal expansion can be significantly reduced by pouring the concrete in small slabs rather than in one continuous strip.

The size of the resin particles also has an impact on their rinsing characteristics. Fine-mesh resins require significantly less rinse water after regeneration than coarse resins. Three to five BV of rinse water are typically required to rinse the acid or caustic regenerant chemicals from the resin. Most of the material is rinsed within the first 0.5 BV, corresponding to the interstitial volume, the remainder being required to remove the last traces, which slowly diffuse out of the resin micropores. This rinse water is normally combined with the spent regenerant and discharged, usually as waste. Removal of the last traces of chemicals from inside the resin beads is a diffusion-limited process. Not surprisingly, by using finer

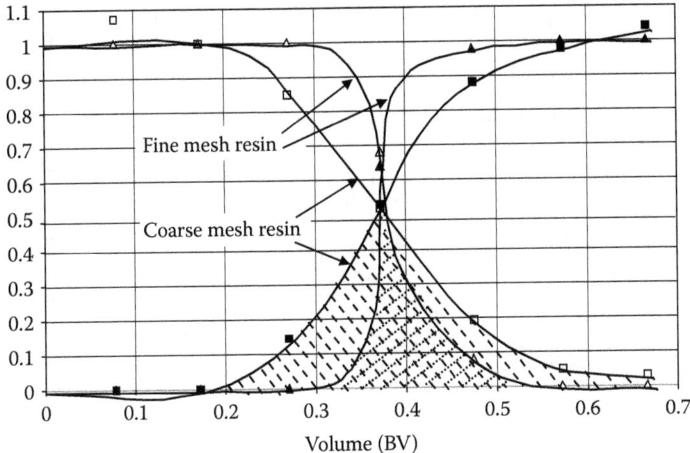

FIGURE 7.7 Effect of resin particle size when sulfuric acid is pumped into the bottom of a 60-cm bed of cation resin and then displaced from the column by admitting water to the top.

resins, this rinse-water requirement can be reduced by 50% or more, resulting in a significant reduction in waste volume.

This issue is particularly important in process applications when treating or producing concentrated solutions. While in water treatment, hundreds of BV of solution are treated each cycle; in some process applications where concentrated solutions are involved, less than 1 BV may be processed each cycle. Examples of such applications would include chromatographic separations such as ion exclusion or acid retardation. In such cases the fluid must be passed in and out of the column with a minimum of dilution or loss. Figure 7.7 shows the effect of resin particle size when sulfuric acid is pumped into the bottom of a 60-cm bed of cation resin and then displaced from the column by admitting water to the top. Assuming we wish to have no overvolume (i.e., collect the same volume out of the column that was first admitted), the cross-hatched area represents the quantity of acid that would be lost each cycle. With the coarse-mesh material (20 to 50 mesh), the equivalent of 0.087 BV would be lost each cycle compared to 0.0337 BV with the fine-mesh (80 to 120 mesh) resin. This reduction would be very significant — 43% loss compared to 17% loss, for a process such as ion exclusion where a net of only about 0.2 BV of solution are processed each cycle.

7.4.6 Flow Distribution

Perhaps the most frequently asked questions concerning short-bed ion exchange relate to flow distribution. Conventional wisdom, acquired through decades of practical experience, is that one must have a bed height of at least 60 to 100 cm in order to achieve an even flow distribution through the bed. The reason for this

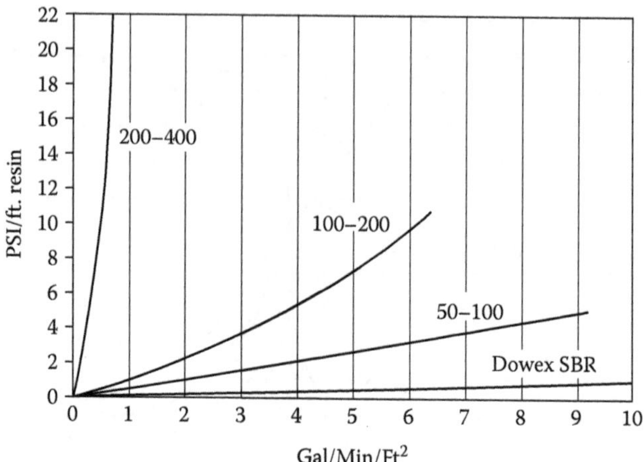

FIGURE 7.8 Pressure drop across resins of various particles sizes.

is that the major mechanism for flow distribution in an ion exchange bed is pressure drop. The so-called choking effect will cause increased pressure drop if the liquid flow attempts to go through one section of the bed at a higher velocity than another. This higher back pressure will force the liquid to find another path, thereby evening out the flow. If the bed is too short, or if the overall flow is too low, there will be insufficient pressure drop to evenly distribute the fluid.

This pressure drop can be improved by increasing the bed height or by using finer resin. Figure 7.8 shows the pressure drop across resins of various particle sizes. Since we are generally in the laminar flow region, the pressure drop varies approximately inversely with the square of the particle diameter. In other words, a 4-cm-high bed of 0.2-cm diameter resin will have about the same pressure drop as 100 cm of 1 mm resin at the same flow.

This assumes, of course, that the resin bed is uniformly packed. If we have a reasonable amount of pressure drop, any flow maldistribution will be the result of packing irregularities. Regions of less-packed resin in the bed will offer lower resistance to flow and allow fluid to pass more rapidly at the same pressure drop.

Achieving uniform bed packing is difficult in ion exchange because the resins are constantly changing volume, and this has a major impact on flow distribution.[13] Swelling of the resin tends to flatten the concentration profiles since, as the flow in a localized area advances ahead of the main flow, swelling of the particles reduces the size of interstices between the particles and increases resistance to flow. Resin shrinkage, on the other hand, even in a packed resin bed, can result in flow channeling (see Figure 7.9). The liquid tends to accelerate in areas where local flow has advanced ahead of the main flow and shrinks the resin in its path. This is particularly a problem near the walls of the column where the resin mass tends to shrink away from the walls, thus producing channels where the flow advances farther and farther ahead of the main flow. This is potentially even more of an issue in a short bed.

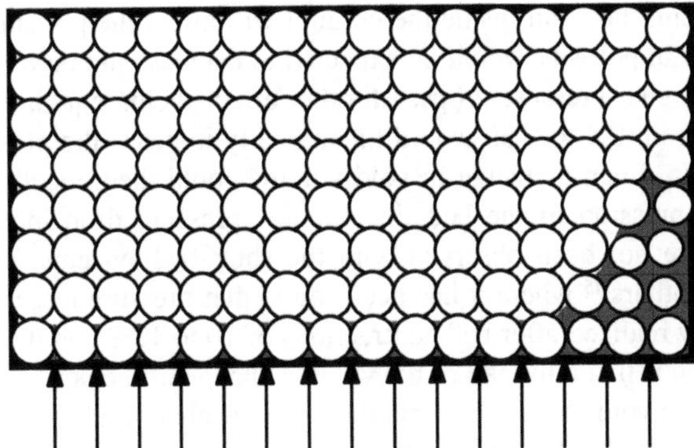

FIGURE 7.9 Resin shrinkage, even in a packed resin bed, can result in flow channeling.

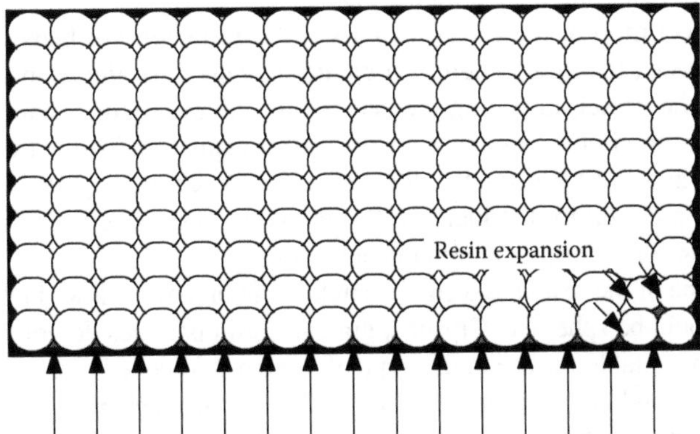

FIGURE 7.10 Any flow channels that develop due to localized resin shrinkage are immediately closed off by expansion of the surrounding resin.

In the Recoflo system, even flow distribution is achieved by compressing the resin bed inside the vessel.[14] The resin remains under compression at all times. Any flow channels that develop due to localized resin shrinkage are immediately closed off by expansion of surrounding resin (see Figure 7.10). On the other hand, since there is no free space in the column, gross movement of the resin cannot occur so that exchange profiles move up and down the bed in a reproducible manner as the process cycles. The potential benefits of countercurrent regeneration can therefore be fully realized.

There are several ways to compress the resin inside the column. The easiest way is to mound up the resin over the top flange when the column is assembled. The resin is then compressed when the column is reassembled. Alternatively, the resin can be packed in concentrated brine, which causes the resin to shrink. It then swells again when the column is rinsed with water after it has been reassembled. Success has also been achieved by employing a positive displacement pump to pump the resin into an assembled column.

Compressing the resin inside the column will affect the porosity of the bed. This reduction in porosity would be expected to decrease the permeability of the bed, causing higher pressure drops. All other effects being equal, decreasing the bed porosity from 0.4 to 0.35 would theoretically increase the pressure drop by a factor of 1.75. Such a change in void volume could theoretically be achieved by a 5% compression of the bed. In practice, pressure drop increases of this magnitude have not been observed with the short-bed systems. This has been confirmed by others,[15] where it has been found that pressure losses of this magnitude are only realized after bed contractions of 10 to 12% and then only rarely. One conclusion is that only part of the volume change of the bed is brought about by a reduction in void space. The remaining volume change must be by decreasing the size of the beads, that is, by forcing water out. The net effect is that it is possible to compress a resin bed to a much greater extent than one would expect without encountering excessive pressure drops. A side benefit is that the bed will contain more resin and have a higher capacity.

The aspect ratio of the bed is also a significant factor. For beds with a large aspect ratio (i.e., tall beds), there is a greater tendency for the resin to "lock" to the wall of the vessel, impeding the bed's ability to evenly distribute changes in resin swelling and compression over the entire resin bed. The extremely small aspect ratio in a short bed renders the wall effect much less significant, allowing the bed compression to distribute uniformly as the resin swells and shrinks due to changes in ionic form or fluid concentration.

The shape of the resin has a significant bearing on how it flows in the bed. The particles should be spherical. Crushed, granular resin particles do not flow and are generally not suitable for use in this type of system. Moreover, crushed resins tend to be unstable and are much more prone to fracture than spherical bead resins.

Resin breakage due to excessive compression is a potential issue, particularly if the resin is caused to swell too much inside the column. Expanding resins can generate enormous forces that, in addition to fracturing resin, can damage column internals or crack vessel walls. There are several mitigating factors that help to avoid this problem. First of all, as discussed above, fine resins are much stronger than conventional resins and have a surprising amount of elasticity. Secondly, as also discussed above, due to the low loadings, the resins change ionic form to only a small extent each cycle so the reversible swelling is quite small. In practice, mechanical resin attrition or equipment damage due to resin compression or swelling has not been found to occur under normal circumstances.

7.5 SHORT-BED COLUMN DESIGN

A typical short-bed column design is shown in Figure 7.11.[14] Note that there are no flow nozzles, strainers, or pipe laterals inside the column.

The resin (46) occupies the space formed between the two surfaces of a flanged pipe spool (58). Against each flange surface (60, 62), and adjacent to the resin is a fine-mesh cloth screen (68) that serves to retain the resin inside the column. Outside of the screens are flow distribution plates (54, 56). Flow channels

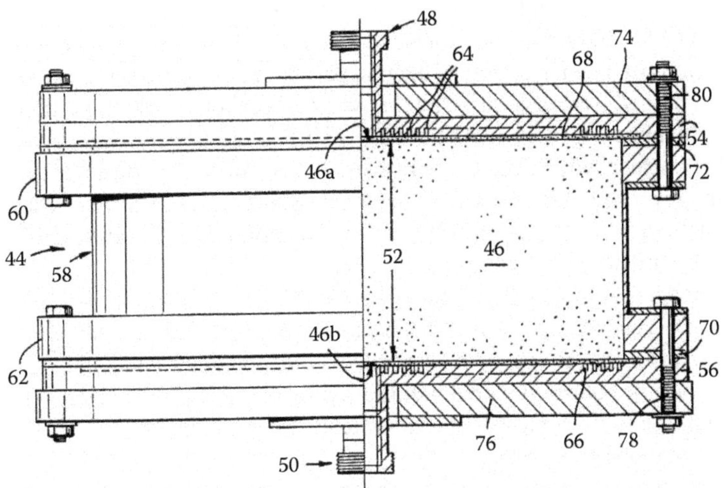

FIGURE 7.11 A typical short-bed column design.

(64, 66) are machined into these flow distribution plates. The flow distribution plates, which are normally fabricated in plastic, are reinforced with flat steel cover plates (74, 76). Fluid is admitted to the column through the inlet port (48). The fluid passes through the radial flow channels and around circumferential flow channels over the entire surface of the bed. The fluid then passes through the screen into the resin bed. After leaving the resin bed, the fluid passes through the other screen, flow channels, and exit port (50) in a similar manner.

Since the bed height is defined by the length of the exchange zone, it normally remains constant for any specific process application. Scale-up is achieved by increasing the bed height only, not the diameter. While most of the process development work has been done on columns 5 to 10 cm in diameter, commercial systems have been constructed as large as 274 cm in diameter. The fact that the results obtained are virtually identical regardless of column diameter would confirm the efficacy of the flow distribution system.

It is essential to avoid flexure of the distribution plates under pressurized conditions since this would relieve the resin compression. This is somewhat of a challenge for large diameter columns. Various reinforcement designs have been developed to prevent flexure while minimizing the thickness of the steel cover plate.

7.6 PREFILTRATION

With conventional cocurrent ion exchange systems, a small amount of solids accumulating within the resin bed can often be removed by regular backwashing. Often, no prefiltration is provided with these systems. It is not possible to backwash a packed-bed ion exchange system, however, because there is no freeboard in the column to take up bed expansion. Moreover, to do so would obviate the advantages offered by countercurrent regeneration.

The fine resins employed in the short-bed system increase the tendency of the resin bed to filter out any suspended material in the feed. This is due not only to the improved sieving action of the fine particles; ion exchange resins also have highly charged surfaces that are known to adsorb colloidal particles. The extremely high surface areas of fine-mesh resin beds are very effective at removing very fine suspended material. These particles are then very difficult to remove from the resin bed. This is perhaps the single major negative aspect of the short-bed system. In order to be successful, careful attention must be paid to prefiltration ahead of a short-bed ion exchange system.

Although various methods have been developed to help alleviate this situation,[16,17] the general consensus is that packed-bed ion exchange systems in general are more prone to fouling with suspended solids than conventional systems and must always be equipped with a prefiltration system. It is fair to say that the level of filtration required for a short-bed ion exchange system is similar to that required for a reverse osmosis system.

To address this requirement, a number of different filter designs have been applied. Generally speaking, cartridge filters are not economical unless the level of suspended solids is extremely low. Multimedia depth filters with coagulation have been utilized with satisfactory results for water-treatment applications. In particular, a novel multimedia filter called a *micromedia filter,* which utilizes fine-mesh filter media, has been shown to provide adequate filtration.[18] In some cases, ultrafiltration or microfiltration membranes have been installed. The cost of membrane filtration systems has been reduced appreciably over the past few years, so this approach is now often viable. Of course, if the ion exchange unit is being used as a polisher after a reverse osmosis unit, the problem is nonexistent.

In the event that dirt does accumulate in the bed, it is usually necessary to remove the resin from the bed to clean it. Fortunately, because of the small quantity of resin involved, this chore is not too onerous. A typical Recoflo resin bed can generally be disassembled, manually unloaded, cleaned, reloaded, and reassembled in less than one 8-h shift. Cleaning is accomplished by slurrying the resin in a small tank, usually with a small amount of nonionic detergent. The resin is allowed to settle and the supernatant containing the fine suspended contaminants is decanted to waste.

Automated systems that can pump out the resin from the column, clean it, and pump it back into the column have been designed and utilized in those cases where frequent cleanings are required.

7.7 APPLICATION OF SHORT-BED ION EXCHANGE

The Recoflo system has seen extensive application since it was first commercialized in 1973. Over 1500 units have been installed in over 50 different countries. The initial application was in the recovery and purification of chemicals from industrial wastewater. This market was targeted initially since it was felt that there would be less resistance to acceptance of the new technology compared to the traditional water-treatment applications. More recently, it has been applied to water-treatment applications such as demineralization and softening. Although no systems are operating on sweeteners, this should be an excellent field of application.

7.7.1 Chromic Acid Recovery

The first commercial Recoflo success, a process for recovery of chromic acid from electroplating rinse waters, was developed and commercialized in 1973. This process has been applied in hundreds of plating operations, ranging from small job shops to large, continuous-strip, "tin free-steel" plating operations.[19]

For this process, three resin beds are employed. Dilute chromic acid rinse waters typically containing 0.1 to 1 g/L CrO_3 as well as trace cationic contaminants such as Fe^{+++}, Cr^{+++}, Cu^{++}, and Ni^{++} are first passed through a 15-cm-high cation exchange bed for removal of the contaminants and then through a 7.5-cm-high anion exchange bed for removal of chromate ions along with sulfate and fluoride electroplating catalysts. The resulting demineralized water is recycled to the plating line for rinsing parts.

After loading of the resins, the cation bed is regenerated with sulfuric acid, the spent regenerant passing to waste. The anion bed is regenerated with sodium hydroxide, yielding a sodium chromate solution, which is passed immediately through a second 30-cm-high cation exchange bed, where the sodium ions are exchanged for hydrogen. This results in the production of a moderately concentrated ($[CrO_3]$ = 60 to 80 g/L) high-purity chromic acid solution. Although this is significantly more concentrated than can be produced by a conventional ion exchange system, in most cases it is necessary to install a small evaporator on the plating bath to remove sufficient water to allow recycling of the recovered chromic acid.

7.7.2 Chromic Acid Purification

Cation exchange can be used to remove cationic contamination such as iron and trivalent chromium directly from chromium plating solutions. Once loaded, the resin is regenerated with sulfuric acid.

Although this is one of the first applications for ion exchange in the plating industry dating back to 1956,[20] it was not until a simple, compact, short-bed unit called Chromapur® (Eco-Tec Inc.) was developed that the process became widely applied. A typical unit is shown in Figure 7.12. Since the first unit was installed in the mid 1970s, about 100 of these units have been installed around the world.

A significant issue with this application is oxidation of the resin by the chromic acid feed. Oxidation causes de-cross-linking and swelling of the resin. It has been found that if the plating solution is diluted to about $[CrO_3]$ = 150 g/L, the resin life is acceptable. This dilution has proven not to be a problem since evaporative losses from the plating bath are sufficient to compensate for the water addition.

7.7.3 Nickel Salt Recovery

Using strong acid cation exchange resin, Recoflo has been extensively used to recover nickel from electroplating rinse waters.[21] The rinse solution is passed

FIGURE 7.12 Typical short-bed unit called Chromapur (Eco-Tec Inc. With permission.).

through a 30-cm-high resin bed and the nickel is exchanged for hydrogen. The dissolved organic electroplating addition agents, which tend to have a negative charge, as well as the borate, chloride, and sulfate, all pass through into the effluent to waste and are thus removed from the system. Note that the water is not recycled in this application.

After loading, the resin is regenerated with acid, producing a recovered nickel salt solution. If sulfuric acid is used, nickel sulfate will be produced. If a mixed nickel sulfate-chloride product is preferable, the resin can be regenerated with a sulfuric-hydrochloric acid mixture.

Unfortunately, an excess of acid must be used to effect the regeneration, which produces a recovered nickel product with a pH less than 1, which is totally unusable. This amount of acid exceeds the normal acid makeup requirement for the plating bath and would drop the bath pH to unacceptably low levels. Chemical neutralization of this excess acidity is not permissible since it would introduce contaminants such as sodium or calcium to the plating bath.

The "acid retardation" process is used to remove the residual free acid from the recovered nickel sulfate stream, thus raising the pH to approximately 3. Acid is eluted from the acid retardation resin with water, so that it can be recovered for use in subsequent regenerations. The acid retardation process is discussed in more detail below. Typical results for this process are shown in Table 7.1.

The same process has been employed on a much larger scale for recovery of nickel and zinc sulfate from large, continuous-strip electrogalvanizing rinse waters.[22]

TABLE 7.1
Typical Results: Nickel Electroplating Rinse-Water Recovery

	[Ni] (g/L)	[Na] (g/L)	pH
Feed	3.95	0.375	N/A
Waste	0.001	0.147	N/A
Eluate	46.1	2.55	3.1

7.7.4 Selective Copper Recovery from Printed Circuit Board Wastewaters

Chelating resins bearing functional groups such as iminodiacetic acid and picolylamine show a high selectivity for transition metal ions, especially copper. Such resins have been shown to be very effective in treating rinse waters from electroless copper plating operations in printed circuit board shops, where the copper is complexed with chelating agents such as quadrol and ethylenediaminetraacetic acid (EDTA).[23,24] The copper sulfate that is recovered upon regeneration with sulfuric acid is typically sent to a small electrowinning system where metallic copper is recovered from the eluate. This is very attractive, since conventional waste-treatment systems based upon simple pH neutralization and precipitation are not effective on these types of wastewaters due to metal complexation. Using this type of process, copper levels can be reduced to levels of less than 0.02 mg/L.

7.7.5 Acid Retardation

Hatch and Dillon introduced a process called *acid retardation* in 1968.[25] Although this process uses anion exchange resins, the process is not really ion exchange. Strong mineral acids such as sulfuric, nitric, or hydrochloric acid are sorbed by the resin, while metal salts of these acids are excluded. When a contaminated acid solution is passed through a bed of resin, a deacidified metal-salt solution is collected as a waste or by-product. Purified acid is reclaimed by stripping the resin with water. The challenge is to minimize the quantity of water used for stripping. The Recoflo process proved ideal for this process, since only a very small volume of solution can be treated and produced each cycle. A large number of these acid purification units, called APU® (Eco-Tec Inc.), have been installed since being first commercialized in 1977.

7.7.6 Sulfuric Acid Aluminum Anodizing and Steel Pickling

The first successful commercialization of the APU was in the aluminum finishing industry.[26] Sulfuric acid solutions at a concentration of 15 to 25% w/w are used for anodizing and become contaminated with dissolved aluminum through use.

TABLE 7.2
Typical APU Results: Sulfuric Acid Anodizing Recovery

	$[H_2SO_4]$ (g/L)	[Al] (g/L)
Feed (bath)	183.8	12.2
Product (recycle)	175	4.2
Byproduct (waste)	13	12

By operating an APU in a continuous bypass arrangement directly on the anodizing bath, it is possible to extend bath life indefinitely by continuously removing an aluminum sulfate by-product. This process has actually become a *de facto* standard in the industry. Typical performance for this application is shown in Table 7.2. The same process has also been used for purification of sulfuric acid carbon steel pickle liquors, but to a lesser extent.

7.7.7 Aluminum Etchants

In many automotive and appliance applications, aluminum is bright etched with concentrated (80%) phosphoric acid solutions. The rate of drag-out of this viscous acid on the work from the process bath is appreciable. A phosphoric acid recovery system was developed that combines cation exchange with the APU acid retardation process.[27] This was also an ideal application for Recoflo; because of the high contaminant level, very small volumes of concentrated solutions are treated and produced. The flow sheet for this process is shown in Figure 7.13.

Rinse water from the first of a series of counterflow rinses is collected at a phosphoric acid concentration of typically 15 to 30% w/w. This is fed to a Recoflo cation exchange unit where aluminum contaminants are exchanged for hydrogen ions. The purified, dilute phosphoric acid is then concentrated by vacuum evaporation to about 80% H_3PO_4 and recycled back to the process bath. The cation exchanger is regenerated with sulfuric acid; however, because of the high selectivity for the trivalent aluminum cation, an excess of sulfuric acid of several times the stoichiometric quantity is required. This spent regenerant is then passed to an APU that reclaims purified sulfuric acid for reuse on the cation exchanger. An aluminum sulfate byproduct solution is generated as waste. This process provides very attractive economics and has been installed at 19 major aluminum bright dip operations in the United States over the past 20 years.

Dilute solutions of hydrochloric acid along with other acids such as nitric, phosphoric, and acetic at a concentration of about 3 to 4% (w/w) are used for etching aluminum lithographic plates as well as aluminum electrolytic capacity foils. A cation exchange–acid retardation process similar to that used for phosphoric acid recovery has been extensively employed for these applications.[28] Because the process acid is relatively dilute, the cation exchanger operates directly

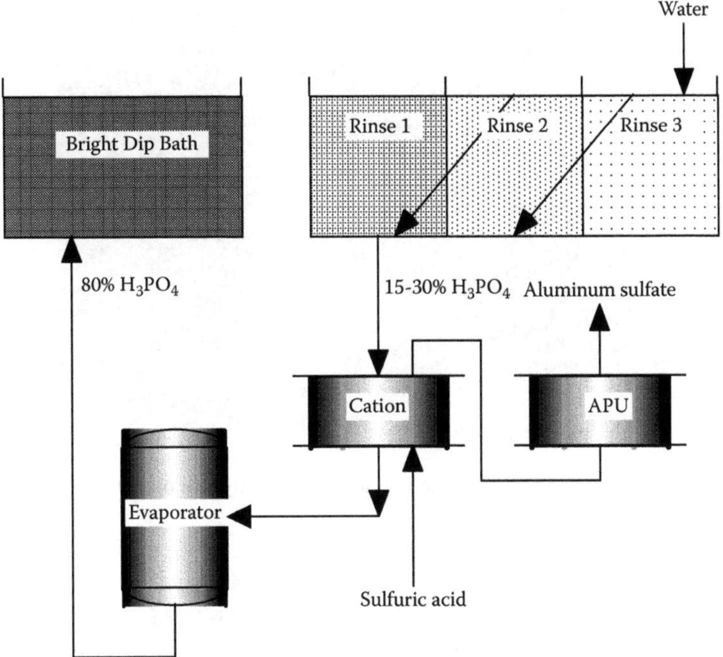

FIGURE 7.13 Flow sheet for bright etching aluminum with concentrated (80%) phosphoric acid solutions.

on the etch bath to continuously remove aluminum contamination. The cation exchanger is regenerated with hydrochloric acid. For these applications, the acid retardation unit employed for regenerant recovery is integrated with the cation exchanger in a unit called a DPU. A typical DPU installed on a large capacity foil operation in Japan is shown in Figure 7.14. This large unit is equipped with 244-cm (96") diameter resin beds.

7.7.8 Stainless Steel Pickle Liquors

One of the most successful applications of the APU is purification of nitric-hydrofluoric acid stainless steel pickle liquors. Of course, this system is only capable of recovering the so-called free acid. Nitric and hydrochloric acids that have actually reacted with the oxide scale or metal substrate are not recoverable. Although recovery efficiency of free acid with the APU is normally better than 95%, it is estimated that purchases of nitric acid can be reduced by typically 50%, while hydrofluoric acid savings are somewhat less.[29] Nevertheless, paybacks on capital investment for these systems are very attractive. Almost as important as the direct economic savings through reduced acid purchases and neutralization costs are the process benefits associated with consistent pickle bath performance.[30] In many instances, a major driving force is the reduction of nitrate discharges in

FIGURE 7.14 A typical DPU installed on a large-capacity foil operation in Japan.

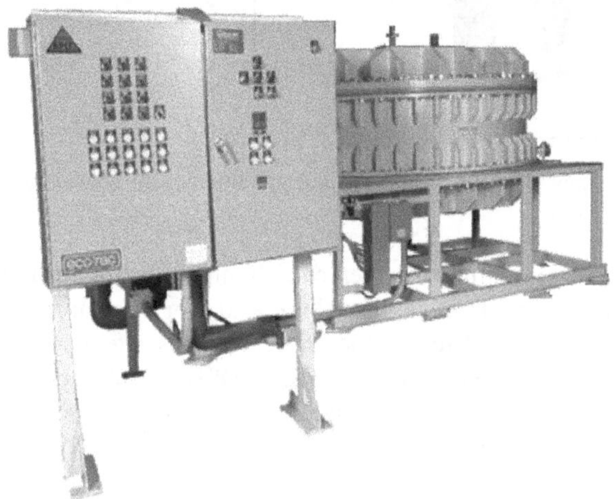

FIGURE 7.15 A large APU with a resin bed 274 cm (108 in) in diameter by 60 cm (24 in) high.

the mill effluent. Today, most major stainless steel continuous strip plating operations in the world utilize this technology. It is estimated that more than 100 systems of this type have been installed. A large APU with a resin bed 274 cm (108") in diameter by 60 cm (24") high is shown in Figure 7.15.

A major consideration in this application is the stability of the ion exchange resin in the nitric acid solutions being treated. Ion exchange resin manufacturers warn against possible violent chemical oxidation reactions between nitric and ion exchange resins with possible safety concerns. No problems have been experienced with these systems, since the concentration and temperature of the nitric acid as well as the resin contact time are carefully controlled and limited.

7.7.9 Metal Refinery Electrolyte Bleeds

In electrolytic copper refineries, high-purity copper is electrolytically recovered from a solution containing about 50 g/L copper and 150 g/1 sulfuric acid.[31] Metallic contaminants such as nickel, antimony, and bismuth build up, eventually to the point where electrolyte must be purged to limit the level of contamination. For example, at Falconbridge's Kidd Creek copper refinery in Timmins, Ontario, Canada, electrolyte was bled at a rate of 67 m^3/day to control nickel buildup. An APU was commissioned in 1995 to recover the sulfuric acid values from the decopperized electrolyte bleed stream. The recovered acid is recycled back to the tank house, while the deacidified nickel salt stream is neutralized with soda ash to recover nickel carbonate, which is sold. Several other refineries have installed APUs for similar applications.

Electrolytic zinc refineries often have a similar problem as copper refineries due to buildup of contaminants such as manganese and magnesium in the sulfuric acid electrolyte. The APU has been successfully evaluated on a pilot plant scale for this application, although to date no full-scale systems have been installed.

7.7.10 Pulp and Paper Industry

A number of new short-bed ion exchange applications have been developed utilizing a process called *ion retardation* to facilitate closure of the kraft chemical pulping process. The ion retardation process, which was originally developed over 40 years ago, is quite analogous to the acid retardation process discussed above. Ion retardation uses an amphoteric resin wherein a polyacrylic acid cation exchanger is impregnated inside a strong base anion resin. As in acid retardation, the resin is regenerated with only water.

In a patented[32] process called PDP (precipitator dust purification), Recoflo ion retardation units are used to remove chloride contaminants from sodium sulfate that is collected from the electrostatic precipitator of the kraft recovery boiler. The chloride is stripped from the resin with only water and then discharged to waste. The purified sodium sulfate is recycled back to the kraft pulping process. Typical results are shown in Table 7.3.

TABLE 7.3
Typical Chloride Removal Results in a PDP System

	Relative Volume (L/L Feed)	[Na_2SO_4] (g/L)	[NaCl] (g/L)
Feed	1	198	18.3
Waste	1	11.0	17.5
Purified salt	1	188	0.31
Percent removal		5.6%	95.6%
Mass balance		0.3%	–2.4%

In another patented[33] process called GLS (green liquor splitter), sulfide ions are removed from green pulping liquor. The desulfurized sodium carbonate can then be causticized with lime to make sodium hydroxide that can be used in the bleach plant. Again, the ion retardation resin is regenerated with water only. The sulfide is recycled back to the pulping process.

Although these processes have not yet been fully commercialized, extended pilot plant trials have been completed at mills in Canada with excellent results.[34]

7.7.11 Amine Purification

Alkanolamines such as methyldiethanolmamine (MDEA) are extensively used in the petrochemical industry to remove sulfide from gas. These alkanolamines are regenerated with steam but tend to accumulate salts such as oxalates, sulfate, chlorides, and thiocyanates, which are not steam-strippable. Eventually, the alkanolamines must be discarded because of accumulation of these so-called heat stable salts.

Recoflo units called Amipur® (Eco-Tec Inc.) are now extensively utilized to remove these heat stable salts by OH-cycle anion exchange, thereby extending the life of the alkanolamine solutions. Choice of the anion resin is critical, as some of the ions tend to foul the resin. A typical Amipur unit is shown in Figure 7.16.

7.7.12 Water Demineralization

Water demineralization was and still is a major application for ion exchange, notwithstanding the major inroads that have been made by reverse osmosis (RO) into this market over the past 10 to 20 years. The major advantage that ion exchange has over RO is a much lower waste volume and the ability to produce much higher purity water. On the other hand, its major disadvantage vis-à-vis RO is its use of chemicals for regeneration and the resulting chemical waste. Among other advantages, the short-bed approach can improve the competitive position of ion exchange by further reducing the volume of waste and the quantity of chemicals required for regeneration.

The basic Recoflo demineralization unit utilizes a two-bed, strong acid–strong base configuration. During the on-stream or service cycle, feed water is passed through the cation and anion beds in the usual manner. A typical on-stream resistivity profile of the demineralized product water is shown in Figure 7.17. The length of the on-stream cycle depends on the feed total dissolved solids (TDS) concentration. For example, for a feed water containing [TDS] = 150 mg/L, the on-stream time would only be about 10 min. This is indeed short compared to a conventional demineralizer, which may remain in service for a day or more.

The average water quality of the product water is somewhat dependent on the feed TDS. For waters greater than [TDS] = 170 mg/L, average conductivities of less than 1 µs/cm can readily be achieved. For [TDS] = 50 mg/L, a conductivity of less than 0.1 µs/cm (> 10 MΩ cm) can be ensured, while for [TDS] = 10 mg/L a conductivity of less than 0.067 µs/cm (> 15 MΩ cm) is obtained. Perhaps more

FIGURE 7.16 A typical Amipur unit.

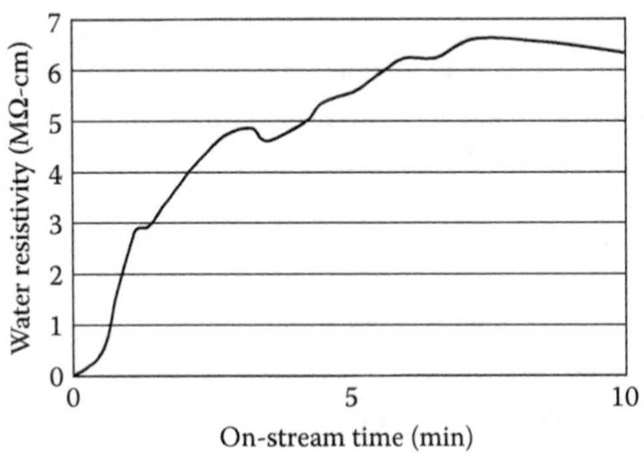

FIGURE 7.17 A typical on-stream resistivity profile of the demineralized product water.

significantly for boiler makeup applications, silica levels of less than 10 μg/L can be guaranteed. According to usual water-treatment practice, a mixed-bed polisher is usually utilized to ensure purity of this level. By using Recoflo, the necessity of employing mixed-bed ion exchange polishers is eliminated.

After the resins beds have become loaded, they are regenerated. The complete regenerant dilution and feed systems are all located on the same compact skid as the resin beds. The spent regenerant chemicals from each bed can be collected in a small tank where they self-neutralize. Since the volume of spent regenerants is only a few BV and the BV are so small, the neutralization tank is appreciably smaller than with a conventional demineralizer. This represents a major space savings in the plant.

After regeneration, final rinse-to-quality is achieved by recirculating the contents of the two resin beds using the feed pump. This minimizes the quantity of wastewater that is produced and ensures that the final product water quality is acceptable.

The total off-stream time, including regeneration and rinsing of both beds as well as recirculation rinse, is only about 7 min. Because the total off-stream time for a conventional plant is about 4 h, a duplicate, standby unit is usually supplied to ensure continuous service. With the Recoflo system, a small water surge tank with only a few minutes' residence time eliminates the need for redundancy.

In order to produce demineralized water with a conductivity of less than 0.1 μs (> 10 MΩ cm), a mixed-bed ion exchange unit is always used to polish the water produced by a primary two-bed demineralizer or reverse osmosis unit. Conventional wisdom says that it is not possible to achieve this quality using separate beds, even using counterflow packed-bed technology. While mixed beds do the job, they have a well-deserved reputation for being unreliable and maintenance intensive. According to a well-known manufacturer of mixed-bed systems, "statistically on a per capita basis, the call-out rate for problems on mixed beds was about 20 times higher than that for two-bed plants." He went on to say, "Mixed beds were more liable to organic fouling and other wasting diseases and that automatic mixed beds, in particular, needed an amount of tender loving care that seemed somewhat out of proportion to the results being achieved."[35]

The main limitation to water purity from a two-bed demineralizer is sodium leakage. Because selectivity for sodium ions on cation resin is very poor, it is very difficult to remove the last little bit of sodium from the feed water. Moreover, rinsing the last traces of NaOH from the anion bed after regeneration is difficult, particularly as the resin ages.

By incorporating a second Recoflo cation "polisher" bed immediately following the anion bed, it is possible to produce "mixed-bed quality" water using a three-bed system configuration. The polisher bed picks up sodium leakage from the primary cation bed as well as any residual caustic left in the anion bed after regeneration. In addition, any trimethylamine coming off the anion resin is also removed by this bed. Figure 7.18 shows the resistivity of demineralized water produced by the laboratory pilot plant after the primary cation-anion beds and then again after the polishing cation bed. The unit was fed tap water from the

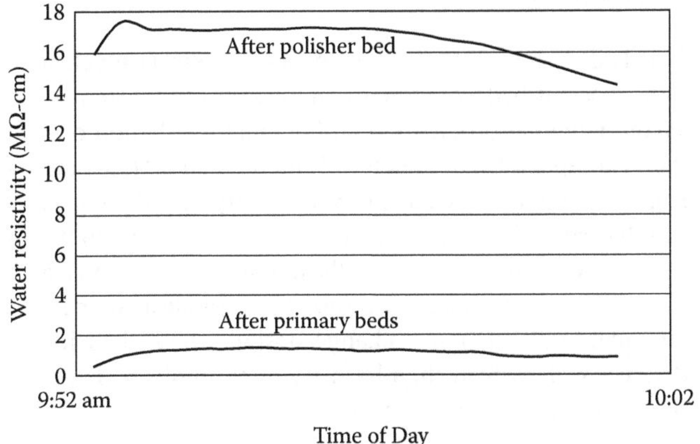

FIGURE 7.18 Resistivity of demineralized water produced by the laboratory pilot plant after the primary cation/anion beds and then again after the polishing cation bed.

TABLE 7.4
Recoflo Three-Bed Demineralization Field Results

Description	Water Quality Requirement	Recoflo Three-Bed Results
Conductivity	< 0.1 μS/cm	< 0.06 μS/cm
Sodium	< 3 ppb	< 1.2 ppb
Chloride	< 3 ppb	< 1.3 ppb
Sulfate	< 3 ppb	0.8 ppb
Silica	< 3 ppb	1.5–2.5 ppb
TOC	< 300 ppb	100 ppb

city mains at a feed concentration of approximately 170 mg/L. The polisher bed was not regenerated over this period.

Recently, two Recoflo three-bed units were installed at a power generation plant in the U.S. Midwest. The Recoflo units, one of which is shown in Figure 7.2, are equipped with two 152 cm (60") diameter primary beds and a 122 cm (48") diameter cation polisher. The primary beds are 15 cm (6") in height, while the polisher bed is 7.5 cm (3") high. The on-stream flow rate is 136 m^3/h (600 gpm). Raw water is withdrawn from Lake Michigan and after filtration is fed directly to the unit. The typical analysis of product water from this unit and the minimum quality requirement are shown in Table 7.4. The feed water contains a TDS of about 160 mg/L and a total organic carbon (TOC) of 2 mg/L.

The primary beds (cation and anion) of the unit are regenerated approximately four times per hour as described above. The polisher cation bed is regenerated once per day.

7.8 CONCLUSION

A viable short-bed system such as the Recoflo system is much more (or less) than just another packed-bed, countercurrent ion exchange system with short beds. In order to avoid potential problems such flow maldistribution or damage to ion exchange resin and apparatus, a number of interrelated features are incorporated into the design. Although the technology is ideally suited to a wide variety of applications, it has some limitations, particularly with respect to pretreatment for removal of suspended solids. Providing these issues are properly addressed, the short-bed ion exchange system has been shown to provide a number of advantages over "conventional" tall-bed ion exchange designs. Short-bed ion exchange is now considered a well-proven, highly efficient technology. It is bound to see increased application in the coming years.

REFERENCES

1. Smith, J.H., Renouf, P.W., and Crossen, M., 50 years in separate beds, in *Proceedings of the 45th International Water Conference,* Pittsburgh, 1984.
2. Tsouris, C.P. and Joseph, V., Process intensification: has its time finally come? *Chem. Eng. Prog.,* 6, 50, 2003.
3. McGarvery, F.X, Hauser, E.W., and Kiefer, B., Hydraulic properties of ion exchange resin, in *Proceedings of the 45th International Water Conference,* Pittsburgh, 1984.
4. Brown, C.J. and Fletcher, C.J., The Recoflo short bed ion exchange process, in *Ion Exchange for Industry,* Streat, M., Ed., Ellis Horwood, Chichester, 1988, p. 392.
5. Brown, C.J. and Fletcher, C.J., Water deionization by Recoflo short bed ion exchange, in *Proceedings of the 47th International Water Conference,* Pittsburgh, 1986.
6. Jackson, D. et al., Short bed demineralizer technology at Transalta Energy Corporation, in *Proceedings of the 65th International Water Conference,* Pittsburgh, 1994.
7. Hunter, R.F., Ph.D. thesis, University of Toronto, 1963.
8. Hunter, R.F., Process of Removing a Component from a Fluid, U.S. Patent No. 3,386,914, 1968.
9. Hunter, R.F., Process for Operating Fixed Beds of Active Media, U.S. Patent No. 3,385,788, 1968.
10. Sabzali, J. and Michaud, C.F., A shortcut to higher regeneration efficiency with shallow shell resins, in *Proceedings of the IEX 2000 Ion Exchange at the Millennium,* Greig, J.A., Ed., Imperial College Press, London, 2000.
11. Reichenberg, D., Properties of ion-exchange resins in relation to their structure. III. Kinetics of exchange, *J. Am. Chem. Soc.,* 1953.
12. Dowex™ Fine Mesh Spherical Ion Exchange Resins for Pharmaceutical and Fine Chemical Column Separations, September 2004, *Dow Liquid Separations,* Form 177-01509-904, http://www.dow.com/liquidseps/prod/sp_fine.htm
13. Hochmuller, K., Shock test for the determination of the resistance of ion exchange resins to osmotic and hydromechanical stress, in *Ion Exchange Technology,* Naden, D. and Streat, M., Eds., Ellis Horwood, Chichester, p. 472.

14. Byrne, E.B. and Lapidus, L., Concentration profiles in packed-bed ion-exchange systems, *J. Am. Chem. Soc.*, 77, 6506, 1955.
15. Brown, C.J., Fluid Treatment Process and Apparatus, U.S. Patent No. 4,673,507, 1987.
16. Golden, L.S. and Irving, J., Resin structure: mechanical strength, in *Theory and Practice of Ion Exchange,* Society of Chemical Industry, Cambridge, 1976.
17. Hoehn, K., Medete, A., and Rice, D., How to handle suspended solids within ion exchange resin (IER) packed bed systems, in *Proceedings of the 61st International Water Conference*, Pittsburgh, 2000.
18. Puropack packed bed technology, in *Purolite Engineering Manual*, Purolite Corporation, 1999.
19. Brown, C.J., and Sheedy, M., Method and Apparatus for Increasing Filter Contaminant Loading Capacity, U.S. Patent No. 7,045,067, 2006.
20. Brown, C.J., Chromic acid recovery by reciprocating flow ion exchange, in *Proceedings of the American Electroplater's Society 2nd Continuous Strip Plating Symposium*, Williamsburg, VA.
21. Costa, R.L., Treatment of Hexavalent Chromium Solutions, U.S. Patent No. 2,733,204, 1956.
22. Brown, C.J., A better way to recover nickel, *Prod. Finishing*, 52, 11, 54, 1988.
23. Brown, C.J., Bauer, G.A., and Hughes, C.R., Metal recovery from electrogalvanizing process streams, in *Proceedings of the AESF 6th Continuous Strip Plating Symposium*, 1990.
24. Brown, C.J., Metal recovery by ion exchange and electrowinning, *Proceedings of the 4th International Forum on Electrolysis in the Chemical Industry*, Electrosynthesis Corp., Fort Lauderdale, FL, 1990.
25. Brown, C.J., Process for Removal of Copper from Solutions of Chelating Agent and Copper, U.S. Patent No. 4,666,683, 1987.
26. Hatch, M.J. and Dillon, J.A., *Ind. Eng. Chem. Proc. Des. Dev.*, 2, 4, 253,1963.
27. Brown, C.J., Purification of sulfuric acid anodizing solutions, *Plating Surf. Finish.*, 1979.
28. Brown, C.J., Recovery of phosphoric acid by ion exchange and evaporation, in *Proceedings of the Surfin '84 American Electroplater's Society Technical Conference*, 1984.
29. Brown, C.J., Recovery of aluminum etch and anodizing solutions, in *Proceedings of the 169th Meeting of the Electrochemical Society*, Boston, 1986.
30. Brown, C.J., Recovery of stainless steel pickle liquors: purification vs. regeneration, in *Proceedings of the CISA International Steel Congress*, Beijing, 2002.
31. Brown, C.J., Productivity improvements through recovery of pickle liquors with the APU process, *Iron Steel Eng.*, 1990.
32. Brown, C.J., Control of copper electrolyte impurities using the Eco-Tec APU®, *CIM Bull.*, 98, 1088, 2005.
33. Paleologou, M., Thompson, R., Berry, R., Sheedy, M., and Brown, C., Method and Apparatus for Removing Sodium Chloride from Pulping Chemicals Using an Amphoteric Ion-Exchange Resin, U.S. Patent No. 5,992,171, 1999.
34. Thompson, R., Paleologou, M., Berry, R., Brown, C., and Sheedy, M., Process for the Separation of Sulphides from Pulping Liquors Using an Amphoteric Ion-Exchange Resins, U.S. Patent No. 5,992,171, 1999.
35. Paleologou, M. et al., Short-bed ion-exchange technologies for kraft pulp mill system closure, in *Proceedings of the 66th International Water Conference*, Orlando, 2005.

Contents of Other Volumes

Volumes 1–4, 6: Out of print

Volume 5

NEW INORGANIC ION EXCHANGERS
A. Clearfield, G. H. Nancollas, and R. H. Blessing

APPLICATION OF ION EXCHANGE TO ELEMENT SEPARATION AND ANALYSIS
F. W. E. Strelow

PELLICULAR ION EXCHANGE RESINS IN CHROMATOGRAPHY
Csaba Horvath

Volume 7

INTERPHASE MASS TRANSFER RATES OF CHEMICAL REACTIONS WITH CROSSLINKED POLYSTYRENE
Gabriella Schmuckler and Shimon Goldstein

INFLUENCE OF POLYMERIC MATRIX STRUCTURE ON PERFORMANCE OF ION-EXCHANGE RESINS
V. A. Davankov, S. V. Rogozhin, and M. P. Tsyurupa

SPECTROSCOPIC STUDIES OF ION EXCHANGERS
Carla Heitner-Wirguin

ION-EXCHANGE MATERIALS IN NATURAL WATER SYSTEMS
Michael M. Reddy

THE THERMAL REGENERATION OF ION-EXCHANGE RESINS
B. A. Bolto and D. E. Weiss

Volume 8

METAL EXTRACTION WITH HYDROXYOXIMES
Richard J. Whewell and Carl Hanson

ELECTRICAL PHENOMENA IN SOLVENT EXTRACTION
Giancarlo Scibona, Pier Roberto Dansei, and Claudio Fabiani

EXTRACTION WITH SOLVENT-IMPREGNATED RESINS
Abraham Warshawsky

SOLVENT EXTRACTION OF ELEMENTS OF THE PLATINUM GROUP
Lev M. Gindin

SOLVENT EXTRACTION FROM AQUEOUS-ORGANIC MEDIA
Jiri Hala

Volume 9

ION-EXCHANGE PROCESSES USED IN THE PRODUCTION OF ULTRAPURE WATER REQUIRED IN FOSSIL FUEL POWER PLANTS
Calvin Calmon

A SYSTEMATIC APPROACH TO REACTIVE ION EXCHANGE
Gilbert E. Janauer, Robert E. Gibbons, Jr., and William E. Bernier

ION-EXCHANGE KINETICS IN SELECTIVE SYSTEMS
Lorenzo Liberti and Roberto Passino

SORPTION AND CHROMATOGRAPHY OF ORGANIC IONS
G. V. Samsonov and G. E. Elkin

THERMODYNAMICS OF WATER SORPTION OF DOWEX 1 OF DIFFERENT CROSS-LINKING AND IONIC FORM
Zoya I. Sosinovich, Larissa V. Novitskaya, Vladimir S. Soldatov, and Erik Högfeldt

DOUBLE-LAYER IONIC ADSORPTION AND EXCHANGE ON POROUS POLYMERS
Frederick F. Cantwell

HUMIC-TRACE METAL ION EQUILIBRIA IN NATURAL WATERS
Donald S. Gamble, Jacob A. Marinsky, and Cooper H. Langford

Volume 10

SOLVENT EXTRACTION OF INDUSTRIAL ORGANIC SUBSTANCES FROM AQUEOUS STREAMS
C. Judson King and John J. Senetar

LIQUID MEMBRANES
Richard D. Noble, J. Douglas Way, and Annett L. Bunge

MIXED SOLVENTS IN GAS EXTRACTION AND RELATED PROCESSES
Gerd Brunner

INTERFACIAL PHENOMENA IN SOLVENT EXTRACTION
Valery V. Tarasov and Gennady A. Yagodin

SYNERGIC EXTRACTIONS OF ZIRCONIUM (IV) AND HAFNIUM (IV)
Jiri Hala

Volume 11

CHEMICAL THERMODYNAMICS OF CATION EXCHANGE REACTIONS: THEORETICAL AND PRACTICAL CONSIDERATIONS
Steven A. Grant and Philip Fletcher

A THREE-PARAMETER MODEL FOR SUMMARIZING DATA IN ION EXCHANGE
Erik Högfeldt

DESCRIPTION OF ION-EXCHANGE EQUILIBRIA BY MEANS OF THE SURFACE COMPLEXATION THEORY
Wolfgang H. Höll, Matthias Franzreb, Jürgen Horst, and Siefried H. Eberle

SURFACE COMPLEXATION OF METALS BY NATURAL COLLOIDS
Garrison Sposito

A GIBBS-DONNAN-BASED ANALYSIS OF ION-EXCHANGE AND RELATED PHENOMENA
Jacob A. Marinsky

INFLUENCE OF HUMIC SUBSTANCES ON THE UPTAKE OF METAL IONS BY NATURALLY OCCURING MATERIALS
James H. Ephraim and Bert Allard

Volume 12

HIGH-PRESSURE ION-EXCHANGE SEPARATION IN RARE EARTHS
Liquan Chen, Wenda Xin, Changfa Dong, Wangsuo Wu, and Sujun Yue

ION EXCHANGE IN COUNTERCURRENT COLUMNS
Vladimir I. Gorshkov

RECOVERY OF VALUABLE MINERAL COMPONENTS FROM SEAWATER BY ION-EXCHANGE AND SORPTION METHODS
Ruslan Khamizov, Dmitri N. Muraviev, and Abraham Warshawsky

INVESTIGATION OF INTRAPARTICLE ION-EXCHANGE KINETICS IN SELECTIVE SYSTEMS
A. I. Kalinitchev

EQUILIBRIUM ANALYSIS OF COMPLEXATION IN ION EXCHANGERS USING SPECTROSCOPIC AND DISTRIBUTION METHODS
Hirohiko Waki

ION-EXCHANGE KINETICS IN HETEROGENEOUS SYSTEMS
K. Bunzl

EVALUATION OF THE ELECTROSTATIC EFFECT ON METAL ION-BINDING EQUILIBRIA IN NEGATIVELY CHARGED POLYION SYSTEMS
Tohru Miyajima

ION-EXCHANGE EQUILIBRIA OF AMINO ACIDS
Zuyi Tao

ION-EXCHANGE SELECTIVITIES OF INORGANIC ION EXCHANGERS
Mitsuo Abe

Volume 13

EXTRACTION OF SALTS BY MIXED LIQUID ION EXCHANGERS
Gabriella Schmuckler and Gideon Harel

ACID EXTRACTION BY ACID-BASE-COUPLED EXTRACTANTS
Aharon M. Eyal

HOST-GUEST COMPLEXATION AS A TOOL FOR SOLVENT EXTRACTION AND MEMBRANE TRANSPORT OF (BIO)ORGANIC COMPOUNDS
Igor V. Pletnev and Yuri A. Zolotov

NEW TECHNOLOGIES FOR METAL ION SEPARATIONS: POLYETHYLENE GLYCOL BASED-AQUEOUS BIPHASIC SYSTEMS AND AQUEOUS BIPHASIC EXTRACTION CHROMATOGRAPHY
Robin D. Rogers and Jianhua Zhang

DEVELOPMENTS IN SOLID–LIQUID EXTRACTION BY SOLVENT-IMPREGNATED RESINS
José Luis Cortina and Abraham Warshawsky

PRINCIPLES OF SOLVENT EXTRACTION OF ALKALI METAL IONS: UNDERSTANDING FACTORS LEADING TO CESIUM SELECTIVITY IN EXTRACTION BY SOLVATION
Bruce A. Moyer and Yunfu Sun

Volume 14

POLYMER-SUPPORTED REAGENTS: THE ROLE OF BIFUNCTIONALITY IN THE DESIGN OF ION-SELECTIVE COMPLEXANTS
Spiro D. Alexandratos

RECOVERY OF VALUABLE SPECIES FROM DISSOLVING SOLIDS USING ION EXCHANGE
Jannie S. J. van Deventer, P. G. R. de Villiers, and L. Lorenzen

POLYMERIC LIGAND-BASED FUNCTIONALIZED MATERIALS AND MEMBRANES FOR ION EXCHANGE
Stephen M. C. Ritchie and Dibakar Bhattacharyya

BIOSORPTION OF METAL CATIONS AND ANIONS
Bohumil Volesky, Jinbai Yang, and Hui Niu

SYNTHESIS AND APPLICATION OF FUNCTIONALIZED ORGANO-CERAMIC SELECTIVE ADSORBENTS
Lawrence L. Tavlarides and J. S. Lee

ENVIRONMENTAL SEPARATION THROUGH POLYMERIC LIGAND EXCHANGE
Arup K. SenGupta

IMPRINTED METAL-SELECTIVE ION EXCHANGER
Masahiro Goto

SYNTHESIS AND CHARACTERIZATION OF A NEW CLASS OF HYBRID INORGANIC SORBENTS FOR HEAVY METALS REMOVAL
Arthur D. Kney and Arup K. SenGupta

Volume 15

AN INTEGRATED METHOD FOR DEVELOPMENT AND SCALING UP OF EXTRACTION PROCESSES
Baruch Grinbaum

DESIGN OF PULSED EXTRACTION COLUMNS
Alfons Vogelpohl and Hartmut Haverland

PURIFICATION OF NICKEL BY SOLVENT EXTRACTION
Kathryn C. Sole and Peter M. Cole

TREATMENT OF SOILS AND SLUDGES BY SOLVENT EXTRACTION IN THE UNITED STATES
Richard J. Ayen and James D. Navratil

THE DESIGN OF SOLVENTS FOR LIQUID–LIQUID EXTRACTION
Braam van Dyk and Izak Nieuwoudt

EXTRACTION TECHNOLOGY FOR THE SEPARATION OF OPTICAL ISOMERS
André B. de Haan and Béla Simándi

REGULARITIES OF EXTRACTION IN SYSTEMS ON THE BASIS OF POLAR ORGANIC SOLVENTS AND USE OF SUCH SYSTEMS FOR SEPARATION OF IMPORTANT HYDROPHOBIC SUBSTANCES
Sergey M. Leschev

DEVELOPMENTS IN DISPERSION-FREE MEMBRANE-BASED EXTRACTION–SEPARATION PROCESSES
Anil Kumar Pabby and Ana-Maria Sastre

Volume 16

ADSORPTION AND ION-EXCHANGE PROPERTIES OF ENGINEERED ACTIVATED CARBONS AND CARBONACEOUS MATERIALS
Michael Streat, Danish J. Malik, and Basudeb Saha

ENTROPY-DRIVEN SELECTIVE ION EXCHANGE FOR HYDROPHOBIC IONIZABLE ORGANIC COMPOUNDS (HIOCs)
Ping Li and Arup K. SenGupta

ION-EXCHANGE ISOTHERMAL SUPERSATURATION: CONCEPT, PROBLEMS, AND APPLICATIONS
Dmitri N. Muraviev and Ruslan Khamizov

METAL SEPARATION BY pH-DRIVEN PARAMETRIC PUMPING
Wolfgang H. Höll, Randolf Kiefer, Cornelia Stöhr, and Christian Bartosch

SELECTIVITY CONSIDERATIONS IN MODELING THE TREATMENT OF PERCHLORATE USING ION-EXCHANGE PROCESSES
Anthony R. Tripp and Dennis A. Clifford

ION-EXCHANGE KINETICS FOR ULTRAPURE WATER
Dennis F. Hussey and Gary L. Foutch

Index

A

Absolute permeabilities, 319
Acetalization reaction, 55, 81
Acetals, 47
Acid retardation, 392, 393, 394
Acidic catalysts, 77
Acidic organophosphorous, 343
Activation energy, 71, 253, 266
Activity coefficient, 63
Adhesive tendency, 142
Adsorption, 66
Adsorption parameters, 68
Adsorptive chromatographic reactors, 71
Aerobic oxidation, 228
Agglomeration of nanocrystals, 204
Agglomeration tendency, 126
Aggregation of nanoparticles, 260
Air suspension coating process, 125
Alginate, 131,158, 170
Aliphatic aldehyde, 48
Alkaline-earth metals, 178
Alkoxide, 187
Aluminum etchants, 394
Amine purification, 398
Amphilic properties, 162
Anion chromatograph, 322
Anion-cation pair, 332
Anomalous or non-Fickian diffusion, 29
Anticancer drugs, 138
Antiinflammatory drugs, 117
APU® (Eco-Tec Inc.), 393, 395
Arrhenius equation, 71
Arsenic, 323
Asymmetric hydrogenation, 195, 205
Asymmetric reaction, 195
Asymmetric selectivity, 227
Axial dispersion plug flow model, 77
Azeotropic distillation, 56
    chitin, 157
    helix structure, 166
    keratin, 166
    type and -type chitosan, 340

B

Bacteria chitosan, 372
Basket and paddle, 115
BET (Brunauer-Emmet-Teller) isotherm, 211
Bidisperse catalysts, 74
Bimetallic catalyst, 221, 238, 239
Bimodals, 25
Binary adsorptive experiments, 75
Binary catalyst, 218
Binary-metal colloids, 218
Binuclear phthalocyanine derivatives, 235
Bioadhesive system, 140
Bioavailability, 132
Biomass, 340
Biopolymers for supported catalysis, 155
Biopolymers, 181, 182
Bipolymer-supported catalyst, 269
Bjerrum length, 8
Borderline Lewis acid, 366
Borderline Lewis base, 366
Bridge model, 180
Bronsted-Lowry base, 178
Buccal absorption, 141
Bulk condition, 327
Bulk modulus, 21, 30
BV(bed volume), 384
    chitin, 157
    Keratin, 166

C

Carbonylation, 239, 240
Carbopol, 141
Carboxylate, 170
Carboxylation, 172
Carboxymethyl cellulose (CMC), 190, 193, 222, 224
Carrageenan, 168, 169
Casein, 167
Catalytic converter, 50, 362
Catalytic distillation, 60
Catalytic hydrogenation, 201, 256
Catalytic oxidation, 228
Cationic surfactant, 242
Cellulose based materials, 188
Cellulose membrane, 189
Cellulose, 160
Cetane number, 53
Chain–chain interaction, 29

Chain-length distribution, 9
Chelate complex, 214
Chelating agents, 393
Chelating cage, 230
Chelating resins, 393
Chelation, 191
Chelation of amine, 247
Chemical reactors, 3
Chemisorbed, 76, 228
Chemisorption, 199
Chemoselective hydrogenation, 223
Chiral catalysts, 153
Chitin, 341
Chitosan hollow fiber, 256
Chitosan, 155, 340
Choking effect, 386
Chromatogram, 357
Chromatographic reactor, 79, 82
Chromic acid, recovery, 391
CLC, 342, 343, 360, 365
Cloud temperature, 55
CMC (*See* Carboxymethyl cellulose)
Coacervation method, 133
Cocatalysts, 240
Coflow (cocurrent) regeneration, 376
Coimmobilization, 190
Collagen, 161, 165
Colloidal gold, 234
Colloid-supported catalysis, 185
Colloid-supported catalyst, 154
Column redundancy, 378
Combustion chamber, 54
Concentration polarization factor, 306
Concentration polarization, 306, 307
Conditioning of the catalyst, 170
Conductivity, 311
Configurational entropy, 4
Conversion yield, 264
Coordination number, 180
Cooxidants, 239
Copolymerization, 74
Copper-hydroquinone, 191
Coprecipitaion, 183, 222, 230, 234
Cosolvent effect, 26
Coulombic interaction, 310
Coulometrically controlled, 243
Coumarin, 47
Countercurrent moving bed, 82
Countercurrent regeneration, 381
Countercurrently, 380
Counterdiffusion, 195
Counterflow (countercurrent) regeneration, 376
Counterion, 105
Counterion condensation, 6
Counterion condensation theory, 8
Coupled transport, 296, 334
Coupled transport mechanism, 300, 302, 305
Coupling of carbon dioxide, 246
Cross-coupling reaction, 245
Cross-linking, 108
Cross-linking of chitosan, 157
Cross-linking process, 247
Cross-linking treatment, 262
Catalyst:cocatalyst, 246
Cycle times, 378
Cyclopropanation, 237

D

Damkohler number, 94
DD (degree of deacetylation), 173
Deactivation, 50, 215, 219
Dearomatization, 249
DEE (1,1-diethoxyethane), 46, 60, 79, 91
Degree of ionization, 121
Dehalogenation, 225, 249
Deliquescence, 110
Demineralization, 377, 398
Desorbent, 79, 90
Desorption, 66
Dextrorotatory, 166
Dielectric constant, 266, 298
Diffraction angle, 200
Diffusion, 29
Diffusion coefficient, 30
Diffusion limitation, 269
Diffusion path length, 120
Diffusion properties, 260
Diffusive migration, 71
Dihydronicotinamide, 227
Di-isopropyl ether (DIPE), 51
Dismutation, 85
Dispersible tablets, 140
Dispersion stability, 218
Distribution coefficient, 171
Distribution ratio, 351
Dithiocarbamate, 370
Divalent cations, 294
DME (1,1-dimethoxyethane), 46, 60, 79, 91
Donnan approach, 179
Donnan co-ion exclusion effect, 315
Donnan potential, 123, 124
Double-decomposition process, 105
DPU, 395
Drug delivery, 105
Drug delivery vehicle, 131
Drug displacement, 112

Drug release, 122
Drug resin complex, 113, 128
Drug stabilization, 141
Drug waste, 132
DTC-type lipophilic chitosan, 370
DTPA, 347
DTPA-type chitosan, 348, 350
Dynamic binary adsorption, 93

E

EDTA, 347, 393
EDTA type chitosan, 348, 350
Effective diffusion coefficient, 116
Effective diffusivity, 69
Effectiveness factor, 88
Egg-box model, 131
Elastic and mixing models, 14
Elastic properties, 2
Electrolyte permeation, 328
Electrolyte transport, 302, 305
Electron density, 346
Electroneutrality, 315
Electroplating addition agents, 392
Electrostatic affinity, 123
Electrostatic attraction, 156
Electrostatic interaction, 176
Electrostriction, 27, 35
Electrosynthesized, 243
Eley-Rideal mechanism, 228
Eluent stream, 83
Eluting agent, 357
Elution, 78, 310, 321
Elution profile, 349, 360
Elution time, 329
Emblematic reactions, 241
Emulsification solvent evaporation method, 133
Enantiometer, 227
Enantiometric excess, 205, 216, 218
Enantioselectivity, 206, 212, 237, 238
Enantioseparation, 157
Encapsulation, 161, 182, 191, 196
Encapsulation medium, 197
Enthalpic interaction, 39
Entropy, 3, 4, 9, 15
Enzymatic catalyst, 151
Equilibrium adsorption constant, 67
Equilibrium-limited reversible reaction, 84
Equilibrium swelling, 22
Equivalent conductance, 295, 318
ESR (electron spin resonance), 212
Esterification, 57
Etherification, 57
Ethyl tertiary butyl ether (ETBE), 51
Eudragit-coated resin particles, 126
EXAFS (Extended x-ray adsorption fine structure), 212
Exchange kinetics, 382
Exchange zone, 380
Extract purity, 90

F

Fick's law, 32
Fickian diffusion, 29, 40
Fickian diffusion coefficient, 32, 34
Film diffusion, 116, 134, 252
Fixed bed adsorptive reactor, 74
Flat-leaf membrane, 326
Flat-leaf NF, 302
Flat-leaf test, 305
Flat-leaf test cell, 312
Flexure of the distribution plates, 389
Floating beads, 132
Flory elastic model, 9
Flory-Huggins interaction, 7, 37
Flory-Huggins model, 31
Fluoride, 323
Fouling, 326
Fourier transform infrared, 171
Fourier transform infrared spectra, 76
Fractional extension of chains, 10
Free radical polymerization, 236
Functional group, 269

G

G:M ratio, 186
Gas chromatograph, 112
Gas chromatography, 14, 61
Gastric drug delivery, 107
Gastric mucoadhesion, 135
Gastric retentive device, 132
Gastroretentive formulation, 136
Gaussian chain length distribution, 9
Gaussian model, 20
Gel-bead preparation, 187
Gel transition, 27
Gelatin, 161
Gibbs energy, 5
Glass and gel transition, 28
Glass transition, 18, 20, 27, 35
Glass transition temperature, 113
Glassy, 15
Glassy polymers, 18
Glassy state, 25

GLC (chiral liquid chromatography), 207
GLS (green liquor splitter), 398
Glucosamine, 344
Guluronate zipped ribbons, 159
Guluronic units (high G/M ratio), 159
Gusler-Cohen expression, 10

H

Half reaction, 249
Haloacetic acids, 323
Hard and soft acids and bases, 177
Hard Lewis acid, 371
Helicobacter pylori colonization, 135
Hemoperfusion, 143
HEPES, 345
Heterogeneization, 153, 203
Heterogeneous catalysis, 154
Heterogeneous catalysts, 189, 191, 197, 215
Heterogenized, 232
Heteropolymer, 159, 170
High-octane fuel, 50
High optical resolution, 238
High performance liquid chromatography (HPLC), 85, 112, 207
Higuchi matrix-diffusion-controlled model, 117
Homogeneous catalysts, 197
Homovalent ions, 321
HPMC matrix tablets, 122
HPMC tablets, 128
HR-TEM (High-resolution transmission electron microscopy), 199
Hydrated ionic radii, 306, 309, 316, 318
Hydrated ionic radius, 295
Hydrated radii, 334
Hydrating properties, 119
Hydration in the inner-sphere, 179
Hydration reactions, 237
Hydroformylation reaction, 215, 226
Hydrogels, 15
Hydrogen bonds, 238
Hydrogenation catalysts, 190
Hydrogenation kinetics, 253
Hydrogenation rate, 225
Hydrogenation ratio, 223
Hydrogenation reaction, 194, 215, 216, 219
Hydrogenation transfer reactions, 247
Hydrogenolysis, 224
Hydrolyzed species, 175
Hydrophilic properties, 156
Hydrophilicity, 214, 242
Hydrophobic, 161, 330
Hydrophobicity, 25
Hydroxo-complexes, 171
Hydroxylation, 215
Hydroxylation reactions, 239

I

IDA (iminodiacetic acid), 347
IDA-type chitosan, 350
IER as therapeutics, 143
Immobilizing functional groups, 347
Implantation device, 143
Impregnating agent, 127
Impregnation, 196
Impregnation with metals, 222
Impregnation-reduction, 217
Inactivated, 232
Inactivation mechanism, 191
Inactive dimmers, 232
Infrared, 114
Initiation mechanism, 235
Interdiffusion coefficients, 306, 308, 319, 324, 331, 334
Intermolecular chelation, 179
Internal energy, 4
Intraparticle diffusion, 198, 209, 255
Intraparticle mass transfer, 190
Intriclinic and monoclinic phases, 160
Inverse Langevin function, 10
Ion chromatograph, 313, 318, 320
Ion exchange capacity, 113
Ion exchange microsphere, 139
Ion exchange selectivity, 295, 310, 312, 334
Ion exchange separation factor, 309, 319
Ion exclusion or acid retardation, 385
Ion pair formation, 298
Ion pair mechanism, 296, 302, 305, 306
Ion pairs, 296, 306
Ion retardation, 397
Ionic strength, 123
Ionotropic gelation, 186, 188, 202
Isomerization, 230
Isomerized products, 223
Isopiestic, 23
Isothermal glass transition, 40

K

Katchalsky's equation, 157
Keck coupling, 244
Keratin, 165
Keratinous materials, 48
Kinetic constant, 68
Kinetic model, 65

Knudsen diffusion, 69
Kraft pulping, 397

L

Lagrangian coordinate system, 30
Langmuir adsorption model, 67
Langmuir-Hinshelwood equation, 208
Langmuir-Hinshelwood-Hougen-Watson model, 66
Lanthanide metals, 178
Leachability, 154
Levenberg-Marquadrt method, 70
Lewis acid/base parameters, 177
Ligand exchange, 193
Ligand substitution, 236
Linear driving force (LDF), 87
Lipophilic, 344
Lipophilic character, 242
Lipophilic chitosan, 340
Lipophilicity, 144
Liquid crystalline phase, 48
Liquid lattice model, 6
Liquid–liquid equilibrium, 23
Liquid–liquid extraction, 57
Liquid phase mass transfer coefficient, 299
Loading cycle, 380

M

Macroporous, 191
Macroporous resin, 11, 14
Macroreticular resin, 57, 207
Macroreticular strong cation exchanger, 11
Mannuronate, 159
Mass coordinate system, 30
Mass spectrometry, 112
Mass-transfer zone, 380
Maximum contaminant level (MCL), 330
Maxwell-Stefan diffusion coefficient, 31
Maxwell-Stefan diffusion model, 40
Maxwell-Stefan model, 38
MDEA (methyldiethanolmamine), 398
Membrane reactors, 84
Membrane water interface, 294, 305, 308
Mesoporous, 191
Mesoporous resin, 11
Metal impregnation, 226
Metal refinery, 397
Methyl *tert*-butyl ether (MTBE), 51
Methylcellulose, 193, 219
Microemulsions, 48
Microencapsulation, 125
Micromedia filter, 390
Micronization, 110
Micropore–macropore model, 74
Microwave irradiation, 244
Mixed-bed quality, 400
Mixing Gibbs energy, 6
Molar fraction, 63, 95
Molecular diffusion, 69
Monodentate Lewis base, 212
Monolayer coverage, 210
Monomeric units, 201
MOPAC 93, 346
MTPC, 368
Multicomponent adsorption, 75
Multicomponent solution, 324
Multifunctional reactor, 60, 71
Multivalent species, 321
Myelosuppression, 139

N

Nanocomposites, 201, 243
Nanofiltration, 294
Nanosized metal catalysts, 185
Nenrst-Planck equation, 319
Neutral ion pairs, 296
Neutral polymer, 236
Neutralization tank, 400
Nickel salt recovery, 391
Nicotine, 109
Nitrate, 320, 323
Nominal cross-link density, 11
Nominal degree of neutralization, 18
Nonaqueous, 2
Nonfloating IER beads, 133
Nontransition metals, 180
NO_x, 47
Nuclear magnetic resonance (NMR), 178

O

OCFEM (orthogonal collocation infinite elements method), 91
Octane number, 50
ODAEs (ordinary differential and algebraic equations), 91
OH radical scavenger, 227
Oligomers, 162, 180
Operating capacity, 378
Opthalmic drug delivery, 142
Optical yield, 219, 226
Optimum complexation, 121
Oral-dosage, 132

Oral drug delivery, 104
Order of permeation, 322
Organophosphorous compounds, 350
Osmotic pressure, 22
Osmotic shock, 384
Osmotic stability, 384
Oxine, 359
Oxine type chitosan, 359
Oxygenates, 49, 50

P

Paddle method, 115
Particle diffusion, 116, 134
Particle porosity, 88, 95
Particulate matter (PM), 52
Partitioning coefficient, 299
Patented processes for acetal production, 58
PDP (precipitator dust purification), 397
Peclet number, 94
Pectins, 168
Pendant model, 180
Perchlorate, 320, 323
Perfumery, 47
Permeability, 310, 315
Permeating electrolyte, 332
Permitivity, 8
Peroxo-bridged dimmers, 213, 232
Peroxy-bonded, 191
Pervaporation, 57
Pharmaceutical field, 104
Pharmaceutical grade ion-exchange resin, 108
Phase diagram, 24, 25, 26
Photochemical activation, 197
Photoelectron pulses, 204
Photoirradiation, 181
Physisorbed, 228
Pickle liquors, 395
Plasticization, 20
Plasticizers, 47
PMAA (Polymethylacrylic acid), 222
Poisson ratio, 21
Polisher, 398
Polyacetals, 47
Polyamino acids, 190
Polycondensation, 182
Polyelectrolyte gel, 10
Polyelectrolytes, 179, 182
Polymer dissolution, 182
Polymer-enhanced ultrafiltration (PEUF), 183
Polymeric hydrogel, 144
Polymeric ligands, 240
Polymerization, 57
Polymorphism, 111
Polysaccharide based materials, 155
Pore diffusion, 68
Pore size, 266
Pore-transport model, 318
Porogen, 189
Porosity of the bed, 388
Powdered ion exchange resin, 135
Prefiltration, 389, 390
Process intensification, 377
Protonated amine, 170, 247
Protonation, 172
Purification of Rhodium, 362
Pycnometer, 13
Pyridine functional group (PMC), 365

Q

Quasi-irreversible sorption, 175
Quaternarization, 161

R

Radiolytic treatment, 197
Raffinate, 85
Raffinate purity, 90
Raffinate streams, 83
Rate-limiting step, 383
Rate of permeation, 142
Reactive distillation, 56
Reactive extraction, 84
Reactive separation, 59
Reciprocating flow ion exchange, 377, 378
Recoflo, 379, 387
Reduction potential, 231
Reductive hydrogenation, 250
Reformulating diesel fuel, 54
Regenerant consumption, 382
Regioselectivity, 212, 242
Relative permeability, 312
Relative permeation, 295, 310, 316
Relative selectivity, 320
Relaxation-controlled, diffusion, 29
Renkin equation, 260
Residence time, 266
Resin breakage, 388
Resin shrinkage, 386
Resinate, 104, 107
Resinate (drug-resin complex), 109
Reverse osmosis (RO), 294, 400
Reynolds number, 95
RFG (Reformulated gasoline program), 51
Rubbery polymer, 2, 13

S

Sacrificial anode, 243
Salen complex, 191
Salt passage, 305
Salt permeability coefficient, 325, 327, 329, 330
Salt permeation experiments, 326
Salting-out effect, 25
Schmidt number, 95
Selective separation, 79
Selectivity, 228, 310
SEM (scanning electron microscopy) microphotographs, 257
SEM-EDAX (scanning electron microscopy coupled with x-ray diffraction analysis), 248
Separation factor, 310, 311
Shear moduli, 15
Shear modulus, 9, 10, 12, 35, 40
Sherwood number, 95
Short-bed, 386, 390, 402
Short-bed system, 388
Shrinking and swelling kinetics, 33
Sigmoidal release system, 142
Simulated moving bed reactor (SMBR), 46, 57, 81, 85, 91, 93
Solid-state nuclear magnetic resonance, 112
Solubilization, 265
Solute flux, 298
Solution diffusion, 307, 318
Solution-diffusion model, 296
Solvent dielectric constant, 319
Solvent extraction reagents, 372
Solvent flux, 298
Solvent polymer friction coefficient, 30
Solvent-polymer interactions, 7
Solvent–solvent interactions, 3, 7
Solvent uptake, 13
Sonogashira coupling, 244
Sorption, 25
Sorption kinetics, 172
Sorption sites, 219
Spatial arrangements of ligands, 229
Speciation, 175
Specific diffusivity, 295
Specific surface area, 232
Spectrometric analysis, 172
Spirally wounded membrane (SWM), 302
Spray solvent, 126
Square-planar coordination, 214
SST (shallow shell technology), 383
Stability constants, 356
Stabilizing treatment, 257
Starch-polysulfoxane, 194
Static compression method, 12
Stereoselectivity, 153, 198, 237
Stereospecific arrangement, 269
Steric effect, 245, 251
Steric hindrance, 173, 247
Stoichiometric coefficient, 88, 95
Strong electrolyte, 324
Sulfonic acidic ion exchange resin, 60
Supercritical conditions, 211
Supramolecular organization, 167
Surface coordination number, 201
Surface diffusion, 68
Surface reaction, 66
Sustained release formulation, 125
Suzuki coupling reaction, 244
Swelling, 119, 386
Swelling and shrinking kinetics, 35
Swelling equilibria, 3
Swelling kinetics, 40
Swelling pressure, 3, 5, 9
Swelling ratio, 13, 14, 15, 18, 19, 20, 21, 27, 34, 38, 40, 76
Swollen network, 40
Synergistic effect, 348

T

Tablet disintegration, 142
Tanins, 181
Taste masking, 109, 140
TDS (total dissolved solids), 398
Template role, 360
Ternary precursor complex, 231
Tert-butyl isopropyl ether (IPTBE), 51
Tertiary amyl ether (TAME), 51
Thermostatic bath, 75
Thermostatic water pump, 61
Thickness of the membrane, 325
Thiol or sulfhydryl groups, 165
Time of complexation, 119
TMBR (True moving bed reactor), 89, 91
TMC, 368
TOC (total organic carbon), 401
Tortuosity, 95
Tortuosity factor, 69
Trace anions, 308
Trace species, 323
Transdermal drug delivery, 144
Transmembrane pressure, 301, 312, 316, 326
Transmission electron microscopy (TEM), 199, 249
Transport of electrolytes, 295
Transport phenomena, 68

Tripolyphosphate, 198
True moving bed (TMB), 83
Turn over frequency (TOF), 204, 228, 249, 253, 264, 266, 269
Two-bed demineralizer, 400

U

Unassociated compounds, 70
Unsaturated coordinative state, 232
UV irradiation, 197

V

Van't Hoff equation, 61
Vanillin, 47
Vapor-liquid equilibrium, 27
Variable order kinetic equation, 250
Vitamin B_{12}, 109
Volumetric density, 187

W

Water uptake selectivity, 25
WAXS (wide-angle x-ray scattering anaysis), 199
Wurster process, 125

X

XPS (x-ray photoelectron spectroscopy) analyses, 249, 259
XPS (x-ray photoelectron spectroscopy), 181, 204
X-ray diffractograms, 203

Z

Zeolites, 77
Zero-order release profile, 130
Zero proton condition (pH_{zpc}), 172
Zero-valent state, 204